QUALITY OF SERVICE AND RESOURCE ALLOCATION IN WIMAX

Edited by **Roberto C. Hincapie**
and **Javier E. Sierra**

Quality of Service and Resource Allocation in WiMAX
Edited by Roberto C. Hincapie and Javier E. Sierra

Published by InTech
Janeza Trdine 9, 51000 Rijeka, Croatia

Publishing Process Manager Tajana Jevtic

Technical Editor Teodora Smiljanic

Cover Designer InTech Design Team

Additional hard copies can be obtained from orders@intechopen.com

Quality of Service and Resource Allocation in WiMAX,
Edited by Roberto C. Hincapie and Javier E. Sierra
p. cm.
978-953-307-956-1

For EU product safety concerns:
IN TECH d.o.o., Prolaz Marije Krucifikse Kozulić 3, 51000 Rijeka, Croatia,
info@intechopen.com or visit our website at intechopen.com.

Meet the editors

Professor Roberto C. Hincapie received his PhD degree in Telecommunications Engineering from the Universidad Pontificia Bolivariana, Medellin, Colombia in 2009. Currently, he is an Associate Professor at the same university, where he is also the director of GIDATI research group. His current research interests are Wireless Mesh Networks, resource allocation and OFDM wireless technologies such as WiMAX and LTE. He has taught courses in Traffic Engineering, Simulation and Design of Telecommunication systems. He has participated in several publications in indexed journals and addressed a number of key international conferences.

Professor Javier E. Sierra received his PhD degree in Telecommunications Engineering from the Universidad Pontificia Bolivariana, Medellin, Colombia in 2009. Currently, he is a professor at the Engineering Department of the same university. He works for GIDATI research group and Universidad Pontificia Bolivariana. He has participated in various national and international conferences (IEEE, ACM), and has received awards, including the award for the best paper presented. He is also an author of several publications in indexed journals. His biography has been included in the book Who's Who in the World, 2010 edition, and Who's Who in Science and Engineering by Marquis Who's Who in the World. His research interests include wireless networks, network optimization, traffic grooming and optical transport networks.

Contents

Preface

This book has been prepared to present state of the art on WiMAX Technology. It has been constructed with the support of many researchers around the world, working on resource allocation, quality of service and WiMAX applications. Such many different works on WiMAX, show the great worldwide importance of WiMAX as a wireless broadband access technology.

This book is intended for readers interested in resource allocation and quality of service in wireless environments, which is known to be a complex problem. All chapters include both theoretical and technical information, which provides an in depth review of the most recent advances in the field for engineers and researchers, and other readers interested in WiMAX.

In the first section, readers will find chapters on resource allocation techniques, such as scheduling, call admission control, frequency reuse and cross-layer techniques. The second section presents the evaluation of various models for ensuring the QoS for applications running on WiMAX networks. Finally in the third section, applications for WiMAX are presented, with wireless mesh networks based on multi-hop and relay architectures.

Roberto C. Hincapie, PhD & Javier E. Sierra, PhD
Universidad Pontificia Bolivariana, Medellín,
Colombia

Part 1

Scheduling and Resource Allocation Algorithms

1

Scheduling Mechanisms

Márcio Andrey Teixeira
and Paulo Roberto Guardieiro
[1]*Federal Institute of Education, Science and Technology of São Paulo,*
[2]*Faculty of Electrical Engineering, Federal University of Uberlândia,*
Brazil

1. Introduction

The WiMAX technology, based on the IEEE 802.16 standards (IEEE, 2004) (IEEE, 2005), is a solution for fixed and mobile broadband wireless access networks, aiming at providing support to a wide variety of multimedia applications, including real-time and non-real-time applications. As a broadband wireless technology, WiMAX has been developed with advantages such as high transmission rate and predefined Quality of Service (QoS) framework, enabling efficient and scalable networks for data, video, and voice. However, the standard does not define the scheduling algorithm which guarantees the QoS required by the multimedia applications. The scheduling is the main component of the MAC layer that helps assure QoS to various applications (Bacioccola, 2010). The radio resources have to be scheduled according to the QoS parameters of the applications. Therefore, the choice of the scheduling algorithm for the WiMAX systems is very important. There are several scheduling algorithms for WiMAX in the literature, however, studies show that an efficient, fair and robust scheduling algorithm for WiMAX systems is still an open research area (So-in et al., 2010) (Dhrona et al., 2009) (Cheng et al., 2010).

The packets that cross the MAC layer are classified and associated with a service class. The IEEE 802.16 standards define five service classes: Unsolicited Grant Service (UGS), extended real-time Polling Service (ertPS), real-time Polling Service (rtPS), non real-time Polling Service (nrtPS) and Best Effort (BE). Each service class has different QoS requirements and must be treated differently by the Base Station. The scheduling algorithm must guarantee the QoS for both multimedia applications (real-time and non-real-time), whereas efficiently utilizing the available bandwidth.

The rest of the chapter is organized as follows. Section 2 presents the features of the WiMAX MAC layer and of the WiMAX scheduling classes. The main components of the MAC layer are presented. Then, the key issues and challenges existing in the development of scheduling mechanisms are shown, making a link between the scheduling algorithm and its implementation. Section 3 provides a comprehensive classification of the scheduling mechanisms. Then, the scheduling mechanisms are compared in accordance with the QoS requirement guarantee. Section 4 describes the scheduling algorithms found in the literature in accordance with the classification of the scheduling mechanisms provided in the Section

3. Then, the performance evaluation of these algorithms is made. Section 5 presents a synthesis table of the main scheduling mechanisms and highlights the main points of each of them. Section 6 does the final consideration of this chapter.

2. WiMAX MAC scheduling and QoS: Issues and challenges

The major purpose of WiMAX MAC scheduling is to increase the utilization of network resource under limited resource situation. In the WiMAX systems, the packet scheduling is implemented in the Subscriber Station (uplink traffic) and in the Base Station (downlink and uplink traffic). The Figure 1 shows the packets scheduling in the Base Station (BS) and in the Subscriber Station (SS) (Ma, 2009).

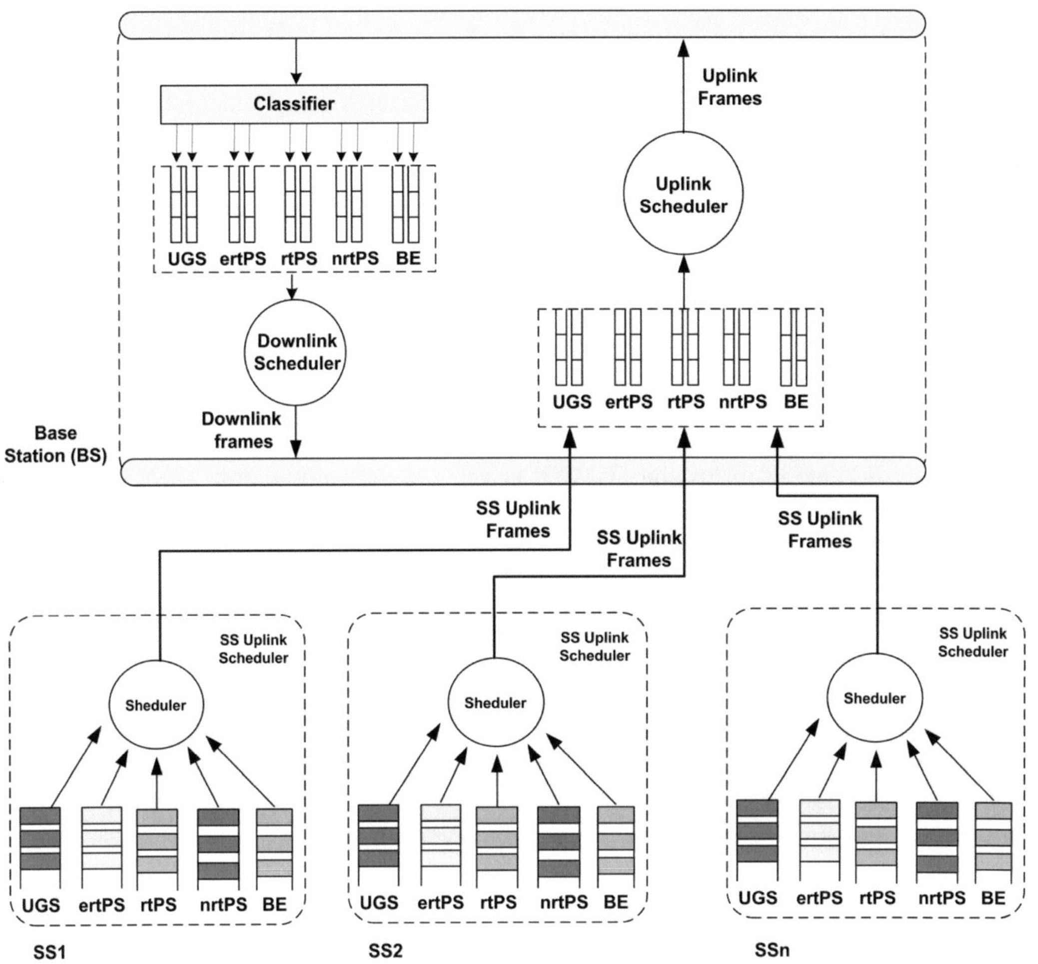

Fig. 1. Packet scheduling in the BS and in the SS (Ma, 2009).

In the downlink scheduling, the BS has complete knowledge of the queue status and the BS is the only one that transmits during the downlink subframe. The data packets are broadcasted to all SSs and an SS only picks-up the packets destined to it. The uplink scheduling is more complex than downlink scheduling. In the uplink scheduling, the input queues are located in the SSs and are hence separated from the BS. So, the BS does not have any information about the arrival time of packets in the SSs queues.

2.1 The uplink medium access

The BS is responsible for the whole medium control access for the different SSs. The uplink medium access is based on request/grant mechanisms. Firstly, the BS makes the bandwidth allocation so that the SSs can send their bandwidth request messages before the transmitting of data over the medium. This process is called polling. The standard defines two main request/grant mechanisms: unicast polling and contention-based polling. The unicast polling is the mechanism by which the BS allocates bandwidth to each SS to send its BW-REQ messages. The BS performs the polling periodically. After this, the SSs can send its BW-REQ messages as a stand-alone message in response to a poll from the BS or it can be piggy-backed in data packets. The contention-based polling allows the SSs to send their bandwidth requests to the BS without being polled. The SSs send BW-REQ messages during the contention period. If multiple request messages are transmitted at the same time, collisions may occur. There are other mechanisms that the SSs can use to request uplink bandwidth such as multicast polling, Channel Quality Indicator Channel (CQICH) (Lakkakorpi & Sayenko, 2009) etc. Depending on the QoS and traffic parameters associated with a service, one or more of these mechanisms may be used by the SSs. A comparison of these mechanisms is presented in (Chuck, 2010).

The choice of the bandwidth request and grant mechanisms has an impact directly on the scheduling delay parameter. The scheduling delay parameter corresponds to the time interval between when the bandwidth is requested and when it is allocated. The scheduling algorithms try to minimize this interval time in order to meet the time constraints of delay-sensitive applications. Moreover, because the standard gives a choice among several bandwidth request mechanisms, it is important for each scheduling mechanism solution to define its own bandwidth request strategy.

2.2 The WiMAX scheduling classes

The packets that cross the MAC layer are classified in connections. At the MAC, each connection belongs to a single service class and is associated with a set of QoS parameters that quantify its characteristics. The standard defines five QoS classes (Li et al., 2007):

- The Unsolicited Grant Service (UGS) receives unsolicited bandwidth to avoid excessive delay and has higher transmission priority among the other services. This service supports constant bit rate (CBR) or fixed throughput connections such E1/T1 lines and voice over IP (VoIP). The BS uplink scheduler offers fixed size uplink (UL) bandwidth (BW) grants on a real-time periodic basis. The QoS specifications are: Maximum sustained rate, Maximum latency tolerance, Jitter tolerance.
- The extended real-time Polling Service (ertPS) also receives unsolicited bandwidth to avoid excessive delay. However, the ertPS service can send bandwidth request messages to change the allocated resource. This service is designed to support real-time multimedia applications that generate, periodically, variable size data packets such as VoIP services with silence suppression. The BS uplink scheduler offers real-time uplink

bandwidth request opportunities on a periodic basis, similar to UGS, but the allocations are made in a dynamic form, not fixed. The QoS specifications are: Maximum sustained rate, Minimum reserved rate, Maximum latency tolerance, Jitter tolerance, Traffic priority.

- The real-time Polling Service (rtPS) uses unicast polling mechanism and receives from BS periodical grants in order to send its BW-REQ messages. This service is designed to support variable-rate services (VBR) such as MPEG video conferencing and video streaming. The BS uplink scheduler offers periodic uplink bandwidth request opportunities. The QoS specifications are: Maximum sustained rate, Minimum reserved rate, Maximum latency tolerance, Jitter tolerance and Traffic priority.
- The non-real time Polling Service (nrtPS) can use contention request opportunities or unicast request polling. However, the nrtPS connections are polled on a regular basis to assure a minimum bandwidth. So, the BS uplink scheduler provides timely uplink bandwidth request opportunities (in order of a second or less) (IEEE, 2005). This service is designed to support applications that do not have delay requirements. The QoS specifications are: Maximum sustained rate, Minimum reserved rate and Traffic priority.
- The Best Effort (BE) service can use unicast or contention request opportunities. However, the BS uplink scheduler does not specifically offer any uplink bandwidth opportunity. This service does not have any QoS requirements.

The Table 1 shows a comparison of WiMAX service classes. Adapted from (So-in et al., 2010).

Service Class	Pros	Cons
UGS	No overhead. Meets guaranteed latency for real-time service	Bandwidth may not be utilized fully since allocations are granted regardless of current need.
ertPS	Optimal latency and data overhead efficiency	Needs to use the polling mechanism (to meet the delay guarantee) and a mechanism to let the BS know when the traffic starts during the silent period.
rtPS	Optimal data transport efficiency	Requires the overhead of bandwidth request and the polling latency (to meet the delay guarantee)
nrtPS	Provides efficient service for non-real-time traffics with minimum reserved rate	N/A
BE	Provides efficient service for BE traffic	No service guarantee; some connections may starve for a long period of time.

Table 1. Comparison of WiMAX Service classes (So-in et al., 2010).

The scheduling algorithm must guarantee the QoS for both multimedia applications (real-time and non-real-time), while efficiently utilizing the available bandwidth. However, the scheduling algorithm for the service classes is not defined by the IEEE 802.16 standards.

2.3 The scheduling and the link adaptation

The design of scheduling algorithms in WiMAX networks is highly challenging because the wireless communication channel is constantly varying (Pantelidou & Ephremides, 2009). The key issue to meet the QoS requirements in the WiMAX system is to allocate the resources among the users in a fair and efficient way, especially for video and voice transmission. However, the amount of allocated resources depends on the Modulation and Coding Schemes (MCSs) used in the physical layer. The aim of the MCSs is to maximize the data rate by adjusting transmission modes to channel variations. The WiMAX supports a variety of MCSs and allows for the scheme to change on a burst-by-burst basis per link, depending on channel conditions. The Figure 2 shows the processing units at MAC and PHY (Liu et al., 2006).

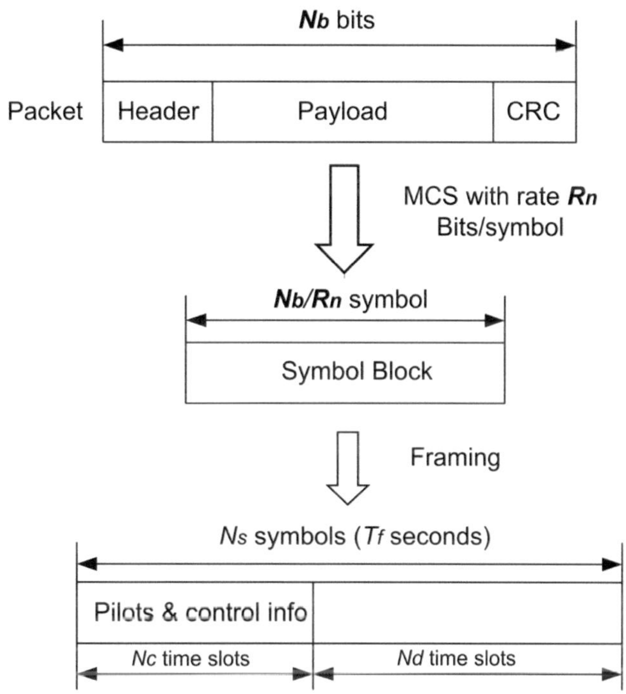

Fig. 2. Processing units at MAC and PHY (Liu et al., 2006).

The MCS is determined in accordance with the Signal-to-Noise Ratio (SNR) and depends on two values:

- The minimum entry threshold: represents the minimum SNR required to start using more efficient MCS.

- The mandatory exit threshold: represents the minimum SNR required to start using a more robust MCS.

The Table 2 shows the values of the receiver SNR assumptions which are proposed in Table 266 of IEEE 802.16e amendment of the standard (Aymen & Loutfi, 2008).

Modulation	Codification rate	SNR(dB)
BPSK	1/2	3.0
QPSK	1/2	6.0
	3/4	8.5
16QAM	1/2	11.5
	3/4	15.0
64QAM	2/3	19.0
	3/4	21.0

Table 2. Values of the SNR (Aymen & Loutfi, 2008).

The link adaptation mechanism allows the making of an adaptive modification of the burst profiles, adapting the traffic to a new radio condition. However, a new issue emerges: how to make an efficient scheduling of the SSs, located in different points away from the BS, sending data to different burst profiles, in accordance with the MCSs used for data transmission. This issue is important because the scheduler must guarantee the application's QoS requirements and allocate the resources in a fair and efficient way.

2.3.1 The WiMAX system capacity

The WiMAX system capacity determines the amount of data that can be delivered to and from the users (Dietze, 2009). There are several ways of quantifying the capacity of a wireless system. The traditional way of quantifying capacity is by calculating the data rate per unit bandwidth that can be delivered in a system. The OFDM symbol is a basic parameter used to calculate the data rate. The expression (1) is used to calculate the data rate (Nuaymi, 2007):

$$Data\,Rate = \left(\frac{Number\,of\,uncoded\,bits\,per\,OFDM\,symbol}{OFDM\,symbol\,time} \right) \quad (1)$$

$$Data\,Rate = \frac{Nsc \times d \times c}{\left[NFFT/(BW \times n) \right] \times (1+G)} \quad (2)$$

Where:

- *Nsc*: is the number of subcarriers used for useful data transmission. In OFDM PHY, 192 subcarriers are used for useful data transmission whereas the total number of subcarriers is equal to 256.
- *d*: represents the number of bits per symbol of modulation. This number depends on the MCS used.
- *c*: represents the code rate of the Forward Error Correction (FEC).

- NFTT : represents the total number of subcarriers. For the OFDM PHY, the total number of subcarriers is equal to 256.
- *BW*: represents the channel bandwidth;
- *n*: represents the sampling factor;
- *G*: represents the ratio of the guard time to the useful symbol time.

Given the values of BW = 7MHz, n = 8/7, d = 4 (16QAM modulation), c = 3/4 and G = 1/16, the data rate is computed as following (Nuaymi, 2007):

$$Data\,Rate = \frac{192 \times 4 \times (3/4)}{\left[NFFT/(BW \times n)\right] \times (1+G)} \quad (3)$$

$$Data\,Rate = \frac{192 \times 4 \times (3/4)}{\left[256/(7MHz \times (8/7))\right] \times (1+1/16)} = 16.94\,Mb/s \quad (4)$$

The Table 3 shows the data rates for different MCSs and G values (Nuaymi, 2007).

G Ratio	BPSK 1/2	QPSK 1/2	QPSK 3/4	16-QAM 1/2	16-QAM 3/4	64-QAM 2/3	64-QAM 3/4
1/32	2.92	5.82	8.73	11.64	17.45	23.27	26.18
1/16	2.82	5.65	8.47	11.29	16.94	22.59	25.41
1/8	2.67	5.33	8.00	10.67	16.00	21.33	24.00
1/4	2.40	4.80	7.20	9.60	14.40	19.20	21.60

Table 3. Data rates for different MCSs and G values (Nuaymi, 2007).

As it can be seen in the Table 3, the highest order modulations offer a larger throughput. However, in a practical use, not all users receive adequate signal levels to reliably decode all modulations. Users that are close to the BS are assigned with the highest order modulation, while users that are far from BS use lower order modulations for communications to ensure that the data are received and decoded correctly. This implies that the BS needs to allocate more resources for these users aiming at maintaining the same throughput as the users that use the highest order modulation. This issue must be taken into account in the scheduling development, in order to maximize the resources in a function of the number of users at the access networks and the modulation types used.

3. The WiMAX scheduling mechanisms

The scheduling mechanism plays an important role in the provisioning of QoS for the different types of multimedia applications. The WiMAX resources have to be scheduled according to the QoS requirements of the applications. Therefore, the application performance depends directly on the scheduling mechanism used. In the last few years, the scheduling mechanism research has been intensively investigated. However, recent studies show that an efficient, fair and robust scheduler for WiMAX is still an open research area, and the choice of a scheduling algorithm for WiMAX networks is still an open question. Since the scheduling is a very active field, we cannot describe all the algorithms proposed for WiMAX. However, we present a study of some proposals for WiMAX.

3.1 The WiMAX scheduling mechanisms classification

There are several proposals about WiMAX scheduling mechanisms. In a general way, these proposals can be classified in: Point-to-Multipoint (PMP) scheduling mechanisms and Mesh scheduling mechanisms. Moreover, some scheduling works are focused on downlink scheduling, others on uplink scheduling, and others on both scheduling (downlink and uplink). The Figure 3 shows the general classification of WiMAX scheduling mechanisms.

Taking into account the classification shown in the Figure 3, the scheduling mechanisms are classified in three categories (Dhrona et al., 2009):

- Homogeneous.
- Hybrid.
- Opportunistic algorithms.

The three categories of scheduling mechanisms have the same aims which are to satisfy the QoS requirements of the applications. What differs one category from the other are the characteristics of the scheduling algorithms employed in the scheduling mechanism and the number of algorithms used to ensure QoS for the service classes.

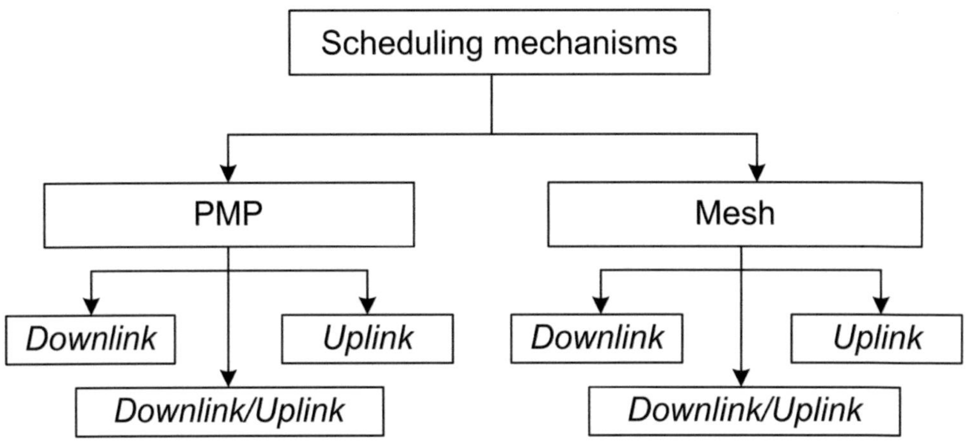

Fig. 3. General classification of WiMAX scheduling mechanisms

The homogeneous category uses scheduling algorithms which were originally proposed for wired networks, but are used in WiMAX to satisfy the QoS. Generally, these algorithms do not address the issue of link channel quality. Some examples of these algorithms are: Round Robin (RR) (Cheng, 2010), Weighted Round Robin (WRR) (Sayenko et al., 2006), Deficit Round Robin (DRR) (Shreedhar &Varghese, 1995), Earliest Deadline First (EDF) (Andrews, 2005), Weighted Fair Queuing (WFQ) (Cicconetti, 2006) etc.

The hybrid category employs multiple legacy schemes in an attempt to satisfy the QoS requirements of the multi-class traffic in WiMAX networks. Some of the algorithms in this category also address the issue of variable channel conditions in WiMAX. Some examples of these algorithms are: EDF+WFQ+FIFO (Karim et al., 2010), EDF+WFQ (Dhrona et al., 2009), adaptive bandwidth allocation (ABA) (Sheu & Huang, 2011) etc.

The opportunistic category refers to algorithms that exploit variations in channel conditions in WiMAX networks. This technique is known as cross-layer algorithms. Some examples of these algorithms are: Temporary Removal Scheduler (TRS) (Ball, 2005), Opportunistic Deficit Round Robin (O-DRR) (Rath, 2006) etc.

The authors in (So-in et al., 2010) classify the scheduling algorithms into two categories:

- Algorithms that use the physical layer.
- Algorithms that do not use the physical layer.

Furthermore, algorithms that do not use the physical layer are divided into two groups:

- Intraclass.
- Interclass.

The authors in (Msadaa et al., 2010) also classify the algorithms into three categories: algorithms based on packet queuing, algorithms based on optimization strategies and cross-layer algorithms. The scheduling strategy based on queuing packet has the same characteristics of the algorithms developed for wired networks. This category is divided into two groups: one layer structure which is shown in the Figure 4 and the multi-layer structure, illustrated in the Figure 5.

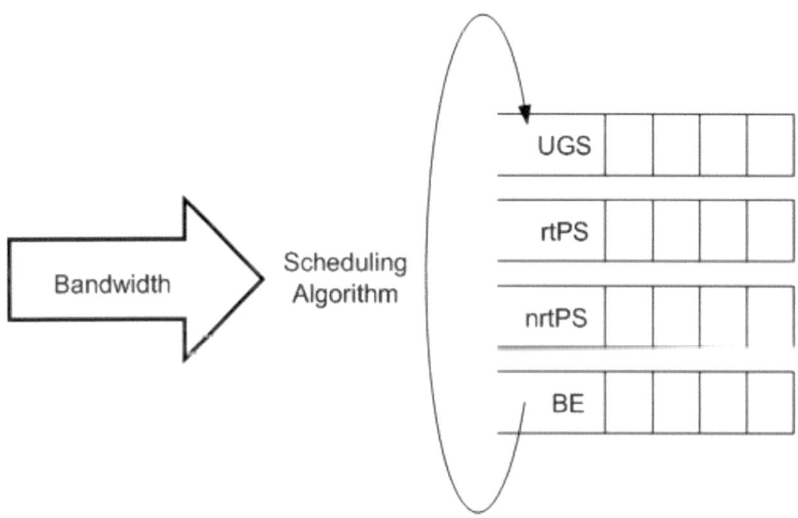

Fig. 4. One layer scheduling structure (Msadaa et al., 2010).

In the one layer scheduling structure, only a single scheduling algorithm is used for all service classes. For example, in (Sayenko et al., 2008), it was proposed that a scheduling solution based on the RR approach. In this case, the authors consider that there is very litte time to do the scheduling decisions, and a simple one-layer scheduling structure is a better solution than a multi-layer scheduling structure.

In the multi-layer scheduling structure, two or more steps are used in the scheduling which defines a multi-layer scheduling. The authors in (Wongthavarawat & Ganz, 2003) were the first to introduce this scheduling structure model. The multi-layer structure, shown in the

Figure 5, combines the strict priority policy among the service classes, and an appropriate queuing discipline for each service class.

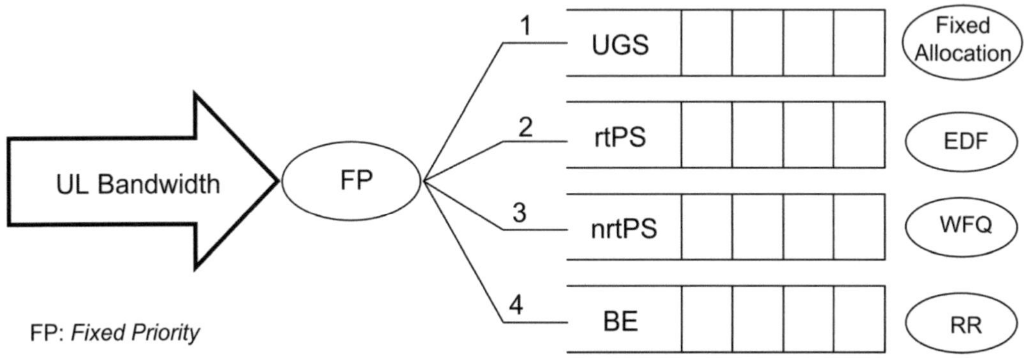

Fig. 5. Multi-layer scheduling structure (Msadaa et al., 2010).

The Table 4 summarizes the existing classification in the literature on scheduling mechanisms and exemplifies some scheduling algorithms that have been evaluated for WiMAX networks.

	Proposes					**Scheduling Algorithms**
	(Dhrona et al., 2009)	(So-in et al., 2010)		(Msadaa et al., 2010)		
Classification	Heterogeneous	Channel Anware	Intra class	Packet Queuing derived strategies	One Layer structure	RR, WRR,DRR
					Hierarquical structure	EDF, WFQ
	Hybrid		Inter class	Optimization based strategies		EDF +WFQ+FIFO, EDF + WFQ, WRR+RR+PR DFPQ
	Opportunistic		Cross-layer approach algorithms			TRR, O-DRR, mSIR, mmSIR

Table 4. Classification of scheduling mechanisms.

4. The uplink scheduling algorithms

The uplink scheduling algorithms executed at BS for uplink traffic have to make complex decisions, because it does not have queue information. So, the main focus of this section is on these scheduling algorithms. We made the choice of the main algorithms found in the literature and we distinguished them among the scheduling categories described above.

4.1 The homogeneous scheduling algorithms

Some homogeneous scheduling algorithms are based on the RR scheduler. The RR scheduler is the simplest algorithm that distributes the equal bandwidth to the SSs. However, it does not support the QoS requirements for different traffic classes, such as delay and jitter. In order to improve the RR algorithm for WiMAX systems, some proposes based on RR scheduler were made and they can be found in the literature.

4.1.1 Weighted Round Robin (WRR) scheduling

The WRR scheduling is an extension of RR scheduling. This algorithm has been implemented and evaluated in (Dhrona et al., 2009). The algorithm is executed at the beginning of every frame at the BS. At this moment, the WRR algorithm determines the allocation of bandwidth among the SSs based on their weights. So, the authors assign weight to each SS with respect to its Minimum Reserved Traffic Rate (MRTR) as follows:

$$Wi = MRTRi \Bigg/ \sum_{j=1}^{n} MRTRj \tag{5}$$

where Wi is the weight of SSi and n the number of SSs

4.1.1.1 Performance evaluation

The WRR algorithm was evaluated by means of simulation study described in the reference (Dhrona et al., 2009). The main parameters used were: OFDM PHY layer, symbol duration time of 12.5 μs and the channel bandwidth of 20 MHz. The authors have observed that the WRR algorithm does not perform well when the traffic contains variable sized packets. The algorithm will not provide a good performance in the presence of variable size packets.

4.1.2 Deficit Round Robin (DRR) scheduler

This algorithm is a variation of RR. A fixed quantum (Q) of service is assigned to each SS flow (i). When an SS is not able to send a packet, the remainder quantum is stored in a deficit counter (DC_i). The value of the deficit counter is added to the quantum in the following round. When the length of the packet (L_i) waiting to be sent is less than the deficit counter DC_i the head of the queue (Q_i) is dequeued and the value of the (DC_i) is decremented by L_i. The algorithm is flexible enough as it allows provision of quantum of different sizes depending on the QoS requirements of the SSs. However, the DRR algorithm requires accurate knowledge of packet size (L_i), very complex in its implementation.

4.1.3 Earliest Deadline First (EDF)

The EDF was originally proposed for real-time applications in wide area networks (Khan et al., 2010). This algorithm assigns a deadline to each packet and allocates bandwidth to the SS that has the packet with the earliest deadline (Hussain et al., 2009). The deadlines can be assigned to the packets of the SSs based on the SS's maximum delay requirement. Since each SS specifies a value for the maximum latency parameter, the arrival time of a packet is added to the latency to form the tag of the packet. The EDF algorithm is suitable for SSs belonging to the UGS and rtPS scheduling services.

4.1.4 Performance evaluation of DRR and EDF algorithms

The performance of the DRR and EDF algorithm were evaluated in (Karim et al., 2010). However, the authors consider only the downlink resource allocation. The simulation configuration and the parameters follow the performance evaluation parameters specified in Mobile WiMAX systems Evaluation Document and WiMAX Profile. The results showed that the EDF algorithm introduces unfairness when a under loaded. The DRR algorithm is fair and gives a better performance than EDF algorithm.

4.1.5 Weighted Fair Queuing (WFQ) scheduler

The WFQ scheduler assigns finish times to the packets. So, the packets are selected in increasing order according to their finish times. The finish times of the SS packets are calculated based on the size of the packets and the weight assigned to the SS. The WFQ was also evaluated in (Dhrona et al., 2009). The algorithm results in superior performance as compared to the WRR algorithm in the presence of variable size packets. However, the disadvantage of the WFQ algorithm is that it does not consider the start time of a packet.

4.2 Heterogeneous and opportunistic scheduling algorithms

The heterogeneous scheduling algorithm category is used as the combination of legacy scheduling algorithms. An important aspect of heterogeneous algorithms is the allocation of bandwidth among the traffic classes of WiMAX. Some of the algorithms in this category also address the issue of variable channel conditions in WiMAX.

4.2.1 Adaptive Bandwidth Allocation (ABA)

An adaptive bandwidth allocation (ABA) model for multiple traffic classes was proposed in (Sheu & Huang, 2011). In order to promise the quality of real-time traffic and allow more transmission opportunity for other traffic types, the ABA algorithm first serves the UGS connections. Then, polling bandwidth is allocated for rtPS service to meet their delay constraints and for the nrtPS to meet their minimum throughput requirements. For the BE service, the ABA algorithm will prevent it from starvation. The ABA algorithm assigns initial bandwidth, UGS, rtPS, nrtPS and BE, based on the requested bandwidth of UGS, the required minimum bandwidth of rtPS and nrtPS and the queue length of BE service respectively. If remaining bandwidth exists, the ABA then assigns extra bandwidth for the rtPS, nrtPS and BE services.

4.2.1.1 Performance evaluation

The analytical results of the ABA algorithm were obtained by running on the MATLAB software. These results were validated through the simulator developed by the authors written in Visual C/C++. The results showed that the ABA algorithm meet the delay constraints of rtPS and the minimum throughput requirements of nrtPS, while it endeavors to avoid any possible starvation of BE traffic.

4.2.2 EDF, WFQ and FIFO scheduling algorithms

The authors in (Karim et al., 2010) have combined three scheduling algorithms. It is used the strict priority mechanism for overall bandwidth allocation. The EDF scheduling algorithm is used for SSs of ertPS and rtPS classes. The WFQ algorithm is used for SSs of nrtPS class and FIFO is used for SSs of BE class. The bandwidth distribution among the traffic classes is executed at the beginning of every frame whereas the EDF, WFQ and FIFO algorithms are executed at the arrival of every packet. This algorithm was evaluated in (Sayenko et al., 2008). A drawback of this algorithm is that lower priority SSs will essentially starve in the presence of a large number of higher priority SSs due to the strict priority of overall bandwidth allocation.

4.2.3 EDF and WFQ

It was proposed in (Dhrona et al., 2009) a hybrid algorithm that uses the EDF scheduling algorithm for SSs of ertPS and rtPS classes and WFQ algorithm for SSs of nrtPS and BE classes. Although the details of overall bandwidth allocation are not specified, it is not done in a strict priority manner, but a fair manner is used to allocate the bandwidth among the classes. At the arrival of every packet the EDF and WFQ algorithms are executed.

4.2.3.1 Performance Evaluation of (EDF+WFQ+FIFO) and (EDF+WFQ) scheduling algorithms

The performance analysis of the scheduling scheme above described is performed by simulations (Dhrona et al., 2009). The main parameters of the simulation are the following: the air interface is WirelessMAN-OFDM, the channel bandwidth is 20 MHz, the OFDM symbol duration is 12.5 μs.

- EDF+WFQ+FIFO: This solution shows superior performance for SSs of ertPS classes in relation to average throughput, average delay and packet loss when the concentration of real-time traffic is high.
- EDF+WFQ: This algorithm is limited by the allocation of bandwidth among the traffic classes.

4.2.4 EDF and Connection Admission Control (CAC) scheme

A scheduling scheme which combines the CAC mechanism and the EDF algorithm was proposed in (Wu, 2010). This solution aims at the scheduling of rtPS class and uses the EDF algorithm to reduce the average latency. The Figure 6 shows the flowchart about the scheduling scheme.

In the proposed scheduling solution, the BS verifies if the requested service is rtPS class or not, when an SS asks the BS for a connection request. If the connection request is the rtPS, the CAC will judge if the connection can be admitted or not. The connection request will be admitted if the sum of available bandwidth and the collected bandwidth from BE service is greater than the Maximum Sustained Traffic Rate (MSTR) of rtPS class. In this case, the Minimum Reserved Traffic Rate (MRTR) will be taken into account to determine if the request can be admitted. Once admitted the rtPS connection, the BS will schedule the rtPS service class according to the EDF algorithm.

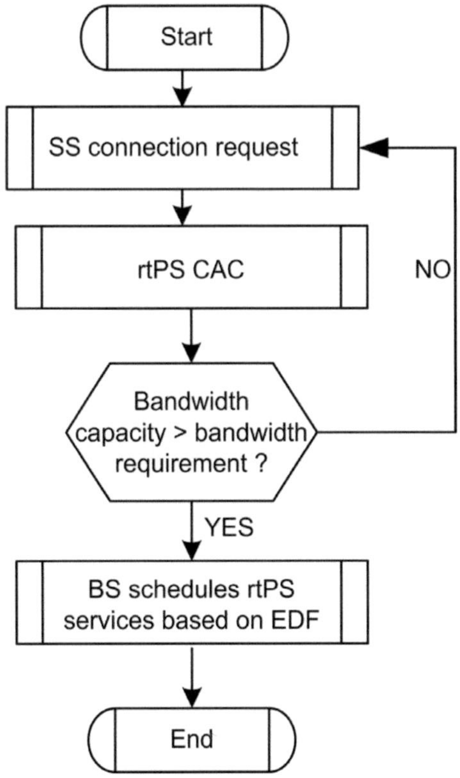

Fig. 6. Flowchart of the scheduling algorithm (Wu, 2010).

4.2.4.1 Performance evaluation

The performance analysis of the scheduling scheme above described is performed by simulations. The main parameters of the simulation are the following: the PHY layer is OFDM, the MCS is 64QAM 3/4, and the frame duration is 5 ms. The compared performance metrics are latency, jitter, throughput and the rate of packet loss. The results showed that the scheduling scheme can reduce the average latency and achieve the QoS.

4.3 Cross layer approach algorithms

4.3.1 Temporary Removal Scheduler (TRS)

The TRS scheduler makes the scheduling list in accordance with the SSs that have Signal Interference Ratio (SIR) greater than a preset threshold (Ball, 2005). When the radio conditions are poor then the scheduler suspends the packet call from the scheduling list for an adjustable time period *Tr*. The scheduling list contains all the SSs that can be served at the next frame. When the *Tr* expires, the suspended packet is checked again, and if the radio conditions are still poor the packet is suspended for another time period *Tr*. This process is repeated *L* times, where *L* is equal to consecutive suspend procedure. The TRS scheduler can be combined with the Round Robin (RR) and maximum Signal to Interface Ratio (mSIR) schedulers. When TRS is combined with RR the whole radio resources are divided by the number of subscribers in the list, and all the subscribers will get resources equitably.

4.3.2 UAF WiMAX scheduler

The UAF_WiMAX scheduler was developed in (Khan et al., 2010). The scheduler serves the SSs with minimum signal to interface ratio, taking into account the already allocated resources to the SSs which have greater signal to interface ratio. When the BS provides periodical unicast request polling to the subscribers, the subscribers respond with their required bandwidth request that is equal to its uplink data connection queue. The UAF_WiMAX scheduler first chains the SSs in the scheduling list that have bandwidth request. The UAF_WiMAX scheduler identifies the SSs packets call-power depending upon the radio conditions.

4.3.3 Performance evaluation of the UAF and TRS schedulers

The UAF scheduler was evaluated by means of simulation study, and the results were compared with the TRS scheduling algorithm (Khan et al., 2010). The main parameters used were: OFDM PHY layer, channel bandwidth is 5MHz, the frame duration is 20 ms, and the MCSs used are 64QAM 3/4, 64QAM 2/3, 16QAM 3/4, 16QAM 1/2, QPSK 3/4, QPSK 1/2.The results showed that the UAF_WiMAX scheduler has less mean sojourn time when compared to TRS combined with mSIR scheduler. The UAF scheduler serves the SSs with the minimum signal to interface ration when it already has resources to the SSs having a greater signal to interface ratio.

4.3.4 maximum Signal-to-Interference Ratio (mSIR)

The authors in (Belghith et al., 2010) make a comparison of WiMAX scheduling algorithms and propose an enhancement of the maximum Signal-to-Interference Ratio (mSIR) scheduler, called modified maximum Signal-to-Interference Ratio (mmSIR) scheduler. The mSIR scheduler serves those SSs having the highest Signal-to-Interference Ratio (SIR) at each frame. However, SSs having slightly smaller SIR may not be served and then the average delay to deliver data increases. To solve this problem, the authors proposed a solution where the BS only serves the SSs that do not have unicast request opportunities in the same frame.

4.3.4.1 Performance evaluation

The mSIR scheduler was evaluated by means of simulation study in (Belghith et al., 2010), where three scenarios were used: pessimistic, optimistic and realistic. In the pessimistic scenario, bad radio conditions are considered. All SSs use the most robust MCS (BPSK 1/2). In the optimistic scenario, ideal radio conditions are considered. All the SSs use the most efficient MCS (64QAM 3/4). In the realistic scenario, random radio conditions are considered. Hence, the SSs may have different MCSs (64QAM 3/4, 64QAM 2/3, 16QAM 3/4, 16QAM 1/2, QPSK 3/4, QPSK 1/2). The air interface is OFDM and the channel bandwidth is 7 MHz. The mmSIR scheduler was compared with two schedulers: RR and Prorate. The mmSIR had good spectrum efficiency in all scenarios.

4.3.5 Adaptive scheduler algorithm

An adaptive scheduling packets algorithm for the uplink traffic in WiMAX networks is proposed in (Teixeira & Guardieiro, 2010). The proposed algorithm is designed to be

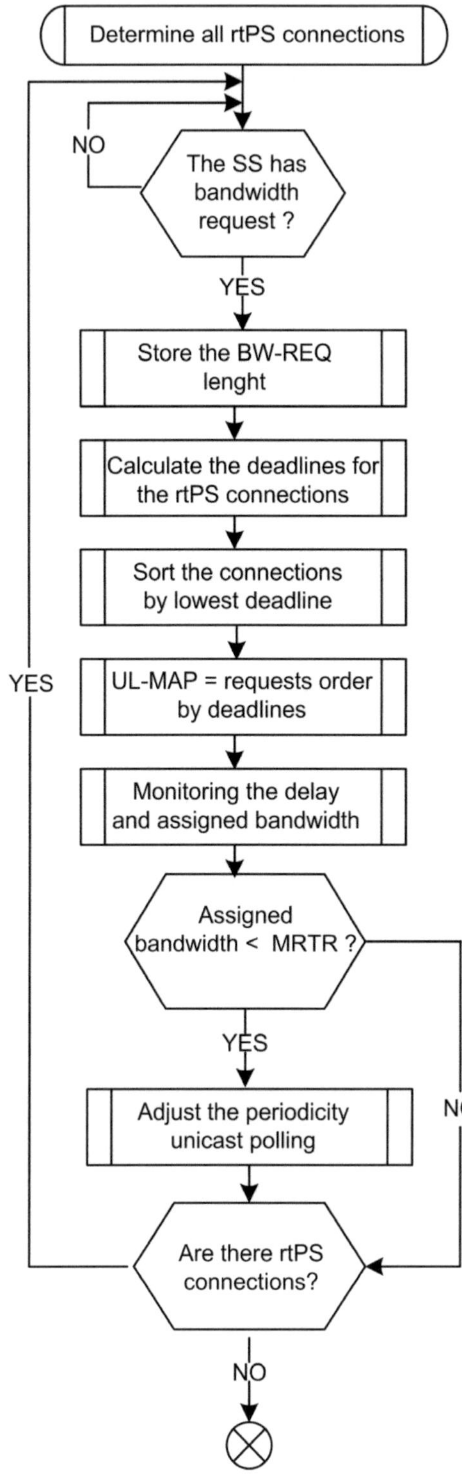

Fig. 7. Flowchart of the adaptive scheduling scheme.

completely dynamic, mainly in networks that use various MCSs. Moreover, a method which interacts with the polling mechanisms of the BS was developed. This method controls the periodicity of sending unicast polling to the real-time and non-real-time service classes, in accordance with the QoS requirements of the applications. The Figure 7 shows a flowchart of the adaptive scheduling scheme.

The scheduler monitors the average delay of the rtPS service and the minimal bandwidth assiged to the rtPS and nrtPS service classes. The limited maximum delay is guaranteed for the rtPS service through the use of a new deadlines based scheme. The deadlines calculation is made by using the following parameters: the information about the MCSs used for the sending packets between the SS and the BS; the information about the bandwidth request messages sent by the SSs; and the queuing delay of each bandwidth request message in the BS queue. Once the deadlines are calculated, they are assigned to the rtPS connections. Thus, the scheduler defines the transmission order of the rtPS connections based on the lowest deadline. Moreover, the scheduler also ensures the minimal bandwidth for rtPS and nrtPS services in accordance with the minimum bandwidth requirement per connection, the amount of bytes received in a current period, and the amount of backlogged requests (in bytes).

4.3.5.1 Performance evaluation

The adaptive scheduling algorithm was evaluated by means of modeling and simulation in environments where various MCSs were used and also in environments where only one type of MCS was used. The main parameters of the simulation are the following: OFDM PHY layer, frame duration is 20 ms, MCSs used are 64QAM 3/4, 64QAM 2/3, 16QAM 3/4, 16QAM 1/2, QPSK 3/4, QPSK 1/2. The performance of the adaptive scheduling algorithm was compared with RR and WRR algorithms in (Teixeira & Guardieiro, 2010), where it showed a better performance.

5. A synthesis of scheduling mechanisms

The goals of the scheduling mechanisms are basically to meet QoS guarantees for all service class, to maximize the throughput, to maintain fairness, to have less complexity and to ensure the system scalability. There are several scheduling mechanisms in the literature, however, each one with its own characteristics.

The homogenous and hybrid scheduling mechanisms do not explicitly consider all the required QoS parameters of the traffic classes in WiMAX. The algorithms consider only some of the parameters which are not sufficient since the scheduling classes have multiple QoS parameters such as the rtPS class that requires delay, packet loss and throughput guarantee. The WRR, WFQ and hybrid (EDF+WFQ) algorithms provide a more fair distribution of bandwidth among the SSs. The WFQ and WRR algorithms attempt to satisfy the minimum reserved traffic rate (MRTR) of the SSs by assigning weights to the SSs based on their MRTR. The worst case delay bound guaranteed by the WFQ algorithm can be sufficient for the UGS connections but not for ertPS and rtPS connections.

The algorithms such as EDF and hybrid (EDF+WFQ+FIFO) indicate superior performance for SSs of ertPS and rtPS classes with respect to average throughput, average delay and packet loss when the concentration of real-time traffic is high. However, these algorithms will also result in starvation of SSs of nrtPS and BE classes.

The cross-layer algorithm takes into account some QoS requirements of the multi-class traffic in WiMAX such as the average delay, average throughput and the channel quality. The mSIR scheduler serves those SSs having the highest SIR at each frame. However, SSs having slightly smaller SIR may not be served and then the average delay to deliver data

Scheduling Mechanisms	Possibility of use for WiMAX	DL	UL	Comments	Algorithm parameters
WRR	Yes	Yes	Yes	Not provide good performance in the presence of variable size packets.	Static weights.
DRR	Yes	Yes	No	Requires accurate knowledge of packet size, being very complex in its implementation.	Fixed quantum.
EDF	Yes	Yes	Yes	Allocates bandwidth to the SS that has the packet with the earliest deadline. Needs to know the arrival time of the packets.	Dealines (can be the arrival time - send time of the packets in some cases).
ABA	Yes	Yes		Initially assigns bandwidth for UGS, rtPS, nrtPS and BE serice classes based. If remaining bandwidth exists, the ABA then assigns extra bandwidth for the rtPS, nrtPS and BE services.	UGS bandwidth requirement.
EDF, WFQ, FIFO	Yes	Yes	Yes	SSs with lower priority will starve in the presence of a large number of higher priority SSs due to the strict priority.	Weights for WFQ, Deadlines for EDF.
WFQ, EDF	Yes	Yes	Yes	Limited by the allocation of bandwidth among the traffic classes.	Weights for WFQ, deadlines for EDF.
EDF + CAC	Yes	Yes	Yes	Can reduce the average latency and achieve the QoS.	MRTR, deadline.

Table 5a. Synthesis of some scheduling mechanisms.

Scheduling Mechanisms	Possibility of use for WiMAX	DL	UL	Comments	Algorithm parameters
TRF	Yes	Yes	Yes	The scheduler makes the scheduling list in accordance with the subscribers that have SIR greater than a preset threshold.	Removal time (Tr).
UAF	Yes	Yes	Yes	Serves the SSs with minimum signal to interface ratio, taking into account the already allocated resources to the SSs which have greater signal to interface ratio.	SIR.
mSIR	No	Yes	Yes	SSs having a poor SIR may be scheduled after an excessive delay	SIR.
Adaptive Scheduler	Yes	Yes	Yes	Controls the periodicity of sending unicast polling to the real-time and non-real-time service classes, in accordance with the Quality of Service requirements.	Polling interval, SIR, MCS.

Table 5b. Synthesis of some scheduling mechanisms.

increases. The cross-layer algorithms do not exploit all characteristics of WiMAX system. On the other hand, the optimization scheduler mechanisms take into account the characteristics of WiMAX system, for example, polling mechanism, backoff optimization, overhead optimization and so on. For example, the adaptive scheduler uses a cross-layer approach where it makes the scheduling in accordance with the MCSs and interacts with the polling mechanisms of the BS. Scheduling mechanisms, cross-layer and optimization mechanisms are still an open ongoing research topic. The Tables 5a and 5b show a synthesis of deployment of some important scheduling mechanisms.

6. Conclusion

In this chapter, we present the state of the art of WIMAX scheduling mechanisms. Firstly, we we present the features of the WiMAX MAC layer and of the WiMAX scheduling classes. The main components of MAC layer are also presented. After that, we present the key issues and challenges existing in the development of scheduling mechanisms. A classification of the scheduling mechanisms was also made. So, we present a synthesis table of the scheduling mechanisms performance where we highlight the main points of each of them. All the proposed WiMAX algorithms could not be studied in this chapter, but we have shown some relevant proposals.

The adaptive scheduling algorithm proposed in (Teixeira & Guardieiro, 2010) makes the scheduling in accordance with the MCSs and interacts with the polling mechanisms of the BS. Its evaluation shows a good performance in the realistic and optimistic scenarios. The mSIR and mmSIR scheduling algorithms were evaluated in (Belghith et al., 2010) and show good spectrum efficiency in the realistic scenarios where random radio conditions are considered. The scheduling algorithms such as WRR, WFQ and EDF+WFQ were evaluated in (Dhrona et al., 2009) and show a fair distribution of bandwidth among the SSs. However, the performance evaluation of these algorithms was made considering only the optimistic scenarios.

7. References

Andrews, M. (2000). Probabilistic end-to-end delay bounds for earliest deadline first scheduling. *Proceedings of IEEE Computer Communication Conference.* ISBN 0-7803-5880-5, Israel. February, 2000.

Belghith, A., Norovjav, O., & Wang, L. (2010). WiMAX spectrum efficiency: considerations and Simulation results. *Proceedings of 5th International Symposium on Wireless Pervasive Computing (ISWPC).* ISBN 978-1-4244-6855-3, Modena, May, 2010.

Bacioccola, A., et. al (2010). IEEE 802.16: History, status and future trends. *Computer Communications,* Vol. 33, No. 2, (February, 2010), pp. 113-123.

Belghith, A. Loutfi, N. (2008). WiMAX capacity estimations and simulation results. *Proceedings of 67th IEEE Vehicular Technology Conference.*ISBN 978-1-4244-1644-8, Singapore, May, 2008.

Ball, C. F. et al. (2005). Performance Analysis of Temporary Removal Scheduling applied to mobile WiMAX Scenarios in Tight Frequency Reuse. *Proceedings of The 16th Annual IEEE International Symposium on Personal Indoor and Mobile Radio Communications (PIMRC'2005).* ISBN 9783800729098, Berlin, September, 2005.

Cicconetti, C. et al. (2006). Quality of Service Support in IEEE 802.16 Networks. *IEEE Network,* Vol. 20, No. 2, (April 2006) pp. 50-55, ISSN 0890-8044.

Cheng, S. T., Hsieh, M. T., Chen, B. F. (2010). Fairness-based scheduling algorithm for time division duplex mode IEEE 802.16 broadband wireless access systems. *IET Communications,* Vol. 4, No. 9, (April, 2010) pp. 1065 - 1072, ISSN 1751-8628.

Chuck, D., Chen, K. Y., Chang, J. M. (2010). A Comprehensive Analysis of Bandwidth Request Mechanisms in IEEE802.16 Networks. *IEEE Transactions on Vehicular Technology,* Vol. 59, No. 4, (May, 2010) pp. 2046-2056, ISSN 0018-9545.

Dhrona, P., Abu, N. A., Hassanein, H. S. (2009). A performance study of scheduling algorithms in Point-to-Multipoint WiMAX networks, *Computer Communications,* vol. 32, (February, 2009) pp. 511-521.

Dhrona, P., Abu, N. A., Hassanein, H. S. (2008). A performance study of scheduling algorithms in Point-to-Multipoint WiMAX networks, *IEEE Conference on Local Computer Networks (LCN 2008),* ISBN 978-1-4244-2412-2, Montreal, October 2008.

Dietze, k., Hicks, T., Leon, G. (2009). WiMAX System Performance Studies, Available from <www.edx.com/files/WiMaxPaperv1.pdf>.

Hussain, M. A. R. et al. (2009). Comparative Study of Scheduling Algorithms in WiMAX. *Proceedings of National Conference on Recent Developments in Computing and its Applications (NCRDCA ´09)*. ISBN 978-93-80026-78-7, New Delhi, India, August 2009.

IEEE 802.16 Working Group. IEEE Standard for Local and Metropolitan Area Networks - Part. 16: Air Interface for Fixed Broadband Wireless Access Systems. October, 2004.

IEEE Standard 802.16e-2005,Part16. IEEE Standard for Local and Metropolitan Area Networks- Part 16: Air Interface for Fixed and Mobile Broadband Wireless Access Sys-tems, 2005.

Karim, A. et al. (2010). Modeling and Resource Allocation for Mobile Video over WiMAX Broadband Wireless Networks. *IEEE Journal on Selected Areas in Communications* ,Vol. 28, No. 3, (April 2010) pp. 354-365. ISSN 0733-8716.

Khan, M. A. S. et al. (2010). Performance Evaluation and Enhancement of Uplink Scheduling Algorithms in Point to Multipoint WiMAX Networks. *European Journal of Scientific Research*. Vol.42 No.3 (2010), pp.491-506.

Lakkakorpi, J., Sayenko, A. (2009). Uplink VoIP Delays in IEEE 802.16e Using Differents ertPS Resumption Mechanisms, *Proceedings of the Third International Conference on Mobile Ubiquitous Computing, Systems, Services and Technologies*. ISBN 978-0-7695-3834-1, Slieme, Malta, October, 2009.

Li, B. et al. (2007). A Survey on Mobile WiMAX. *IEEE Communications Magazine*, Vol. 45 No 12, (December, 2007) pp. 70-75, ISSN 0163 6804.

Liu, Q. et al. (2006), A Cross-Layer Scheduling Algorithm With QoS Support in Wireless Networks. *IEEE Transactions on Vehicular Technology*, Vol. 55, No 3, (May 2006) pp. 839-847, ISSN 0018-9545.

Ma, M. (2009). *Current Technology Developments of WiMAX Systems* (ed. 1), Springer. ISBN 978-1-4020-9299-2, Singapore.

Msadaa, I. C., Camara, D., Filali, (2010). Scheduling and CAC in IEEE 802.16 Fixed BWNs: A Comprehensive Survey and Taxonomy. *IEEE Communications Surveys & Tutorials*. Vol. 12, No. 4, (May, 2010) pp. 459 - 487, ISSN 1553-877X.

Nuaymi, L. (2007). WiMAX Physical Layer, In: *WiMAX: Technology for Broadband Wireless Access*, John Wiley & Sons, ISBN 9780470028087.

Pantolidou, A. and Ephremides, A. (2009). The Scheduling Problem in Wireless Networks. *Journal of Communications and Networks*, Vol. 11, No 5, (October, 2009) pp. 489-499. ISSN 1976-5541.

So-in, C., Jain, R., Tamimi, A. (2010). Scheduling in IEEE 802.16e Mobile WiMAX Networks: Key Issues and a Survey. *IEEE Journal on Selected Areas in Communications (JSAC)*, Vol. 27, No. 2, (February, 2009), pp. 156-151, ISSN 0733-8716.

Sayenko, Alanen, A. Hamalainen, O. (2008). Scheduling solution for the IEEE 802.16 base station. *Internation Journal of Computer Networks*. Vol. 52, No. 1, (January 2008) pp. 96 - 115. ISSN 1985-4129.

Sayenko, A. et al. (2006). Comparison and analysis of the revenue-based adaptive queuing models. *Computer Networks*, vol. 50, No. 8 (June 2006) pp. 1040-1058.

Sheu, T. L.; Huang, K. C.; (2011). Adaptive bandwidth allocation model for multiple traffic classes in IEEE 802.16 worldwide interoperability for microwave access networks. *IET Communications*, VoL. 5, No. 1, (January 2011) pp. 90 - 98, ISSN 1751-8628.

Shreedhar, M.; Varghese, G. (1995). Efficient fair queueing using deficit round robin. *IEEE/ACM Transactions on Networking*, Vol. 4, pp. 375 – 385, june, 1996. ISSN 1063-6692.

Teixeira, M. A., Guardieiro, P. R. (2010). A Predictive Scheduling Algorithm for the Uplink Traffic in IEEE 802.16 Networks. *Proceeding of The 12th IEEE International Conference on Advanced Communication Technology (ICACT)*,ISBN 978-1-4244-5427-3, February 2010.

Wu, C. F. (2010). Real-time scheduling for Multimedia Services in IEEE 802.16 Wireless Metropolitan Area Network. *Information Technology Journal*, Vol. 9, No. 6, (June 2010) pp. 1053-1067.

Wongthavarawat, K. & Ganz, A. (2003). Packet Scheduling for QoS Support in IEEE 802.16 Broadband Wireless Access Systems. *International Journal of Communications Systems*. Vol. 16, No. 1, (February, 2003) pp. 81-96. ISSN: 1074-5351.

2

A Comprehensive Survey on WiMAX Scheduling Approaches

Lamia Chaari, Ahlem Saddoud,
Rihab Maaloul and Lotfi Kamoun
*Electronics and Information Technology Laboratory,
National School of Engineering of Sfax (ENIS),
Tunisia*

1. Introduction

The institute of Electrical and Electronics IEEE 802.16 standard is a real revolution in wireless metropolitan area networks (wireless MANs) that enables high-speed access to data, video and voice services. The IEEE 802.16 is mainly aimed at providing broadband wireless access (BWA). Thus, it complements existing last mile wired networks such as cable modem and xDSL. Its main advantage is fast deployment which results in cost saving.

WiMAX networks are providing a crucial element in order to satisfy on-demand media with high data rates. This element is the QoS and service classes per application. In Broadband Wireless communications, QoS is still an important criterion. So the basic feature of WiMAX network is the guarantee of QoS for different service flows with diverse QoS requirements. While extensive bandwidth allocation and QoS mechanisms are provided, the details of scheduling and reservation management are left not standardized. In fact, the standard supports scheduling only for fixed-size real-time service flows. The scheduling of both variable-size real-time and non-real-time connections is not considered in the standard. Thus, WiMAX QoS is still an open field of research and development for both constructors and academic researchers. The standard should also maintain connections for users and guarantee a certain level of QoS. Scheduling is the key model in computer multiprocessing operating system. It is the way in which processes are designed priorities in a queue. Scheduling algorithms provide mechanism for bandwidth allocation and multiplexing at the packet level.

In this chapter, we proposed a survey on WiMAX scheduling scheme in both uplink and downlink traffic. The remainder of this chapter is organized as follows: Section 2 presents the QoS support in WiMAX networks, and section 3 presents scheduling mechanisms classifications. In section 4, we discuss channel-unaware and channel aware schedulers proposed for both uplink and downlink. We present the relay WiMAX schedulers in section 5. Section 6 presents a comparative study. Finally, we conclude the chapter in section 7.

2. Quality of services provisioning in WiMAX networks

2.1 Services and parameters

In WiMAX (Jeffrey,2007)(Labiod & Afifi, 2007)(Shepard,2006)(Nuaymi, 2007), a service flow is a MAC transport service provided for transmission of uplink, downlink traffic, and is a key concept of the QoS architecture. Each service flow is associated with a unique set of QoS parameters, such as latency, jitter throughput, and packet error rate. The various service flows admitted in a WiMAX network are usually grouped into service flow classes, each identified by a unique set of QoS requirements. This concept of service flow classes allows higher-layer entities at the subscriber station (SS) and the base station (BS) to request QoS parameters in globally consistent ways. The WiMAX networks is a connection-oriented MAC in that it assigns traffic to a service flow and maps it to MAC connection using a Connection ID (CID). In this way, even connectionless protocols, such as IP and UDP, are transformed into connection-oriented service flows. The connection can represent an individual application or a group of applications sending with the same CID. A service flow is a unidirectional flow of packets that is provided a particular QoS. The SS and BS provide this QoS according to the QoS parameter set defined for the service flow. Each data service is associated with a set of QoS parameters that quantify its behavior aspects. These parameters are managed through a series of MAC management messages referred to as DSA, DSC, and DSD. The DSA messages create a new service flow. The DSC messages change an existing service flow. The DSD messages delete an existing service flow. An SS wishing to either create an uplink or downlink service flow sends a request to the BS using a DSA-REQ message. The BS checks the integrity of the message and, if the message is intact, sends a message received (DSX-RVD) response to the SS. The BS checks the SS's authorization for the requested service and whether the QoS requirements can be supported, generating an appropriate response using a DSA-RSP message. The SS concludes the transaction with an acknowledgment message (DSA-ACK). An SS that needs to change a service flow definition performs the following operations. The SS informs the BS using a DSC-REQ. The BS checks the integrity of the message and, if the message is intact, sends a message received (DSX-RVD) response to the SS. The BS shall decide if the referenced service flow can support this modification. The BS shall respond with a DSC-RSP indicating acceptance or rejection. In the case when rejection was caused by presence of non-supported parameter of non-supported value, specific parameter may be included into DSC-RSP. The SS reconfigures the service flow if appropriate, and then shall respond with a DSC-ACK. Any service flow can be deleted with the DSD messages. When a service flow is deleted, all resources associated with it are released. This mechanism allows an application to acquire more resources when required. Multiple service flows can be allocated to the same application, so more service flows can be added if needed to provide good QoS.

Five services are supported in the mobile version of WiMAX: Unsolicited Grant Service (UGS), Real-Time Polling Service (rtPS), Extended Real-Time Polling Service (ErtPS) , non-real-time polling service (nrtPS), and Best Effort (BE). Each of these scheduling services has a mandatory set of QoS parameters that must be included in the service flow definition when the scheduling service is enabled for a service flow. These are summarized in Table 1.

QoS Category	Applications	QoS Specifications
UGS Unsolicited Grant Service	VoIP	-Maximum Sustained Rate -Maximum Latency Tolerance -Jitter Tolerance
rtPS Real-Time Polling Service	Streaming Audio or Video	-Minimum Reserved Rate -Maximum Sustained Rate -Maximum Latency Tolerance -Traffic Priority
ErtPS Extended Real-Time Polling Service	Voice with Activity Detection (VoIP)	-Minimum Reserved Rate -Maximum Sustained Rate -Maximum Latency Tolerance -Jitter Tolerance -Traffic Priority
nrtPS Non-Real-Time Polling Service	File Transfer Protocol (FTP)	-Minimum Reserved Rate -Maximum Sustained Rate -Traffic Priority
BE Best-Effort Service	Data Transfer, Browsing, Web etc.	Maximum Sustained Rate -Traffic Priority

Table 1. WiMAX applications and QoS specifications

2.2 Functional elements

Based on the IEEE 802.16e specification (Standard, 2006), the proposed QoS functional elements includes call admission control (CAC), scheduling and bandwidth allocation.

2.2.1 Bandwidth allocation schemes

During initialization and network entry, the BS assigns up to three dedicated CID to each SS in order to provide the SS the ability to sends and receives control messages. The SS can send the bandwidth request message to the BS by numerous methods. In the IEEE 802.16 standard, bandwidth requests are normally transmitted in two modes: a contention mode and a contention-free mode (polling). In the contention mode, the SSs send bandwidth-requests during a contention period, and the BS using an exponential back-off strategy resolves contention. In the contention-free mode, the BS polls each SS, and an SS in reply sends its BW-request. The basic intention of unicast polling is to give the SS a contention-free opportunity to tell the BS that it needs bandwidth for one or more connections In addition to polling individual SSs, the BS may issue a broadcast poll by allocating a request interval to the broadcast CID, when there is insufficient bandwidth to poll the stations individually.

Similarly, the standard provides a protocol for forming multicast groups to give finer control to contention-based polling. SSs with currently active UGS connections may set the

PM bit (bit PM in the Grant Management subheader) in a MAC packet of the UGS connection to indicate to the BS that they need to be polled to request bandwidth for non-UGS connections. Variable bandwidth assignment is possible in rtPS, nrtPS and BE services, whereas UGS service needs fixed and dedicated bandwidth assignment. The BS periodically in a fixed pattern offers bandwidth for UGS connections so UGS connections do not request bandwidth from the BS. Each connection in an SS requests bandwidth with a BW Request message, which can be sent as a stand-alone packet or piggybacked with another packet. A bandwidth request can be incremental or aggregate. An incremental bandwidth request means the SS asks for more bandwidth for a connection. An aggregate bandwidth request means the SS specifies how much total bandwidth is needed for a connection. Most requests are incremental, but aggregate requests are occasionally used so the BS can efficiently correct its perception of the SSs needs.

Furthermore, the IEEE 802.16 MAC accommodates two classes of SS, differentiated by their ability to accept bandwidth grants simply for a connection or for the SS as a whole. Both classes of SS request bandwidth per connection to allow the BS uplink-scheduling algorithm to properly consider QoS when allocating bandwidth. With the grant per connection (GPC) class of SS, bandwidth is granted explicitly to a connection, and the SS uses the grant only for that connection. With the grant per SS (GPSS) class, SSs are granted bandwidth aggregated into a single grant to the SS itself. GPC is more suitable for few users per subscriber station. It has higher overhead, but allows a simpler SS GPSS is more suitable for many connections per terminal. It is more scalable, and it reacts more quickly to QoS needs. It has low overhead, but it requires an intelligent SS.

Based on the methods by which the SS can send the bandwidth request message to the BS, bandwidth allocation mechanisms can be classified according table 2.

2.2.2 Call Admission Control

Researchers have characterized CAC as the decision maker for the network. When a subscriber station SS send a request to the base station (BS) with a certain QoS parameters for a new connection, the BS will check whether it can provide the required QoS for that connection. If the request was accepted, the BS verifies whether the QoS of all the ongoing connections can be maintained. Based on this it will take a decision on whether to accept or reject the connection. The process described above is called as CAC mechanism. The basic components in an admission controller are performance estimator which is used to obtain the current state of the system; resource allocator uses this state to reallocate available radio resource. Then the admission control decision is made to accept or reject an incoming connection. A connection is admitted: if there is enough bandwidth to accommodate the new connection. The newly admitted connection will receive QoS guarantees in terms of both bandwidth and delay and QoS of existing connections must be maintained (Chou et al,2006). A more relaxed rule would be considered to limit admission control decision (to reject) to applications with real-time hard constraints, for example, IP telephony and video conferencing. For other requests (e.g: video streaming, web browsing) if there are insufficient resources, one can provide throughput less than requested by them. A simple admission control decision can be evident: if there are enough available resources in the BS so new connections are admitted else it will be rejected. However, a simple admission

Type	QoS Classes	Mechanisms
Unsolicited request	UGS and ertPS	- Periodically allocates bandwidth at setup stage: - No overhead and meet guaranteed latency for real-time service - Exhausted bandwidth if it is granted and the flow has no packets to send.
Poll-me bit (PM)	UGS	-Asks BS to poll non UGS connections implicitly in MAC header No overhead but Still needs the unicast polling
Piggybacking	ertPS, rtPS, BE & nrtPS	- Piggyback BWR over any other MAC packets being sent to the BS. - Do not need to wait for poll, Less overhead; 2 bytes vs. 6 bytes Grant management.
Bandwidth stealing	nrtPS and BE	- Sends BWR instead of general MAC packet - BWR (6 bytes = MAC header) - Do not need to wait for poll
Contention region (WiMAX)	ertPS, nrtPS and BE	- MSs use contention regions to send BWR ➔ Need the backoff mechanism - Overhead Adjustable - Reduced polling overhead
Codeword over CQICH	ertPS	- Specifies codeword over CQICH - Makes use of CQI channel - Limit number of bandwidth on CQICH
CDMA code based BWR (Mobile WiMAX)	nrtPS and BE	- MS chooses one of the CDMA request codes from those set aside for bandwidth requests. - Six sub channels over 1 OFDM symbol for up to 256 codes - Reduced polling overhead compared to contention region - Results in one more frame delay compared to contention region
Unicast Polling	ertPS, rtPS, nrtPS and BE	- BS polls each MS individually and periodically. - Guarantees that MS has a chance to ask for bandwidth - More overhead BWR (6 bytes per MS) periodically
Multicast, Broadcast and Group Polling	ertPS, nrtPS and BE	- BS polls a multicast group of MSs. - BWR (6 bytes) per multicast ➔ Reduced polling overhead - Some MSs may not get a chance to request bandwidth; need contention resolution technique.

Table 2. Taxonomy of Bandwidth request mechanisms

control is not efficient to guarantee QoS of different types of connections and in the same time, it can affect the performance of IEEE 802.16 network. An important question might be asked: What is the decision of the call admission module when no resources are available for new flows? The answer must be a solution to avoid dropping and blocking new connection requests when it is possible. These solutions are presented in the proposals described below. We present a classification and a description of CAC algorithms proposed in the literature for PMP (Point-to- Multipoint) mode. We classify CAC proposals into two classes. The first category is called *"with degradation"*; it is based on decreasing the resources provided to existing connections in the purpose to allow a new service flow to be accepted in the network. In the second policy named *"without degradation"*, it is forbidden to adopt any strategy of degradation in order to maintain the QoS of existing connections. Figure1 shows a diagram with the topics used in the classification.

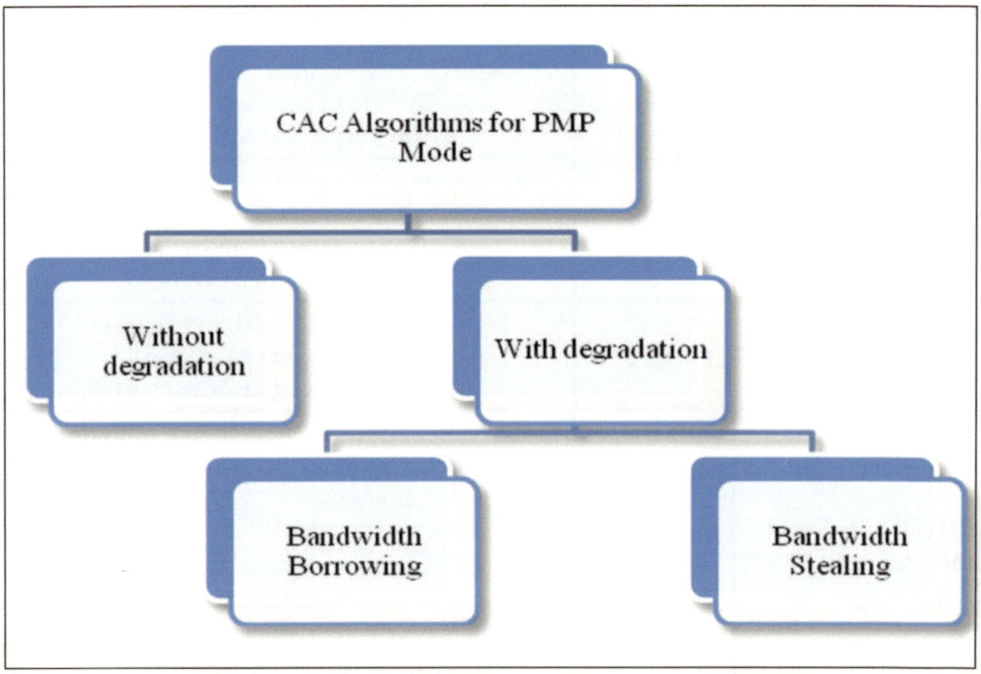

Fig. 1. Proposed classification for WiMAX CAC algorithms in PMP mode

First, *"without degradation"* policy is more flexible than the second one as it offers more opportunities and chance for new requests to be accepted and to get the possible resources when it is necessary. Second, CAC schemes based on degradation strategies are unfortunately less conservative and not simple.

We classify the CAC scheme based on degradation policy in two sub-classes: based on bandwidth borrowing mechanism, or bandwidth stealing. The main concept of these CAC schemes is to decrease the resources afforded to existing connections in order to support requests of a new service flow and to satisfy their demand.

We have regrouped and compared the most related CAC proposals in table 3.

Proposals	QoS Parameter (min/max)	Analytical validation	Token Bucket policy	Bandwidth estimation	Degradation strategy
(H.Wang et al, 2005)	Max Bandwidth Utilization	Markov	N	N	borrowing
(Zhu & Lu, 2006)	Max Bandwidth Utilization	Markov	N	N	borrowing
(Kalikivayi et al, 2008)	Delay guarantee	Markov	S	S	borrowing
(Kitti & Aura, 2003)	Delay guarantee	N	S	N	N
(Wang et al, 2007)	Min blocking probability	N	N	N	borrowing
(Tzu-Chieh et al, 2006)	Delay guarantee	Markov chain	S	N	stealing
(Shafaq et al, 2007)	Min blocking probability	N	N	S	N
(Chandra & Sahoo, 2007)	Delay & jitter	N	N	S	N
(Yu et al, 2009)	Delay, Min blocking probability	Markov chain	N	N	N
(Rango et al, 2011)	throughput, average delay and jitter	N	N	S	N
(Shida & Zisu, 2008)	Max throughput of all flows and decrease the delay of the VBR	N	N	S	N

N: Not supported S: supported

Table 3. CAC in IEEE 802.16 PMP Mode: A Comparative table

An admission control module in BS (J.Chen et al, 2005a) (Carlos,2009) has as input a Dynamic Service Addition (DSA; essentially a new connection), Dynamic Service Change (DSC) or a Dynamic Service Deletion (DSD) requests, either. These need to be considered in terms of a set of predefined QoS parameters. It also needs to know the current resource state of the network, which it can only determine by consulting the Scheduler. With that information, it applies the particular CAC algorithm and informs the scheduler of whether a request has been admitted or not. Most of the scheduling algorithm presented in literature assumes a simple CAC is present but this is inappropriate in some cases. Since both CAC and scheduling handle, the QoS a proper CAC algorithm is needed in order to guarantee the promised QoS. Sometimes CAC and scheduling algorithm working on different criteria can interfere, which necessitate CAC algorithms that works in an independent manner from the scheduling algorithm based on bandwidth and delay prediction (Castrucci et al,2008).

2.2.3 MAC scheduling services

In WiMAX network, a service flow is a MAC transport service provided for transmission of uplink, downlink traffic, and is a key concept of the QoS architecture. Each service flow is associated with a unique set of QoS parameters, such as latency, jitter throughput, and packet error rate. The various service flows admitted in a Mobile WiMAX network are usually grouped into service flow classes. This concept of service flow classes allows higher-layer entities at the SS and the BS to request QoS parameters in globally consistent ways. A service flow is a unidirectional flow of packets that is provided a particular QoS. The SS and BS provide this QoS according to the QoS Parameter Set defined for the service flow.

A service flow is partially characterized by the following attributes: (Standard, 2004)

- Service Flow ID: An SFID is assigned to each existing service flow. The SFID serves as the principal identifier for the service flow in the network. A service flow has at least an SFID and an associated direction. The SFID identifies a services which in turn identifies the right of the IEEE 802.16 SS to certain system resources, and also defines which of user's packets will be mapped to the corresponding MAC connection
- CID: Mapping to an SFID that exists only when the connection has an admitted or active service flow.
- "ProvisionedQoSParamSet": A QoS parameter set provisioned: When a service level was set up (neither reserved nor allocated).
- "AdmittedQoSParamSet": Defines a set of QoS parameters for which the BS (and possibly the SS) is reserving resources. The principal resource to be reserved is bandwidth.
- "ActiveQoSParamSet": Defines a set of QoS parameters defining the service actually being provided to the service flow. Only an Active service flow may forward packets.
- Authorization Module: A logical function within the BS that approves or denies every change to QoS Parameters and Classifiers associated with a service flow.

Scheduling is the main component of the MAC layer that assures QoS to various service classes. The MAC scheduling Services are adopted to determine which packet will be served first in a specific queue to guarantee its QoS requirement. In fact, the scheduler works as a distributor in order to allocate the available resources among SSs. Thus, an efficient scheduling algorithm could enhance the QoS provided by IEEE 802.16 network. As well, scheduling architecture should ensure good use of bandwidth, maintain the fairness among users, and satisfy the requirements of QoS. It is important to mention that Scheduling algorithms can be implemented in the BS as well as in the SSs. Those are implemented at the

BS have to deal with both uplink and downlink traffics. Therefore, there are three different schedulers: two at the BS schedule the packet transmission in downlink and uplink sub frame and the latter at the SS for uplink to apportion the assigned BW to its connections.

In order to indicate the allocation of transmission intervals in both uplink and downlink, in each frame, the signaling messages UL-MAP and DL-MAP are broadcasted at the beginning of the downlink sub frame. The scheduling decision for the downlink traffic is relatively simple as only the BS transmits during the downlink sub frame and the queue information is located in the BS. While, an uplink scheduler at the BS must synchronize its decision with all the SSs.

We describe a better understanding of some specific factors that should be considered in the scheduling policy as follows:

- **QoS requirements**: An efficient scheduling algorithm could enhance the QoS specification of the different types of service classes as it is mentioned in table1.
- **Fairness**: Besides assuring the QoS requirements, the bandwidth resources should be shared fairly between users. Thus, fairness represents one of the most challenging problems in the scheduling approaches.
- **Channel Utilization**: It is the fraction of time used to transmit data packets. It is almost equal to the channel capacity in PMP communications. A scheduling mechanism has to check that resources are not allocated to SSs that do not have enough data to send, thus resulting in wastage of resources.
- **Complexity**: The scheduling algorithm must be simple, fast and should not have a prohibitive implementation complexity as it serves different service classes in various constraints.
- **Scalability**: It is the capacity to handle a growing number of flows. A scheduling algorithm should efficiently operate as the number of connections increases.
- **Cross-layer design**: A scheduling algorithm should take into account the characteristics of different layers (e.g. the adaptive modulation and coding (AMC) scheme). It is significant to consider the burst profile in such scheduling policy in order to improve system performance.

2.3 A QoS framework

A novel design paradigm, the so-called cross-layer optimization, is one of the most promising issues of research for the improvement of wireless communication systems (Zhang & Chen, 2008). Cross-layer operation can be formulated conceptually as the selection of strategies across multiple layers such that the resultant interlayer operation is optimized. Each layer has optimal schemes under given states, such as channel condition and QoS parameters, and the combination of schemes selected in all layers results in optimized interlayer operation. In this section, we elaborate architecture for integrated QoS control with respect to cross-layer design. The IEEE 802.16 uses the PMP centralized MAC architecture where the BS scheduler controls all the system parameters (radio interface). It is the role of the BS scheduler to determine the burst profile and the transmission periods for each connection; the choice of the coding and modulation parameters are decisions that are taken by the BS scheduler according to the quality of the link and the network load and demand. Therefore, the BS scheduler must monitor the received carrier-to-interference-plus-noise-ratio (CINR) values (of the different links) and then determine the bandwidth requirements of each station taking into consideration the service class for this connection and the quantity of traffic required. Figure2 shows the BS scheduler operation based on cross layer approach.

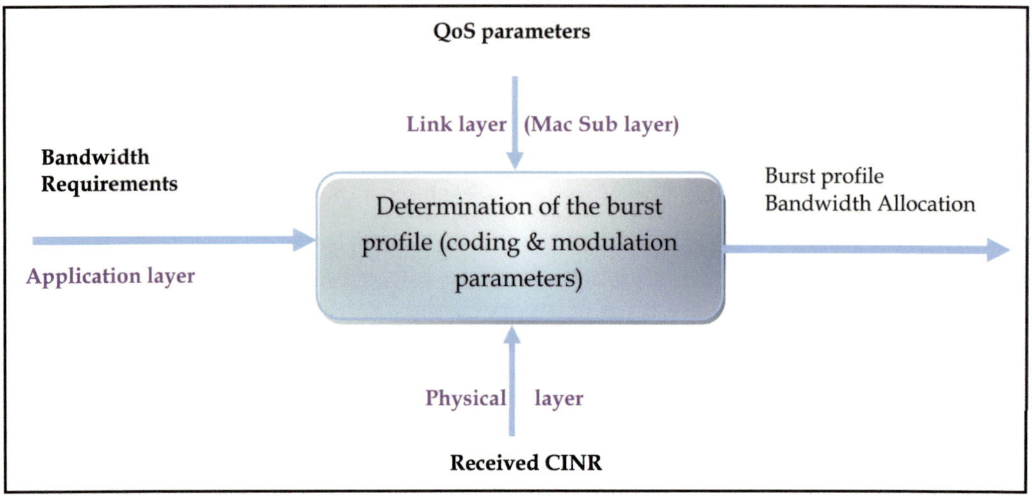

Fig. 2. Burst profile parameter

In figure 3, we give an idea about the architecture of the IEEE 802.16 QoS platform of the BS and SSs to support multimedia services.

This chapter emphasis especially in relationship between modules and the control information flows to provide cross-layer operation. In the downlink, all decisions related to the allocation of bandwidth to various SSs are made by the BS on a per CID basis. As MAC PDUs arrive for each CID, the BS schedules them for the PHY resources, based on their QoS requirements. Once dedicated PHY resources have been allocated for the transmission of the MAC PDU, the BS indicates this allocation to the SS, using the DL-MAP message. While the scheduler independently builds the DL-MAP and UL-MAP, the CAC needs to closely consult these in order to determine the available resources and consequently, whether to admit or deny a connection of a particular traffic type. Frames arriving at the BS were previously scheduled on the UL-MAP to be either BW requests or data PDUs to be forwarded on the DL or data PDUs destined for the BS itself. A BW request must be taken up by the CAC that decides whether to admit the request and, if so, will pass this information to the centralized scheduler.

The UL packets from the upper layer are classified into service flows by a packet classifier within the SS, and the SS requests BW according to the UL grant/scheduling type. From the amount of BW requested, the BS estimates the queue status information of each SS. In IEEE 802.16 systems, all resources are managed by the BS, thus the BS performs channel- and QoS-aware scheduling, on the basis of measured UL channel information, the negotiated QoS parameter and estimated queue status.

In the uplink, the SS requests resources by either using a stand-alone bandwidth-request MAC PDU or piggybacking bandwidth requests on a generic MAC PDU, in which case it

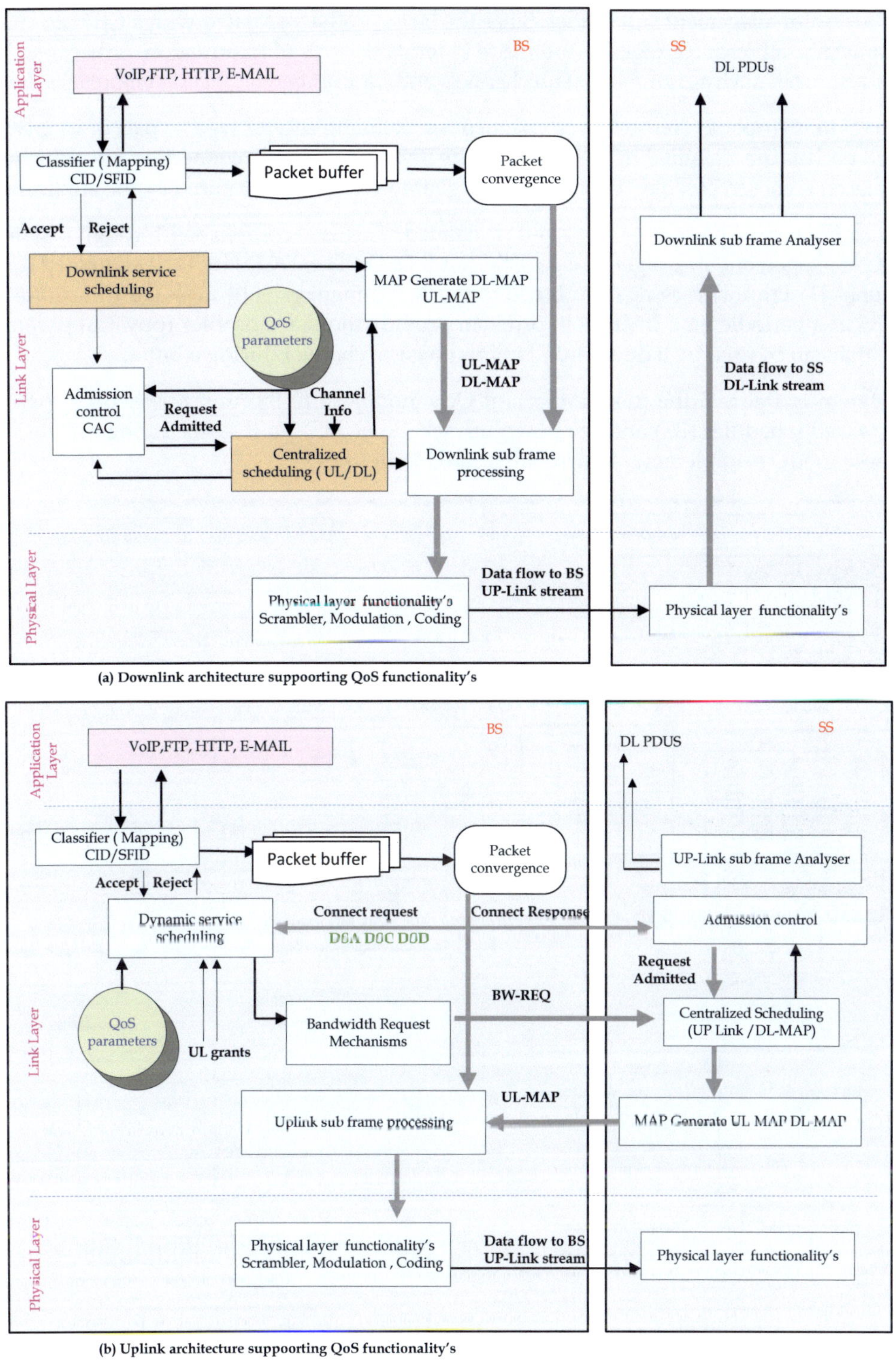

Fig. 3. QoS support for multimedia services in IEEE 802.16

uses a grant-management sub header. Since the burst profile associated with a CID can change dynamically, all resource requests are made in terms of bytes of information, rather than PHY layer resources, such as number of sub channels and/or number of OFDM symbols.

Each SS to BS (uplink) connection is assigned a scheduling service type as part of its creation. When packets are classified in the Convergence Sublayer (CS), the connection into which they are placed is chosen based on the type of QoS guarantees that are required by the application.

Service flows may be created, changed or deleted. This is accomplished through a series of MAC management messages: DSA, DSC and DSD. The DCD/UCD (Downlink/Uplink Channel Descriptor) message are broadcasted MAC management message transmitted by the BS at a periodic time interval in order to provide the burst profiles (physical parameter sets) that can be used by a downlink/Uplink physical channel during a burst.

As shown in figure 3 the most important QoS modules are the uplink scheduler (SS), the centralized scheduler (BS) and the downlink scheduler (BS), so the scheduling architectures of those modules implementation are illustrated in figure 4.

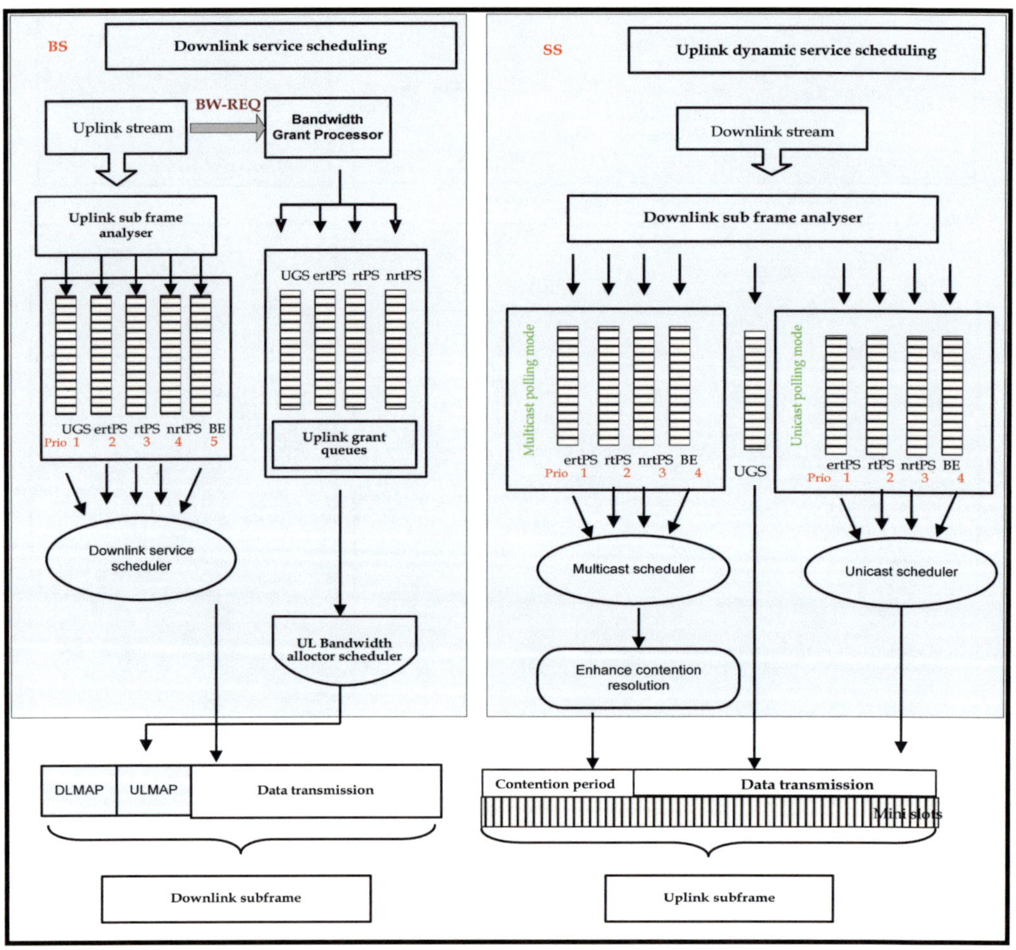

Fig. 4. Scheduling architecture in BS and SS using TDD mode

The WiMAX MAC layer uses a scheduling service to deliver and handle SDUs and MAC PDUs with different QoS requirements. A scheduling service uniquely determines the mechanism the network uses to allocate UL and DL transmission opportunities for the PDUs. When packets are classified in the convergence sublayer, the connection into which they are placed is chosen based on the type of QoS guarantees that are required by the application.

3. Scheduling mechanisms classification

In the research literature, we find an important number of studies focus on mechanisms for packet scheduling in WiMAX networks (Kitti & Aura, 2003)(Sonia & Hamid, 2010)(Ridong et al, 2009)(G.Wei et al, 2009). We classify the scheduling methods proposed in the literature of IEEE 802.16 as is shown in figure 5. The scheduling algorithms used in WiMAX network could be originally designed for wired network in order to satisfy the QoS requirements. Therefore, these algorithms do not take into account WiMAX channel characteristics. The Schedulers of this kind is belonging to the channel unaware scheduling category. But the scheduling algorithm which takes into account the variability of channel characteristics can be categorized as channel aware scheduler. The objective of the following sections is to provide a comprehensive survey on the scheduling research works proposed for WiMAX. These works are described according to the above taxonomy illustrated by the figure5.

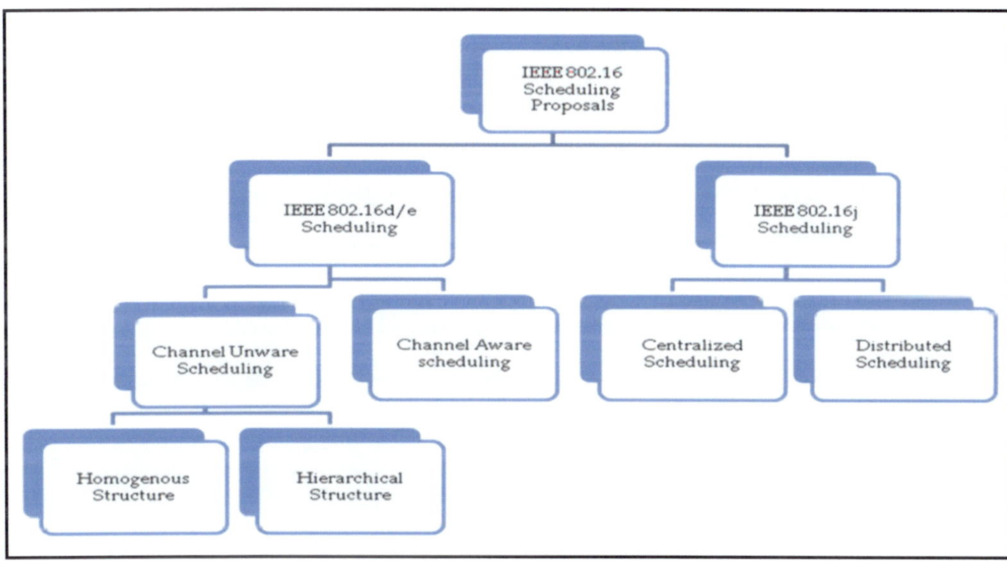

Fig. 5. Proposed classification for WiMAX Scheduling algorithms

4. IEEE 802.16e/d scheduling

4.1 Channel unaware scheduling

The algorithms belonging to this class are classical schedulers. The algorithms applied in both homogenous and hierarchical structures were originally designed for wired networks but are used in WiMAX in order to satisfy the QoS requirements. Therefore, the algorithms of this category do not consider the WiMAX channel conditions such as the channel error and loss rates.

4.1.1 Homogenous structures

Uplink homogeneous schedulers

This category of scheduling is based on simple algorithms such as Earliest Deadline First (EDF)(S.Ouled et al, 2006), Round Robin (RR), Fair Queuing (FQ), and their derivatives. A modified version of the Deficit Round Robin (DRR) is proposed in (Elmabruk et al, 2008), as a scheduling algorithm to ensure the QoS in the IEEE 802.16. The authors try to preserve the available simplicity in the original DRR design which provides O(1) complexity. The proposed scheme has one queue for both UGS and Unicast polling, and one queue for BE and a list of queues for rtPS and nrtPS. Each queue in the list represents one connection as shown in figure 6. The list is updating in each frame by adding new queues and removing the empty queues from the list. The bandwidth requirement is calculated depending on the traffic type by using the maximum sustained traffic rate r_{max} and the minimum reserved traffic rate r_{min}. Each queue in the list is related with a deficit counter variable to determine the number of the requests to be served in the round and this is incremented in every round by a fixed value called Quantum, which is computed as follow:

$$\text{Quantum} = \sum_{K=0}^{Ki} r_{min}(i, K) \quad (1)$$

Where r_{min} is the minimum reserved traffic rate and Ki is the total connections for the ith class of the service flow. An extra queue has been introduced to store a set of requests whose deadline is due to expire in the next frame.

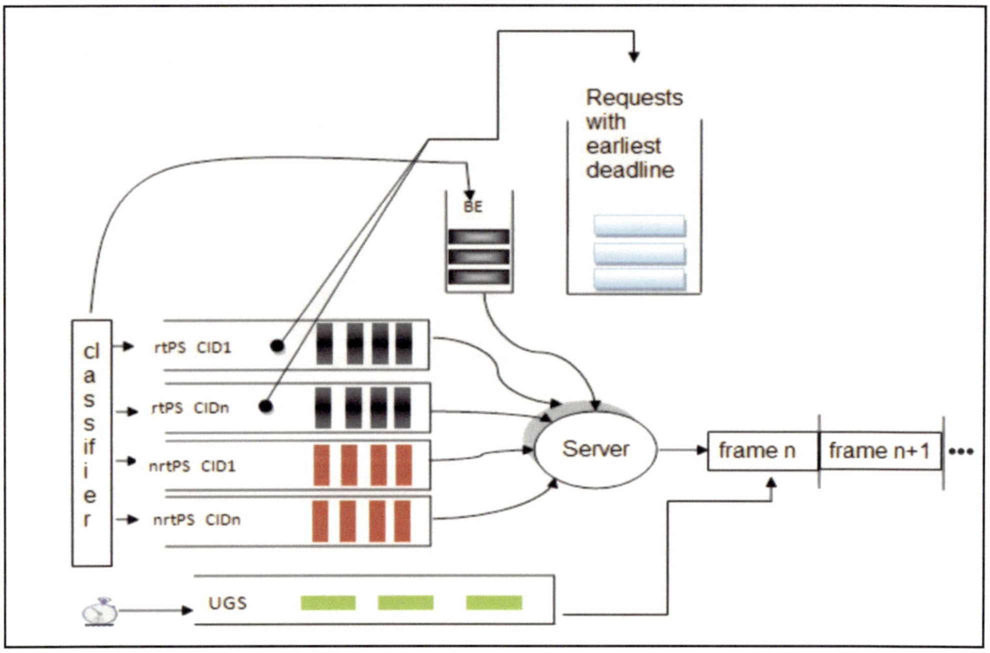

Fig. 6. Scheduler architecture proposed in (Elmabruk et al, 2008)

Every time the scheduler starts the scheduling cycle, this queue will be filled by all rtPS requests, which are expected to miss their deadline in the next frame. In the proposed scheme, it is assumed that the deadline of a request should be equal to the sum of the arrival time of the last request sent by the connection and its maximum delay requirement. In the next scheduling cycle, the scheduler will check if there are any request has been added to this extra queue. If so, the scheduler will then serve this queue after the UGS and polling queue. Once the extra queue becomes empty and there are available BW in the UL_MAP, the scheduler will continuing serving the PS list, using DRR with priority for rtPS, followed by nrtPS. For BE, the remaining bandwidth will assigned using FIFO mechanism.

In (Chirayu & Sarkar, 2009), authors propose an enhancement to the EDF principle to ensure that low priority traffic would not starved. Since the EDF tends to starve the BE traffic in presence of high number of rtPS packets. The WiMAX frame is divided into time slots, and SS are required to transmit packets in these slots, the original packets generated at the application level are fragmented to ensure that these packets fit into and can be transmitted in a time slot. When a packet is fragmented, the last fragmented packet might be of any length from 1 byte to the maximum size, which can be transmitted in a slot. If the last fragment contains lesser number of bytes than the maximum allowable fragment size, then they can stuff a part of a BE packet into this empty section. In this way, two or three such empty slots might be enough to transmit a complete BE packet to the BS. Thus, the chance that BE traffic will find an empty spaces to be transmitted is increase even there more rtPS traffic.

Downlink homogeneous schedulers

Since homogenous algorithms cannot assure the QoS guarantee for different service classes, a limited number of studies focused on this category of scheduling. RR and WRR (Cicconetti et al, 2006)(Sayenko et al, 2008) are applied in IEEE 802.16 networks in order to schedule the downlink traffic. RR algorithm allocates fairly the resources for users even they have nothing to transmit, so it may be non-conserving work scheduler and does not take into account the QoS characteristics. In WRR algorithm, the weights are assigned to adjust the throughput and latency requirements. Variants of RR such as DRR (Cicconetti et al, 2007) are applied for downlink packet scheduling in order to serve the variable size packet. The major advantage of the RR variants is their simplicity; their complexity is O (1).

In (Kim & Kang, 2005) and (Ku et al, 2006), the authors proposed a packet scheduling scheme called DTPQ (Delay Threshold-based priority Queuing) where both real time (RT) and non-real time (NRT) services are supported. The purpose of the proposed DTPQ scheduling scheme aims to maximize the number of users in the system and increasing the total service revenue. The main important parameters taken into account in this scheduling policy is the weight of both RT and NRT services denoted by ω_{RT} and ω_{NRT} respectively. The downlink packet scheduling scheme proposed in (Kim & Kang, 2005) does not address how the delay threshold can be set while an adaptive version of DTPQ scheme is implemented in (Ku et al, 2006). In fact, the delay threshold is updated based on the variation of the weighted sum of the delay for the most urgent RT users and average data rate for RT users.

4.1.2 Hierarchical structures

Uplink hierarchical schedulers

In (Kitti & Aura, 2003), authors introduce a hierarchical structure of bandwidth allocation for IEEE 802.16 systems. Figure 7 shows a sketch of the proposed implementation UPS (Uplink Packet Scheduling). In the first level, the entire bandwidth is distributed in a strict priority manner. UGS has the highest priority, then rtPS, nrtPS, and finally BE. So inter class fairness is not achieved in presence of large number of the higher priority packets. In the second level, different mechanisms are used to control the QoS for each class of service flow.

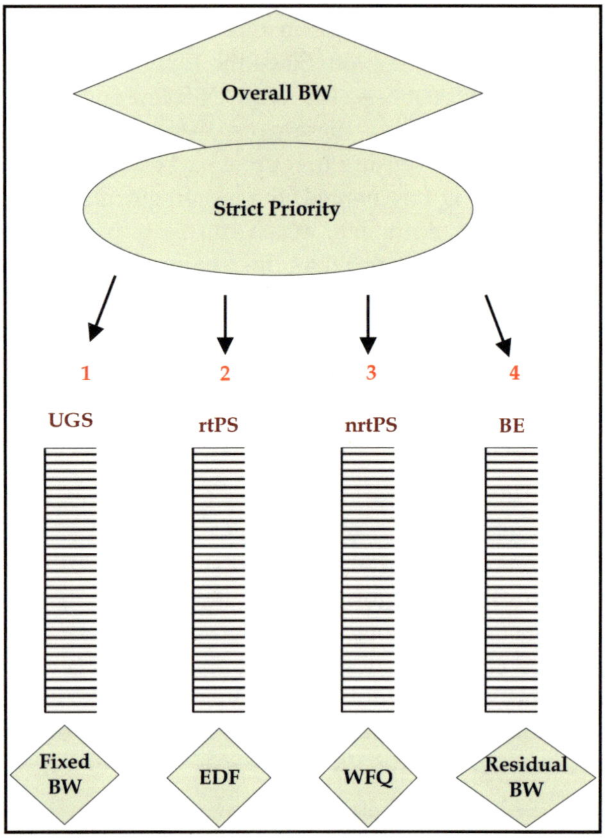

Fig. 7. Hierarchical structure proposed in (Kitti & Aura, 2003)

The uplink packet scheduler allocates fixed bandwidth to the UGS connections. Earliest deadline first (EDF) is used to schedule rtPS service flows, in which packets with the earliest deadline are scheduled first. The nrtPS service flows are scheduled using the weight fair queuing (WFQ) based on the weight of the connection. The remaining bandwidth is equally allocated to each BE connection. The UPS solution is composed of three modules: information, scheduling database and service assignment modules. Here is a brief description of the different of UPS:

- At the beginning of each time frame, the Information Module collects the queue size information from the BW-Requests received during the previous time frame. The

Information Module will process the queue size information and update the Scheduling Database Module.

- The Service Assignment Module retrieves the information from the Scheduling Database Module and generates the UL-MAP.
- BS broadcasts the UL-MAP to all SSs in the downlink subframe.
- BS's scheduler transmits packets according to the UL-MAP received from the BS.

Authors in (Tsu-Chieh et al, 2006) present an uplink packet scheduling with call admission control mechanism using the token bucket. Their proposed CAC is based on the estimation of bandwidth usage of each traffic class, while the delay requirement of rtPS flows shall be met. Each connection is controlled by token rate r_i and bucket size bi. Then, they find an appropriate token rate by analyzing Markov Chain state and according to delay requirements of connections. In their Uplink Packet Scheduling Algorithm, they adopt Earliest Deadline First (EDF) mechanism proposed in (Kitti & Aura, 2003). There is a database that records the number of packets that need to be sent during each frame of every rtPS connection. The disadvantage of this mechanism is that depends on the estimation model that is used.

In (Yanlei & Shiduan, 2005), authors propose a hierarchical packet scheduling model for WiMAX uplink by introducing the "soft-QoS" and "hard-QoS" concepts as shown in figure 8. The rtPS and nrtPS traffic are classified as soft-QoS because their bandwidth requirement varies between the minimum and maximum bandwidth available for a connection. UGS traffic is classified as hard-QoS. The model is able to distribute bandwidth between BE and other classes of traffic efficiently and guarantees fairness among the QoS-supported traffic (UGS, rtPS and nrtPS).

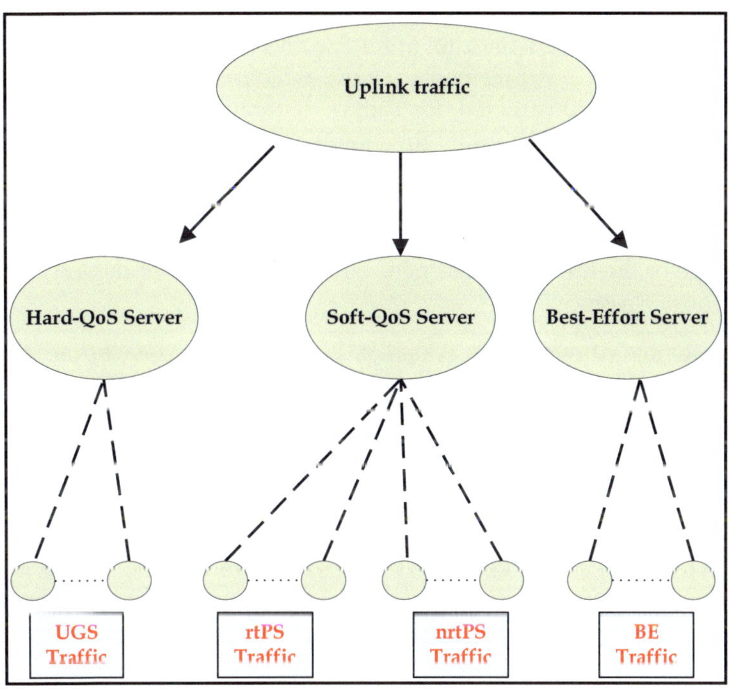

Fig. 8. Hierarchical structure as proposed in (Yanlei & Shiduan, 2005)

The packet-scheduling algorithm comprises of four parts:

1. hard-QoS server scheduling
2. soft-QoS server scheduling
3. best-effort server scheduling
4. Co-scheduling among the above three servers

The four servers implement WFQ (Weighted Fair Queuing) in their queues, for the first three servers a virtual finish time for each packet has to be calculated. The weight must be the weight of the packet and the packet having the smallest time is put at the head of the queue. The co-scheduling server calculates a virtual finish time too but here the weight should be the weight of the queue and the packet with the smallest time is served firstly.

A new distributed uplink Packet scheduling algorithm is proposed in (Sonia & Hamid, 2010). When uplink capacity cannot satisfy the required resource of connections, the traffic of one or some user terminals from user terminals in the overlapping cells are selected for transferring to the neighboring under-loaded cells. The algorithm is described as follow:

In the first step, the service assignment module, as proposed in (Kitti & Aura, 2003), calculates the uplink free capacity and resources required for each connection of a user terminal using the information saved in the scheduling database.

In the second step using the information calculated in the previous step and the traffic characteristics of the scheduling services of the user terminals, the BS checks the uplink free capacity in each time frame. If the free capacity is not enough to be allocated to necessary connections, the BS concludes that a handover is needed.

Authors in (Ridong et al, 2009) propose a utility-based dynamic bandwidth allocation algorithm in IEEE 802.16 networks to minimize the average queuing delay. The utility function is introduced as a supplementary unit, which is related to the average queuing delay of each SS node, is constructed, for QoS consideration, weight factors are introduced for different type of services. The utility function is expressed as follows:

$$U_{i,k}\left(B_{i,k}^{alloc}\right) = 1 - e^{\frac{-1}{\alpha_k \times \overline{D_{i,k}}}} \qquad (2)$$

The disadvantage of the hierarchical structure is the starvation of the lower priority classes by the high priority classes.

In order to avoid this drawback, in (Chafika, 2009), authors develop an algorithm called courteous algorithm that consists of servicing the lower priority traffic without affecting the high priority traffic. Authors analyze two queues c1 and c2, which related respectively to rtPS and nrtPS classes. Packets of the c1 class have priority Pr1, while those of class c2 have priority Pr2.

Four conditions must be satisfied before applying the courteous algorithm in order to serve packet of class c1 before those of c2. These conditions are as follow:

1. Pr1>Pr2
2. $\eta_1(t) < \omega_1$
3. $\eta_2(t) > \omega_2$
4. $\tau_2 < \xi_1$

The first condition is that the priority of queue c1 is higher than that of queue c2. In the second one ω1 represents the tolerated threshold of packet loss rate for class c1 traffic, and η_1 represents the packet loss probability at time t for the class c1, which must not reach the value of ω1. The third condition relates to the probability of packet loss for class c2, which is η_2 at time t just before the application of the courteous algorithm. η_2 is the factor that determines if class c2 traffic needs more bandwidth and ω_2 represents the tolerated threshold of packet loss rate for class c2 traffic. Thus, if η_2 is greater than ω_2, then the packets of this class require to be served. In the fourth condition τ_2 is the time that required to service class c2 packets and should not exceed the tolerated waiting time ξ1 of packets of class c1. The main idea of the courteous algorithm consists in substituting service of packet of high priority with service to lower priority traffic whenever possible. This scheduling scheme is recommended when nrtPS traffic is important with respect to rtPS traffic. One more advantage of this proposal is that it improves indirectly the overall traffic since it contributes to the reduction of the packet loss rate.

Downlink hierarchical schedulers

In (Xiaojing, 2007), an Adaptive Proportional Fairness (APF) scheduling algorithm, was proposed, which is designed to extend the PF scheduling algorithm to a real time service and provides a satisfaction of various QoS requirements. The proposed APF try to differentiate the delay performance of each queue based on the Grant per Type-of- Service (GPTS) principle. The introduced priority function for queue i is defined as:

$$\mu_i(t) = \frac{r_i(t)}{R_i(t) / C_i(t)M_i(t)} \qquad (3)$$

Where $r_i(t)$ is the current data rate, $R_i(t)$ denotes an exponentially smoothing average of the service rate received by SS i up to slot t. $M_i(t)$ is the minimum rate requirement, $C_i(t)$ is the number of connections of the ith queue. Each queue corresponds to one QoS class, respectively. The queue having the highest value of $\mu_i(t)$ is served first. So the priority can be respectively UGS, ertPS, rtPS, nrtPS, and BE. In fact, the packets with minimum deadline or latency are measured at the highest priority level.

In (N.Wei et al, 2011), the authors proposed a QoS priority and fairness scheduling scheme for downlink traffic which guarantees the delay requirements of UGS, ertPS and rtPS service classes. The proposed mechanism is a two-level scheduling scheme that intends to maximize the BE traffic throughput. Firstly, a strict priority between service classes is adapted in the first level as follows UGS > ertPS > rtPS > nrtPS > BE. Secondly, a fixed-size data is granted periodically for UGS service class, an Adaptive Proportional Fairness (APF) scheduling is applied for both rtPS and ertPS service classes, and a Proportional Fairness (PF) scheduling is used for nrtPS and BE service classes.

A comparative study in (Y.Wang et al, 2008) is presented, compared with RR, and PF schemes, APF algorithm outperforms in service differentiation and QoS provisioning. APF is flexible to the system size in terms of the number of I accommodated users.

The priority order applied may starve some connections of lower classes. In (J.Chen et al, 2005b), a Deficit Fair Priority Queue (DFPQ) is introduced in order to reduce the problem of lower priority classes' starvation. A DFPQ is deployed in the first layer with counter to serve different types of service flows in both uplink and downlink. The counter is deceases according to the size of the packets. The scheduler moves to the next class when the counter returns to zero. Three different scheduling algorithms are used for each traffic class in the second layer. The proposed scheme is as shown in figure 9 EDF for rtPS traffics, WFQ for nrtPS class and RR for BE class. A DFPQ is better than the strict priority scheduling in order to achieve the fairness among classes.

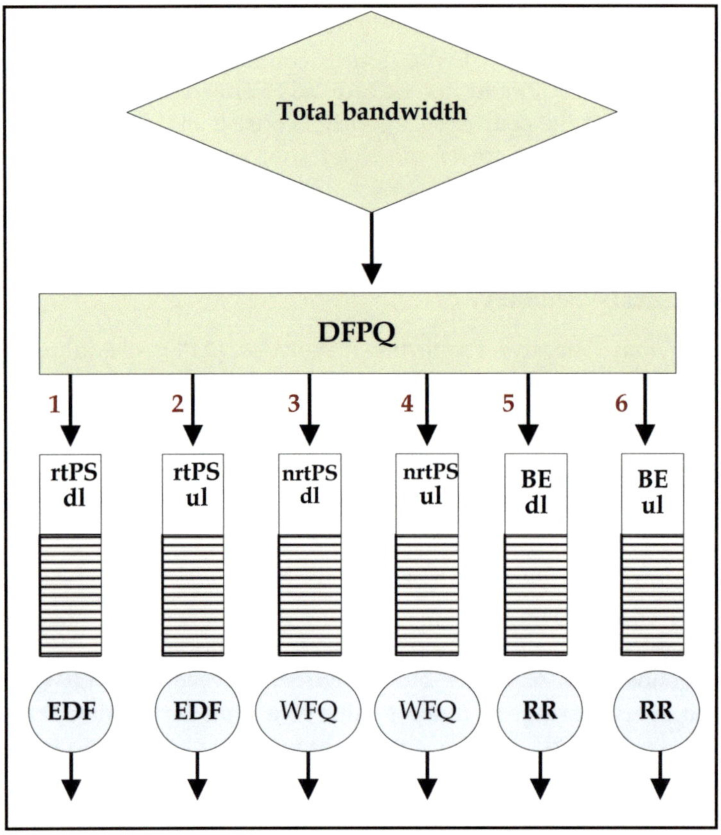

Fig. 9. Deficit Fair Priority Queue (DFPQ) as proposed in (J.Chen et al, 2005b)

4.2 Channel aware scheduling

Uplink aware schedulers

This category is also called opportunistic scheduling algorithms that is proposed for WiMAX and exploit variation in channel quality giving priority to users with better channel quality, while attempting to satisfy the QoS requirements of the multi-class traffic. A cross layer scheduling is proposed in (G.Wei et al, 2009) designed for WiMAX uplink, considering the states of queues, the channel conditions and the QoS requirements of service classes, authors propose a cross layer designed scheduling algorithm called DMIA (Dynamic MCS and Interference Aware Scheduling Algorithm) which can dynamically adapt the varying modulation and coding scheme (MCS) and the interferences in wireless channel. The objective is to maximize the total throughput, while satisfying the QoS requirement of different service classes. So, it is a constrained optimization problem.

Frequently, the cross layer algorithms formulate the scheduling problem as an optimization problem.

The DMIA proposed in (G.Wei et al, 2009) is designed to two-stage. On the first one, the dynamic bandwidth values are set for the five service classes. Therefore, the algorithm can prevent the high priority traffics from occupying too much bandwidth resources, and adjust the amount of scheduling data according to the varying MCS. On the second stage, different connections belong to the same service class will be scheduled according to the priority functions.

In (Liu et al, 2005), the authors propose a priority- based scheduler at the MAC layer for multiple connections with divers QoS requirements, where each connection employs adaptive modulation and coding (AMC) scheme at the physical layer. The authors define a priority function that integrates in its formulation the delay of HOL packet and the minimum required bandwidth. Each non-UGS connection admitted in the system is assigned with a priority, which is updated dynamically based on the channel quality and on its service class. The number of time slots allocated per frame to UGS connections is fixed. The proposed scheduler enjoys flexibility since it does not depend on any specific traffic or channel model. Besides, in (Pratik, 2007) authors have chosen to evaluate the performances of the proposed cross layer in (Liu et al, 2005), their evaluation indicates high frame utilization as it indicates poor performance with respect to average throughput, average delay and fairness.

A Cross-Layer Scheduling Algorithm based on Genetic Algorithm (CLSAGA) under the Network Utility Maximization (NUM) concepts is proposed in (Jianfeng et al, 2009). Adaptive modulation and coding (AMC) scheme and QoS category index of each service flow jointly decide the weights of utility functions to calculate the scheduling scheme of MAC layer. The genetic algorithm can be used to solve optimization problems. The cross layer diagram is shown in figure 10.

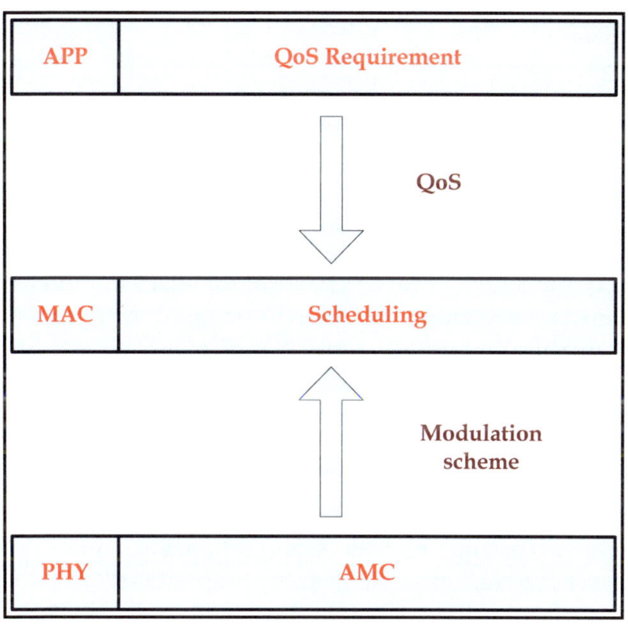

Fig. 10. Cross layer diagram as proposed in (Jianfeng et al, 2009)

Downlink aware schedulers

In (Hongfei et al, 2009), the authors proposed a practical cross-layer framework for downlink scheduling with multimedia traffic called CMA (Connection-oriented Multistate Adaptation) illustrated in figure 11. A multisession MBS scheduling in multicast/Broadcast (MC/BC)-based WiMAX is taken into account in the proposed scheme. The authors adopt the service-oriented design on per-service-flow carrying multisession MBS. The framework performs simultaneous adaptations across protocol stacks on source coding, queue prioritization, flow queuing, and scheduling. CMA achieves the lowest variance value with the fastest convergence curve and lowest max-min variations, which mean that it can provide SSs with better throughput equality in a short time.

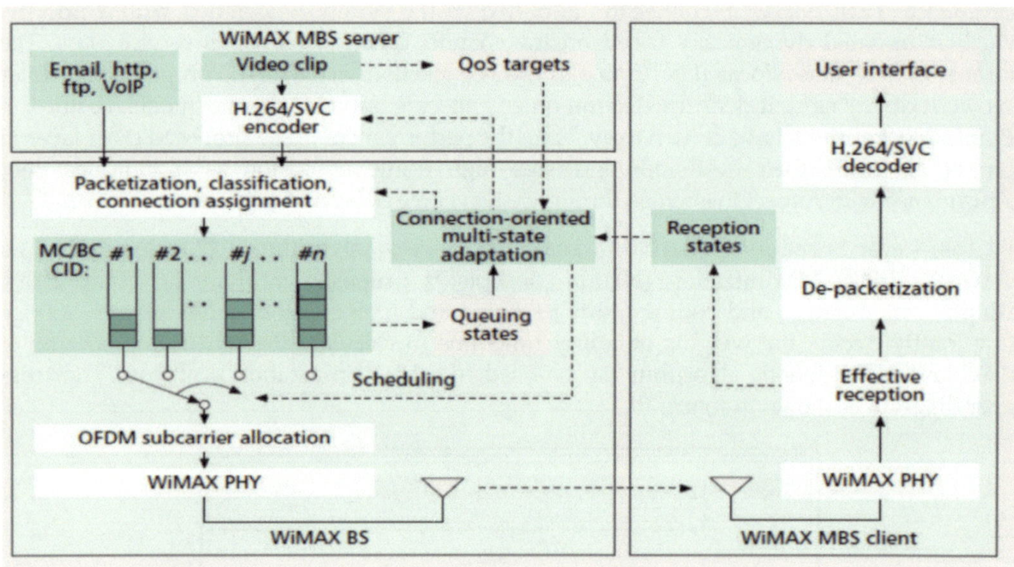

Fig. 11. CMA scheduling framework proposed in (Hongfei et al, 2009)

In (Vishal et al, 2011), the authors proposed a resource allocation mechanism for downlink OFDMA, which aims to maximize the total throughput with lesser complexity while maintaining rate proportionally between users. The BS allocates sub carries to existing users and the number of bits per OFDMA symbol from each user to be transmitted on each sub carries.

The main steps of the proposed mechanism are described as follows:

- Calculate number of subcarriers assigned to each user.
- Assign subcarriers to each user to achieve proportionality.
- Assign total power P_k to each user maintaining proportionality.
- Assign $P_{k,n}$ for each users subcarriers subjected to P_k constraints. Where $P_{k,n}$ is the power assigned per subcarrier per user.

5. IEEE 802.16j scheduling

Unlike in single hop networks, in a Mobile Multihop Relays (MMR) system (Standard, 2008), service scheduling more complicated. Because the BS need to discover if all the RSs (relay stations) in the path to the SS have sufficient resources to support the QoS request. The discovery procedure begins with the BS sending a DSA request message to its subordinate RS. Then the RS sends its own DSA request message to its subordinates RSs and so on until the access RS is reached.

There are two different options of scheduling in MMR networks: centralized and distributed. In the first option, the BS performs the scheduling of all nodes, while in the second option; the relay stations have certain autonomy and can make scheduling decisions for nodes in communications.

An IEEE 802.16j frame structure is divided into access and relay zones, as well as the uplink sub frame that is divided into access zone and relay zone. The IEEE 802.16j standard defines two kinds of relay:

1. Transparent relays: Access zone is used by SS to transmit on access links to the BS and RSs. The relay zone is used by RSs to transmit to their coordinates RSs or BS. This kind of relay operates only in centralized scheduling mode within the topology of maximum two hops.
2. Nontransparent relays: This mode introduces the multihop scenario. There are two ways of transmitting and receiving the frame. The first one is to include multiple relay zones in a frame and relays can alternately transmit and receive in the different zones. The second one is to group frames together into a multi-frame and coordinate a repeating pattern in which relays are receiving or transmitting in each relay zone.

There are three cases of SS/BS communication:

i. The SS is connected to BS directly.
ii. The SS is connected to the BS via a transparent relay.
iii. The SS is connected to the BS via one or more nontransparent relays.

5.1 Distributed scheduling

Uplink relay schedulers

In (Debalina et al, 2010), authors propose a heuristic algorithm, OFDMA Relay Scheduler (ORS) algorithm, for IEEE 802.16j networks. The ORS algorithm is used to schedule traffic for every SS/RS in each scheduling period. A scheduling period consists of an integral number of frames. The ORS scheduler works for all three cases of SS/BS communication and it consists of two main parts:

- Frame division and Bandwidth Estimation:

The frame relay zone is divided into even and relay zone to maintain the half-duplex nature of the node. So, nodes are labeled alternately even or odd. Even nodes transmit in even relay zone and odd nodes transmit in odd relay zone. The BS is assigned an even label. Thus, the children of the BS are labeled odd.

For the Bandwidth Estimation, if the BS obtains information about the CINR (Carrier to Interference-plus-Noise-) and RSSI (Received Single Strength Indication) values, it can determine the data rate used by the sub channel. Therefore, if the BS does not know about the CINR and RSSI values, then the ORS algorithm compute the lower bound of the network capacity by assuming all the slots available for data transmission are modulated at the most robust and least rate.

- Slot Scheduling:

The ORS heuristic schedules slots for a particular service class to all the nodes in a zone before considering the next zone. The proper zone where the slots for a particular node will be allocated is based on whether the child is a MS or RS and the label of an RS. The node is then allocated slots based on the best available sub channels, which are picked for scheduling the link based on their CINR and RSSI values.

The proposed ORS in (Debalina et al, 2010) addresses adaptive zone boundary computation, determination of schedule for prioritized traffic based on traffic demand while incorporating frequency selectivity within a zone and adapting to changing link conditions in IEEE 802.16j networks.

Downlink relay schedulers

In (Yao et al, 2007), a factor-graph-based low-complexity distributed scheduling algorithm in the downlink direction is proposed. The proposed algorithm manages excellent performance by exchanging weighted soft-information between neighboring network nodes to obtain a series of valid downlink transmission schedules that lead to high average values and low standard deviations in packet throughputs.

A factor graph consisting of agent nodes, variable nodes, and edges is a graphic representation for a group of mutually interactive local constraint rules. Soft-information indicates the probability that each network link will be activated at each packet slot. The proposed scheme consists of three main parts are described as follows:

- **Factor Graph Modeling and Sum-Product Algorithm:** the factor graph model is constructed in order to model the example network scenario and to specify all the local constraint rules enforced by each agent node. The local rules specified by the BS denoted by *B*, two relays *R*1 and *R*2, and four MSs *M*1, *M*2, *M*3 and *M*4.
- **Calculation and Transportation of Soft-Information**: Four iterative steps are implemented in this part. In step1, an initialization of soft-Information is done to indicate the transfer from a node to another one. The second step processes the passage of the soft-Information from a variable node to one of the agent node. The third step assigns weights to the soft-information of a local transmission pattern according to the network traffic condition. Finally, in the fourth step, the stop criterion is set. The proposed algorithm sets the maximum number of iterations at 10. If the number of iterations exceeds this number, the algorithm will stop and the procedure will restart from the initialization step.
- **A Feasible-Weighting Scheme:** A heuristic and feasible weighting scheme is defined. Weight is assigned for each local transmission pattern differently in order to increase

resource utility. Thus, the information required for the weighting scheme must be locally achievable.

5.2 Centralized scheduling

Uplink relay schedulers

A traffic adaptive uplink-scheduling algorithm for relay station is proposed in (Ohym & Dong, 2007). It focuses on the system transparency in IEEE 802.16j. The aims of this algorithm are to minimize the end-to-end delay and signaling overhead and to avoid the resource waste. Authors in (Ohym & Dong, 2007) consider two main strategies: one is elaborated for the real time service flows, and the other is elaborated for the non real time service flows. The first strategy has to allocate resources for RS based on the bandwidth request information of the Mobile Station (MS) is defined as MS-REQ since it use the bandwidth request information of the MS. BS allocates bandwidth for uplink data transmission at each frame based on the bandwidth request information of only MSs without any bandwidth request of the RS. The MS-REQ is as follow:

While any service flow exists

1. BS allocates bandwidth for MS and RS at a time

2. MS transmits data to RS

3- a) On receiving the data from MS, RS transmits the received data by using pre-allocated resource

3- b) If the data from MS is broken, RS transmits nothing

End

Allocating bandwidth for relay station in advance may generate resource waste. If the data is broken between the MS and the RS, the pre-allocated bandwidth is not used and the bandwidth efficiency cannot be maximized only with MS-REQ method. Thus, this scheduling strategy is suitable for the case of light traffic load.

The second strategy is defined as TR-QUE. It has to allocate resources for RS based on the direct bandwidth request of RS. The relay station queues the received data from mobile stations according to the existing scheduling classes: UGS, rtPS, ertPS, nrtPS, and BE. Then, the RS regards each queue as each service flow. The TR-QUE is detailed as follow:

While any service flow exists

1. BS firstly allocates bandwidth only for MS

2. MS transmits data and RS receives and queues the data

3. RS requests bandwidth for successful data

4. BS allocates bandwidth for RS and RS transmits the data

End

It is the optimized scheduling solution for RS in term of bandwidth efficiency. This scheduling strategy is suitable for the case of heavy traffic load. Since The RS does not waste resource even if some part of data-packets from the mobile stations to the relay station are broken due to poor channel. However, the delay performance cannot optimize only by this strategy. In order to optimize both the delay and bandwidth requirements, authors propose a hybrid method Hyb-REQ that uses MS-REQ method for real time traffic and TR-QUE method for non-real time traffic, respectively. The Hyb-REQ algorithm is defined as follow:

While any service flow exists

If non-real time service flow

1. BS firstly allocates bandwidth only for MS
2. MS transmits data and RS receives and queues the data
3. RS requests bandwidth for successful data
4. BS allocates bandwidth for RS and RS transmits the data

If real time traffic service flow

5. BS allocates bandwidth for MS and RS at a time
6. MS transmits data to RS

7- a. In case of success, RS transmits the received data
7- b. In case of error, RS transmits the queued data in step 2

End

When some real time service packets are broken between the MS and the RS, Hyb-REQ transmits some part of non-real time data packets queued at the RS without bandwidth request by using the pre-allocated bandwidth, which was supposed to be wasted. The Hyb-REQ scheduling improves the delay requirement for the real time service traffic using and maximizes the throughput for the non real time service. So the proposal algorithm tends to satisfy the QoS dependent on the traffic.

Downlink relay schedulers

In (Gui, 2008), two relay-assisted scheduling schemes are defined, in which the RS assists the BS in its scheduling decision and therefore it is possible for the BS to exploit CSI (Channel State Information) on the access links without those of the relay links from all the users directly. Authors consider a set of K mobile users, uniformly distributed in a cell, served by a single base station with M relay stations, in which each mobile device intends to receive its NRT data from the BS, possibly by multi-hop routing. Each user rightly predicts its own downlink channel state information and feedback information, combined with the information of the quality of service (e.g. throughput and delay) that each user has achieved so far, is used to calculate the priorities by certain scheduling algorithm at the BS side. For each time slot, either a mobile terminal or a relay terminal with the highest priority is selected by BS for the transmission of the data packets. Figure 12 describes the packet scheduler structure proposed.

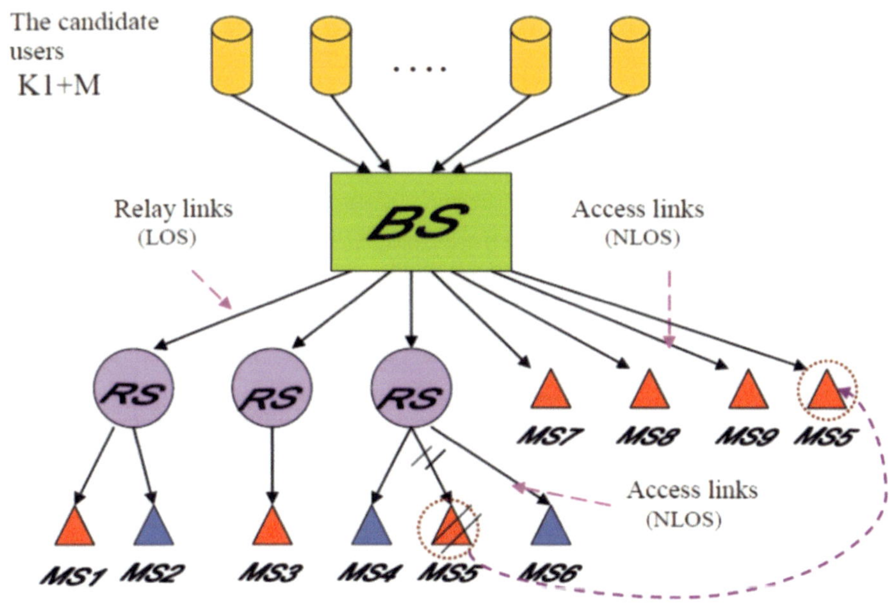

Fig. 12. Packet scheduler structure proposed in (Gui, 2008)

6. Comparative and synthesis study

The Table 4, as shown below presents a comparative analysis of the QoS Scheduling Algorithms in PMP mode.

Category	Traffic	Scheduling Proposal	Strength	Limitation	QoS aspects
Homo-genous	Uplink	(Elmabruk et al, 2008)	Simple	Unsuitable for uplink traffic	Attempt to satisfy all classes
		(Chirayu & Sarkar, 2009)	Does not starve the BE traffics	Introduce overheads	Throughput for NRT and delay for RT classes
	Downlink	(Cicconetti et al, 2006) (Sayenko et al, 2008)	Enhance the QoS satisfaction	Does not consider the channel behavior	2 types of class (rtPS, BE)
		(Kim & Kang, 2005)	Maximize the number of SS and increase the total revenue	Does not address the delay setting	Delay for RT classes and throughput for NRT
		(Ku et al, 2006)	Maximize the number of SS and increase the total revenue and The delay threshold is updated	Unstable	Maximize throughput while maintaining delay

Category	Traffic	Scheduling Proposal	Strength	Limitation	QoS aspects
Hiera-rchical	Uplink	(kitti & Aura, 2003)	Satisfy the major QoS requirements	Complex and unfair	Delay for RT traffics and throughput for NRT traffics
		(Tzu-Chieh, 2006)	Satisfy the major QoS requirements	Complex and need estimation model	Satisfy delay requirements
		(Yanlei & Shiduan, 2005)	Satisfy the main QoS requirements	Complex	3 types of service (UGS, rtPS, nrtPS)
		(Sonia & Hamid, 2010)	QoS guarantee	Complex and handover process	Attempt to serve all types of connections
		(Ridong et al, 2009)	Minimize the average queuing delay	Unfair	Delay requirements for RT classes
		(Chafika, 2009)	Fair and satisfy the QoS requirements	Complex and need mathematical model	Serve the lower priority traffic
	Downlink	(Xiaojing, 2007)	Performs throughput, fairness, and frame utilization	low average delay	All types of service
		(N.Wei et al, 2011)	increase the network throughput and lower delay	Does not support the radio channel	All types of traffic
		(J.Chen et al, 2005b)	Provides more fairness to the system	Complex implementation	All types of traffic
Aware schedulers	Uplink	(G.Wei et al, 2009)	Address the channel state condition and try to satisfy the QoS requirements	Complex	Maximize the total system throughput
		(Liu et al, 2005)	Use the AMC scheme and try to satisfy the QoS constraints	Complex	Respect to average throughput and average delay
		(Jianfeng et al, 2009)	Genetic algorithm implementation	Complex	Balances priorities between mobile stations
	Downlink	(Hongfei et al, 2009)	Viable end-to-end architecture	Complex	Delay, throughput and fairness
		(Vishal et al, 2011)	Low complexity adaptive resource allocation	Starve the lower priority traffic	Maximize total throughput

Table 4. IEEE 802.16d/e proposed methods comparison based on the proposed classification

The Table 5, as shown below presents a comparative analysis of QoS scheduling algorithms, which are dedicated to support the relay mode.

Category	Traffic	Scheduling Proposal	Strength	Limitation	QoS considered
Distributed	Uplink	(Debalina et al, 2010)	Adaptive computation to the channel conditions	complex	All types of service
	Downlink	(Yao et al, 2007)	Increase data packet throughput And increase resource utility, avoid collision	Complex Does not address the delay constraint	-
Centralized	Uplink	(Ohyun & Dong, 2007)	Enhancement of delay and bandwidth requirements	Complex	All types of service
	Downlink	(Gui, 2008)	overhead is avoided	Complex	NRT services

Table 5. IEEE 802.16j proposed methods comparison based on the proposed classification

7. Conclusion

In this chapter, we have provided an extensive survey of recent WiMAX proposals that provide and enhance QoS. All the relevant QoS functionality's such as bandwidth allocation, scheduling, admission control, physical modes and duplexing for WiMAX are deeply discussed. Call Admission Control (CAC) is an important QoS component in WiMAX networks as it has a strong relationship with QoS parameters such as delay, dropping probabilities, jitter and scalability. Therefore, we present a classification and a description of CAC algorithms proposed in the literature for PMP mode. We describe, classify, and compare CAC proposals for PMP mode. Although many CAC scheme has be introduced in the literature, there is stillroom for improvement CAC mechanism.

The QoS platform designers need to be familiar with WiMAX characteristics. So in this chapter, we have present cross-layer designs of WiMAX/802.16 networks. A number of physical and access layer parameters are jointly controlled in synergy with application layer to provide QoS requirements. Most important QoS key concepts are identified. Relations and interactions between QoS functional elements are discussed and analyzed with cross layer approach consideration.

Moreover, scheduling is a main component of the MAC layer that assures QoS to various service classes. Scheduling algorithms implemented at the BS has to deal with both uplink and downlink traffics. An understanding classification of the uplink and downlink scheduling in the IEEE 802.16 networks is described in details. We present a survey of some scheduling research in literature for WiMAX fixe, mobile, and relay. In order to give a comparative study between the proposals mechanisms, we draw two summary tables showing the strength, the limitation and QoS observed aspect of each scheduling method proposed for fixed, mobile and relay WiMAX network. We have discussed the approaches and key concepts of different scheduling algorithms which can be useful guide for further research in this field.

As the scheduling in WiMAX wireless network is a challenging topic, future works should include advanced investigations on scheduling algorithms under different CAC schemes and bandwidth allocation mechanisms.

Furthermore, we intend to evaluate the behavior and the efficiency of some scheduling and CAC modules for the mobile and the relay WiMAX networks under full saturation condition and to provide a mathematical analysis combined with extensive simulations.

8. References

A. Sayenko, O. Alanen, and T. Hamaainen. (2008). Scheduling solution for the IEEE 802.16 base station, Int. *J. Computer and Telecommunications Networking*, vol. 52, pp. 96-115, Jan. 2008.

C. Cicconetti, L. Lenzini, placeE. Mingozzi, and C. Eklund. (2006). Quality of service support in IEEE 802.16 networks, *IEEE Network*, vol. 20, pp. 50-55

C.Cocconetti, A.Erta, L.Lenzini and E.Mingozzi. (2007). Performance Evaluation of IEEE802.16 MAC for QoS Support, *IEEE Transactions on Mobile Computing*, vol.6,no1,pp.26-38

C.-M. Chou, B.-J. Chang, and Yung-Fa Huang. (2006). Dynamic Polling Access Control for High Density Subscribers in Wireless WiMAX Networks, *Proceedings of the 2006 Taiwan Area Network Conference (TANET'06)*, Hualien, Taiwan, Nov. 2006

Carlos Valencia, (2009). Scheduling Alternatives for Mobile WiMAX End-to-End Simulations and Analysis, Master of Applied Science in Technology Innovation Management, Carleton University Ottawa, Canada, K1S 5B6

Chafika TATA, (2009). *Algorithme de Courtoisie : optimisation de la performance dans les réseaux WiMAX fixes*, Master thesis 2009.

Chirayu Nagaraju and Mahasweta Sarkar, (2009). A Packet Scheduling To Enhance Quality of Service in IEEE 802.16, *Proceedings of the World Congress on Engineering and Computer Science* Vol IWCECS 2009.

D.H. Kim and C. G. Kang. (2005). Delay Threshold-based Priority Queuing Packet Scheduling for Integrated Services in Mobile Broadband Wireless Access System,

Proceedings of the IEEE Int. Conf. High Performance Computing and Communications, Kemer-Antalya, Turkey, pp.305-314.

Debalina Ghosh, Ashima Gupta, Prasant Mohapatra. (2009). Adaptive Scheduling of Prioritized Traffic in IEEE 802.16j Wireless Networks, *Proceedings of the IEEE International Conference on Wireless and Mobile Computing, Networking and Communications, (WIMOB'09)*, pp 307 - 313

Elmabruk. Laias, Irfan Awan, et Pauline.ML.Chan. (2008). An Integrated Uplink scheduler In IEEE 802.16, *Proceedings of the second UKSIM European Symposuim on Computer Modeling and Simulation*

F.Rango, A.Malfitano and S.Marano, (2011). GCAD: A Novel Call Admission Control Algorithm in IEEE 802.16 based Wireless Mesh Networks, JOURNAL OF NETWORKS, VOL. 6, NO. 4, p 595-606

Gui Fang. (2008). Performance Evaluation of WiMAX System with Relay-assisted Scheduling, Master Thesis 2008

H. LABIOD, H. AFIFI, WI-FI, (2007). *ZIGBEE AND WIMAX*, 327 pages, 978-1-4020-5396-2 (HB) ISBN 978-1-4020-5397-9 (e-book), Published by Springer 2007.

H. Zhu and K. Lu. (2006). Above Packet Level Admission Control and Bandwidth Allocation for IEEE 802.16 Wireless MAN, Simulation Modeling Practices and Theory, 24 November 2006

H.Wang, W. Li, and D. P. Agrawal, (2005). Dynamic admission control and QoS for 802.16 wireless MAN, in Wireless Telecommun. Symp. pp. 60–66

Hongfei Du, Jiangchuan Liu, and Jie Liang. (2009). Downlink Scheduling For Multimedia Multicast/Broadcast Over Mobile WiMAX: Connection-Oriented Multistate Adaptation, *IEEE Wireless Communications*, August 2009

IEEE Draft Standard P802.16j/D5, (2008). Part 16: Air Interface for Fixed and Mobile Broadband Wireless Access Systems -Multihop Relay Specification," May 2008.

IEEE Standard for Local and metropolitan area networks. (2006). Part 16: Air Interface for Fixed and Mobile Broadband Wireless Access Systems" Amendment 2: Physical and Medium Access Control Layers for Combined Fixed and Mobile Operation in Licensed Bands, 28 February 2006.

IEEE Standard. (2004). Local and metropolitan area networks Part 16: Air Interface for Fixed Broadband Wireless Access Systems", 3 Park Avenue, New York, NY 10016-5997, USA, IEEE Computer Society and the IEEE Microwave Theory and Techniques Society Sponsored by the LAN/MAN Standards Committee, Print: SH95246, PDF: SS95246

J. Chen, W. Jiao and H. Wang. (2005). A service flow management strategy for IEEE 802.16 broadband wireless access systems in TDD mode, *Proceedings of IEEE International Conference on Communications*, Seoul, Korea, vol. 5, pp.3422-3426.

J. M. Ku, S. K. Kim, S. H. Kim, S. Shin, J. H. Kim, and C. G. Kang. (2006). Adaptive delay threshold-based priority queuing scheme for packet scheduling in mobile broadband wireless access system, *Proceedings of the IEEE Wireless Communication and Networking Conf.*, Las Vegas, NV, vol. 2, pp. 1142-1147.

J.Chen, Wenhua Jiao and Qian Guo. (2005). An integrated QoS control architecture for IEEE 802.16 broadband wireless access systems, *Proceedings of the Global Telecommunications Conference*, St. Louis, MO, pp. 6-11, ISBN: 0-7803-9414-3

Jeffrey G. Andrews, (2007). *Fundamentals of WiMAX Understanding Broadband Wireless Networking,* ISBN 0-13-222552-2,478 pages, Prentice Hall Communications Engineering and Emerging Technologies Series, text printed in the United States in Westford, Massachusetts. First printing, February 2007

Jianfeng Song, Jiandong Li, Changle Li, (2009). A Crosss-Layer Scheduling Algorithm based on Genetic Algorithm, *Proceeding of the seventh Annual Communication Networks and services Research Conference.*

K.Yu, X.Wang, S.Sun, L. Zhang, X.Wu, (2009). A Statistical Connection Admission Control Mechanism for Multiservice IEEE 802.16 Network, Proceeding of 69th IEEE International conference on Vehicular Technology Conference, VTC Spring 2009

Kalikivayi Suresh, Iti Saha Misra and Kalpana Saha. (2008). Bandwidth and delay guaranteed call admission control for QoS provisioning in IEEE 802.16e Mobile WiMAX, *Proceedings of the GLOBECOM*, 2008

L. Wang, F. Liu, Y. Ji, and N. Ruangchaijatupon. (2007). Admission control for non-preprovisioned service flow in wireless metropolitan area networks, *Proceedings of the Fourth European Conference of Universal Multiservice Netw. ECUMN '07*, 2007, pp. 243–249

Loutfi Nuaymi, (2007). *WiMAX: Technology for Broadband Wireless Access*, John Wiley & Sons 2007 (310 pages), ISBN:9780470028087

M. Castrucci, F. Delli Priscoli, C. Buccella, V. Puccio, I. Marchetti. (2008). Connection Admission Control in WiMAX networks, *Proceedings of the 17th ICT Mobile and Wireless Communications Summit (ICTMobileSummit)*, Stockholm Sweden, June 2008

Ohyun Jo and Dong-Ho Cho. (2007). Traffic adaptive uplink scheduling scheme for relay station in IEEE 802.16 based multi-hop system", *Proceedings of the 66th Conference on Vehicular Technology, (VTC-2007), pp 1679 - 1683*

Pratik Dhrona. (2007). *A Performance Study of Uplink Scheduling Algorithms in Point to Multipoint WiMAX Networks*, Master thesis, December 2007.

Q.Liu, X.Wang and G.Giannakis. (2005). Cross-layer scheduler design with QoS support for wireless access networks, *Proceedings of International Conference on Quality of Service in Heterogeneous Wired/Wireless Networks*

Ridong Fei, Kun Yang, Shumao Ou, et Wenyoung Wang. (2009). A Utility-based Dynamic Bandwidth Allocation Algorithm in IEEE 802.16 NETWORKS to Minimize Average Queuing Delay, *Proceedings of the International Conference on Communication and Mobile Computation 2009.*

S. Chandra and A. Sahoo, (2007). An efficient call admission control for IEEE 802.16 networks, *Proceedings of 15th IEEE LAN/MAN Workshop, LANMAN 2007*, pp. 188–193.

S. Ould Cheikh El Mehdi, W. Fawaz, Ken Chen. (2006). Une nouvelle approche pour la gestion de flux temps réels basée sur l'algorithme EDF, *Colloque Francophone sur l'ingénierie des Protocoles CFIP'06*

Shafaq B. Chaudhry, Ratan K. Guha.(2007). Adaptive connection admission control and packet scheduling for QoS provisioning in mobile wiMAX, *Proceedings of IEEE International Conference on Signal Processing and Communications (ICSPC 2007)*, Dubai, United Arab Emirates, 2007

Shepard, Steven, (2007). *WiMAX Crash Course*, ISBN: 0072263075, EAN: 9780072263077, pages 339 edition McGraw-Hill, 1/6/2006

Shida Luo, Zisu Li, (2008). An Effective Admission Control for IEEE802.16 WMN, *Proceedings of the second International Symposium on Intelligent Information Technology Application*, 2008

Sonia Nazari and Hamid Beigy. (2010). A New Distributed Uplink Packet Scheduling Algorithm In WiMAX Newtorks, *Proceedings of the Second International Conference Future Computer and Communication (ICFCC'10)*

Tzu-Chieh Tsai, Chi-Hong and Chuang-Yin Wang. (2006). CAC and Packet Scheduling Using Token Bucket for IEEE 802.16 Networks", *Journal of communications*, vol 1, NO 2

W. Kitti, and G. Aura. (2003). Packet Scheduling for QoS Support in IEEE 802.16 Broadband Wireless Access Systems, *International Journal on Communication Systems*. Vol.16, Issue1, pp: 81-96

Wankhede Vishal A, Jha Rakesh, Upena Dalal, (2011). Resource Allocation for Downlink OFDMA in WiMAX Systems, *Proceedings of the International Conference on Communication, Computing and Security*, ICCCS'11, India, February 2011

Wei Gan, Jiqing Xian, Xianzhong Xie, et Juan Ran. (2009). A Cross-layer Designed Scheduling Algorithm for WiMAX Uplink, *Proceedings of the Ninth International Conference on Electronic Measurement & Instruments ICEMI'2009*

Wei Nie, Houjun Wang and Jong Hyuk Park. (2011). Packet Scheduling with QoS and Fairness for Downlink Traffic in WiMAX Networks, *Journal of Information Processing Systems*, Vol.7, No.2, June 2011 DOI:10.3745/JIPS.2011.7.2.261

Xiaojing Meng, (2007). An Efficient Scheduling for Diverse QoS Requirements in WiMAX, Master thesis 2007.

Y. Wang, S. Chan, M. Zukerman, and R.J. Harris. (2008). Priority-Based fair Scheduling for Multimedia WiMAX Uplink Traffic, *Proceedings of the IEEE International Conference on Communications*, Beijing, China, pp. 301-305

Y.Zhang Hsiao-Hwa Chen, (2008). *MOBILE WiMAX Toward Broadband Wireless Metropolitan Area Networks*, ISBN 978-0-8493-2624-0, edition Auerbach, 2008

Yanlei Shang et Shiduan Cheng, (2005). An Enhanced Packet Scheduling Algorithm for QoS Support in IEEE 802.16 Wireless Network, *Proceedings of the third International Conference on Networking and Mobile Computing (ICCNMC'05)*, China, pp.652-661.

Yao-Nan Lee, Jung-Chieh Chen, Yeong-Cheng Wang, and Jiunn-Tsair Chen. (2007). A Novel Distributed Scheduling Algorithm for Downlink Relay Networks", *IEEE TRANSACTIONS ON WIRELESS COMMUNICATIONS*, VOL. 6, NO. 6

3

Scheduling Mechanisms with Call Admission Control (CAC) and an Approach with Guaranteed Maximum Delay for Fixed WiMAX Networks

Eden Ricardo Dosciatti[1,2,3], Walter Godoy Junior[1,2,3] and Augusto Foronda[2,3]
[1]*Graduate School of Engineering and Computer Science (CPGEI),*
[2]*Advanced Center in Technology of Communications (NATEC),*
[3]*Federal University of Technology Parana (UTFPR),*
Brazil

1. Introduction

The major challenge for the second decade of this century is the implementation of access to high-speed internet, known as broadband internet. With the popularization of access to the global network, it is evident that there will be a reduction of physical barriers to the transmission of knowledge, as well as in transaction costs, and will be instrumental in fostering competitiveness, especially for developing countries. But, wired access to broadband internet has a very high cost and is sometimes impracticable, since the investment needed to deploy cabling throughout a region often outweighs the reduces the service provider's financial gains. One of the possible solutions in reducing the costs of deploying broadband access in areas where such infrastructure is not present is to use wireless technologies, which require no cabling and reduce both implementation time and cost of deployment (Gosh et al., 2005).

Motivated by the growing need ubiquitous high speed access, ie, the use of computers everywhere, embedded in the structures of our lives, wireless technology is an option to provide a cost-effective solution that may be deployed quickly and with easily, providing high bandwidth connectivity in the last mile, ie, in places where business or residential customers, also known as the tail of the distribution network of services. As wireless networks have an ease of deployment and low maintenance cost, ease of configuration and mobility of their devices, there are challenges that must be overcome in order to further advance the widespread use of this type of network.

Thus, the IEEE (Institute of Electrical and Electronics Engineers) has developed a new standard for wireless access, called IEEE 802.16 (802.16-2004, 2004). Also known as WiMAX (Worldwide Interoperability for Microwave Access), it is an emerging technology for next generation wireless networks which supports a large number of users, both mobile and nomadic (fixed), distributed across a wide geographic area. Furthermore, this technology provides strict QoS (Quality of Service) guarantees for data, voice and video applications (Camargo et al., 2009). As a service provider, WiMAX will create new alternatives for applications such as telephony, TV broadcasts, broadband Internet access for residential users, and commercial, industrial and university centers. This is a new market niche that is revolutionizing telecom companies and interconnection equipment manufacturers (Eklund et al., 2002). Moreover, WiMAX enables broadband connection in areas which are inaccessible or lacking in infrastructure, since it requires no installation or complex physical connections via cables and traditional technologies (WiMAX Forum, 2011). The increasing deployment of wireless infrastructure is enabling a variety of new applications that require flexible, but also robust, support by the network, such as multimedia applications including video streaming and VoIP (Voice over Internet Protocol), among others, which demand real-time data delivery (Sun et al., 2005).

Based on these assumptions and considering that the standard leaves open certain issues related to network resources management and mechanisms for packet scheduling, is that several researchers have presented proposals to resolve issues related to scheduling mechanisms and QoS architectures for Broadband Wireless Access (BWA). However, many of these solutions only address the implementation or addition of a new QoS architecture to the IEEE 802.16 standard.

Thus, this work presents a new scheduler with call admission control to a WiMAX Base Station (BS). An analytical model, based on Latency-Rate (LR) server theory (Stiliadis & Varma, 1998) is developed, from which an ideal frame size, called Time Frame (TF), is estimated, with guaranteed delays for each user. At the same time, the number of stations allocated in the system is maximized. In this procedure, framing overhead generated by the MAC (Medium Access Control) and PHY (Physical) layers is considered when is calculated the duration of each time slot. After the developed this model, a set of simulations is presented for constant bit rate (CBR) and variable bit rate (VBR) streams, with performance comparisons between situations with different delays and different TFs. The results show that an upper limit on the delay may be achieved for a wide range of network loads, optimizing the bandwidth.

This work is structured as follows: in Section 2, a brief overview of IEEE 802.16 standard is presented. In Sections 3, 4 and 5 we present the concepts about the operation of scheduling mechanisms, call admission control and QoS, respectively. In Section 6 the related research in this area are discussed. Our proposal for a new scheduler, with its implementation and evaluation, is explained in Section 7. Finally, in Section 8 the conclusion are briefly described.

2. Overview of IEEE 802.16 fixed standard

The basic topology of an IEEE 802.16 network includes two entities that participate in the wireless link: Base Stations (BS) and Subscriber Stations (SS), as shown in Figure 1 (Dosciatti et al., 2010). The BS is the central node, responsible to coordinate communications and provide connectivity to the SSs. BSs are kept in towers distributed so as to optimize network coverage area, and are connected to each other by a backhaul network, which allows SSs to access external networks or exchange information between themselves.

Networks based on the IEEE 802.16 standard can be structured in two schemes. In PMP (Point-to-MultiPoint) networks, all communication between SSs and other SSs or external networks takes place through a central BS node. Thus, traffic flows only between SSs and the BS, as shown in Figure 1. In Mesh mode, SSs communicate with each other without the need for intermediary nodes; that is, traffic can be routed directly through SSs. So, all stations are peers which can act as routers and forward packets to neighboring nodes (Akyildiz & Wang, 2005). This work only considers the PMP topology, since it is implemented by first-generation WiMAX devices, and also due to the strong trend towards its adoption by Internet providers because it allows them to control network parameters in a centralized manner, without the need to recall all SSs (WiMAX Forum, 2011).

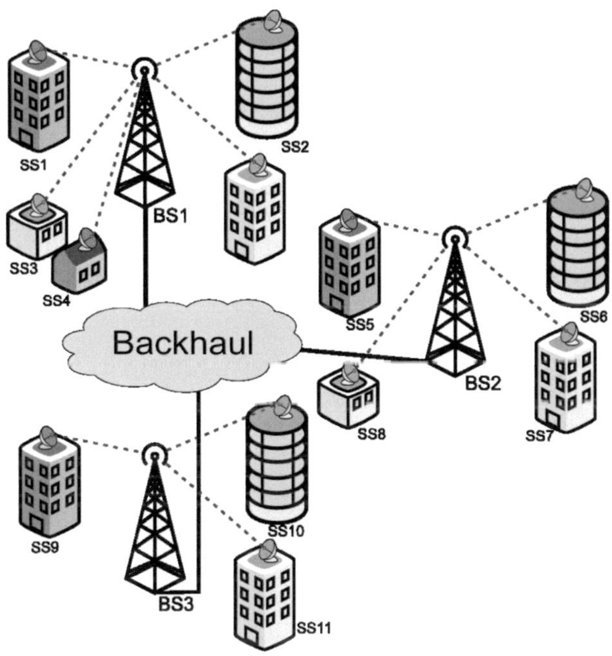

Fig. 1. IEEE 802.16 network architecture

Although it is referred to as fixed pattern, IEEE 802.16 allows stations to provide customers with low-speed mobility. A feature missing in this pattern and that justifies its designation as fixed is the possibility to perform handoffs/handovers, which allow a client station to switch to another base station without to lose connectivity. In this case, SSs are instead called mobile stations (MSs). The functionality of handoff/handover was included in the IEEE 802.16 standard in early 2006 with the publication of the IEEE 802.16e (802.16e-2005, 2006), which quickly received the name of "IEEE 802.16 mobile".

WiMAX is designed to leverage wireless broadband metropolitan area networks and, it is obtained performance comparable to traditional cable and xDSL technology, with the following main advantages:

- The ability to provide services in areas with poor infrastructure deployment;
- Elimination unnecessary expenses with facilities;
- The ability to overcome physical boundaries, such as walls or buildings;
- High scalability; and
- Low update and maintenance costs.

WiMAX technology can to reach a theoretical maximum distance of 50 km (Tanenbaum, 2003). Data transmission rates can vary from 50 to 150 Mbps, depending on channel frequency bandwidth and modulation type (Intel, 2005). Communication between a BS and SSs occurs in two different channels: uplink (UL) channel, which is directed from SSs to the BS, and downlink (DL) channel, which is directed from the BS to SSs. DL data is transmitted by broadcasting, while in UL access to the medium is multiplexed. UL and DL transmissions can to operate in different frequencies using Frequency Division Duplexing (FDD) mode or at different times using Time Division Duplexing (TDD) mode.

In TDD mode, the channel is segmented in fixed-size time slots. Each frame is divided into two subframes: a DL subframe and an UL subframe. The duration of each subframe is dynamically controlled by the BS; that is, although a frame has a fixed size, the fraction of it assigned to DL and UL is variable, which means that the bandwidth allocated for each of them is adaptive. Each subframe consists of a number of time slots, and thus both the SSs and the BS must be synchronized and transmit the data at predetermined intervals. The division of TDD frames between DL and UL is a system feature controlled by the MAC layer. Figure 2 (Wongthavarawant & Ganz, 2003) shows the structure of a TDD frame. In this work, the system was operated in TDD mode with the OFDM (Orthogonal Frequency Division Multiplexing) air interface, as determined by the standard.

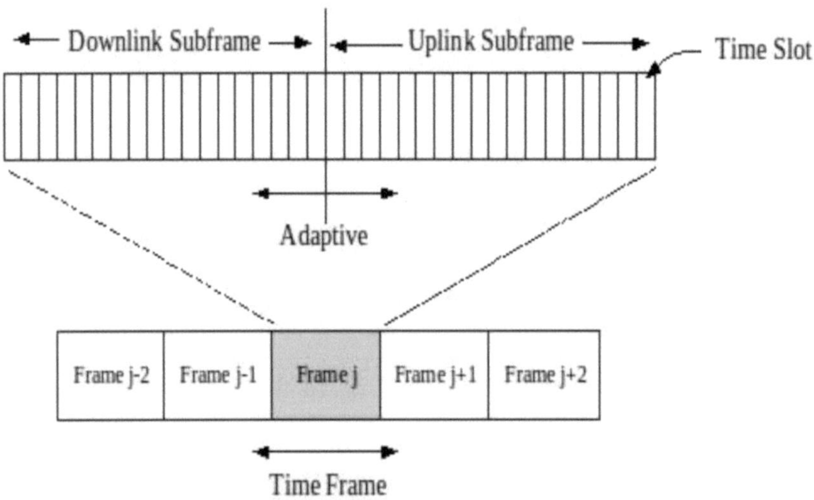

Fig. 2. IEEE 802.16 frame structure

Figure 3 (Hoymann, 2005) shows an example OFDM frame structure in TDD mode. As seen earlier, each frame has a DL subframe followed by an UL subframe. In this structure, the system supports frame-based transmission, in which variable frame lengths can be adopted. These subframes consists of a fixed number of OFDM symbols. Details of the OFDM symbol structure may be found in (Gosh et al., 2005).

The DL subframe starts with a long preamble (two OFDM symbols) through which SSs can synchronize with the network and check the duration of the current frame. Instantly after the DL long preamble, the BS transmits the Frame Control Header (FCH), which consists of an OFDM symbol and is used by SSs to decode MAC control messages transmitted by the BS.

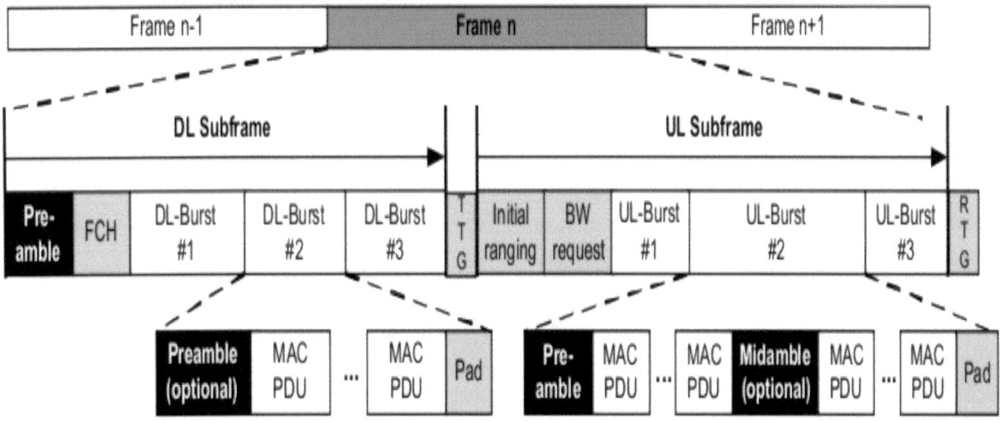

Fig. 3. OFDM frame structure with TDD

The UL subframe consists in contention intervals for initial raning and bandwidth request purposes and one or several UL transmission bursts, each from a different SS. The initial ranging slots allows an SS to enter the system, by adjusted its power level and frequency offsets and correctness of its time offset. Bandwidth request slots are used by SSs to transmit bandwidth request headers.

Two gaps separate the DL and UL subframes: the Transmit/Receive Transtion Gap (TTG) and the Receive/Transmit Transition Gap (RTG). These gaps allow the BS to switch from transmit to receive mode, and vice versa.

3. Scheduling mechanisms

Scheduling mechanisms were intentionally left outside the scope of the IEEE 802.16 standard. The diversity of service offered combined with scheduling mechanisms is an important area for differentiation in the development of research in both industry and academia. However, some concepts are common to all implementations, and some ideas were intended even while not explicitly made a part of the standard.

The scheduling mechanism in the WiMAX MAC layer is designed to efficiently deliver broadband data services such as voice, video and other data related to change of broadband wireless channel. Scheduling is the main component of the MAC layer that helps assure QoS to various service classes. The scheduler is located at each BS to enable rapid response to traffic requirements and channel conditions. So, the scheduler works as a distributor to allocate the resources among SSs. The scheduling mechanism is provided for both DL and UL traffic. The allocated resource can be defined as the number of slots and then these slots are mapped into a number of subchannels (each subchannel is a group of multiple physical subcarriers) and time duration (OFDM symbols). These allocated resources are delivered in MAP messages at the beginning of each frame. Therefore, the resource allocation can be changed from frame to frame in response to traffic and channel conditions. The amount of resources in each allocation can range from one slot to the whole frame. The MAC scheduler handles data transport on a connection-by-connection basis. Each connection is associated with a single scheduling mechanism that is determined by a set of QoS parameters that quantify aspects of its behaviour.

Scheduling has also been studied intensively in many disciplines, such as CPU task scheduling in operating systems, service scheduling in a client-server model, and events scheduling in communication and computer networks. Thus a lot of scheduling algorithms have been developed. However, compared with the traditional scheduling problems, the problem of scheduling at the MAC layer of WiMAX networks is unique and worth study by four reasons described below.

1. The total bandwidth in a WiMAX network is adaptive by Adaptive Modeling and Coding (AMC) and is deployed at the physical layer and the number of bytes each time slot can carry depends on the coding and modulation scheme.
2. Multiple service types have been defined and their QoS requirements need to be satisfied at the same time. How to satisfy various QoS requirements of different service types simultaneously has not been addressed by any other wireless access standard before.
3. The time complexity of WiMAX scheduling algorithm must be simple, since in real-time services require a fast response from the central controller in the BS.
4. The frame boundary in the WiMAX MAC layer also serves as the scheduling boundary, which makes the WiMAX scheduling problem different from the continuous time scheduling problem.

To implement a scheduler, the following aspects, as defined in IEEE 802.16, must be taken into consideration:

- The distribution of resources should be made based on the bandwidth requests sent by the SSs and QoS parameters of each connection, and different connections use the same type of service, different values for the same QoS parameter may occur.
- Bandwidth allocation should allow not only the transmission of data, but also the transmission of bandwidth requests in accordance with the request mechanism established for each type of service.
- All QoS parameters defined by the standard should be guaranteed.

In addition, the scheduler must efficiently use the available bandwidth so that a greater number of users can be admitted, thus resulting in high levels of network utilization.

Although the scheduler is implemented at the MAC layer, the technology used at the PHY layer can influence its project. When used a WirelessMAN-SC (Single Carrier) PHY, there is only one carrier frequency and the whole is given to an SS. This PHY layer requires line of sight (LOS) communication. Rain attenuation and multipath also affect reliability of the network at these frequencies. To allow non-line of sight (NLOS) communication, IEEE 802.16 designed the Orthogonal Frequency Division Multiplexing (OFDM) PHY, popularly known as IEEE 802.16d and used in this work, and designed for wireless fixed stations. In this PHY layer, multiple subcarriers form a physical slot, but because they are transparent to the MAC layer, the subcarriers can be seen as a logical channel from the point of view of the scheduler. However, each subchannel may use different modulations so that different SSs may have different transmission rates. How OFDM is a multicarrier transmission in which thousands of subcarriers are transmitted, each user is given complete control of all subcarriers. The scheduling decision is simply to decide what time slots should be allocated to each SS. For mobile users, it is better to reduce the number of subcarriers and to have higher signal power per SS. Therefore, multiple users are allowed to transmit on different subcarriers at the same time slot. The scheduling decision is then to decide which subcarriers and what time slots must be allocated to a given user. This combination of time division and frequency division multiple access in conjunction with OFDM is called Orthogonal Frequency Division Multiple Access (OFDMA). Then, the OFDMA requires allocation of resources in two dimensions: frequency and time. In other words, the scheduler must decide not only on the allocation of slots, but also of subcarriers for each user. Since more than one user can use the channel at the same instant of time, it is essential that the scheduling algorithms consider the characteristics of the OFDMA physical layer. Figure 4 (So-In et al., 2009) illustrates a schematic view and the differences between the three types of the 802.16 PHYs, discussed above. The details of these interfaces can be found in (802.16-Rev2/D2, 2007).

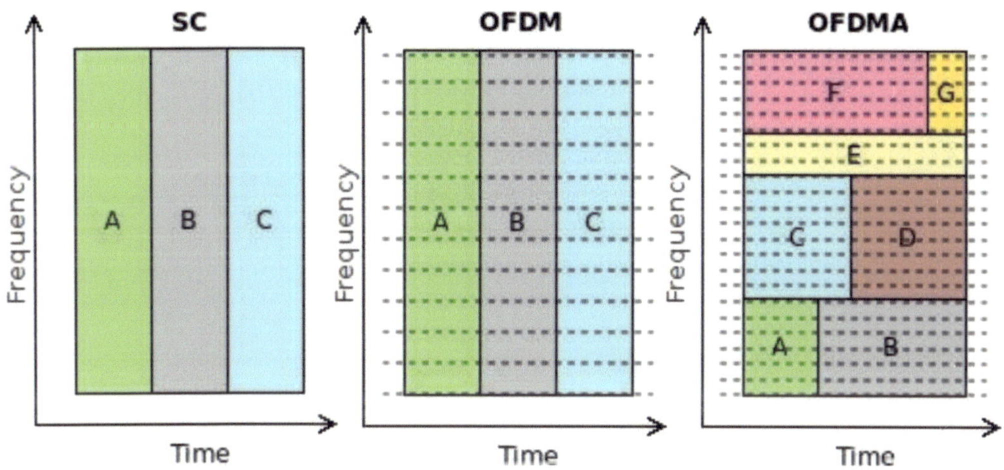

Fig. 4. IEEE 802.16 PHYs: SC, OFDM and OFDMA (the letters A, B, C, D, E, F and G represent different users)

The scheduler for WirelessMAN-SC can be fairly simple because only time domain is considered. The entire frequency channel is given to the BS. For OFDM, it is more complex since each subchannel can be modulated differently, but it is still only in time domain. On the other hand, both time and frequency domains need to be considered for OFDMA. The OFDMA scheduler is the most complex because each SS can receive some portions of the allocation for the combination of time and frequency so that the channel capacity is efficiently utilized.

A scheduling mechanism needs to consider the allocations logically and physically. Logically, the scheduler should calculate the number of slots based on QoS service classes. Physically, the scheduler needs to select which subchannels and time intervals are suitable for each user. The goal is to minimize power consumption, to minimize bit error rate and to maximize the total throughput.

The scheduling mechanism in IEEE 802.16 includes the scheduling of downlink traffic, carried out by BS, and the scheduling of uplink traffic, performed by two schedulers, one at the BS and another at the SSs as shown in Figura 5. To perform the allocation of resources, the schedulers use information about the QoS requirements and the status of the queues of the connections.

At the BS, packets from the upper layer are put into different queues. However, the optimization of queue can be done and the number of required queues can be reduced. Then, based on the QoS parameters and some extra information such as the channel state condition, the DL-BS scheduler decides which queue to service and how many service data units (SDUs) should be transmitted to the SSs. Since the BS controls the access to the medium, the second scheduler (the UL-BS scheduler) makes the allocation decision based on the bandwidth requests from the SSs and the associated QoS parameters. Finally, the third scheduler is at the SS. Once the UL-BS grants of the bandwidth for the SS are allocated, the SS scheduler decides which queues should use that allocation. While the requests are per connections, the grants are per subscriber and the subscriber is free to choose the appropriate queue to service. The SS scheduler needs a mechanism to allocate the bandwidth in an efficient way, what is Call Admission Control and is treated in the next section.

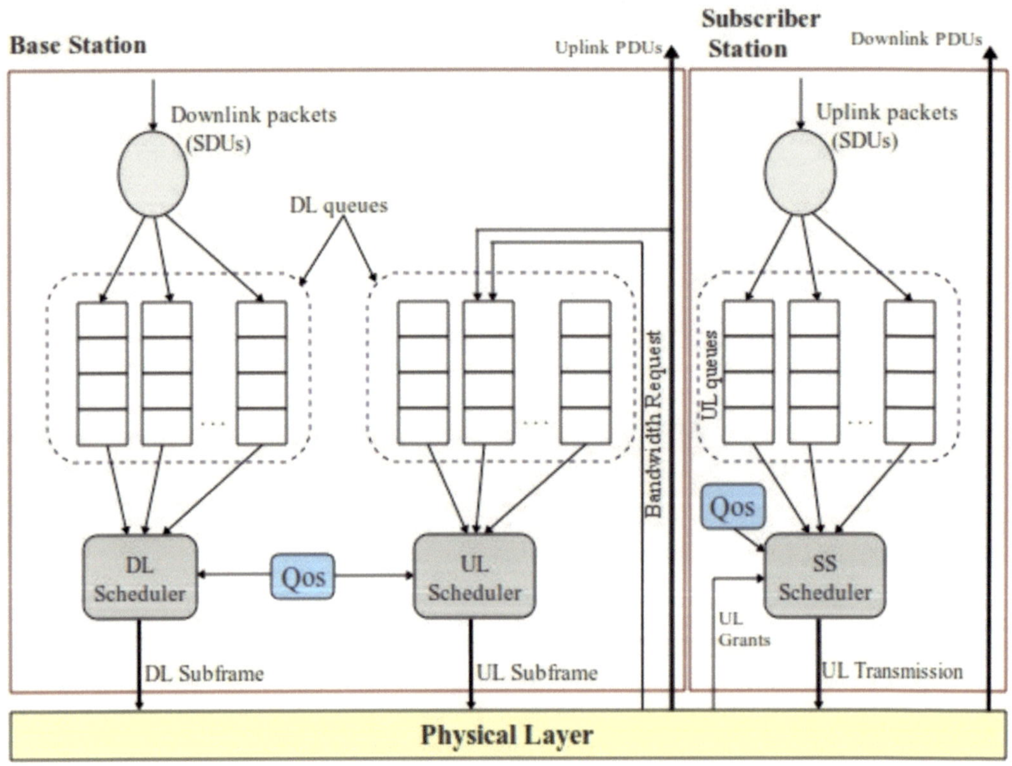

Fig. 5. Components of the schedulers at BS and SSs

4. Call Admission Control - CAC

Typically, a Call Admission Control (CAC) procedure is also implemented at the BS that ensures the load supplied by the SSs can be handled by the network.

While the scheduling mechanism ensures that the required amount of resources is allocated to the connections, so that the QoS requirements are met, the admission control mechanism limits the number of connections to the network so the network is not overloaded by many users.

Whenever an user wants to establish a new connection, a request sent to the BS for the admission control mechanism to decide whether the new connection may or may not be accepted. To make this decision, the admission control must ensure that there are sufficient resources to meet the QoS requirements of the new connection without compromising the minimum QoS requirements of ongoing connections.

The choice of admission control policy to be adopted in an IEEE802.16 network is strongly associated with the scheduling mechanism used. For example, by the adoption of an admission control mechanism that estimates the resources available from the difference between the total capacity of the link and the sum of the minimum rate requirements of already admitted connections, you should ensure that the scheduler will not allocate more than the minimum rate for a connection when other connections have not yet met their minimum requirements. In addition, the integration of scheduling and admission control can result in simple solutions, because when one the mechanisms is able to guarantee the fulfillment of a requirement for the same QoS guarantees need not be implemented by other mechanism.

Thus, CAC restricts the access to the network in order to prevent network congestion or service degradation for already accepted users. It can prevent the system from be overloaded. CAC has been characterized as the decision maker for the network as is shown in Figure 6.

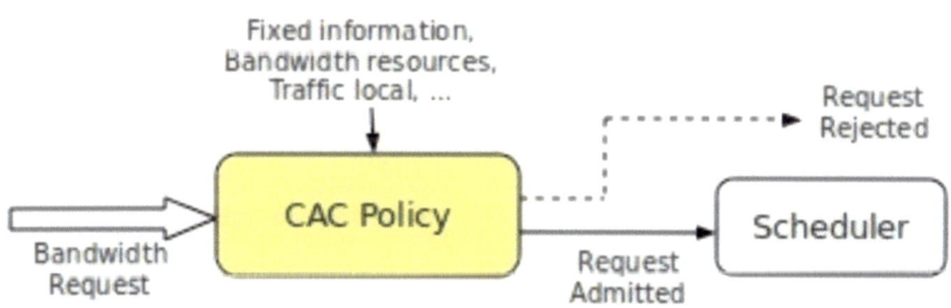

Fig. 6. CAC policy process

However, before the user is allocated in the CAC policies, a bandwidth request is required. Thus, to design a CAC algorithm, we must also worry about the bandwidth allocation algorithm. For this, the following points should be highly valued (Mei et al., 2010):

- The fairness of the bandwidth allocation: Different terminals carry different data transmission business, so the bandwidth requirement of different business is varying.

This means that to allocate the bandwidth, QoS should not simply be the average allocation. Meanwhile, it would be not appropriate if bandwidth quantity between terminals varies dramatically.

- Data transmission delay: Adjust the delay time of data transmission of terminals or connections through allocation of bandwidth, and to keep the delay in a reasonable range tolerated by terminals or connections.
- The data throughput of the system.
- Combined with protocol construction of WiMAX, the ways of the bandwidth allocation request and the existence mechanism of QoS assurance mechanism.

In this work, we only consider the PMP mode architecture of IEEE 802.16 BWA networks, where transmission only occurs between a BS and SSs and the BS controls all the communications between BS and SSs. The connection can be either downlink (from BS to SS) or uplink (from SS to BS) as it is depicted in Figure 1. In PMP architecture, two modes are defined: Grant-Per-Connection (GPC) and Grant-Per-Subscriber-Station (GPSS). Under GPC, the CAC algorithm considers each individual connection arriving from an SS, while for GPSS each SS manages admission of its own individual connections before sending a single bandwidth (BW) request to the BS.

The IEEE 802.16 standard does not define policies for admission control, which has encouraged researchers from academia and industry to investigate solutions to this problem. There are already several proposals in the literature, such as (Chen et al., 2005; Guo et al., 2007; Masri et al., 2009; Rong et al., 2008; Wang H et al.,2007; Wang L et. al. 2007). In Section 7 is introduced a new solution to the problem.

5. Quality of Service - QoS

IEEE 802.16 can support multiple communication services (data, voice, video) with different Quality of Service (QoS) requirements organized into different connections. The MAC layer of IEEE 802.16 standard defines mechanisms to provide QoS and control of data transmission between the BS and SSs. The main QoS mechanism is the association of packets that pass through the MAC layer to a service flow. The service flow is a MAC layer service that provides unidirectional message transport. During connection establishment, these service flows are created and activated by the BS and SSs. Each service flow must define its own set of QoS parameters, among them maximum delay, minimum bandwidth and type of service scheduling.

Within this context, the IEEE 802.16 standard defines four service classes associated with traffic flows, and each such class has different QoS requirements, which are fulfilled has a scheduler allocated bandwidth to the SSs under a set of rules (802.16-2004, 2004). The four service classes defined by standard are:

- Unsolicited Grant Service (UGS): this service is designed to support real-time applications that generate fixed-size data packets on a periodic basis, such as T1/E1 and Voice over IP (VoIP) without silence suppression. UGS does not need the SS to explicity request bandwidth, thus eliminating the overhead and latency associated with bandwidth request. Because UGS connections never request bandwidth, the amount of bandwidth to allocate to such connections is computed by the BS based on the minimum reserved traffic rate defined in the service flow of that connection.

- real-time Polling Services (rtPS): this service is designed to support real-time services that generate variable-size data packets at periodic intervals, such as moving pictures expert group (MPEG) video and VoIP with silence suppression. Unlike UGS connections, rtPS connections must inform the BS of their bandwidth requirements. Therefore the BS must periodically allocate bandwidth for rtPS connections specifically for the purpose to request bandwidth. In this service class, the BS provides unicast polling opportunities for the SS to request bandwidth. The unicast polling opportunities are frequent enough to ensure that latency requirements of real-time services are met. This service requires more request overhead than UGS does but is more efficient for service that generates variable-size data packets or has a duty cycle less than 100 percent.
- non-real-time Polling Services (nrtPS): this service is designed to support delay-tolerant applications such as FTP (File Transfer Protocol) for which a minimum amount of bandwidth is required. Also, this service is very similar to rtPS except that the SS can also use contention-based polling in the uplink channel to request bandwidth. In nrtPS, it is allowable to have unicast polling opportunities, but the average duration between two such opportunities is in the order of few seconds, which is large compared to rtPS. All SSs that are part of the group can also request resources during the contention-based polling opportunity, which can often result in collisions and additional attempts.
- Best-Effort (BE): this service provides very little QoS support and is applicable only for services that do not have strict QoS requirements, such as HTTP (Hypertext Transfer Protocol) and SMTP (Simple Mail Transfer Protocol). Data is sent whenever resources are available and not required by any other scheduling-service classes.

Each of these service classes should be treated differently by the MAC layer packet scheduling mechanism. Thus, each type of application can be included in a class of service. However, the WiMAX standard does not define nor specify a scheduler, and one of the premises needed to guarantee QoS in WiMAX networks is the application of scheduling in both the uplink and downlink directions, which should translate the QoS requirements of SSs to appropriate slot allocation. When the BS makes a scheduling decision, it informs its decision to all SSs using the messages at the beginning of each frame. These messages explicitly define which slots are allocated to each SS in both directions, uplink and downlink.

This work focuses on packet scheduling in the uplink direction, because it guarantees optimization of physical network rate and ensures the delay requested by the user, therefore to maximize the number of users transmitting data in each frame. The classes of services described above, has its own QoS parameters such as minimum throughput requirement and delay/jitter constraints.

6. Related research

Several scheduling algorithms and QoS architectures for Broadband Wireless Access (BWA) have been proposed in the literature, since the standard only provides signaling mechanisms and no specific scheduling and admission control algorithms. However, many of these solutions only address the implementation or addition of a new QoS architecture to the IEEE 802.16 standard.

A scheduling algorithm decides the next packet to be served on the wait list and is one of the mechanisms responsible for the distribution of bandwidth between multiple streams (through the attribution of each stream bandwidth that was needed and available). In these proposals, there are often no analytical models for ensuring maximum delay and to maximize the number of SSs allocated in the system, which are represented accurately by certain performance metrics of the medium access protocol such as delay.

In (Wongthavarawant & Ganz, 2003), a packet scheduler for IEEE 802.16 uplink channels based on a hierarchical queue structure was proposed. A simulation model was developed to evaluate the performance of the proposed scheduler. However, despite presenting simulation results, the authors overlooked the fact that the complexity to implement this solution is not hierarchical, and did not define clearly how requests for bandwidth are made.

In (Chu et al., 2002) authors proposed a QoS architecture to be built into the IEEE 802.16 MAC sublayer, which significantly impacts system performance, but did not present an algorithm that makes efficient use of bandwidth.

In (Cicconetti et al., 2007), authors presented a simulation study of the IEEE 802.16 MAC protocol operating with an OFDM air interface and full-duplex stations. They evaluated system performance under different traffic scenarios, with the variation of a set of relevant system parameters. About the data traffic, it was observed that the overhead due to the physical transmission of preambles increases with the number of stations.

In (Iyengar et al., 2007), a polling-based MAC protocol is presented along with an analytical model to evaluate its performance, considering a system where the BS issues probes in every frame to determine bandwidth requirements for each node. They developed closed-form analytical expressions for cases in which stations are polled at the beginning or at the end of uplink subframes. It is not possible to know how the model may be developed for delay guarantees.

In (Cho et al., 2005), authors proposed a QoS architecture in which the scheduler is based on packet lifetime for each type of flow. In this paper, authors considered the process of data communication between BS and SS from the start, that is, connection and negotiation of traffic parameters such as bandwidth and delay. The proposal features an architecture defined in well-structured blocks, which may make data flows and architecture actions inaccurate. However, in spite to present simulation results, the work neglects performance by not adequately address the functional blocks of the proposed architecture and by not clearly how lifetime is calculated for each packet.

In (Kim & Yeom, 2007), the scheduling algorithm handles traffic with Best Effort (BE), and concludes that it is difficult to estimate the amount of bandwidth required due to dynamic changes in traffic transmission rate. The purpose of this algorithm is to keep fair bandwidth allocation between BE flows and full bandwidth usage. The system measures the transmission rate for each flow and allocates bandwidth based on the average transmission rate.

Finally, in (Maheshwari, 2005) the author presents a well-established architecture for QoS in the IEEE 802.16 MAC layer. The subject of this work is the component responsible for the allocation uplink bandwidth to each SS, although the decision is taken based on the following aspects: bandwidth required by each SS for uplink data transmission, periodic bandwidth needs for UGS flows in SSs and bandwidth required to make requests for additional bandwidth.

Given the limitations exposed above, these works form the basis of a generic architecture, which can be extended and specialized. However, in these studies, the focus is getting QoS guarantees, with no concerns for maximize the number of allocated users in the network. This paper presents a scheduler with admission control of connections to the WiMAX BS. We developed an analytical model based on Latency-Rate (LR) server theory (Stiliadis & Varma, 1998), from which an ideal frame size called Time Frame (TF) was estimated, with guaranteed delays for each user and mazimization of the number of allocated stations in the system. A set of simulations is presented with constant bit rate (CBR) and variable bit rate (VBR) streams and performance comparisons are made for different delays and different TFs. The results show that an upper bound on the delay may be achieved for a large range of network loads with bandwidth optimization.

7. Proposed scheduler: Implementation and evaluation

A minimum acceptable performance level should be sought throughout the development of any system, be it computer-related or not. This requires a measure or gauge of performance in these systems. To accomplish this, there exist design tools that provide the analyst with different metrics and measures. In this context, some features of the system related to the subject were discussed earlier in this work. To achieve this, this section presents an analytical model of the new scheduler and an analytical description of its call admission control.

7.1 System description

Figure 7 (Foronda et al., 2007) illustrates a wireless network that use the new proposed scheduler with call admission control, which is based on a modified LR scheduler (Stiliadis & Varma, 1998) and uses the token bucket algorithm. The basic approach consists of the token bucket to limit incoming traffic and the LR scheduler provide rate allocation for each user. Then, if the rate allocated by the LR scheduler is larger than the token bucket rate, the maximum delay may be calculated. A scheduler that provides guaranteed bandwidth can be modeled as LR scheduler.

The behavior of an LR scheduler is determined by two parameters for each session i: latency θ_i and allocated rate r_i. The latency θ_i of the scheduler may be seen as the worst-case delay and depends on network resource allocation parameters. In the new scheduler with call admission control, the latency θ_i is a TF period, which is the time needed to transmit a maximum-size packet and separation gaps (TTG and RTG) of DL and UL subframes. In the new scheduler, considering the delay for transmitting the first packet, the latency θ_i of is given by

$$\theta_i = T_{TTG} + T_{RTG} + T_{DL} + T_{UL} + \frac{L_{max,i}}{R} \tag{1}$$

where T_{TTG} and T_{RTG} are the DL and UL subframe gap durations, T_{TL} and T_{UL} are the DL and UL subframe durations, $L_{max,i}$ is the maximum packet size and R is the outgoing link capacity.

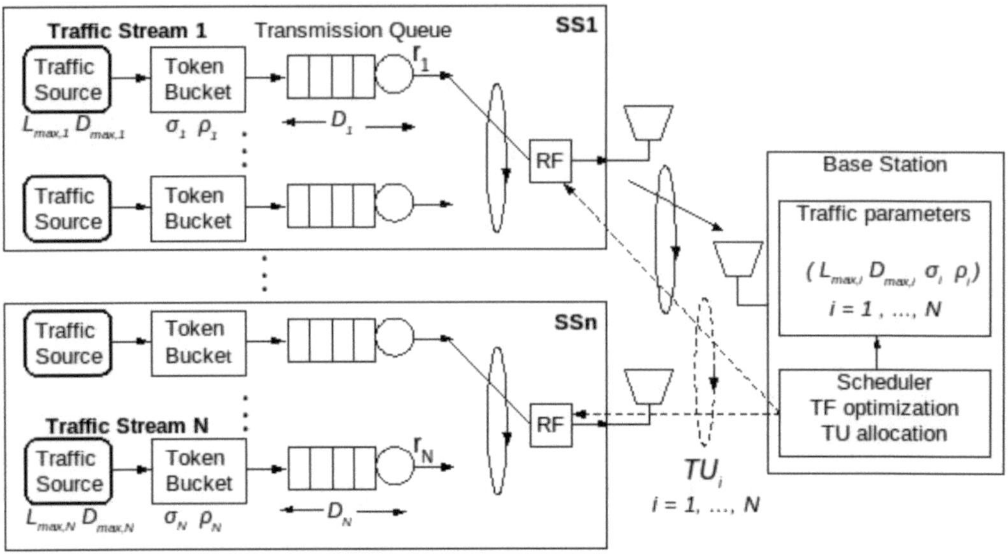

Fig. 7. Wireless network with new scheduler

Now, we show how the allocated r_i for each session i may be determined, and how to optimize TF in order to increase the number of connections accommodated with Call Admission Control (CAC).

7.2 CAC description

An LR scheduler can provide a bounded delay if input traffic is shaped by a token bucket. A token bucket (Gosh et al., 2005) is a non-negative counter which accumulates tokens at a constant rate ρ_i until the counter reaches its capacity σ_i. Packets from session i can be released into the queue only after remove the required number of tokens from the token bucket. In an LR scheduler, if the token bucket is empty, packets that arrive are dropped; however, our model ensures that there will always be tokens in the bucket and that no packets are dropped, as described in next section. If the token bucket is full, a maximum burst of σ_i packets can be sent to the queue. When the flow is idle or running at a lower rate as the token size reaches the upper bound σ_i, accumulation of tokens will be suspended until the arrival of the next packet. We assume that the session starts out with a full bucket of tokens. In our model, we consider IEEE 802.16 standard overhead for each packet. Then, as we will show below, the token bucket size will decrease by both packet size and overhead.

The application using session i declares the maximum packet size $L_{max,i}$ and required maximum allowable delay $D_{max,i}$, which are used by the WiMAX scheduler to calculate the service rate for each session so as to guarantee the required delay and optimize the number of stations in the network. Incoming traffic $A_i t$ from session $i (i = 1, \ldots, N)$ passes through a token bucket inside the user terminal during the time interval $(0, t)$, as shown in Figure 8 (Dosciatti et al., 2010).

This passage of data traffic by the token bucket is bounded by

$$A_i(t) \le \sigma_i + \rho_i t \tag{2}$$

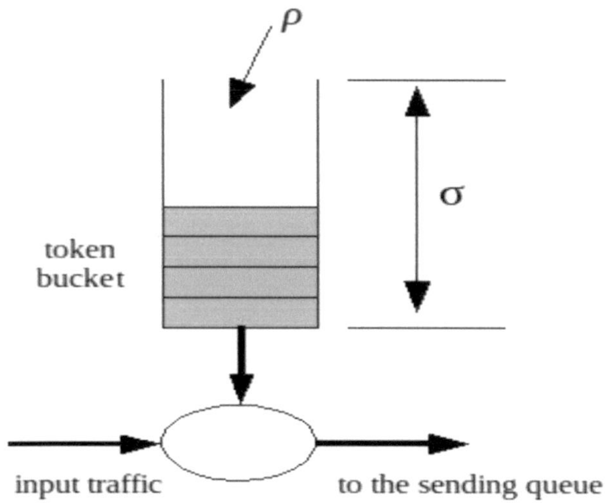

Fig. 8. Input traffic with token bucket

where σ_i the bucket size and ρ_i is the bucket rate.

Then, the packet is queued in the station until it is transmitted via the wireless medium. Queue delay is measured as the time interval between the receipt of the last bit of a packet and its transmission. In the new scheduler with call admission control, queuing delay depends on token bucket parameters, network latency and allocated rate. In (Stiliadis & Varma, 1998) and (Parekh & Gallager, 1993), it is shown that if input traffic $A_i t$ is shaped by a token bucket and the scheduler allocates a service rate r_i, then an LR scheduler can provide a bounded maximum delay D_i:

$$D_i \leq \frac{\sigma_i}{r_i} + \theta_i - \frac{L_{max,i}}{r_i} \tag{3}$$

where σ_i is the token bucket size, r_i is the service rate, θ_i is the scheduler latency, $L_{max,i}$ is the maximum size of a package and, $(\sigma_i / r_i) + \theta_i - (L_{max,i} / r_i)$ is the bound on the delay, D_{bound}.

Equation (3) is an improved bound on the delay for LR schedulers. Thus, the token bucket rate plus the overhead transmission rate must be smaller than the service rate to provide a bound on the delay. D_{bound} should be smaller than or equal to the maximum allowable delay:

$$\frac{\sigma_i}{r_i} + \theta_i - \frac{L_{max,i}}{r_i} \leq D_{max,i} \tag{4}$$

Therefore, three different delays are defined. The first is the maximum delay D_i, the second is the upper bound on the delay D_{bound} and the third is the required maximum allowable delay $D_{max,i}$. The relation between them is $\boldsymbol{D_i \leq D_{bound} \leq D_{max,i}}$. So, the first delay constraint condition of the new scheduler is

$$\frac{\left(\sigma'_i - L'_{max,i}\right)TF}{r'_i TF - \Delta R + L'_{max,i}} + TF + \frac{L'_{max,i}}{R} + T_{TTG} + T_{RTG} \leq D_{max,i} \tag{5}$$

where σ'_i is the token bucket size with overhead, $L'_{max,i}$ is the maximum size of a packet with overhead, *TF* is the time frame, r'_i is the rate allocated by the server with overhead, *R* is the outgoing link capacity, T_{TTG} is the gap between downlink and uplink subframes, T_{RTG} is the gap to between uplink and downlink subframes, $D_{max,i}$ is the maximum allowable delay and Δ is the sum of initial ranging and BW request, which is the uplink subframe overhead and whose value will be discussed when evaluated their performance.

The second delay constraint condition to TF and service rate is that the token bucket rate plus the rate to transmit overhead and a maximum-sized packet that must be smaller than the service rate to place a bound on delay. Thus, the constraint condition is

$$\rho_i + \frac{\Delta R + L'_{max,i}}{TF} \leq r'_i \tag{6}$$

where ρ_i is the bucket rate, Δ is the uplink subframe overhead, *R* is the outgoing link capacity, $L'_{max,i}$ is the maximum packet size with overhead, *TF* is the time frame and r'_i is the rate allocated by the service with overhead.

Figure 9 shows a frame structure with TDD allocation formulas as described by Equation (5). Physical rate, maximum packet size and token bucket size are parameters declared by the application. However, TF and total allocated service rate must satisfy Equation (5).

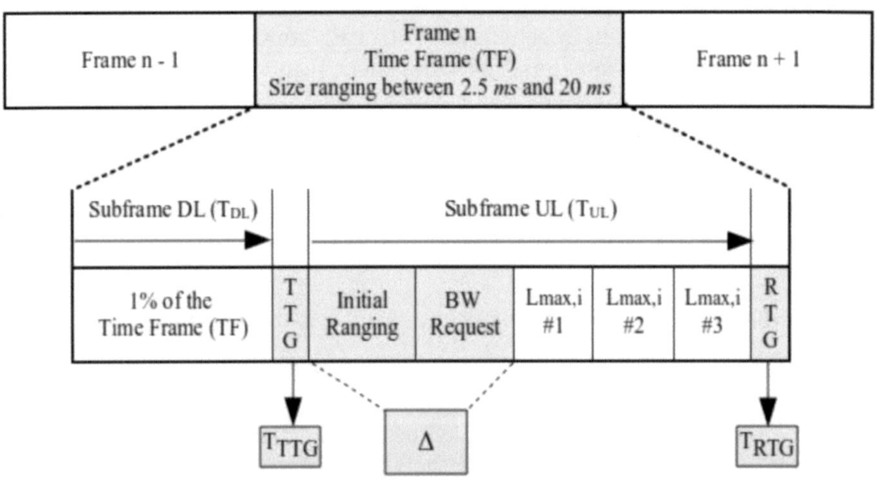

Fig. 9. Frame structure with TDD allocation formulas of Equation (5)

Previous schedulers do not provide any mechanisms to estimate the TF need to place a bound on delay or to maximize the number of stations, because each application requires a TF without the use of criteria to calculate the time assigned to each user. However, TF estimation is important because of a tradeoff. A small TF reduces maximum delay, but increases overhead at the same time. On the other hand, a large TF decreases overhead, but increases delay. Therefore, we must calculate the optimal TF to allocate the maximum number of users under both constraints. The maximum number of users is achieved when the service rate for each user is the minimum needed to guarantee the bound on the delay, D_{bound}.

Different optimization techniques may be used to solve this problem. In this work, we have used a step-by-step approach, which does not change the scheduler's essential operation. We start with a small TF, 2.5 *ms*, calculate r'_i and repeat this process every 0.5 *ms* until we find the minimum r'_i that satisfies both equations.

7.3 Performance evaluation

To analyze the IEEE 802.16 MAC protocol behavior with respect to the new scheduler with call admission control, this section presents numerical results obtained with the analytical model proposed in the previous section. Then, with a simulation tool, the analytical model proposed is validated, showing that the bound on the maximum delay is guaranteed. In this section, two types of delays are treated: required delay, in which the user requires the maximum delay, and the guaranteed maximum delay, which is calculated with the analytical model.

7.3.1 Calculation of optimal time frame

In this work, the duration of downlink subframes is fixed at 1% of the TF because our interest is only in the uplink subframe. In the simulation, after find the optimal number of SSs per frame for each traffic flow, the header value of the uplink subframe is calculated at a rate of 10% of the value of an OFDM symbol. All PHY and MAC layer parameters used in simulation are summarized in Table 1 and can be seen in Figure 9.

Parameter	**Value**
Bandwidth	20 *MHz*
OFDM Symbol Duration	13.89 μs
Delay	5, 10, 15 and 20 *ms*
Δ (Initial Ranging and BW Request) → 9 OFDM Symbols	125.10 μs
TTG + RTG → 1 OFDM Symbol	13.89 μs
UL Subframe (preamble + pad) → 10% OFDM Symbol	1.39 μs
Physical Rate	70 *Mbps*
DL Subframe	1% TF

Table 1. PHY and MAC parameters

Performance of the new scheduler with call admission control is evaluated as the delay requested by the user and assigned stations. Station allocation results, in the system with an optimal TF, limited by the delay requested by the user, are described in sequence.

The first step is define the token bucket parameters, which are estimated in accordance with the characteristics of incoming traffic and are listed on Table 2. It is important to note that the details about the incoming traffic must be known in advance. This is normal for various applications such as audio, CBR and video on demand.

	Audio	VBR video	MPEG4 video
Token Size (bits)	3000	18000	1000
Token Rate (Kb/s)	64	500	4100

Table 2. Token bucket parameters

Thus, the optimal TF value is estimated according to the PHY and MAC layer's parameters (see Table 1), token bucket parameters (see Table 2), required maximum allowable delay, physical rate and maximum packet size. With all parameters defined, and with the constraints set by Equations (5) and (6), described in Section 7.2, the simulation starts with a step-by-step approach, with the value of TF estimated at 2.5 *ms*. The r'_i is calculated and the procedure is repeated every 0.5 *ms* until that the minimum r'_i that satisfies both equations be found.

The graph in Figure 10 shows the optimal TF value, for four delay values required by users (5, 10, 15 and 20 *ms*). For example, in the graph, for a requested delay of 5 *ms*, the optimal TF is 3 *ms*. For a requested delay of 10 *ms*, the optimal TF is 6.5 *ms*. For a requested delay of 15 *ms*, the optimal TF is 10.5 *ms*. Finally, for a requested delay of 20 *ms*, the optimal TF is 15 *ms*.

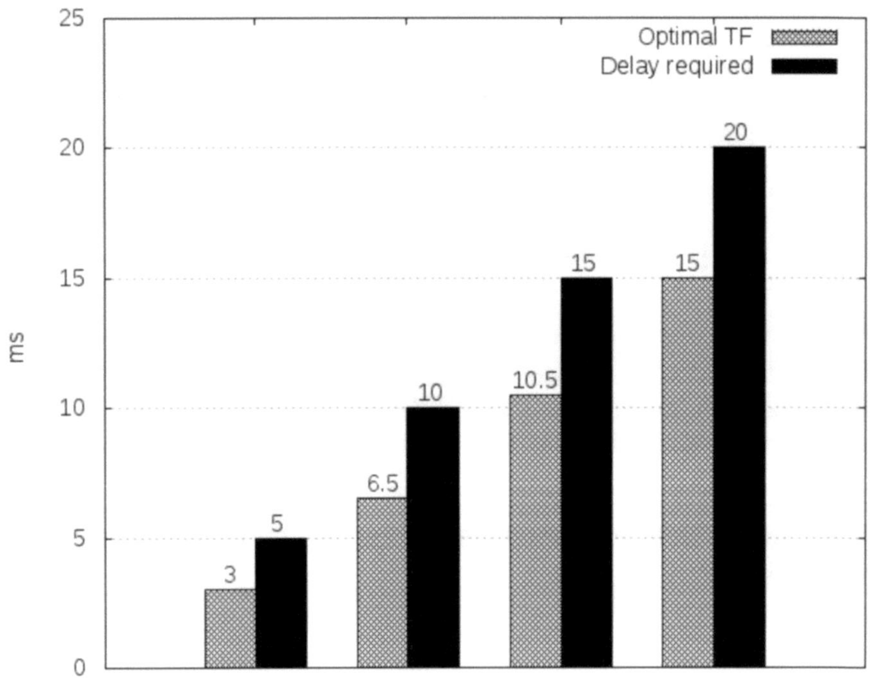

Fig. 10. Optimal TF

Next, we show the number of SSs assigned to each traffic type. The result shows the maximum number of SSs assigned to each range of optimal TF values for each traffic type. It should be noted that three traffic types were used: audio traffic, VBR video traffic and MPEG4 video traffic. For the simulation, the allocation of users is performed by traffic type; i.e., only one traffic at a time will be transmitted within each frame.

As an example, Figure 11 shows that when the user-requested delay is of 20 *ms*, an optimal TF of 15 *ms* is calculated and 50 users can be allocated for audio traffic, or 30 users for VBR video traffic, or 13 users for MPEG4 video traffic.

Two important observations from Figure 11 should be highlighted:

- With a requested delay of 20 *ms*, we cannot choose a TF of less than 15 *ms*, since the restrictions placed by Equation (5) (which regards delay) and Equation (6) (which regards the token bucket) are not respected and thus no bandwidth allocation guarantees exist.
- We also cannot choose a TF greater than 15 *ms*, even though it complies with Equations (5) and (6) with respect to guaranteed bandwidth, because there will be a decrease in the number of users allocated to each traffic flow due to increase of the delay.

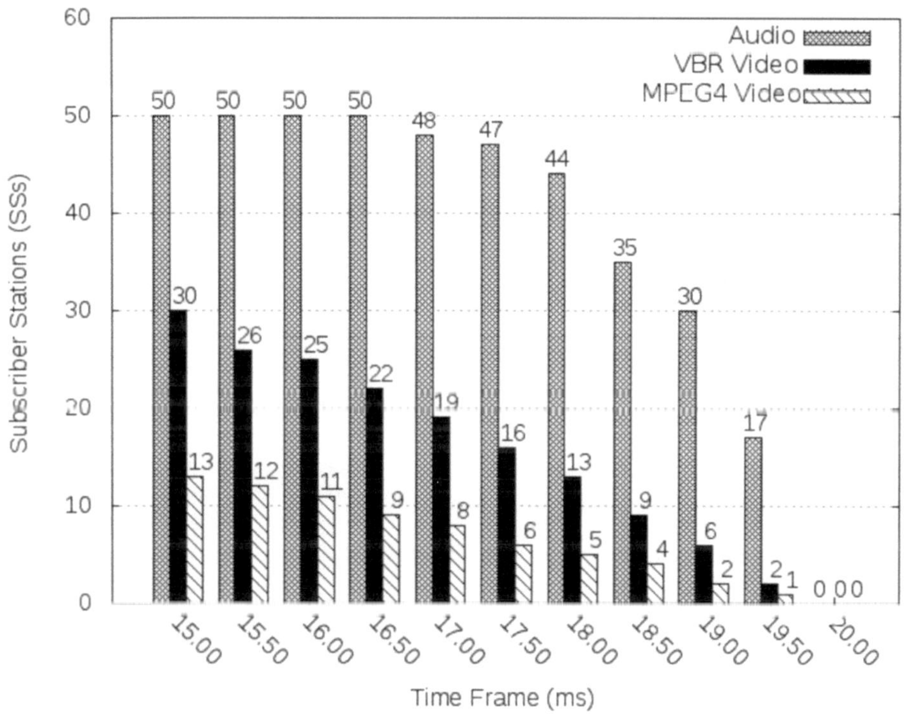

Fig. 11. Number of subscriber stations for 20 *ms* delay

Thus, it is evident that since the IEEE 802.16 standard does not specify an ideal time frame (TF) duration, this approach becomes advantageous because, in addition to comply the restrictions of the analytical model, it optimizes the allocation of users on the system. The same philosophy holds true for other delay values of 5, 10 and 15 *ms*.

7.3.2 Comparison of user allocation and optimal time frame

In this work, an optimal TF was reached, so that the number of SSs in the network may be optimized and a maximum delay may be guaranteed. To make a comparison of the results in this work, Figure 12 shows that, for an audio traffic and a requested delay of 15 *ms*, an optimal TF of 10.5 *ms* is obtained and 41 users can be allocated. When compared to other randomly-chosen TFs, it may be observed that the optimal TF yields a greater number of users.

Thus, when an user requests a delay guarantee, an optimal TF is calculated in order to allocate the largest number of users in a given traffic flow, as seen in the example in Figure 12. It may be noticed, then, that to choose a non-optimal TF will lead to a decreased number of allocated SSs. Therefore, the new scheduler with call admission control proposed herein maximizes the number of SSs and ensures an upper bound on maximum delay, as discussed next.

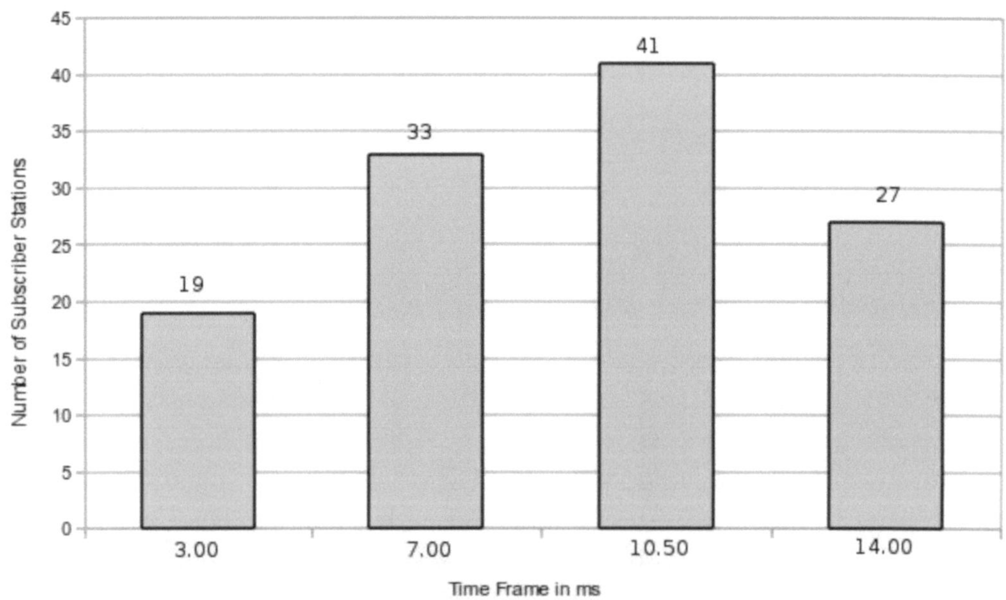

Fig. 12. Users assigned as a function of TF for audio traffic

7.3.3 Guaranteed maximum delay

In this work, only UL traffic is considered. To test the new scheduler's performance, we have carried out simulations of an IEEE 802.16 network consists of a BS that communicates with eighteen SSs, with one traffic flow type by SS and the destination of all flows being the BS, as shown in Figure 13. In this topology, six SSs transmit on-off CBR audio traffic (64 kb/s), six transmit CBR MPEG4 video traffic (3.2 Mb/s) and six transmit VBR video traffic.

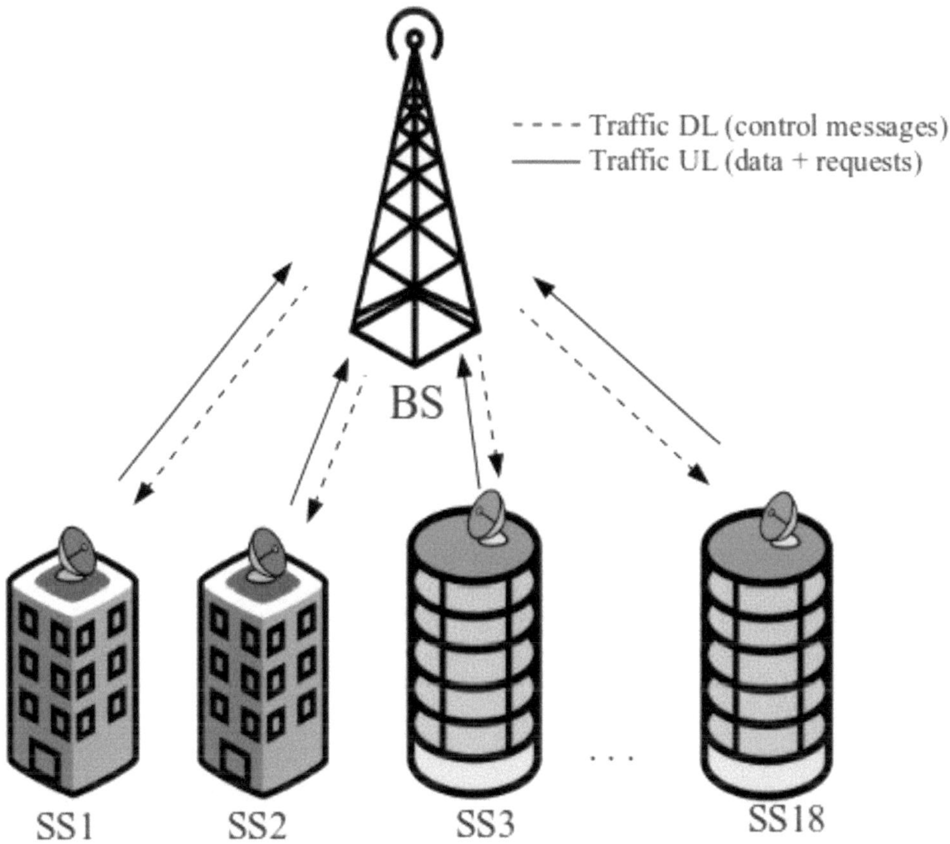

Fig. 13. Simulation scenario

Table 3 summarizes the different types of traffic used in this simulation.

Node	Application	Arrival Period (ms)	Packet size (max)(bytes)	Sending rate (kb/s)(mean)
$1 \rightarrow 6$	Audio	4.7	160	64
$7 \rightarrow 12$	VBR video	26	1024	≈200
$13 \rightarrow 18$	MPEG4 video	2	800	3200

Table 3. Description of traffic types

On Figure 14, with an optimal TF of 3 *ms* and an user-requested delay of 5 *ms*, the average guaranteed maximum delay for audio traffic is 1.50 *ms*. For VBR video traffic, whose packet rate is variable, the average maximum delay is 1.97 *ms*. For MPEG4 video traffic, the average maximum delay is 2.00 *ms*.

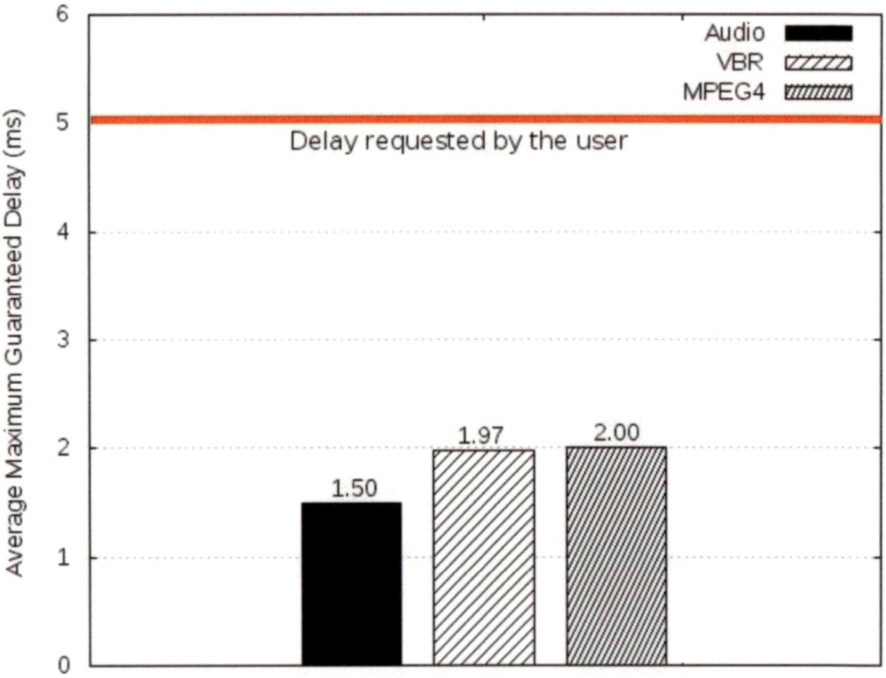

Fig. 14. Guaranteed Maximum Delay

7.3.4 Comparison with other schedulers

The New Scheduler with Call Admission Control was compared to those of (Iyengar et al., 2005), here called Scheduler_1, and (Wongthavarawant & Ganz, 2003), here called Scheduler_2. The comparison was accomplished through the ability to allocate users in a particular time frame (TF).

Table 4 shows the parameters used in the comparisons.

Parameter	Value
Bandwidth	20 *MHz*
OFDM symbol duration	13.89 μs
Delay requested by user	Dependent of each comparison
Maximum data rate	70 *Mbps*
Traffic type	Audio

Table 4. Parameters used in the comparisons

In the graph of Figure 15, we compare the New Scheduler with the Scheduler_1. A maximum delay of 0.12 *ms* was requested by the user, and the duration of each frame (TF) was set at 5 *ms*, as in Scheduler_1. Other parameters are listed in Table 4. In comparison, the New Scheduler allocates 28 users in each frame, while the Scheduler_1, allocates 20 users. Thus, the New Scheduler presents a gain in performance of 40% when compared with the Scheduler_1.

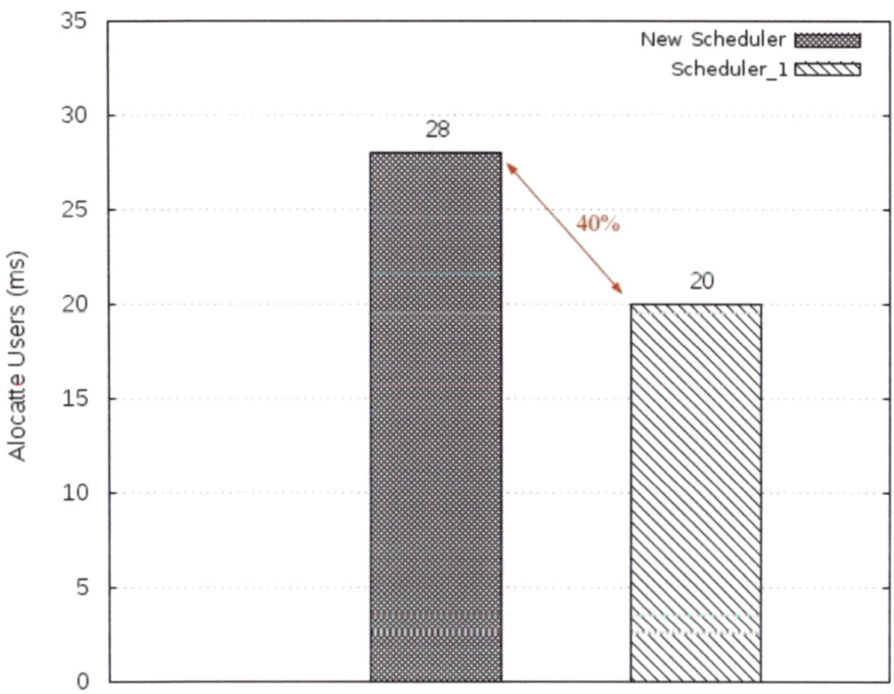

Fig. 15. Comparison between the New Scheduler and Scheduler_1

In the graph of Figure 16, we compare the New Scheduler with the Scheduler_2. A maximum delay of 20 *ms* was requested by the user, and the duration of each frame (TF) was set at 10 *ms*, as in Scheduler_2. Other parameters are listed in Table 4. The comparison was extended by also considering frame duration values of 7.00 *ms*, 8.00 *ms* and 9.00 *ms* to demonstrate the efficiency of the new scheduler. For a TF of 10 *ms*, the New Scheduler allocates 41 users in each frame, while the Scheduler_2 allocates only 33 users. This represents 24.24% better performance for the New Scheduler. Similarly, the New Scheduler also allocates more users per frame in comparison with the Scheduler_2 for all other frame duration values.

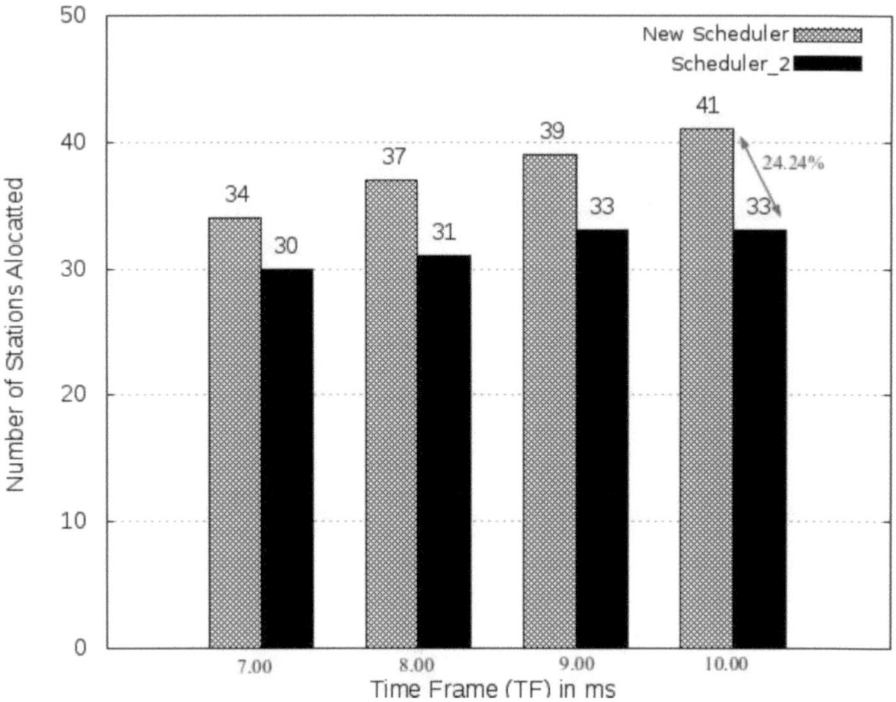

Fig. 16. Comparison between the New Scheduler and Scheduler_2

8. Conclusion

This work has presented the design and evaluation of a new scheduler with call admission control for IEEE 802.16 broadband access wireless networks (known worldwide as WiMAX) that guarantees different maximum delays for traffic types with different QoS requisites and optimizes bandwidth usage. Firstly, we developed an analytical model to calculate an optimal TF, which allows an optimal number of SSs to be allocated and guarantees the maximum delay required by the user. Then, a simulator was developed to analyze the behavior of the proposed system.

To validate the model, we have presented the main results obtained from the analysis of different scenarios. Simulations were performed to evaluate the performance of this model and demonstrated that an optimal TF was obtained with a guaranteed maximum delay in accordance with the delay requested by the user. Thus, the results have shown that the new scheduler with call admission control successfully limits the maximum delay and maximizes the number of SSs in a simulated environment.

9. References

802.16-2004. (2004). *IEEE Standard for Local and Metropolitan Area Networks - Part 16: Air Interface for Fixed Broadband Wireless Access Systems. IEEE Std., Rev. IEEE Std802.16-2004*, IEEE Computer Society, ISBN 0-7381-3986-6, New York, USA.

802.16e-2005. (2006). *IEEE Standard for Local and Metropolitan Area Networks. Amendment 2: Physical and Medium Access Control Layers for Combined Fixed and Mobile Operation in*

Licensed Bands and Corrigendum, IEEE Computer Society, ISBN 0-7381-4857-1, New York, USA.

802.16-Rev2/D2. (2007). *DRAFT Standard for Local and Metropolitan Area Networks.* Part 16: Air Interface for Broadband Wireless Access Systems. IEEE Computer Society, New York, USA.

Akyildiz I. F., & Wang X. (2005). A Survey on Wireless Mesh Networks. *IEEE Communications Magazine,* Vol. 43, No. 9, (September 2005), pp. 523-530, ISSN 0163-6804 .

Camargo E. G., Both C. B., Kunst R., Granville L. Z., & Rochol J. (2009). Uma Arquitetura de Escalonamento Hierárquico para Transmissões Uplink em Redes WiMAX Baseadas em OFDMA. *Proceedings of Brazilian Symposium on Computer Networks and Distributed Systems - SBRC,* pp. 525-538, Recife, Pernambuco, Brazil, May, 2009. (in Portuguese).

Chen J., Jiao W., & Wang H. (2005). A Service Flow Management Strategy for IEEE 802.16 Broadband Wireless Access Systems in TDD Mode. *In IEEE International Conference on Communications (ICC2005),* pp. 3422-3426, ISBN 0-7803-8938-7, Seoul, Korea, May 16-20, 2005.

Cho D-H., Song J-H., Kim M-S., & Han K-J. (2005). Performance Analysis of the IEEE 802.16 Wireless Metropolitan Area Network. *IEEE Computer Society, DFMA'05,* Vol. 1, No. 1, (February 2005), pp. 130-137, ISSN 0-7695-2273-4.

Chu G., Wang D., & Mei S. (2002). A QoS Architecture for the MAC Protocol of IEEE 802.16 BWA System. *In IEEE International Conference on Communications, Circuits, and Systems and West Sino Expositions Proceddings,* pp. 435–439, ISBN 0-7803-7547-5, Tibet Hotel, Chengdu, China, June 29-July 1, 2002.

Cicconetti C., Erta A., Lenzini L., & Mingozzi E. (2007). Performance Evaluation of the IEEE 802.16 MAC for QoS Support. *IEEE Transactions on Mobile Computing,* Vol. 6, No. 1, (January 2007), pp. 26-38, ISSN 1536-1233.

Dosciatti E. R., Godoy-Jr W., & Foronda A. (2010). A New Scheduler for IEEE 802.16 with Delay Bound Guarantee. *The Sixth International Conference on Networking and Services - ICNS '10,* pp. 150-155, ISBN 978-1-4244-5927-8, Cancun, Mexico, March 7-13, 2010.

Eklund C., Marks R. B., Stanwood K. L., & Wang S. (2002). IEEE Standard 802.16: A Technical Overview of the WirelessMAN Air Interface for Broadband Wireless Access. *IEEE Communications Magazine,* Vol. 40, No. 6, (June 2002), pp. 98-107, ISSN 0163-6804.

Foronda A., Higuchi Y., Ohta C., Yoshimoto M., & Okada Y. (2007). Service Interval Optimization with Delay Bound Guarantee for HCCA in IEEE 802.11e WLANs. *IEICE Transactions on Communications,* Vol. E90-B, No. 11, (November 2007), pp. 3158-3169, Print ISSN 0916-8516, Online ISSN 1745-1345.

Gosh A., Wolter D., Andrews J., & Chen R. (2005). Broadband Wireless Access with WiMAX/802.16: Current Performance Benchmarks and Future Potential. *IEEE Communications Magazine,* Vol. 43, No. 2, (February 2005), pp. 129-136, ISSN 0163-6804.

Guo X., Ma W., Guo Z., & Hou Z. (2007). Dynamic Bandwidth Reservation Admission Control Scheme for the IEEE 802.16e Broadband Wireless Access Systems. *In Proceedings of the Wireless Communications and Networking Conference - WCNC'07,* pp. 3420-3425, ISBN 1-4244-0658-7, March 11-15, Hong Kong, 2007.

Hoymann C. (2005). Analysis and Performance Evaluation of the OFDM-based Metropolitan Area Network IEEE 802.16. *Computer Networks,* Vol. 49, No. 3, (October 2005), pp. 341-363, ISSN 1389-1286.

INTEL. (2005). Deploying License-Exempt WiMAX Solutions: White paper. 16 p., January 2005.

Iyengar R., Iyer P., & Sikdar B. (2005). Delay Analysis of 802.16 Based Last Mile Wireless Networks. *Global Telecommunications Conference - GLOBECOM'05 - IEEE*, pp. 3123-3127, ISBN 0-7803-9414-3, St. Louis, Missouri, USA, November 28-December 2, 2005.

Kim S., & Yeom I. (2007). TCP-aware Uplink Scheduling for IEEE 802.16. *In IEEE Communications Letters*, Vol. 11, No. 2, (February 2007), pp. 146-148, ISSN 1089-7798.

Maheshwari S. *An Efficient QoS Scheduling Architecture for IEEE 802.16 Wireless MANs*. Master Degree. K. R. School of Information Technology, Bombay, India, January 2005.

Masri M., Abdellatif S., & Juanole G. (2009). *An Uplink Bandwidth Management Framework for IEEE 802.16 with QoS Guarantees*. Networking 2009 - Lecture Notes in Computer Science, Springer, Vol. 5550, pp. 651-663, ISBN 978-3-642-01399-7_51.

Mei X., Fang Z., Zhang Y., Zhang J., & Xie H. (2010). A WiMAX QoS Oriented Bandwidth Allocation Scheduling Algorithm. *In Second International Conference on Networks Security, Wireless Communications and Trusted Computing - NSWCTC-2010*, pp. 298-301, ISBN 978-0-7695-4011-5, Wuhan, Hubei, China, April 24-25, 2010.

Parekh A., & Gallager R. (1993), A Generalized Processor Control and Topology Management Protocols for Wireless Mobile Networks: the Single-Node Case. *IEEE/ACM Transactions Networking*, Vol. 1, No. 3, (June 1993), pp. 344-357, ISSN 1063-6692.

Rong B., Qian Y., Lu K., Chen H-H.. & Guizani M. (2008). Call Admission Control Optimization in WiMAX Networks. *IEEE Transactions on Vehicular Technology*, Vol. 57, No. 4, (July 2008), pp 2509-2522, ISSN 0018-9545.

So-In C., Jain R., & Tamimi A.-K. (2009). Scheduling in IEEE 802.16e Mobile WiMAX Networks: Key Issues and a Survey. *IEEE Journal on Selected Areas in Communications*, Vol. 27, No. 2 (February 2009), pp. 156-171, ISSN 0733-8716.

Stiliadis D., & Varma A. (1998). Latency-Rate Servers: A General Model for Analysis of Traffic Scheduling Algorithms. *IEEE-ACM Transactions on Networking*, Vol 6, No. 5, (October 1998), pp. 611-624, ISSN 1063-6692.

Sun Y., Sheriff I., Royer E. M. B., & Almeroth K. C. (2005). An Experimental Study of Multimedia Traffic Performance in Mesh Networks. *Proceedings of the International Workshop on Wireless Traffic Measurements and Modeling*, pp. 25-30, ISBN 1-931971-33-1, Seattle, Washington, USA, June 6-8, 2005.

Tanenbaum A. S. (2003). *Computer Networks* (4. ed.), Prentice-Hall, ISBN 0130661023, New Jersey, USA.

Wang H., He B., & Agrawal D. P. (2007). Above Packet Layer Level Admission Control and Bandwidth Allocation for IEEE 802.16 Wireless MAN. *Simulation Modeling Practice and Theory*, Vol. 15, No. 14, (April 2007), pp. 266-382, ISSN 1569-190X.

Wang L., Liu F., Ju Y., & Ruangchaijatupon N. (2007). Admission Control for Non-Preprovisioned Service Flow in Wireless Metropolitan Area Networks. *In Proceedings of the Fourth European Conference on Universal Multiservice Networks – ECUMN'07*, pp. 243-249, ISBN 0-7695-2768-X, Toulouse, France, February 14-16, 2007.

WiMAX Forum. (May 2011). WiMAX Forum, 20.05.2011, Available from http://www.wimaxforum.org.

Wongthavarawant K., & Ganz A. (2003). Packet Scheduling for QoS Support in IEEE 802.16 Broadband Wireless Access Systems. *Internacional Journal of Communications Systems*, Vol. 16, No. 1, (February 2003), pp. 81-96, doi: 10.1002/dac.581.

4

Scheduling Algorithm and Bandwidth Allocation in WiMAX

Majid Taghipoor[1], Saeid MJafari[2] and Vahid Hosseini[3]
[1]*University of Applied Science and Technology Uromieh,*
[2]*Department of Computer and IT Engineering,*
Islamic Azad University of Qazvin, Qazvin,
[3]*Department of Computer and IT Engineering,*
Computer Engineering Deptt., Urmia University,
Iran

1. Introduction

The traditional solution to provide high-speed broadband access is to use wired access technologies, such as cable modem, DSL (Digital Subscriber Line), Ethernet, and fiber optic. However, it is too difficult and expensive for carriers to build and maintain wired networks, especially in remote areas. BWA (Broadband Wireless Access) technology is a flexible, efficient, and cost-effective solution to overcome the problems [1]. WiMAX is one of the most popular BWA technologies today, which aims to provide high speed broadband wireless access for WMANs (Wireless Metropolitan Area Network). [2]

WiMAX provides an affordable alternative for wireless broadband access supporting a variety of applications of different types including video conferencing, non-real-time large volume data transfer, traditional voice/data traffic throughput E1/T1 connection, and web browsing.[1]

Each traffic flow requires different treatment from the network in terms of allocated bandwidth, maximum delay, and jitter and packet loss [3], [5]. Traffic differentiation is thus a crucial feature to provide network-level QoS (Quality of Service). The standard leaves QoS support features specified for WiMAX networks (e.g., traffic policing and shaping, connection admission control and packet scheduling). One of the most critical issues is the design of a very efficient scheduling algorithm which coordinates all other QoS-related functional entities.

The key components in WiMAX QoS guarantee are the admission control and the bandwidth allocation in BS. WiMAX standard defines adequate signalling schemes to support admission control and bandwidth allocation, but does not define the algorithms for them. This absence of definition allows more flexibility in the implementation of admission control and bandwidth allocation.

In this study, we focus on evaluating scheduling algorithms for the uplink traffic in WiMAX. We evaluate a number of WiMAX uplink scheduling algorithms in a single-hop network, which is referred to as PMP (Point Multi Point) mode of WiMAX.

2. Overview of WIMAX

In this section, we discuss the WiMAX, the uniqueness of WiMAX uplink scheduling.

2.1 WiMAX standard

BWA technology promises a large coverage and high throughput. Theoretically, the coverage range can reach 30 miles and the throughput can achieve 75 Mbit/s [1]. Yet, in practice the maximum coverage range observed is about 20 km and the data throughput can reach 9 Mbit/s using UDP (User Datagram Protocol) and 5 Mbit/s using FTP (File Transfer Protocol) over TCP (Transmission Control Protocol)[2].

WiMAX standard has two main variations: one is for fixed wireless applications (covered by IEEE 802.16-2004 standard) and another is for mobile wireless services (covered by IEEE 802.16e standard). The 802.16 standards only specify the PHY (Physical) layer and the MAC (Media Access Control) layer of the air interface while the upper layers are not considered.

The IEEE 802.16 standard specifies a system comprising two core components [6]: the SS (Subscribe Station) or CPE (customer premises equipment) and the BS (Base Station). A BS and one or more SS can form a cell with a P2MP structure. Note that the WiMAX standard also can be used in a P2P (Point to Point) or mesh topology. BS acts as a central entity to transfer all the data from MSs (Mobile Station) in a PMP mode. Transmissions take place through two independent channels: downlink channel (from BS to MS) and uplink channel (from MS to BS). Uplink Channel is shared between all MSs while downlink channels is used only by BS.

To support the two-way communication, either FDD (Frequency Division Duplex) or TDD (Time Division Duplex) can be adopted. In the following discussion, we focus on the popular TDD.

The IEEE 802.16 is connection oriented. Each packet has to be associated with a connection at MAC level. This provides a way for bandwidth request, association of QoS and other traffic parameters and data transfer. All data transmissions are connection-oriented and the connections are classified into four types, namely, UGS (Unsolicited Grant Service), also known as CBR (Constant Bit Rate), rtVR (real-time Variable Bit Rate), nrtVR(non real-time Variable Bit Rate), and BE (Best Effort). Each service related to type of QoS class can has different constraints such as the traffic rate, maximum latency and tolerated jitter. In section 4 we will focus more on QoS the WiMAX technologies.

UGS supports real-time service flows that have fixed-size data packets on a periodic basis. RtVR supports real-time service flows that generate variable data packets size on a periodic basis. The BS provides unicast grants in an unsolicited manner like UGS. Whereas the UGS allocations are fixed in size. NrtVR is designed to support non real-time service flows that require variable size bursts on a regular basis. BE is used for best effort traffic where no throughput or delay guarantees are provided. Those service classes are defined in order to satisfy different types of QoS requirements. However, the IEEE 802.16 standard does not specify the scheduling algorithm to be used. Vendors and operators have to choose the scheduling algorithm(s) to be used.

2.2 WiMAX MAC layer

The 802.16 MAC protocol supports transport protocols such as ATM, Ethernet, and IP, and can accommodate future developments using the specific convergence layer. The MAC also accommodates very high data throughput through the physical layer while delivering ATM-compatible QoS, such as UGS, rtVR, nrtVR, and BE. The 802.16 frame structure enables terminals to be dynamically assigned uplink and downlink burst profiles according to the link conditions. This allows for a tradeoff to occur - in real time - between capacity and robustness. It also provides, on average, a 2x increase in capacity when compared to non-adaptive systems.

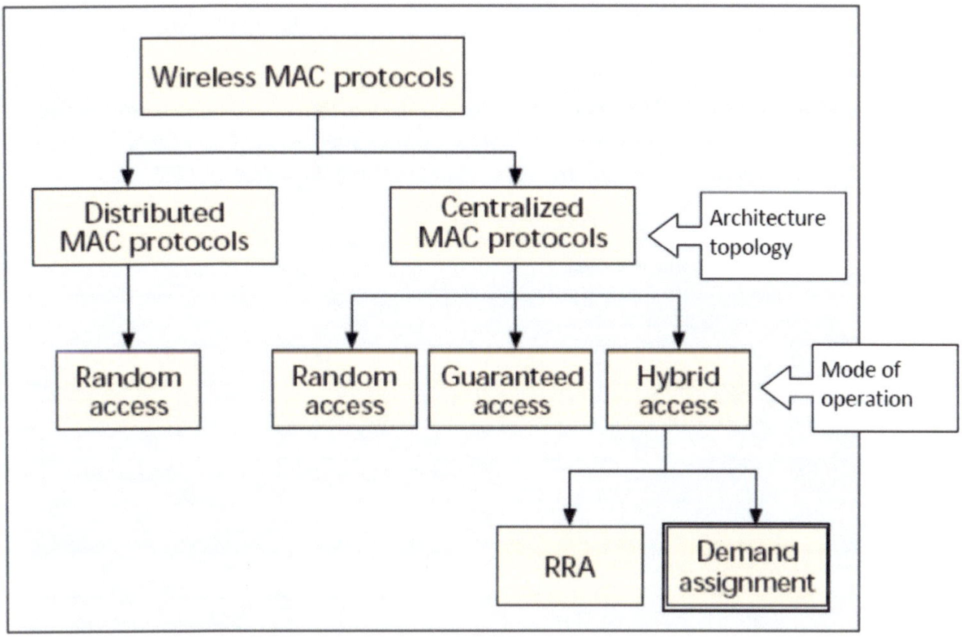

Fig. 1. Wireless MAC protocol classification [25]

According to the architecture topology, there are two main wireless MAC protocols:

1. Distributed MAC Protocols: these protocols are founded on principles of CS and CA, excluding the distributed ALOHA protocol.
2. Centralized MAC Protocols: these protocols are based on communicating with a central entity - in case of cellular mobile communications: the base station. Thus all communications is organized and supervised according to the BS MAC management protocol. [25]

There are three types of wireless MAC protocol types:

1. Random Access Protocols: according to this access protocol, for a node to be able to access the network it should contend for the medium.
2. Guaranteed Access Protocols: unlike the random access protocols, the communication between nodes is made on some predefined rules. This may be either in the form polling the nodes one by one, or by token exchanging.

3. Hybrid Protocols: these type of protocols are more superior to the previously mentioned other two protocols, since they are made out of the top properties of random access protocols and guaranteed access protocols. [25].

Hybrid protocols can be further subdivided into two categories:

a. Random Reservation Access protocol: these are the protocols by the MAC where a periodic reservation of the bandwidth is granted on the reception of a successful request from nodes supported by the central node.
b. Demand Assignment protocol: The MAC allocated bandwidth according to the need of application of the node.

The hierarchy of the wireless MAC protocols classification could be illustrated as it is shown in figure 1.

Therefore, mobile WiMAX MAC protocol could be classified as the demand assignment protocol; knowing that the mobile WiMAX MAC is designed to support QoS and according to the only MAC protocol that guarantees resources is the DA protocol. [25]

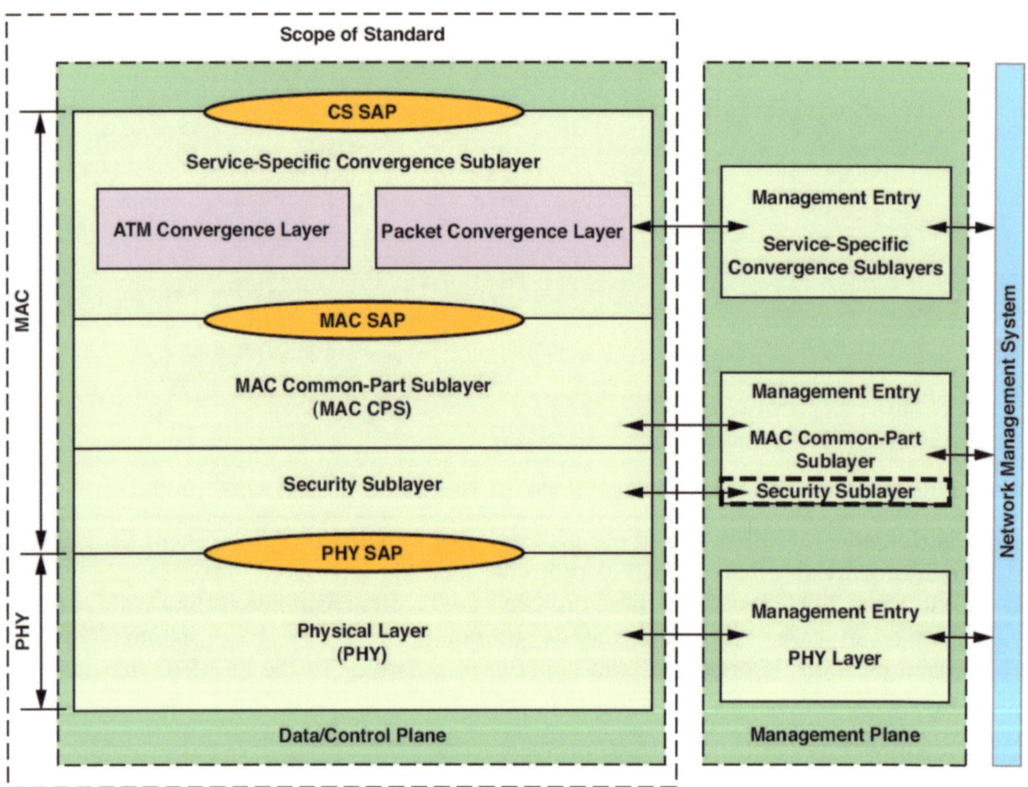

Fig. 2. The 802.16 protocol stack [6]

WiMAX MAC is subdivided into three sub layers with different functionalities. Figure 2 is a basic illustration of the tasks and services that the MAC sub layers are responsible for.

The upper MAC layer is the CS (Convergence Sub layer). This sub layer is responsible mainly for classification and header suppression of the incoming packets from the network layer. The classification is done according to the QoS parameters of the packet. Then each service flow is assigned a service flow identifier number.

The inner layer is called the MAC CSP (common part sub layer) and it is considered the main sub layer of the MAC layer. Finally, the lower MAC layer is called the MAC security sub layer. Functions like support for privacy, user/device authentication are the responsibility of this sub layer.

Figure 3 illustrates the PHY layer with the three sub layers of the MAC layer. The Figure shows the data/control plane only and it regarded as the scope of the standard.

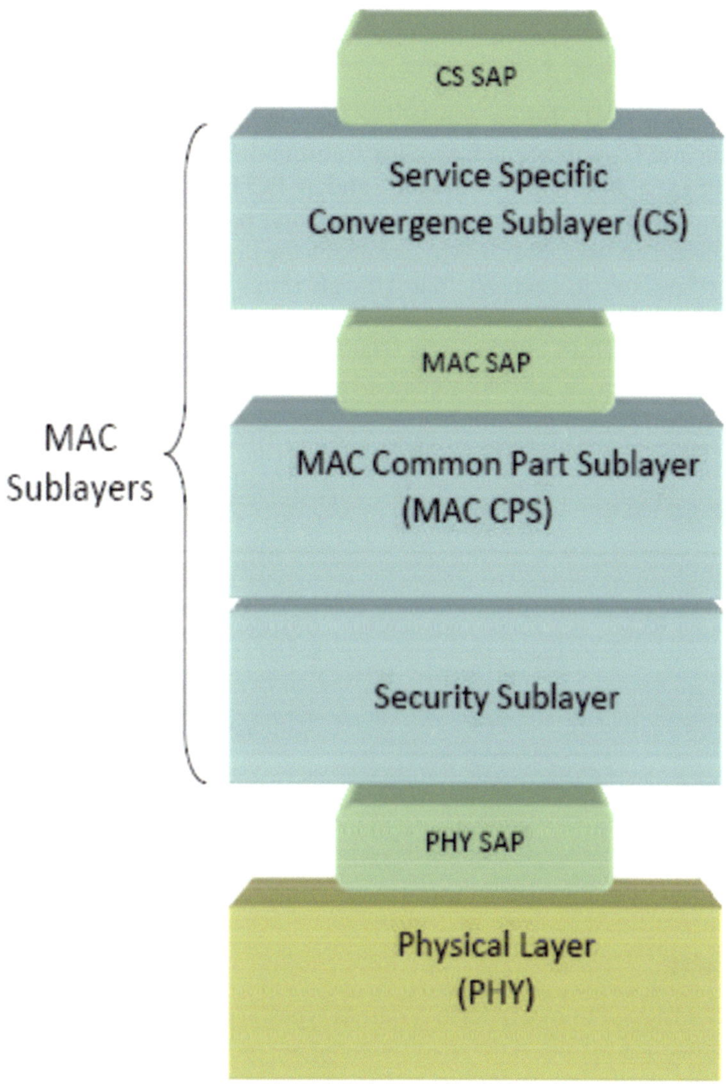

Fig. 3. WiMAX MAC and PHY layers - data/control plane [25]

The 802.16 MAC uses a variable-length PDU (Protocol Data Unit) and other innovative concepts to greatly increase efficiency. Multiple MAC PDUs, for example, may be concatenated into a single burst to save PHY overhead. Multiple SDUs (Service Data Unit) may also be concatenated into a single MAC PDU, saving on MAC header overhead. Fragmentation allows very large SDUs to be sent across frame boundaries to guarantee the QoS. Payload header suppression can be used to reduce the overhead caused by the redundancy within the SDU headers. The 802.16 MAC uses a self-correcting bandwidth request/grant scheme that eliminates any delay in acknowledgements, while allowing better QoS handling than traditional acknowledgement schemes. Depending on the QoS and traffic parameters of their services, terminals have a variety of options available to them for requesting bandwidth. SAP (Service Access Point) is entities located in between the sub layers in order to convert the SDU to PDU. Basically, when PDUs of an upper layer are passed through the SAP to a lower layer, they are considered as SDU for that particular lower layer.

In TDD mode, a WiMAX MAC frame consists of two sub frames, DL-sub frame for downlink transmission and UL-sub frame for uplink transmission, as shown in figure 4. The DL-sub frame comprises a BP (Burst Preamble) and a FCH (Frame Control Header), followed by DL-MAP, UL-MAP, and a number of downlink payload bursts (DL-PL1, …., DL-PLn). As will be discussed later, the UL-MAP contains the uplink scheduling results, i.e., the uplink bandwidth granted to each SS. The UL-sub frame starts with initial ranging contention slots and bandwidth request contention slots, followed by a number of uplink payload bursts (UL-PL1, …, UL-PLn)(Figure 4). The uplink and downlink bursts are not necessarily equal and their length can be adjusted dynamically in order to adapt to the traffic variation. [29]

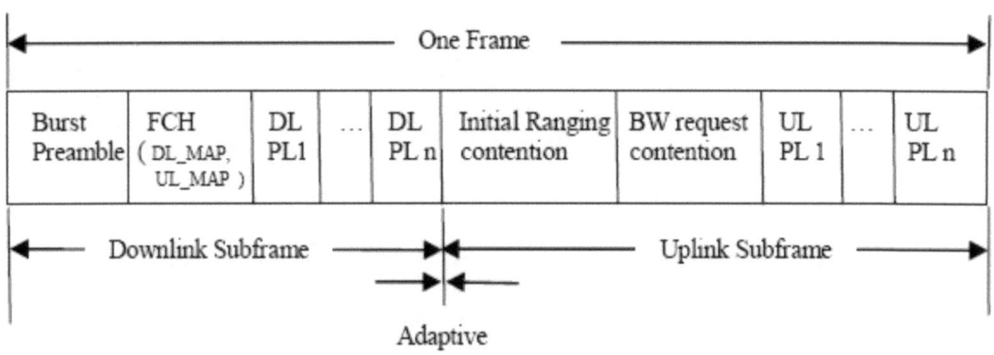

Fig. 4. WiMAX MAC frame structure [29]

3. Problem statement

A WiMAX network is designed to incorporate different types of data streams, and it aims at providing QoS guarantee for all the data streams being served by WiMAX. The WiMAX protocol covers physical layer and MAC layer, and there are several challenges for QoS guarantee in WiMAX.

The IEEE 802.16 standard provides specification for the MAC and PHY layers for WiMAX and there are several challenges for QoS guarantee in WiMAX.

In the physical layer, one challenge is the uncertainty of the wireless channel, which makes the guarantee of broadband wireless data service difficult and renders the static resource allocation scheme unsuitable. In the MAC layer, one challenge is the diversified service types, which requires the WiMAX scheduling scheme to be adaptive to the various QoS parameters of different service types. There have been some studies of the WiMAX MAC scheduling problem [3], [4], [5], and [6].

The key components in WiMAX QoS guarantee are the admission control and the bandwidth allocation in BS. WiMAX standard defines adequate signalling schemes to support admission control and bandwidth allocation, but does not define the algorithms for them. This absence of definition allows more flexibility in the implementation of admission control and bandwidth allocation.

The research problem being investigated here is, after connections are admitted into the WiMAX network, how to allocate bandwidth resources and perform scheduling services, so that the QoS requirements of the connections can be satisfied.

3.1 What is QoS?

QoS refers to the ability of a network to provide improved service to selected network traffic over various underlying technologies including wired-based technologies (Frame Relay, ATM, Ethernet and 802.1 networks, SONET, and IP-routed networks) and wireless-based technologies (802.11, 802.15, 802.16, 802.20, 3G, IMS, etc). In particular, QoS features provide improved and more predictable network service by providing the following services:

- Supporting dedicated bandwidth
- Improving loss characteristics
- Avoiding and managing network congestion
- Shaping network traffic
- Setting traffic priorities across the network

Due to the differences in the wired-based and wireless-based access technologies, the detailed QoS implementations for both tend to be different, however they share common roots. What follows next are the common elements shared between wired-based and wireless-based access methods.

3.2 QoS and scheduling in WiMAX

A high level of QoS and scheduling support is one of the interesting features of the WiMAX standard. These service-provider features are especially valuable because of their ability to maximize air-link utilization and system throughput, as well as ensuring that SLAs (Service-Level Agreements) are met (Figure 5). The infrastructure to support various classes of services comes from the MAC implementation. QoS is enabled by the bandwidth request and grant mechanism between various subscriber stations and base stations. Primarily there are four buckets for the QoS (UGS, rtVR, nrtVR, and BE) to provide the service-class classification for video, audio, and data services, as they all require various levels of QoS requirements. The packet scheduler provides scheduling for different classes of services for a single user. This would mean meeting SLA requirements at the user level. Users can be classified into various priority levels, such as standard and premium.

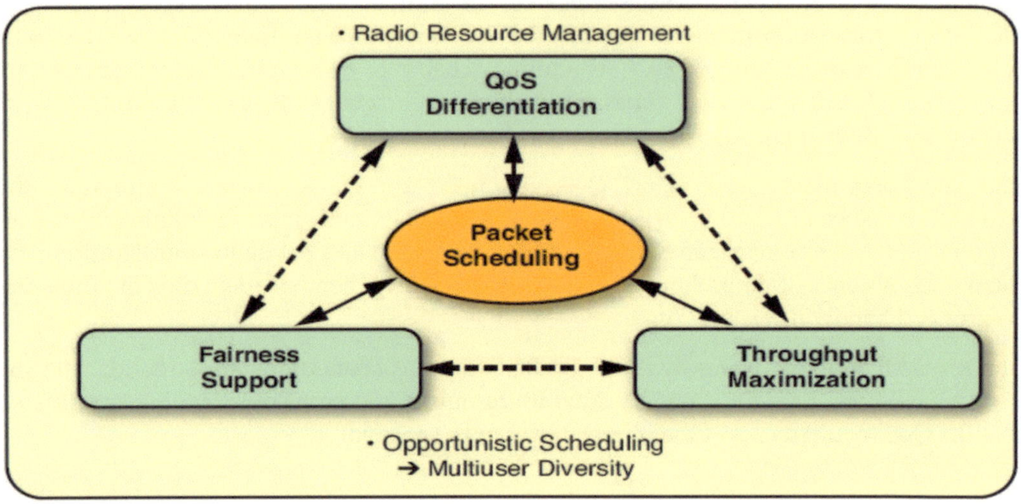

Fig. 5. Packet scheduling, as specified by 802.16 [6]

3.3 Scheduling algorithm and their characteristic

In some cases, separate scheduling algorithms are implemented for the uplink and downlink traffic. Typically, a CAC (Call Admission Control) procedure is also implemented at the BS that ensures the load supplied by the SSs can be handled by the network. A CAC algorithm will admit a SS into the network if it can ensure that the minimum QoS requirements of the SS can be satisfied and the QoS of existing SSs will not deteriorate. The performance of the scheduling algorithm for the uplink traffic strongly depends on the CAC algorithm.

Scheduling has also been studied intensively in many disciplines, such as CPU task scheduling in operating systems, service scheduling in a client-server model, and events scheduling in communication and computer networks. Thus a lot of scheduling algorithms have been developed. However, compared with the traditional scheduling problems, the WiMAX MAC layer scheduling problem is unique and worth study for the following reasons.

First, the total bandwidth in a WiMAX network is adaptive since AMC (Adaptive Modelling and Coding) is deployed in the physical layer and the number of bytes each time slot can carry depends on the coding and modulation scheme. Second, multiple service types have been defined and their QoS requirements need to be satisfied at the same time. How to satisfy various QoS requirements of different service types simultaneously has not been addressed by any other wireless access standard before. Third, the time complexity of the WiMAX scheduling algorithm must be simple since real-time service demands a fast response from the central controller in BS.

Fourth, the frame boundary in the WiMAX MAC layer also serves as the scheduling boundary, which makes the WiMAX scheduling problem different from the continuous time scheduling problem. The above four characteristics make the resource allocation in the WiMAX MAC layer a challenging problem.

While some similarities to the wired world can be drawn, there are certain characteristics of the wireless environment that make scheduling particularly challenging. Five major issues in wireless scheduling are identified in [9]:

- Wireless link variability: Due to characteristics of the channel as well as location of the mobile subscribers.
- Fairness: Refers to optimizing the channel capacity by giving preference to spectrally efficient modulations while still allowing transmissions with more robust modulations (and hence, consuming a major amount of spectrum) to get their traffic through.
- QoS: Particularly for WiMAX, QoS support should be built into the scheduling algorithm to guarantee that QoS commitments are meet under normal conditions as well as under network degradation scenarios.
- Data throughput and channel utilization: Refers to optimizing the channel utilization while at the same time avoiding waste of bandwidth by transmitting over high loss links.
- Power constrain and simplicity: Be considerate of the terminals' battery capacity as well as computational limitations both at the BS and MS.

3.4 Classification scheduling algorithms

Packet scheduling algorithms are implemented at both the BS and SSs. A scheduling algorithm at the SS is required to distribute the bandwidth allocation from the BS among its connections.

The scheduling algorithm at the SS needs to decide on the allocation of bandwidth among its connections. The scheduling algorithm implemented at the SS can be different than that at the BS.

The focus of our work is on scheduling algorithms executed at the BS for the uplink traffic in WiMAX i.e. traffic from the SSs to the BS. A scheduling algorithm for the uplink traffic is faced with challenges not faced by an algorithm for the downlink traffic. An uplink scheduling algorithm does not have all the information about the SSs such as the queue size. An uplink algorithm at the BS has to coordinate its decision with all the SSs where as a downlink algorithm is only concerned in communicating the decision locally to the BS.

In general, the scheduling algorithms can be classified as frame-based scheduling and sorted-based scheduling. Frame-based scheduling algorithms include WRR (Weighted Round Robin)[7], DRR (Deficit Round Robin)[8], etc. Sorted-based scheduling algorithms include WFQ (Weighted Fair Queue)[9], also known as PGPS (Packet-based Generalized Processor Sharing)[10], and a number of variations of WFQ such as WF2Q (Worst Case Fair Queuing)[11], SCFQ (Self-Clock Faire Queuing)[12].

The advantage of frame-based scheduling algorithms is their low computing complexity, while the disadvantage is the significant worst case delay. On the contrary, scheduling algorithm in the WFQ family has better performance in worst case delay, but the algorithm complexity is much higher than that of the frame-based scheduling algorithms.

3.5 Uplink scheduling algorithms

In the coming subsections the fundamental scheduling algorithms will be briefly described

3.5.1 Round Robin

Round Robin as a scheduling algorithm is the most basic and least complex scheduling algorithm. It has a complexity value of O (1) [13].

Basically the algorithm services the backlogged queues in a round robin fashion. Each time the scheduler pointer stop at a particular queue, one packet is dequeued from that queue and then the scheduler pointer goes to the next queue. This is shown in Figure 6.

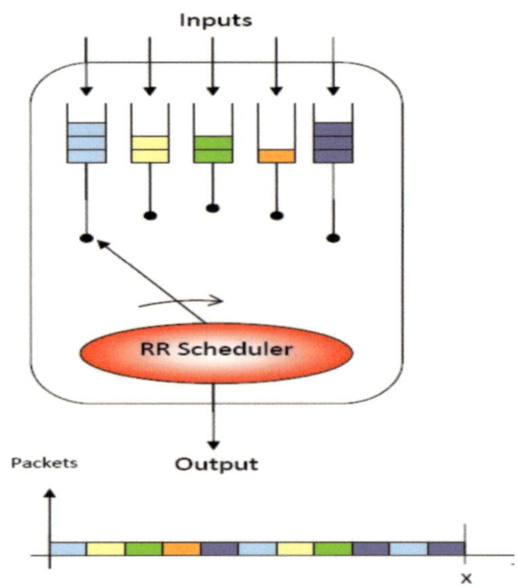

Fig. 6. RR Scheduler

It distributes channel resources to all the SSs without any priority. The RR scheduler is simple and easy to implement. However, this technique is not suitable for systems with different levels of priority and systems with strongly varying sizes of traffic.

3.5.2 Weighted Round Robin

An extension of the RR scheduler, the WRR scheduler, based on static weights.WRR [14] was designed to differentiate flows or queues to enable various service rates. It operates on the same bases of RR scheduling. However, unlike RR, WRR assigns a weight to each queue. The weight of an individual queue is equal to the relative share of the available system bandwidth. This means that, the number of packets dequeued from a queue varies according to the weight assigned to that queue. Consequently, this differentiation enables prioritization among the queues, and thus the SSes. [15]

3.5.3 Earliest deadline first

It is a work conserving algorithm originally proposed for real-time applications in wide area networks. The algorithm assigns deadline to each packet and allocates bandwidth to the SS that has the packet with the earliest deadline. Deadlines can be assigned to packets of a SS based on the SS's maximum delay requirement. The EDF algorithm is suitable for SSs belonging to the UGS and rtVR scheduling services, since SSs in this class have stringent delay requirements. Since SSs belonging to the nrtVR service do not have a delay requirement, the EDF algorithm will schedule packets from these SSs only if there are no packets from SSs of UGS or rtVR class. [16]

3.5.4 Weighted fair queue

It is a packet-based approximation of the Generalized Processor Sharing (GPS) algorithm. GPS is an idealized algorithm that assumes a packet can be divided into bits and each bit can be scheduled separately. The WFQ algorithm results in superior performance compared to the WRR algorithm in the presence of variable size packets. The finish time of a packet is essentially the time the packet would have finished service under the GPS algorithm. The disadvantage of the WFQ algorithm is that it will service packets even if they wouldn't have started service under the GPS algorithm. This is because the WFQ algorithm does not consider the start time of a packet.

3.5.5 Temporary removed packet

The TRS (Temporary Removal Scheduler) involves identifying the packet call power, depending on radio conditions, and then temporarily removing them from a scheduling list for a certain adjustable time period TR. The scheduling list contains all the SSs that can be served at the next frame. When TR expires, the temporarily removed packet is checked again. If an improvement is observed in the radio channel, the packet can be topped up in the scheduling list again, otherwise the process is repeated for TR duration. In poor radio conditions, the whole process can be repeated up to L times at the end of which, the removed packed is added to the scheduling list, independently of the current radio channel condition [18].

The temporary TRS can be combined with the RR scheduler.

The combined scheduler is called TRS+RR. For example, if there are k packet calls and only one of them is temporary removed, each packet call has a portion, equal to $\frac{1}{k-1}$, of the whole channel resources.

3.5.6 Maximum Signal to Interference Ration

The scheduler mSIR (Maximum Signal to Interference Ration) is based on the allocation of radio resources to subscriber stations which have the highest SIR. This scheduler allows a highly efficient utilization of radio resources. However, with the mSIR scheduler, the users with a SIR (Signal to Interference Ratio) that is always small may never be served.[18]

The TRS can be combined with the mSIR scheduler. The combined scheduler is called TRS + mSIR. This scheduler assigns the whole channel resources to the packet call that has the maximum value of the SNR (Signal to Noise Ratio). The station to be served has to belong to the scheduling list.

3.5.7 Reinforcement Learning

The scheduler RL (Reinforcement Learning) is based on the model of packet scheduling described by Hall and Mars [23]. The aim is to use different scheduling policies depending on which queues are not meeting their delay requirements. The state of the system represented by a set of N -1 binary variables {s1: sn-1}, where each variable si indicates whether traffic in the corresponding queue q_i [24].

There is not variable corresponding to the best-effort queue qN, since there is no mean delay requirement for that queue. For example, the state {0; 0; : : : ; 0} represents that all queues have satisfied their mean delay constraint, while (1; 0; : : : ; 0} represents that the mean delay requirements are being satisfied for all queues except q1. Thus, if there are N queues in the system including one best-effort queue, then there are 2^{N-1} possible states. In practice, the number of traffic classes is normally small, e.g., four classes in Cisco routers with priority queuing, in which case the number of states is acceptable.

At each timeslot, the scheduler must select an action a ϵ {a1: aN}, where ai is the action of choosing to service the packet at the head of queue q_i . The scheduler makes this selection by using a scheduling policy Π, which is a function that maps the current state of the system s onto an action a. If the set of possible actions is denoted by A, and the set of possible system states is denoted by S, then Π: S→A.

3.5.8 Hierarchical/hybrid algorithms

Hierarchical/hybrid algorithms build on the fact that scheduling services have different and sometimes conflicting requirements. UGS services must always have their delay and bandwidth commitment met, so simply reserving enough bandwidth for those services and controlling for oversubscription would be enough; rtVR services have little tolerance for delay and jitter, so an algorithm guaranteeing delay commitments would be more suitable; and finally, BE and nrtVR will always be hungry for bandwidth with no considerations for delay, so a throughput maximizing algorithm might be preferred.

While hierarchical refers to two or more levels of decisions to determine what packets to be scheduled, hybrid refers to the combination of several scheduling techniques (EDF for delay sensitive scheduling services such as rtVR and UGS, and WRR for nrtVR and BE for example). There could be hierarchical solutions that are not necessarily hybrid, but hybrid algorithms usually distribute the resources among different service classes, and then different scheduling techniques are used to schedule packets within each scheduling service, making them hierarchical in nature.

A two-tier hierarchical architecture is proposed in [24] for WiMAX uplink scheduling. In the higher hierarchy, strict prioritization is used to direct the traffic into the four queues, according to its type. Then, each queue is scheduled according to a particular algorithm, i.e., fixed allocation for UGS, EDF for rtVR, WFQ for nrtVR, and equal division of remaining bandwidth for BE. Although EDF takes care of the delay requirement of the rtVR, grouping multiple rtVR connections into one queue fails to guarantee the minimum bandwidth requirement of each individual rtVR connection. For example, one rtVR connection with tight delay budget may dominate the bandwidth allocation, resulting in starvation of other rtVR connections.

In [27], the authors use a first level of strict priority to allocate bandwidth to UGS, rtVR, nrtVR and BE services in that order; and then on a second level in the hierarchy, different scheduling techniques are used depending on the scheduling service: UGS, as the highest priority, has pre-allocated bandwidth, EDF is used for rtVR, WFQ for nrtVR, and FIFO for BE. Similarly, explains an algorithm that uses EDF for nrtVR and rtVR classes, and WFQ for nrtVR and BE classes.

In [27], the authors implement a two-level hierarchical scheme for the downlink in which an ARA (Aggregate Resource Allocation) component first estimates the amount of bandwidth required per scheduler class (rtVR, nrtVR, BE and UGS) and distributes it accordingly.

In [28], a SC (Service Criticality) based scheduling is proposed for the WiMAX network, where an SC index is calculated in every SS for each connection and then sent to BS, and BS sorts the SC of all the connections and assigns bandwidth according to the descending order of SC. SC is derived according to the buffer occupancy and waiting time of each connection. If a malicious connection always reports a high SC, or a connection is generating excessive traffic to occupy its sending buffer, this connection will dominate the available bandwidth and affect other connections.

4. Evaluation

This section presents the simulation results for the algorithms scheduling. For testing performance of algorithms, the introduced algorithms are implemented in the NS-2 (Network Simulator) [20] and WiMAX module [21] that is based on the WiMAX NIST module [20].The MAC implementation contains the main features of the 802.16 standard, such as downlink and uplink transmission. We have also implemented the most important MAC signalling messages, such as UL-MAP and DL-MAP, authentication (PKM), capabilities (SBC), REG (Registration), DSA (Dynamic Service Addition), and DSC (Dynamic Service Change). The implemented PHY is OFDM.

Lot size(byte)	Channel coding	modulation
108	3/4	64-QAM
96	2/3	64-QAM
36	3/4	QPSK
24	1/2	QPSK

Table 1. Slot size for OFDM PHY

The current implementation also supports differencing MCSs (Moulding Code Scheme). Table 1 shows present slot size for different modulations and channel coding types.

We present a simulation scenario to study thoroughly the proposed scheduling solution. The scenario will present a multi-service case, in which a provider has to support connections with different 802.16 classes and traffic characteristics.

The purpose of this scenario is to ensure that the scheduler at the BS takes the service class into account and allocates slots based on the QoS requirements and the request sizes sent by SSs. Another purpose is to test that the scheduler at the BS takes the MAC overhead into account. Table 1 presents information about which applications are active at scenario.

Regardless of the simulation scenario, the general parameters of the 802.16 network are the same (see Table 2). There is one BS that controls the traffic of the 802.16 network. The physical layer is OFDM. The BS uses the dynamic uplink/downlink slot assignment for the TDD mode. Both the BS and all SSs use packing and fragmentation in all simulation scenarios. The MAC level uses the largest possible PDU size. ARQ is turned off; neither the BS nor SSs use the CRC field while sending packets.

Value	**Parameter**
OFDM	PHY
7MHz	Bandwidth
400	Frame per Second
TDD	Duplexing mode
OFF	ARQ/CRC

Table 2. WiMAX parameter

We consider a general scenario, where n rtVR and/or nrtVR connections are established. Connection i has an arrival rate of ¸i, a delay budget of i, and a minimum reserved bandwidth of MRRi. For the sake of analytic tractability, we assume that the data arrival forms a Poisson process and all queues have infinite size. Other types of traffic (such as the more practical bursty traffic) are studied through simulations.

The main parameters of the simulation are represented in Table 3. Effects of these parameters are similar over results of all scheduling algorithms. Moreover, producers of this WiMAX module have used these values for testing performance of their simulator.

Parameter	**Value**
Frequency band	5 MHz
Propagation model	Two Ray Ground
Antenna model	Omni antenna
Antenna height	1.5 m
Transmit power	0.25
Receive power threshold	205e-12
Frame duration	20 ms
Cyclic prefix (CP)	0.25
Simulation duration	100 s

Table 3. Main parameters of the simulation

In particular, we consider several comparable scheduling algorithms, including WRR, EDF, and TRS which is a representative WiMAX scheduling algorithm and has been patented and well received).

Besides packet drop rate and throughput that have been studied in analysis, we are also interested in the fairness performance, which is measured by Jain's Fairness Index [22] defined as follows:

$$f(x_1, x_2, \ldots x_n) = \frac{(\sum_{i=1}^{n} x_i)^2}{n \sum_{i=1}^{n} x_i^2} \quad (1)$$

Where xi is the normalized throughput of connection i, and n is the total number of connections. Each SS establishes a number of connections to the BS in our simulation. We consider ten rtVR connections and ten nrtVR connections. Each type of connection is associated with an MRR and a delay budget.

$$Xi = \frac{THi}{MRi} \quad (2)$$

ie, with Thi and MRRi stand for the connection i's actual data rate and reserved data rate, respectively. The Jain's Fairness Index ranges between 0 and 1. The higher the index, the better the fairness. If Thi = MRRi for all i, or in other words, every connection obtains its reserved data rate, then xi = 1 for all i, and Jain's Fairness Index equals 1. All simulations and analytic calculations are done using NS2 simulator.

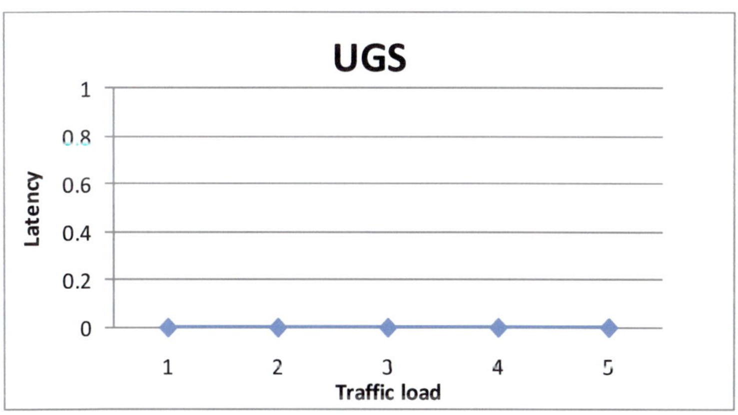

Fig. 7. Latency versus traffic

Figures 7, 8 show delay packets as a function of the traffic load submitted to the network. The data packets are generated by a streaming multimedia application. The diagram of UGS scheduling algorithm by considering delay is linear where its throughput is increasing. As mentioned above, the UGS traffic request is the highest priority. If a packet is available in this type of traffic it will be sent in no time. For accurate performance evaluation, we adopt the WiMAX physical layer standard OFDM_BPSK_1_2 in our simulations. [24]

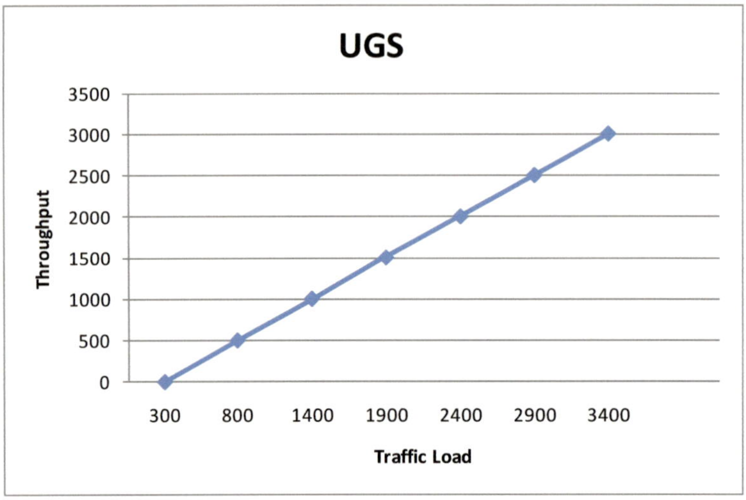

Fig. 8. Throughput versus traffic

The fairness of the scheduling algorithms under bursty traffic is shown in figure 9. As we can see, WRR always maintains almost high fairness, while the fairness of EDF algorithm is the worst among the four algorithms. This is due to the fact that some real time packets rtVR connections are dropped under high burstiness, and thus the throughput of rtVR decreases. [30], [31]

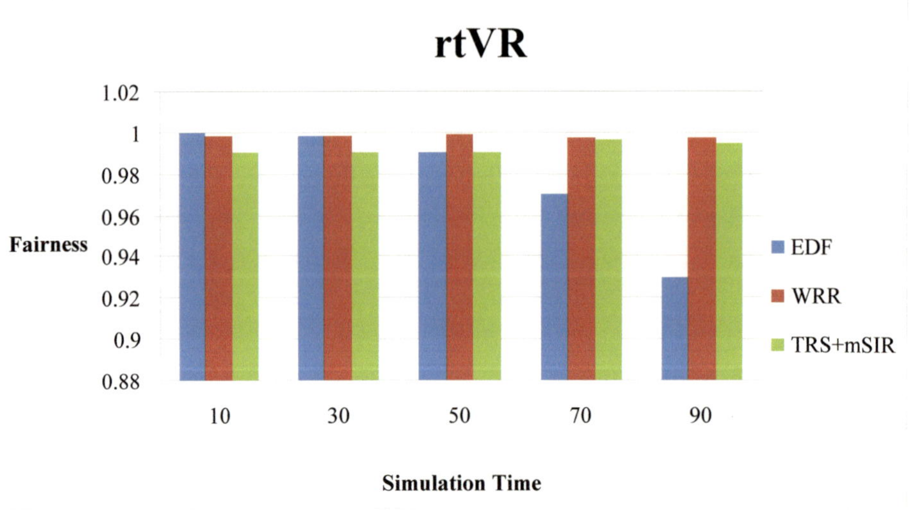

Fig. 9. Fairness versus Simulation Time

Figure 10 shows the latency as a function of rtVR+nrtVR traffic load. We verify that the TRS scheduler provides a decrease in the latency.

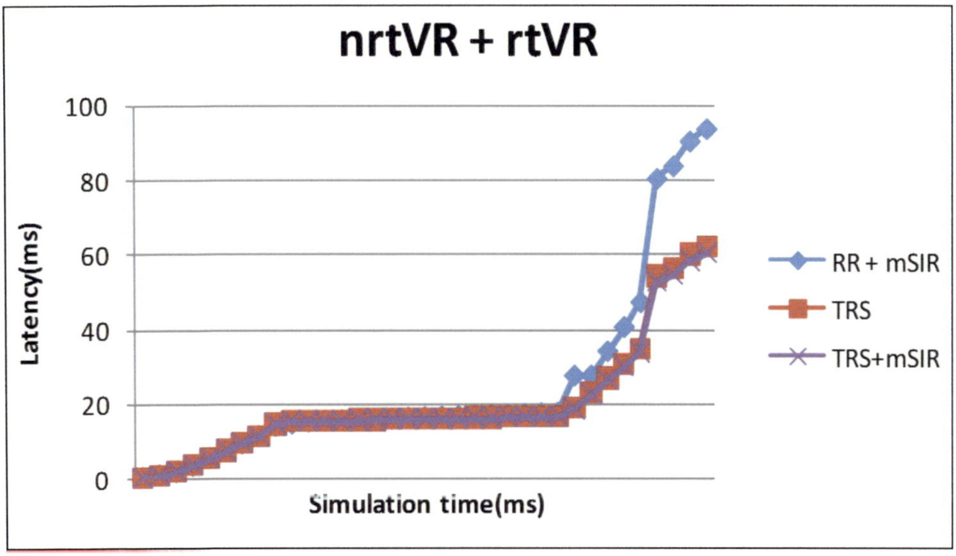

Fig. 10. Latency versus Simulation Time

Figure 11 shows the latency as a function of rtVR traffic load. We verify that the mSIR scheduler provides a decrease in the latency.

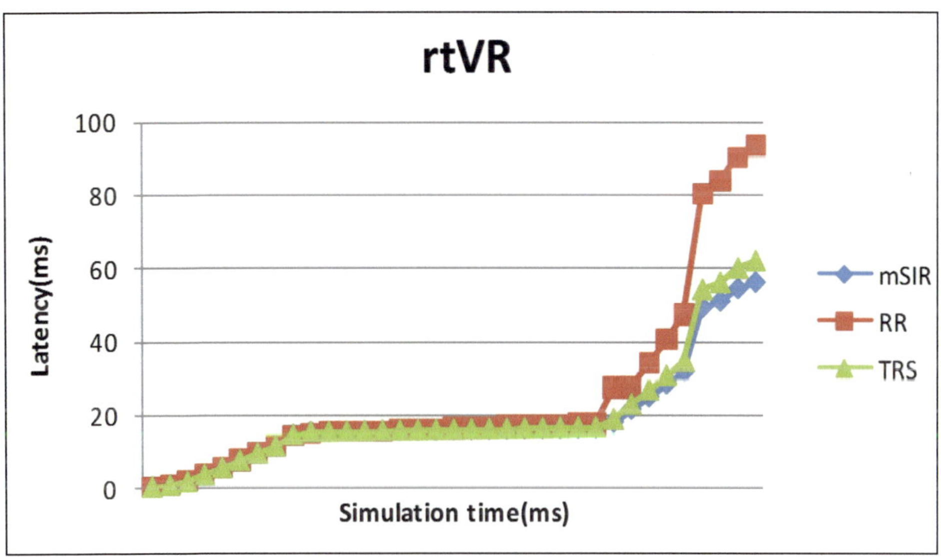

Fig. 11. Latency versus Simulation Time

In figure 12, the protocols have been compared on the base of throughput. As you see, TRS+RR throughput is greater than all.

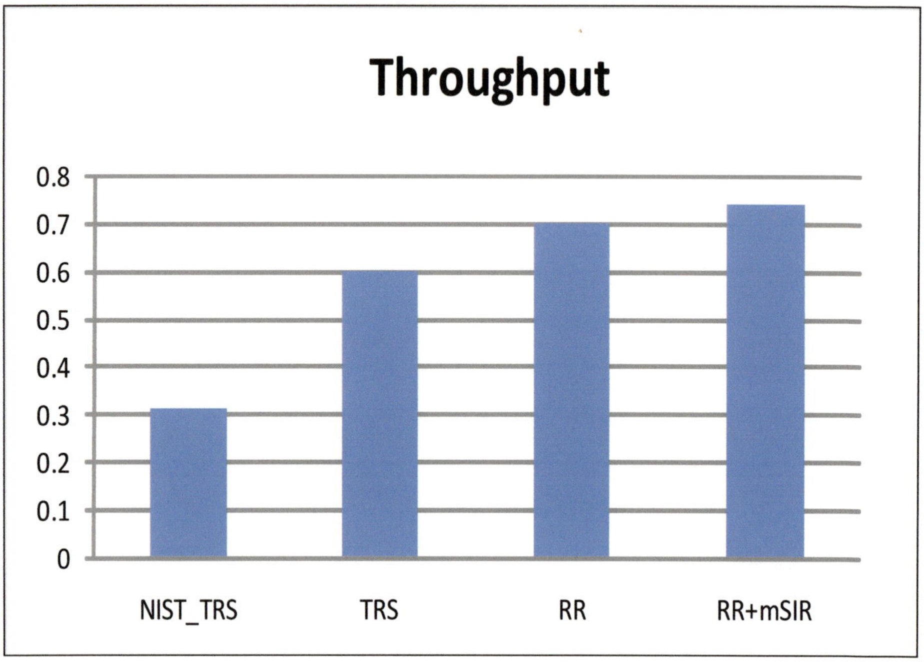

Fig. 12. Throughput

5. References

[1] IEEE Standard for Local and Metropolitan Area Networks—Part 16: Air Interface for Fixed Broadband Wireless Access Systems, 2004, IEEE802.16. Available from : http://www.ieeefor 802.org/16/.

[2] IEEE Standard for Local and Metropolitan Area Networks—Part 16: Air Interface for Fixed Broadband Wireless Access Systems—Amendment 2: Physical and Medium Access Control Layers for Combined Fixed and Mobile Operation in Licensed Bands, 2005, IEEE802.16e.. Available from: http://www.ieee802.org/16/.

[3] Aura Ganz, Zvi Ganz, Kitti Wongthavarawat(18 September 2003). *Multimedia Wireless Networks: Technologies, Standards, and QoS, Prentice Hall Publisher.*

[4] Overcoming Barriers to High-Quality Voice over IP Deployments (2003), Intel Whitepaper.

[5] DiffServ-The Scalable End-to-End Quality of Service Model (August 2005), Cisco Whitepaper.

[6] WiMAX - Delivering on the Promise of Wireless Broadband (Second Quarter 2006), Xcell Journal - Issue 57.

[7] M. Katavenis, S. Sidiropoulos, and C. Courcoubetis. *Weighted Round-Robin Cell Multiplexing in A General-Purpose ATM Switch Chip*, IEEE J. Sel. Areas Commun., vol. 9, no. 8, pp. 1265–1279, Jan. 1991.

[8] M. Shreedhar and G. Varghese (1995), *Efficient Fair Queueing Using Deficit Round Robin*, in Proc. IEEE SIGCOMM, pp. 231–242. 135.

[9] A. Demeres, S. Keshav, and S. Shenker(1989). *Analysis and Simulation of A Fair Queueing Algorithm*, in Proc. IEEE SIGCOMM, pp. 1–12.

[10] A. Parekh and R. Gallager(1992). *A Qeneralized Processor Sharing Approach to Flow Control: The Single Node Case*, in Proc. IEEE INFOCOM , pp. 915–924.
[11] J. Bennet and H. Zhang(1996). *WF2Q: Worst-Case Fair Weighted Fair Queueing, Procceding of IEEE INFOCOM*, 1996, pp. 120–128.
[12] S. Golestani (1994). *A Self-Clocked Fair Queueing Scheme for Broadband Applications*, Procceding of IEEE INFOCOM, pp. 636–646.
[13] R. Jain, lecture notes (2007), A Survey of Scheduling Methods, University of Ohio.
[14] M. Katevenis, S. Sidiropoulos and C. Courcoubetis(1991). *Weighted round robin cell multiplexing in a general purpose ATM switch chip*, Selected Areas in Communications, IEEE Journal on 9(8), pp. 1265_1279.
[15] S. Belenki (2000). *Traffic management in QoS networks: Overview and suggested improvements*, Tech. Rep.
[16] M.Shreedhar and G.Varghese(June 1996). Efficient Fair Queuing using Deficit Round R bin, IEEE/ACM Transactions on Networking, vol. 1, pp. 375-385.
[17] T. Al_Khasib, H. Alnuweiri, H. Fattah and V. C. M. Leung(2005). Mini round robin: enhanced frame_based scheduling algorithm for multimedia networks, IEEE Cmmunications, IEEE International Conference on ICC, pp. 363_368 Vol. 1.
[18] Nortel Networks,Introduction to quality of service (QoS)(September 2008), Nortel NetworksWebsite, 2003. [Online]. Accessed on 1st of September 2008.
[19] C.F. Ball, F. Treml, X. Gaube, and A. Klein(September 2005). Performance Analysis of Temporary Removal Scheduling applied to mobile WiMAX Scenarios in Tight Frequency Reuse, the 16th Annual IEEE International Symposium On Personal Indoor and Mobile Radio Communications, PIMRC 2005, Berlin, 11 - 14.
[20] QoS-included WiMAX Module for NS-2 Simulator. First International Conference on Simulation Tools and Techniques for Communications Networks and Systems, SIMUTools 2008, Marseille,France, March 3-7,2008.
[21] The network simulator ns-2(September 2007). Available from : http://www.isi.edu/nsnam/ns/.
[22] D. M. C. R. Jain andW. Hawe(1984). A Quantitative Measure of Fairness and Discriminationfor Resource Allocation in Shared Systems, dEC Research Report, TR-301.
[23] J. Hall , P. Mars(December 1998). Satisfying QoS with a Learning Based Scheduling Algorithm, School of Engineering, University of Durham,.
[24] M.Taghipoor,G Tavassoli and V.Hosseini(April 2010). Gurantee QoS in WiMAX Networks with learning automata, ITNG 2010 Las Vegas, Nevada, USA. 12-14
[25] Ajay Chandra V. Gummalla, John o. Limb.Wireless Medium Access Control Protocols, IEEE Communications Surveys, 2000.
[26] Q. Liu, X. Wang, and G. Giannakis(May 2006). A Cross-Layer Scheduling Algorithm with QoS Support in Wireless Networks, IEEE Trans. Veh. Tech., vol. 55, no. 3, pp. 839–847.
[27] D. Niyato and E. Hossain (Dec. 2006). Queue-aware Uplink Banwidth Allocation and Rate Control for Polling Service in IEEE 802.16 Broadband Wireless Networks, IEEE Trans. Mobile Comp., vol. 5, no. 8, pp. 668–679.
[28] A. Shejwal and A. Parhar(2007). Service Criticality Based Scheduling for IEEE 802.16 WirelessMAN, in Proc. 2nd IEEE Int. Conf. AusWreless, , pp. 12–18.

[29] H. Chen, thesis (spring 2008). Scheduling and Resource Optimization in Next Generation Hetergeneous Wireless Networks, University of Luoisiana.

[30] Jafari, Saeid. M., Taghipour, M. and Meybodi, M. R.(2011). Bandwidth Allocation in Wimax Networks Using Reinforcement Learning, World Applied Sciences Journal Vol. 15, No. 4, pp. 525-531.

[31] Jafari, Saeid. M., Taghipour, M. and Meybodi, M. R. (2011).Bandwidth Allocation in Wimax Networks Using Learning Automata, World Applied Sciences Journal Vol. 15, No. 4, pp. 576-583.

5

Downlink Resource Allocation and Frequency Reuse Schemes for WiMAX Networks

Nassar Ksairi
HIAST, Damascus
Syria

1. Introduction

Throughout this chapter we consider the *downlink* of a cellular WiMAX network where a number of base stations need to communicate simultaneously with their respective active users [1]. Each of these users has typically a certain Quality of Service (QoS) requirement that needs to be satisfied. To that end, base stations dispose of limited wireless resources (subcarriers and transmit powers) that should be shared between users. They also have some amount of Channel State Information (CSI) about users' propagation channels available, if existent, typically via feedback. The problem of determining the subset of subcarriers assigned to each user and the transmit power on each of these subcarriers is commonly referred to as the *resource allocation problem*. This problem should be solved such that all the QoS exigencies are respected.

Of course, the resource allocation problem has several formulations depending on i) the particular QoS-related objective function which we adopt (*e.g.*, achievable rate, transmission error probability ...) and ii) the channel model that we assume in relation with the available CSI. These CSI-related channel models will be discussed in Section 2 while the different formulations of the resource allocation problem will be covered in Section 4.

Since the set of subcarriers available for the whole WiMAX system is limited, it is typical that some subcarriers are reused at the same time by different base stations. Such base stations will generate multicell interference. Therefore, resource allocation parameters should, in principle, be determined in each cell in such a way that the latter multicell interference does not reach excessive levels. This fact highlights the importance of properly planning the so-called *frequency-reuse scheme* of the network. A frequency-reuse scheme answers the question whether the whole set of subcarriers should be available for allocation in all the cells of the network (meaning better spectral-usage efficiency but higher levels of intercell interference) or whether we should make parts of it exclusive to certain cells (leading thus to less efficiency in spectral usage but to lower levels of interference on the exclusive subcarriers).

Note that frequency-reuse planning is intimately related to resource allocation since it decides the subset of subcarriers that will be available for allocation in each cell of the network. Refer to Sections 4 and 5 for more details.

[1] We assume that the set of active users in the network is determined in advance by the schedulers at the base stations. We also assume that base stations has for each active user an infinite backlog of data to be transmitted

The rest of the chapter is organized as follows. In Section 2, the different kinds of CSI feedback models are presented and their related channel models are discussed. The issue of frequency reuse planning (which is intimately related to cellular resource allocation) is discussed in Section 3. Each different channel model leads to a different formulation of the resource allocation problem. These formulations are addressed in Section 4. Finally, Section 5 deals with the determination of the so-called *frequency-reuse factor*.

2. Feedback for resource allocation: Channel State Information (CSI) and channel models

Consider the downlink of an OFDMA-based wireless system (such as a WiMAX network) and denote by N, K the total number of subcarriers and of active users, respectively. Assume that the subcarriers are numbered from 1 to N and that the active users of the network are numbered from 1 to K. The network comprises a certain number of cells that are indexed using the notation c. Each cell c consists of a base station communicating with a group of users as shown in Figure 1. The signal received by user k at the nth subcarrier of the mth

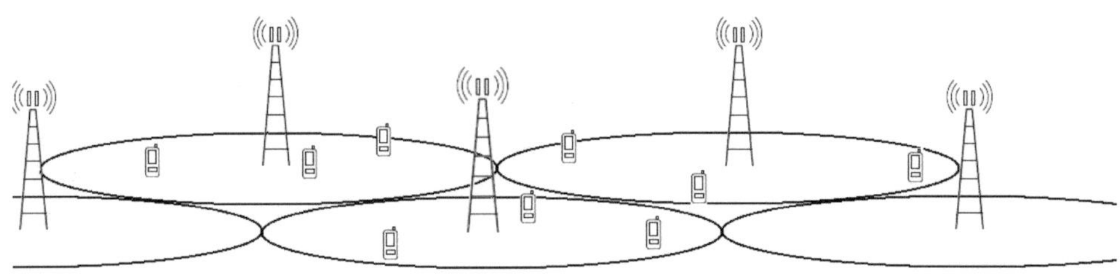

Fig. 1. WiMAX network

OFDM block (m being the time index) is given by

$$y_k(n,m) = H_k^c(n,m)s_k(n,m) + w_k(n,m), \tag{1}$$

where $s_k(n,m)$ is the transmitted symbol and where $w_k(n,m)$ is a random process which is used to model the effects of both thermal noise and intercell interference. Finally, $H_k^c(n,m)$ refers to the (generally complex-valued) coefficient of the propagation channel between the base station of cell c and user k on subcarrier n at time m.

Assume that the duration of transmission is equal to T OFDM symbols *i.e.*, $m \in \{0,1,\ldots,T-1\}$. Denote by $\mathcal{N}_k$ the subset of subcarriers ($\mathcal{N}_k \subset \{1,2,\ldots,N\}$) assigned to user k. The codeword destined to user k is thus the $|\mathcal{N}_k| \times T$ matrix

$$\boldsymbol{S}_k = [\boldsymbol{s}_k(0), \boldsymbol{s}_k(1), \ldots, \boldsymbol{s}_k(T-1)]\,, \tag{2}$$

where each $|\mathcal{N}_k| \times 1$ column-vector $\boldsymbol{s}_k(m)$ is composed of the symbols $\{s_k(n,m)\}_{n\in\mathcal{N}_k}$ transmitted during the mth OFDM block on subcarriers $\mathcal{N}_k$.

Depending on the amount of feedback sent from users to the base stations, coefficients $H_k^c(n,m)$ can be modeled either as deterministic or as random variables. As stated in Section I, different formulations of the resource allocation problem exist in the literature, each formulation being associated with a different model for coefficients $H_k^c(n,m)$. These channels models are summarized in the following subsection.

2.1 Theoretical CSI-related channel models:

The general OFDMA signal model given by (1) does not specify whether the channel coefficients $\{H_k(n,m)\}_{k,n,m}$ associated with each user k are known at the base stations or not. In this chapter, we consider three signal models for these coefficients.

1. **Full CSI: Deterministic channels.**
   In this model, channel coefficients $H_k^c(n,m)$ for each user k are assumed to be perfectly known (thus deterministic) at both the base station side and the receiver side on all the subcarriers $n \in \{0 \ldots N-1\}$. Note that **this assumption implicitly requires that each receiver k feedbacks to the base station the values of the channel coefficients $H_k^c(n,m)$ on all the assigned subcarriers n.**
   For the sake of simplicity, it is also often assumed in the literature that the above deterministic coefficients $H_k^c(n,m)$ remain constant ($H_k^c(n,m) = H_k^c(n)$) during the transmission of a codeword [2] *i.e.*, $\forall m \in \{0,1,\ldots,T-1\}$. Under these assumptions, a transmission to user k at rate R_k nats/sec/Hz is possible from the information-theoretic point of view with negligible probability of error provided that $R_k < C_k$, where C_k denotes the channel capacity associated with user k. If we assume that the noise-plus-interference process $w_k(n,m)$ in (1) is zero-mean Gaussian-distributed[3] with variance σ_k^2, then the channel capacity C_k (in nats/sec/Hz) is given by

$$C_k = \frac{1}{N} \sum_{n \in \mathcal{N}_k} \log\left(1 + P_{k,n} \frac{|H_k^c(n)|^2}{\sigma_k^2}\right),$$

   where $\mathcal{N}_k$, we recall, is the subset of subcarriers ($\mathcal{N}_k \subset \{0,1,\ldots,N\}$) assigned to user k, and where $P_{n,k}$ is the power transmitted by the base station on subcarrier $n \in \mathcal{N}_k$ *i.e.*, $P_{k,n} = \mathbb{E}\left[|s_{k,n}|^2\right]$.
2. **Statistical CSI: Random ergodic (fast-fading) channels.**
   In this model, we assume that coefficients $H_k^c(n,m)$ associated with each user k on any subcarrier n are time-varying, unknown at the base station side and perfectly known at the receiver side. We can thus think of $\{H_k(n,m)\}_m$ as a random process with a certain statistical distribution *e.g.*, Rayleigh, Rice, Nakagami, etc. We also assume that this process undergo **fast fading** *i.e.*, the coherence time of the channel is much smaller than the duration T of transmission of a codeword. It is thus reasonable to model $\{H_k(n,m)\}_m$ as an independent identically distributed (i.i.d) random ergodic process for each $n \in \{0 \ldots N-1\}$. Finally, **we assume that the parameters of the distribution of this process *i.e.*, its mean, variance ..., are known at the base station, typically via feedback.**

---

[2] A codeword typically spans several OFDM blocks *i.e.*, several time indexes m

[3] Eventhough the noise-plus-interference $w_k(n,m)$ is not Gaussian in general, approximating it as a Gaussian process is widely used in the literature (see for instance Gault et al. (2005); S. Plass et al. (2004; 2006)). The reason behind that is twofold: first, the Gaussian approximation provides a lower bound on the mutual information, second it allows us to have an analytical expression for the channel capacity.

Note that since the channel coefficients $\{H_k(n,m)\}_m$ are time-varying in this model, then each single codeword encounters a large number of channel realizations. In this case, it is a well-known result in information theory that transmission to user k at rate R_k nats/sec/Hz is possible with negligible probability of error provided that $R_k < C_k$, where C_k denotes here the channel *ergodic* capacity associated with user k and given (in the case of zero-mean Gaussian distributed noise-plus-interference processes $w_k(n,m)$ with variance σ_k^2) by

$$C_k = \mathbb{E}\left[\log\left(1 + P_{k,n}\frac{|H_k^c(n,m)|^2}{\sigma_k^2}\right)\right].$$

Here, expectation is taken with respect to the distribution of the random channel coefficients $H_k^c(n,m)$.

3. **Statistical CSI: Random nonergodic (slow-fading) channels.**
   In this case, channel coefficients $H_k(n,m) = H_k(n)$ are assumed to be fixed during the whole transmission of any codeword, but nonetheless random and unknown by the base stations. This case is usually referred to as the **slow fading** case. It arises as the best fitting model for situations where the channel coherence time is larger than the transmission duration. We also **assume that the parameters of the distribution of the random variables $H_k(n)$ *i.e.,* their mean, variance ..., are known at the base station via feedback.**

   In contrast to the ergodic case, there is usually no way for the receiver in the nonergodic case to recover the transmitted information with negligible error probability. Assume that the base station needs to send some information to user k at a data rate R_k nats/sec/Hz. The transmitted message (if the transmitted symbols $s_k(n,m)$ are from a Gaussian codebook) can be decoded by the receiver provided that the required rate R_k is less than the mutual information between the source and the destination *i.e.,* provided that $\frac{1}{N}\sum_{n\in\mathcal{N}_k}\log\left(1+P_{k,n}\frac{|H_k(n)|^2}{\sigma_k^2}\right) > R_k$. If the channels realization $H_k(n)$ is such that

$$\frac{1}{N}\sum_{n\in\mathcal{N}_k}\log\left(1+P_{k,n}\frac{|H_k(n)|^2}{\sigma_k^2}\right) \le R_k\,,$$

   then the transmitted message cannot be decoded by the receiver. In this case, user k link is said to be in *outage*. The event of outage occurs with the following probability:

$$P_{O,k}(R_k) \triangleq \Pr\left[\frac{1}{N}\sum_{n\in\mathcal{N}_k}\log\left(1+P_{k,n}\frac{|H_k(n)|^2}{\sigma_k^2}\right) \le R_k\right]. \tag{3}$$

   Probability $P_{O,k}(R_k)$ is commonly referred to as the *outage probability* associated with user k. In the context of communication over slow fading channels as described above, it is of clear interest to minimize the outage probability associated with each user.

It is worth mentioning that the above distinction between deterministic, ergodic and nonergodic channel was originally done in I. E. Telatar (1999) for Multiple-Input-Multiple-Output (MIMO) channels. We present in the sequel the main existing results on resource allocation for each one of the above signal models.

3. Frequency reuse schemes and the frequency reuse factor: Definition and relation to resource allocation

As we stated earlier, management of multicell interference is one of the major issues in cellular networks design and administration. This management is intimately related to the so-called *frequency-reuse scheme* adopted in the network. Indeed, choosing a frequency-reuse scheme means determining the subset of subcarriers that are available for allocation in each cell (or sector) of the network. In some reuse schemes, the whole set of subcarriers is available for allocation in all the cells of the network, while in others some subsets of subcarriers are made exclusive to certain cells and prohibitted for others.

Many reuse schemes have been proposed in the literature, differing in their complexity and in their **repetitive pattern** *i.e.,* the number of cells (or sectors) beyond which the scheme is repleted. In this chapter, we only **focus on three-cell (or three-sector) based reuse schemes**. Indeed, the level of interference experienced by users in a cellular network is related to the value of parameter α defined as

$$\alpha = \frac{\text{number of subcarriers reused by three adjacent cells}}{N}. \tag{4}$$

Where N is the total number of subcarriers in the system. In *sectorized networks i.e.,* networks with 120 °-directive antennas at their base stations (see Figure 2), the definition of α becomes

$$\alpha = \frac{\text{number of subcarriers reused by three adjacent sectors}}{N}. \tag{5}$$

Note that in Figure 2, sectors 3,4,5 form the basic pattern of the reuse scheme that is repleted throughout the network.

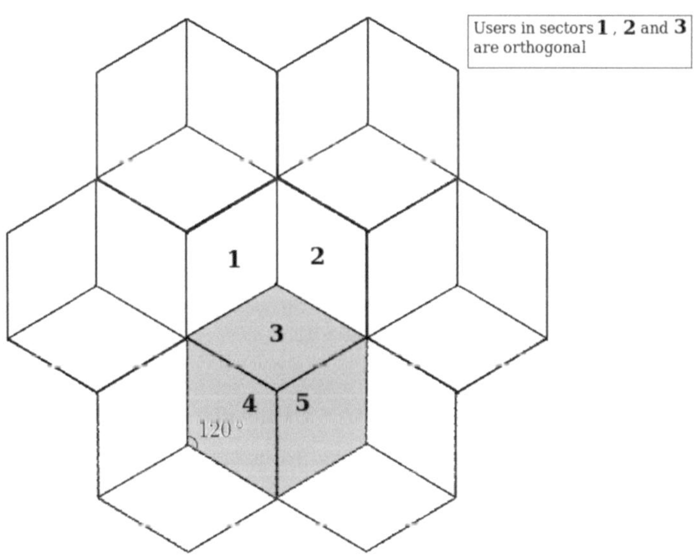

Fig. 2. A sectorized cellular network

Parameter α is called the *frequency reuse factor*. If $\alpha = 1$, then each base station can allocate the totality of the available N subcarriers to its users. This policy is commonly referred to as the *all-reuse scheme* or as the *frequency-reuse-of-one scheme*. Under this policy, all users of the system are subject to multicell interference. If $\alpha = 0$, then no subcarriers are allowed to be used simultaneously in the neighboring cells. This is the case of an *orthogonal reuse scheme*. In such a scheme, users do not experience any multicell interference.

If α is chosen such that $0 < \alpha < 1$, then we obtain the so-called *fractional frequency reuse*, see WiMAX Forum (2006). According to this frequency reuse scheme, the set of available subcarriers is partitioned into two subsets. One subset contains αN subcarriers that can be reused within all the cells (sectors) of the system and is thus subject to multicell interference. The other subset contains the remaining $(1 - \alpha)N$ subcarriers and is divided in an orthogonal way between the different cells (sectors). Such subcarriers are thus protected from interference.

The larger the value of α, the greater the number of available subcarriers for each base station and the higher the level of multicell interference. There is therefore a tradeoff between the number of available subcarriers (which is proportional to α) and the severity of the multicell interference. This tradeoff is illustrated by Figure 3. Generally speaking, the characterization

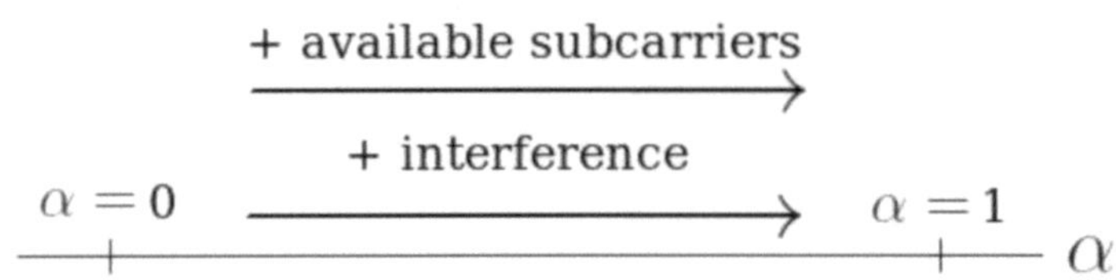

Fig. 3. Tradeoff between interference and number of available subcarriers

of the latter tradeoff is a difficult problem to solve. Most of the approaches used in the literature to tackle this problem were based on numerical simulations. Section 5 is dedicated to the issue of analytically finding the best value of α without resorting to such numerical approaches.

In the sequel, we assume that the frequency-reuse scheme (or the frequency-reuse factor) has already been chosen in advance prior to performing resource allocation. While Section 5 is dedicated to the issue of finding the best value of α.

4. Downlink resource allocation for WiMAX cellular networks

In this section, we give the main existing results on the subject of downlink resource allocation for WiMAX networks. We present the literature on this subject by classifying it with respect to the specific signal models (full-CSI channels, statistical-CSI fast-fading channels, statistical-CSI slow-fading channels).

It is worth noting that many existing works on cellular resource allocation resort to the so-called **single-cell assumption**. Under this simplifying assumption, intercell interference is considered negligible. The received signal model for some user k in cell c on each subcarrier n can thus be written as

$$y(n,m) = H_k^c(n,m)s_k(n,m) + w_k(n,m),$$

where process $w_k(n,m)$ contains **only thermal noise**. It can thus be modeled in this case as AWGN with distribution $\mathcal{CN}(0,\sigma^2)$. Of course, the single-cell assumption is simplifying and unrealistic in real-world cellular networks where intercell interference prevails. However, in some cases one can manage to use the results of single-cell analysis as a tool to tackle the more interesting and demanding multicell problem (see for example N. Ksairi & Ciblat (2011); N. Ksairi & Hachem (2010a)).

4.1 Full-CSI resource allocation (deterministic channels)

Although having full (per-subcarrier) CSI at the base stations is quite unrealistic in practice as we argued in Section 2, many existing works on resource allocation for OFDMA systems resorted to this assumption. We give below the main results in the literature on resource allocation in the case of full CSI, mainly for the sake of completeness.

1) Sum rate maximization

Consider the problem of maximizing the sum of all users achievable rates, first in a **single-cell context** (focus for example on cell c). This maximization should be done such that the spent power does not exceed a certain maximum value and such that the OFDMA orthogonal subcarrier assignment constraint (no subcarrier can be assigned to more than one user) is respected. Recall the definition of $P_{k,n} = \mathbb{E}[|s_k(n,m)|^2]$ for any $n \in \mathcal{N}_k$ as the power allocated to user k on subcarrier n. Let $P_{\max}$ designates the maximal power that the base station is allowed to spend. The maximal sum rate should thus be computed under the following constraint:

$$\sum_{k \in c} \sum_{n \in \mathcal{N}_k} P_{k,n} \leq P_{\max}\,. \tag{6}$$

It is known Tse & Visawanath (2005) that the maximum sum rate is achieved provided that the codeword $\boldsymbol{S}_k = [\boldsymbol{s}_k(0), \boldsymbol{s}_k(1), \ldots, \boldsymbol{s}_k(T-1)]$ of each user is chosen such that

$$\begin{aligned} &\boldsymbol{s}_k(m) \text{ for } m \in \{0,1,\ldots,T-1\} \text{ is an i.i.d process, and} \\ &\boldsymbol{s}_k(m) \sim \mathcal{CN}\left(\mathbf{0}, \operatorname{diag}\left(\{P_{k,n}\}_{n \in \mathcal{N}_k}\right)\right), \end{aligned} \tag{7}$$

where $\boldsymbol{s}_k(m)$, we recall, is the vector composed of the symbols $\{s(n,m)\}_{n \in \mathcal{N}_k}$ transmitted to user k during the mth OFDM symbol. It follows that the maximum sum rate C_{sum} of the downlink OFDMA single cell system can be written as

$$C_{\text{sum}} = \max_{\{\mathcal{N}_k, P_{k,n}\}_{k \in c, n \in \mathcal{N}_k}} \sum_{k \in c} \frac{1}{N} \sum_{n \in \mathcal{N}_k} \log\left(1 + \frac{|H_k^c(n)|^2}{\sigma^2} P_{k,n}\right),$$
subject to subcarrier assignment orthogonality constraint and to (6)

Solving the above optimization problem provides us with the optimal resource allocation which maximizes the sum rate of the system. It is known from Jang & Lee (2003); Tse & Visawanath (2005), that the solution to the above problem is the so-called *multiuser water-filling*. According to this solution, the optimal subcarrier assignment $\{\mathcal{N}_k\}_{k \in c}$ is such that:

Each subcarrier $n \in \{1, 2, \ldots, N\}$ is assigned to the user k_n^* satisfying $k_n^* = \arg\max_k |H_k^c(n)|$.

The powers $\{P_{k_n^*,n}\}_{1 \leq n \leq N}$ can finally be determined by water filling:

$$P_{k_n^*,n} = \left(\frac{1}{\lambda} - \frac{\sigma^2}{|H_{k_n^*}^c(n)|^2} \right)^+ ,$$

where λ is a Lagrange multiplier chosen such that the power constraint (6) is satisfied with equality:

$$\sum_{n=1}^{N} \left(\frac{1}{\lambda} - \frac{\sigma^2}{|H_{k_n^*}^c(n)|^2} \right)^+ = P_{\max} .$$

In a **multicell scenario**, the above problem becomes that of maximizing the sum of data rates that can be achieved by the users of the network subject to a total network-wide power constraint

$$\sum_c \sum_{k \in c} \sum_{n \in \mathcal{N}_k} P_{k,n} \leq P_{\max} . \tag{8}$$

In case the transmitted symbols of all the base stations are from Gaussian codebooks, the sum rate maximization problem can be written as

$$\max_{\{\mathcal{N}_k, P_{k,n}\}_{1 \leq k \leq K, n \in \mathcal{N}_k}} \sum_c \sum_{k \in c} \frac{1}{N} \sum_{n \in \mathcal{N}_k} \log \left(1 + P_{k,n} \frac{|H_k^c(n)|^2}{\sigma_k^2} \right)$$
$$\text{subject to the OFDMA orthogonality constraint and to (8)} , \tag{9}$$

In contrast to the single cell case where the exact solution has been identified, no closed-form solution to Problem (9) exists. An approach to tackle a variant of this problem with per subcarrier peak power constraint ($P_{k,n} \leq P_{\text{peak}}$) has been proposed in Gesbert & Kountouris (2007). The proposed approach consists in performing a decentralized algorithm that maximizes an upperbound on the network sum rate. Interestingly, this upperbound is proved to be tight in the asymptotic regime when the number of users per cell is allowed to grow to infinity. However, the proposed algorithm does not guaranty fairness among the different users.

A heuristic approach to solve the problem of sum rate maximization is adopted in Lengoumbi et al. (2006). The authors propose a centralized iterative allocation scheme allowing to adjust the number of cells reusing each subcarrier. The proposed algorithm promotes allocating subcarriers which are reused by small number of cells to users with bad channel conditions. It also provides an interference limitation procedure in order to reduce the number of users whose rate requirements are unsatisfied.

2) Weighted sum rate maximization

In a wireless system, maximizing the sum rate does not guaranty any fairness between users. Indeed, users with bad channels may not be assigned any subcarriers if the aforementioned multiuser water-filling scheme is applied. Such users may have to wait long durations of time till their channel state is better to be able to communicate with the base station. In order to ensure some level of fairness among users, one can use the maximization of a weighted sum of users achievable rates as the criterion of optimization of the resource allocation.

In a **single-cell scenario** (focus on cell c), the maximal weighted sum rate $C_{\text{weighted sum}}$ is given by:

$$C_{\text{weighted sum}} = \max \sum_{k \in c} \mu_k R_k \,, \tag{10}$$

where R_k is the data rate achieved by user k, and where the maximization is with respect to the resource allocation parameters and the distribution of the transmit codewords $\mathbf{S}_k$. Weights μ_k in (10) should be chosen in such a way to compensate users with bad channel states. As in the sum rate maximization problem, it can be shown that the weighted sum rate is maximized with random Gaussian codebooks *i.e.,* when (7) holds. The optimal resource allocation parameters can thus be obtained as the solution to the following optimization problem:

$$\max_{\{\mathcal{N}_k, P_{k,n}\}_{k \in c, n \in \mathcal{N}_k}} \sum_{k \in c} \frac{\mu_k}{N} \sum_{n \in \mathcal{N}_k} \log\left(1 + P_{k,n} \frac{|H_k(n)|^2}{\sigma^2}\right),$$
$$\text{subject to the subcarrier assignment orthogonality constraint and to (6)}\,.$$

The above optimization problem is of combinatorial nature since it requires finding the optimal set $\mathcal{N}_k$ of subcarriers for each user k. It cannot thus be solved using convex optimization techniques.

For each subchannel assignment $\{\mathcal{N}_k\}_{1 \le k \le K}$, the powers $P_{k,n}$ can be obtained by the so called multilevel water filling Hoo et al. (2004) with a computational cost of the order of $\mathcal{O}(N)$ operations. On the other hand, finding the optimal subcarrier assignment requires an exhaustive search and a computational complexity of the order of K^N operations. The overall computational complexity is therefore $\mathcal{O}(NK^N)$. In order to avoid this exponentially complex solution, the authors of Seong et al. (2006) state that solving the dual of the above problem (by Lagrange dual decomposition for example) entails a negligible duality gap. This idea is inspired by a recent result Yu & Lui (2006) in resource allocation for multicarrier DSL applications.

In a **multicell scenario**, the weighted sum rate maximization problem can be written (in case the transmitted symbols of all the base stations are from Gaussian codebooks) as

$$\max_{\{\mathcal{N}_k, P_{k,n}\}_{c, 1 \le k \le K^c, n \in \mathcal{N}_k}} \sum_c \sum_{k \in c} \frac{\mu_k}{N} \sum_{n \in \mathcal{N}_k} \log\left(1 + P_{k,n} \frac{|H_k^c(n)|^2}{\sigma_k^2}\right),$$
$$\text{subject to the OFDMA orthogonality constraint and to (8)}\,,$$

Here, μ_k is the weight assigned to user k. Since no exact solution has yet been found for the above problem, only suboptimal (with respect to the optimization criterion) approaches exist. The approach proposed in M. Pischella & J.-C. Belfiore (2008) consists in performing resource allocation via two phases: First, the users and subcarriers where the power should be set to zero are identified. This phase is done with the simplifying assumption of uniform power allocation. In the second phase, an iterative distributed algorithm called Dual Asynchronous Distributed Pricing (DADP) J. Huang et al. (2006) is applied for the remaining users under high SINR assumption.

3) Power minimization with individual rate constraints

Now assume that each user k has a data rate requirement equal to R_k in a **signel-cell scenario** (we focus on cell c). The subcarriers $\mathcal{N}_k$ and the transmit powers $\{P_{k,n}\}_{n\in\mathcal{N}_k}$ assigned to user k should thus be chosen such that the following constraint is satisfied:

$$R_k < C_k = \frac{1}{N}\sum_{n\in\mathcal{N}_k} \log\left(1 + P_{k,n}\frac{|H_k(n)|^2}{\sigma^2}\right), \tag{11}$$

and such that the total transmit power is minimal. Here, C_k is the maximal rate per channel use that can be achieved by user k when assigned $\mathcal{N}_k$ and $\{P_{k,n}\}_{n\in\mathcal{N}_k}$. This maximal rate is achieved for each user by using random Gaussian codebooks as in (7). The resource allocation problem can be formulated in this case as follows:

$$\min_{\{\mathcal{N}_k, P_{k,n}\}_{k\in c, n\in\mathcal{N}_k}} \sum_{k\in c}\sum_{n\in\mathcal{N}_k} P_{k,n}$$

such that the subcarrier assignment orthogonality constraint and (11) are satisfied.

Some approaches to solve this combinatorial optimization problem can be found in Kivanc et al. (2003). However, these approaches are heuristic and result in suboptimal solutions to the above problem.

In order to avoid the high computational complexity required for solving combinatorial optimization problems, one alternative consists in relaxing the subcarrier assignment constraint by introducing the notion of subcarrier time-sharing as in Wong et al. (1999). According to this notion, each subcarrier n can be orthogonally time-shared by more than one user, with each user k modulating the subcarrier during an amount of time proportional to $\gamma_{k,n}$. Here, $\{\gamma_{k,n}\}_{k,n}$ are real number from the interval $[0,1]$ satisfying

$$\forall n \in \{1,2,\ldots,N\}, \sum_{k=1}^{K} \gamma_{k,n} \leq 1\,. \tag{12}$$

The rate constraint of user k becomes

$$R_k < \sum_{n=1}^{N} \gamma_{k,n} \log\left(1 + P_{k,n}\frac{|H_k(n)|^2}{\sigma^2}\right). \tag{13}$$

The optimal value of the new resource allocation parameters can be obtained as the solution to the following optimization problem:

$$\min_{\{\gamma_{k,n}, P_{k,n}\}_{1\leq k\leq K, 1\leq n\leq N}} \sum_{k=1}^{K}\sum_{n=1}^{N} \gamma_{k,n} P_{k,n}$$

such that constraints (12) and (13) are satisfied.

The above problem can be easily transformed into a convex optimization problem by a simple change of variables. One can therefore use usual convex optimization tools to find its solution.

Remark 1. *It is worth mentioning here that the assumption of per-subcarrier full CSI at the transmitters is quite unrealistic in practice. First of all, it requires large amounts of feedback messages from the different users to their respective base stations, which is not practically possible in most* *real-world wireless communication systems. Even if the wireless system allows that amount of feedback, it is not clear yet whether the benefit obtained by this additional complexity would outweigh the additional costs due to the resulting control traffic Staćzak et al. (2009). For these reasons, most of the above mentioned resource allocation techniques which assume perfect CSI have not been adopted in practice.*

Remark 2. *So far, it was assumed throughout the previous subsection that all the subcarriers* $\{1, 2, \ldots, N\}$ *are available to the users of each cell* i.e., *a frequency reuse of one is assumed. Resource allocation under fractional frequency reuse is addressed in the next subsection.*

4.2 Average-rate multicell resource allocation in the case of (statistical-CSI fast-fading channels)

Several works such as Brah et al. (2007; 2008); I.C. Wong & B.L. Evans (2009); Wong & Evans (2007) consider the problem of ergodic sum-rate and ergodic weighted sum-rate maximization in WiMAX-like networks. However, these works do not provide analytical solutions to these optimization problems. Instead, they resort to suboptimal (and rather computationally-complex) *duality techniques*. This is why we focus in the sequel on a **special case** of the average-rate resource allocation problem where closed-form characterization of the *optimal* solution has been provided in N. Ksairi & Ciblat (2011); N. Ksairi & Hachem (2010a;b). In particular, we highlight the methodology adopted in these recent works and which consists in using the single-cell results as a tool to solve the more involved multicell allocation problem.

Consider the the **downlink of a sectorized WiMAX cellular system composed of hexagonal cells** as shown in Figure 2. Assume that the **fractional frequency reuse (FFR) scheme** illustrated in Figure 4 is adopted. Due to this scheme, a certain subset of subcarriers $\mathcal{I} \subset \{1, 2, \ldots, N\}$ ($\mathcal{I}$ as in $\mathcal{I}$nterference) is reused in the three cells. If user k modulates such a subcarrier $n \in \mathcal{I}$, process $w_k(n, m)$ will contain both thermal noise and multicell interference. Recall the definition of the reuse factor α given by (5) as the ratio between the number of

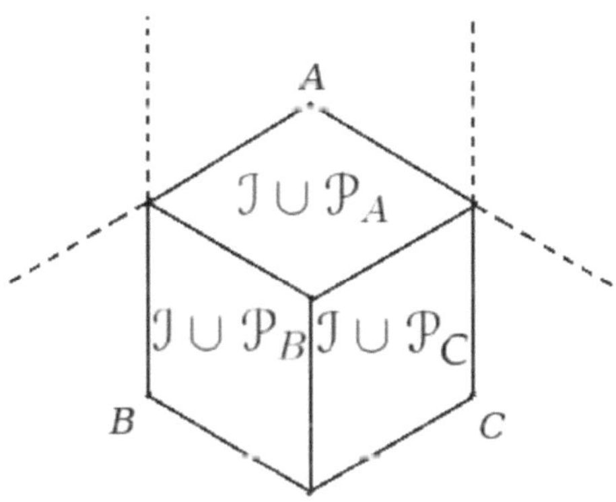

Fig. 4. Frequency reuse scheme

reused subcarriers and the total number of available subcarriers:

$$\alpha = \frac{\text{card}(\mathcal{I})}{N}.$$

Note that $\mathcal{I}$ contains αN subcarriers. The remaining $(1-\alpha)N$ subcarriers are shared by the three sectors in an orthogonal way, such that each base stations c has at its disposal a subset $\mathcal{P}_c$ ($\mathcal{P}$ as in $\mathcal{P}$rotected) of cardinality $\frac{1-\alpha}{3}N$. If user k modulates a subcarrier $n \in \mathcal{P}_c$, then process $w_k(n,m)$ will contain only thermal noise with variance σ^2. Finally,

$$\mathcal{I} \cup \mathcal{P}_A \cup \mathcal{P}_B \cup \mathcal{P}_C = \{0, 1, \ldots, N-1\}.$$

Also assume that **channel coefficients** $\{H_k^c(n,m)\}_{n \in \mathcal{N}_k}$ **are Rayleigh distributed and have the same variance** $\rho_k^c = \mathbb{E}\left[|H_k^c(n,m)|^2\right], \forall n \in \mathcal{N}_k$. This assumption is realistic in cases where the propagation environment is **highly scattering**, leading to decorrelated Gaussian-distributed time-domain channel taps. Under all the aforementioned assumptions, it can be shown that the ergodic capacity associated with each user k **only** depends on the **number of subcarriers** assigned to user k in subsets $\mathcal{I}$ and $\mathcal{P}_c$ respectively, rather than on the **specific** subcarriers assigned to k.

The resource allocation parameters for user k are thus:

i) The sharing factors $\gamma_{k,\mathcal{I}}$, $\gamma_{k,\mathcal{P}}$ defined by

$$\gamma_{k,\mathcal{I}}^c = \text{card}(\mathcal{I} \cap \mathcal{N}_k)/N \qquad \gamma_{k,\mathcal{P}}^c = \text{card}(\mathcal{P}_c \cap \mathcal{N}_k)/N. \tag{14}$$

ii) The powers $P_{k,\mathcal{I}}$, $P_{k,\mathcal{P}}$ transmitted on the subcarriers assigned to user k in $\mathcal{I}$ and $\mathcal{P}_c$ respectively.

We assume from now on that $\gamma_{k,\mathcal{I}}$ and $\gamma_{k,\mathcal{P}}$ can take on any value in the interval $[0,1]$ (not necessarily integer multiples of $1/N$).

Remark 3. *Even though the sharing factors in our model are not necessarily integer multiples of $1/N$, it is still possible to practically achieve the exact values of $\gamma_{k,\mathcal{I}}$, $\gamma_{k,\mathcal{P}}$ by simply exploiting the time dimension. Indeed, the number of subcarriers assigned to user k can be chosen to vary from one OFDM symbol to another in such a way that the* average *number of subcarriers in subsets $\mathcal{I}$, $\mathcal{P}_c$ is equal to $\gamma_{k,\mathcal{I}}N$, $\gamma_{k,\mathcal{P}}N$ respectively. Thus the fact that $\gamma_{k,\mathcal{I}}$, $\gamma_{k,\mathcal{P}}$ are not strictly integer multiples of $1/N$ is not restrictive, provided that the system is able to grasp the benefits of the time dimension. The particular case where the number of subcarriers is restricted to be the same in each OFDM block is addressed in N. Ksairi & Ciblat (2011).*

The sharing factors of the different users should be selected such that

$$\sum_{k \in c} \gamma_{k,\mathcal{I}} \leq \alpha \qquad \sum_{k \in c} \gamma_{k,\mathcal{P}} \leq \frac{1-\alpha}{3}. \tag{15}$$

We now describe the adopted model for the multicell interference. Consider one of the non protected subcarriers n assigned to user k of cell A in subset $\mathcal{I}$. Denote by σ_k^2 the variance of the additive noise process $w_k(n,m)$ in this case. This variance is assumed to be constant w.r.t both n and m. It only depends on the position of user k and the average powers [4] $Q_{B,\mathcal{I}} = \sum_{k \in B} \gamma_{k,\mathcal{I}} P_{k,\mathcal{I}}$ and $Q_{C,\mathcal{I}} = \sum_{k \in C} \gamma_{k,\mathcal{I}} P_{k,\mathcal{I}}$ transmitted respectively by base stations B and C in $\mathcal{I}$. This assumption is valid in OFDMA systems that adopt random subcarrier assignment

[4] The dependence of interference power on only the average powers transmitted by the interfering cells rather than on the power of each single user in these cells is called *interference averaging*

or frequency hopping (which are both supported in the WiMAX standard [5]). Finally, let σ^2 designate the variance of the thermal noise. Putting all pieces together:

$$\mathbb{E}\left[|w_k(n,m)|^2\right] = \begin{cases} \sigma^2 & \text{if } n \in \mathcal{P}_c \\ \sigma_k^2 = \sigma^2 + \mathbb{E}\left[|H_k^B(n,m)|^2\right] Q_1^B + \mathbb{E}\left[|H_k^C(n,m)|^2\right] Q_1^C & \text{if } n \in \mathcal{I} \end{cases} \tag{16}$$

where $H_k^B(n,m)$ (resp. $H_k^C(n,m)$) represents the channel between base station B (resp. C) and user k of cell A at subcarrier n and OFDM block m. Of course, the average channel gains $\mathbb{E}\left[|H_k^A(n,m)|^2\right]$, $\mathbb{E}\left[|H_k^B(n,m)|^2\right]$ and $\mathbb{E}\left[|H_k^C(n,m)|^2\right]$ depend on the position of user k via the path loss model.

Now, let $g_{k,\mathcal{I}}$ (resp. $g_{k,\mathcal{P}}$) be the channel Gain-to-Noise Ratio (GNR) for user k in band $\mathcal{I}$ (resp. $\mathcal{P}_c$), namely

$$g_{k,\mathcal{I}}(Q_{B,\mathcal{I}}, Q_{C,\mathcal{I}}) = \frac{\rho_k}{\sigma_k^2(Q_{B,\mathcal{I}}, Q_{C,\mathcal{I}})} \qquad g_{k,\mathcal{P}} = \frac{\rho_k}{\sigma^2},$$

where $\sigma_k^2(Q_{B,\mathcal{I}}, Q_{C,\mathcal{I}})$ is the variance of the noise-plus-interference process associated with user k given th interference levels generated by base stations B, C are equal to $Q_{B,\mathcal{I}}$, $Q_{C,\mathcal{I}}$ respectively.

The ergodic capacity associated with k in the whole band is equal to the sum of the ergodic capacities corresponding to both bands $\mathcal{I}$ and $\mathcal{P}_A$. For instance, the part of the capacity corresponding to the protected band $\mathcal{P}_A$ is equal to

$$\gamma_{k,\mathcal{P}} \mathbb{E}\left[\log\left(1 + P_{k,\mathcal{P}} \frac{|H_k^A(n,m)|^2}{\sigma^2}\right)\right],$$

where factor $\gamma_{k,\mathcal{P}}$ traduces the fact that the capacity increases with the number of subcarriers which are modulated by user k. In the latter expression, the expectation is calculated with respect to random variable $\frac{|H_k^A(m,n)|^2}{\sigma^2}$. Now, $\frac{|H_k^A(m,n)|^2}{\sigma^2}$ has the same distribution as $\frac{\rho_k}{\sigma^2}Z = g_{k,\mathcal{P}}Z$, where Z is a standard exponentially-distributed random variable. Finally, the ergodic capacity in the whole bandwidth is equal to

$$\begin{aligned} &C_k(\gamma_{k,\mathcal{I}}, \gamma_{k,\mathcal{P}}, P_{k,\mathcal{I}}, P_{k,\mathcal{P}}, Q_{B,\mathcal{I}}, Q_{C,\mathcal{I}}) = \\ &\quad \gamma_{k,\mathcal{I}} \mathbb{E}\left[\log\left(1 + g_{k,\mathcal{I}}(Q_{B,\mathcal{I}}, Q_{C,\mathcal{I}}) P_{k,\mathcal{I}} Z\right)\right] + \gamma_{k,\mathcal{P}} \mathbb{E}\left[\log\left(1 + g_{k,\mathcal{P}} P_{k,\mathcal{P}} Z\right)\right]. \end{aligned} \tag{17}$$

Assume that user k has an *average* rate requirement R_k (nats/s/Hz). This requirement is satisfied provided that R_k is less that the ergodic capacity C_k *i.e.*,

$$R_k < C_k(\gamma_{k,\mathcal{I}}, \gamma_{k,\mathcal{P}}, P_{k,\mathcal{I}}, P_{k,\mathcal{P}}, Q_{B,\mathcal{I}}, Q_{C,\mathcal{I}}). \tag{18}$$

Finally, the quantity Q_c defined by

$$Q_c = \sum_{k \subset c}' (\gamma_{k,\mathcal{I}} P_{k,\mathcal{I}} + \gamma_{k,\mathcal{P}} P_{k,\mathcal{P}}) \tag{19}$$

[5] In WiMAX, one of the types of subchannelization *i.e.*, grouping subcarriers to form a subchannel, is *diversirty permutation*. This method draws subcarriers pseudorandomly, thereby resulting in interference averaging as explained in Byeong Gi Lee & Sunghyun Choi (2008)

denotes the average power spent by base station c during one OFDM block.

$\mathcal{I}$	subset of reused subcarriers that are subject to multicell interference
$\mathcal{P}_c$	subset of interference-free subcarriers that are exclusively reserved for cell c
R_k	rate requirement of user k in nats/s/Hz
C_k	ergodic capacity associated with user k
$g_{k,\mathcal{I}}, g_{k,\mathcal{P}}$	GNR of user k in bands $\mathcal{I}$, $\mathcal{P}_A$ resp.
$\gamma_{k,\mathcal{I}}, \gamma_{k,\mathcal{P}}$	sharing factors of user k in bands $\mathcal{I}$, $\mathcal{P}_c$ resp.
$P_{k,\mathcal{I}}, P_{k,\mathcal{P}}$	power allocated to user k in bands $\mathcal{I}$, $\mathcal{P}_A$ resp.
$Q_{c,\mathcal{I}}, Q_{c,\mathcal{P}}$	power transmitted by base station c in bands $\mathcal{I}$, $\mathcal{P}_A$ resp.
Q_c	total power transmitted by base station c

Table 1. Some notations for cell c

Optimization problem

The joint resource allocation problem that we consider consists in minimizing the power that should be spent by the three base stations A, B, C in order to satisfy all users' rate requirements:

$$\min_{\{\gamma_{k,\mathcal{I}},\gamma_{k,\mathcal{P}},P_{k,\mathcal{I}},P_{k,\mathcal{P}}\}_{k=1..K}} \sum_{c=A,B,C} \sum_{k\in c} \gamma_{k,\mathcal{I}} P_{k,\mathcal{I}} + \gamma_{k,\mathcal{P}} P_{k,\mathcal{P}}$$
$$\text{subject to constraints (15) and (18)}\,. \quad (20)$$

This problem is not convex with repsect to the resource allocation parameters. It cannot thus be solved using convex optimization tools. Fortunately, it has been shown in N. Ksairi & Ciblat (2011) that a resource allocation algorithm can be proposed that is **asymptotically optimal** *i.e.,* the transmit power it requires to satisfy users' rate requirements is equal to the transmit power of an **optimal** solution to the above problem **in the limit of large numbers of users**. We present in the sequel this allocation algorithm, and we show that it can be implemented in a distributed fashion and that it has relatively low computational complexity.

Practical resource allocation scheme

In the proposed scheme we force the users near the cell's borders (who are normally subject to sever fading conditions and to high levels of multicell interference) to modulate uniquely the subcarriers in the protected subset $\mathcal{P}_c$, while we require that the users in the interior of the cell (who are closer to the base station and suffer relatively low levels on intercell interference) to modulate uniquely subcarriers in the interference subset $\mathcal{I}$.

Of course, we still need to define a **separating curve** that split the users of the cell into these two groups of interior and exterior users. For that sake, we define on $\mathbb{R}_+^5 \times \mathbb{R}$ the function

$$(\theta, x) \mapsto d_\theta(x)$$

where $x \in \mathbb{R}$ and where θ is a set of parameters[6]. We use this function to define the separation curves d_{θ^A}, d_{θ^B} and d_{θ^C} for cells A, B and C respectively. The determination of parameters θ^A, θ^B and θ^C is discussed later on. Without any loss of generality, let us now focus on cell A. For

[6] The closed-form expression of function $d_\theta(x)$ is provided in N. Ksairi & Ciblat (2011).

a given user k in this cell, we designate by (x_k, y_k) its coordinates in the Cartesian coordinate system whose origin is at the position of base station A and which is illustrated in Figure 5. In

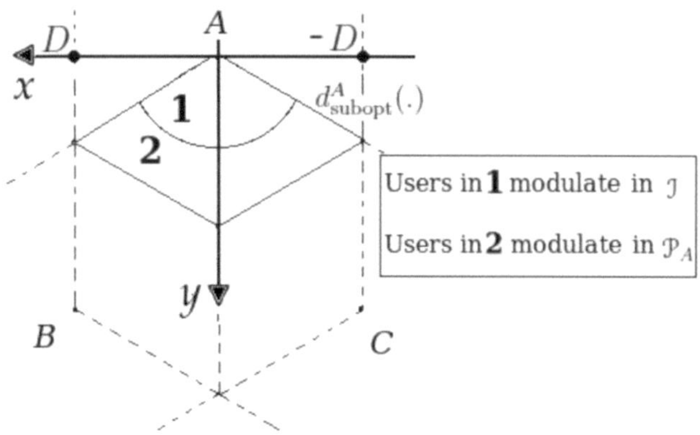

Fig. 5. Separation curve in cell A

the proposed allocation scheme, user k modulates in the interference subset $\mathcal{I}$ if and only if

$$y_k < d_{\theta^A}(x_k)\,.$$

Inversely, the user modulates in the interference-free subset $\mathcal{P}_A$ if and only if

$$y_k \geq d_{\theta^A}(x_k)$$

Therefore, we have defined in each sector two geographical regions: the first is around the base station and its users are subject to multicell interference; the second is near the border of the cell and its users are protected from multicell interference.

The resource allocation parameters $\{\gamma_{k,\mathcal{P}}, P_{k,\mathcal{P}}\}$ for the users of the three protected regions can be easily determined by solving three **independent** convex resource allocation problems. In solving these problems, there is no interaction between the three sectors thanks to the absence of multicell interference for the protected regions. The closed-form solution to these problems is given in N. Ksairi & Ciblat (2011).

However, the resource allocation parameters $\{\gamma_{k,\mathcal{I}}, P_{k,\mathcal{I}}\}$ of users of the non-protected interior regions should be **jointly** optimized in the three sectors. Fortunately, a **distributed iterative** algorithm is proposed in N. Ksairi & Ciblat (2011) to solve this joint optimization problem. This iterative algorithm belongs to the family of **best dynamic response** algorithms. At each iteration, we solve in each sector a single-cell allocation problem given a fixed level of multicell interference generated by the other two sectors in the previous iteration. The mild conditions for the convergence of this algorithm are provided in N. Ksairi & Ciblat (2011). Indeed, it is shown that the algorithm converges for all realistic average data rate requirements provided that the separating curves are carefully chosen as will be discussed later on.

Determination of the separation curves and asymptotic optimality of the proposed scheme

It is obvious that the above proposed resource allocation algorithm is **suboptimal** since it forces a "binary" separation of users into protected and non-protected groups. Nonetheless, it has been proved in N. Ksairi & Ciblat (2011) that this binary separation is **asymptotically**

optimal in the sense that follows. Denote by $Q_{\text{subop}}^{(K)}$ the total power spent by the three base stations if this algorithm is applied. Also define $Q_T^{(K)}$ as the total transmit power of an optimal solution to the original joint resource allocation problem. The suboptimality of the proposed resource allocation scheme trivially implies

$$Q_{\text{subop}}^{(K)} \geq Q_T^{(K)}$$

The asymptotic behaviour of both $Q_{\text{subop}}^{(K)}$ and $Q_T^{(K)}$ as $K \to \infty$ has been studied[7] in N. Ksairi & Ciblat (2011). In the asymptotic regime, it can be shown that the configuration of the network, as far as resource allocation is concerned, is completely determined by i) the **average** (as opposed to **individual**) data rate requirement $\bar{r}$ and ii) a function $\lambda(x,y)$ that characterizes the asymptotic "density" of users' geographical positions in the coordination system (x,y) of their respective sectors. To better understand the physical meaning of the density function $\lambda(x,y)$, note that it is a constant function in the case of uniform distribution of users in the cell area..

Interestingly, one can find values for parameters θ^A, θ^B and θ^C (characterizing the separatin curves d_{θ^A}, d_{θ^B}, and d_{θ^C} respectively) that i) depend only on the average rate requirement $\bar{r}$ and on the asymptotic geographical density of users and ii) which satisfy

$$\lim_{K\to\infty} Q_{\text{subopt}}^{(K)} = \lim_{K\to\infty} Q_T^{(K)} \stackrel{\text{(def)}}{=} Q_T .$$

In other words, one can find separating curves d_{θ^A}, d_{θ^B}, and d_{θ^C} such that the proposed suboptimal allocation algorithm is **asymptotically optimal** in the limit of large numbers of users. We plot in Figure 6 these asymptotically optimal separating curves for several values of the average data rate requirement [8]. The performance of the proposed algorithm *i.e.*, its total

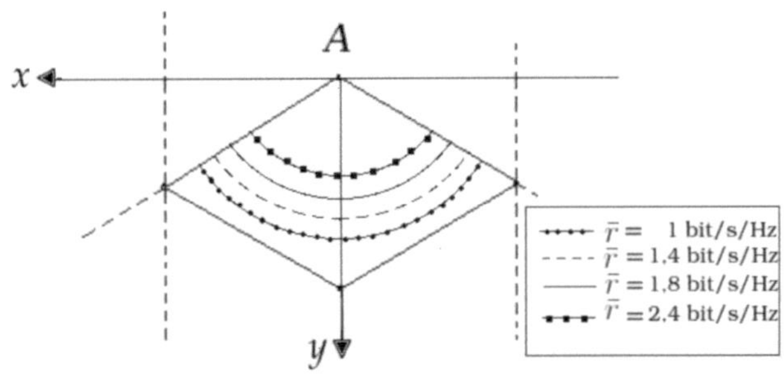

Fig. 6. Asymptotically optimal separating curves

[7] In this asymptotic analysis, a technical detail requires that we also let the total bandwidth B (Hz) occupied by the system tend to infinity in order to satisfy the sum of users' rate requirements $\sum_{k=1}^{K} r_k$ which grows to infinity as $K \to \infty$. Moreover, in order to obtain relevant results, we assume that as K, B tends to infinity, their ratio B/K remains constant

[8] In all the given numerical and graphical results, it has been assumed that the radius of the cells is equal to $D = 500$m. The path loss model follows a Free Space Loss model (FSL) characterized by a path loss exponent $s = 2$. The carrier frequency is $f_0 = 2.4GHz$. At this frequency, path loss in dB is given by $\rho_{dB}(x) = 20\log_{10}(x) + 100.04$, where x is the distance in kilometers between the BS and the user. The signal bandwidth B is equal to 5 MHz and the thermal noise power spectral density is equal to $N_0 = -170$ dBm/Hz.

transmit power when the asymptotically optimal separating curves are used, is compared in Figure 7 to the performance of an all-reuse scheme ($\alpha = 1$) that has been proposed in Thanabalasingham et al. (2006). It is worth mentioning that the reuse factor α assumed for our algorithm in Figure 7 has been obtained using the procedure described in Section 5. It is clear from the figure that a significant gain in performance can be obtained from applying a carefully designed FFR allocation algorithm (such as ours) as compared to an all-reuse scheme. The above comparison and performance analysis is done assuming a 3-sector network. This

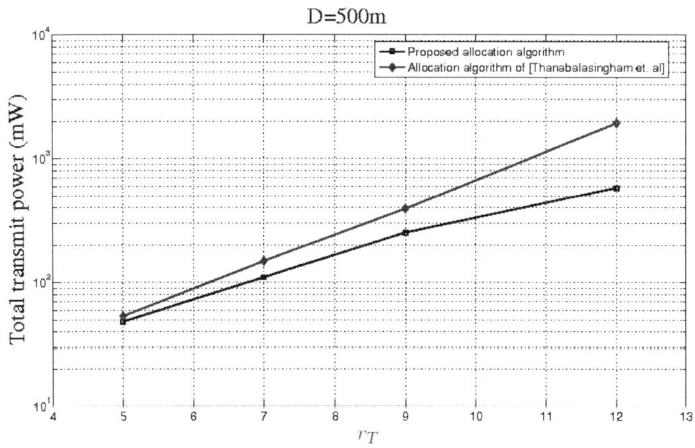

Fig. 7. Performance of the proposed algorithm vs total rate requirement per sector compared to the all-reuse scheme of Thanabalasingham et al. (2006)

assumption is valid provided that the intercell interference in one sector is mainly due to only the two nearest base stations. If this assumption is not valid (as in the 21-sector network of Figure 8), the performance of the proposed scheme will of course deteriorates as can be seen in Figure 9. The same figure shows that the proposed scheme still performs better than an all-reuse scheme, especially at high data rate requirements.

4.3 Outage-based resource allocation (statistical-CSI slow-fading channels)

Recall from Section 2 that the relevant performance metric in the case of slow-fading channels is the outage probability $P_{O,k}$ given by (3) (in the case of Gaussian codebooks and Gaussian-distributed noise-plus-interference process) as

$$P_{O,k}(R_k) \triangleq \Pr\left[\frac{1}{N}\sum_{n\in\mathcal{N}_k}\log\left(1+P_{k,n}\frac{|H_k^c(n)|^2}{\sigma_k^2}\right)\leq R_k\right].$$

Where R_k is the rate (in nats/s/Hz) at which data is transmitted to user k. Unfortunately, no closed-form expression exists for $P_{O,k}(R_k)$. The few works on outage-based resource allocation for OFDMA resorted to **approximations** of the probability $P_{O,k}(R_k)$.

For example, consider the problem of maximizing the sum of users' data rates R_k under a total power constraint $P_{\max}$ such that the outage probability of each user k does not exceed a certain threshold ϵ_k:

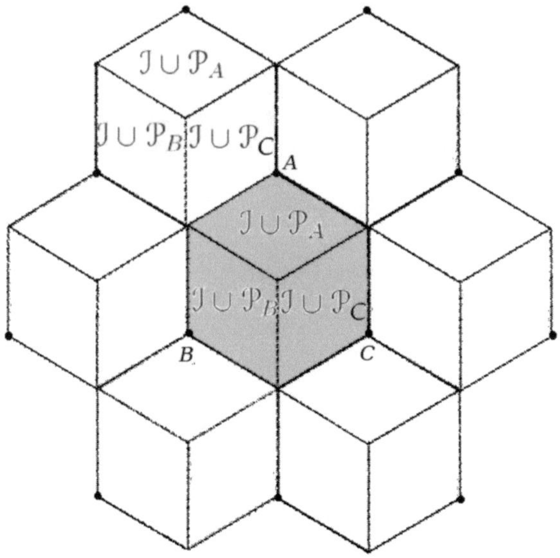

Fig. 8. 21-sector system model and the frequency reuse scheme

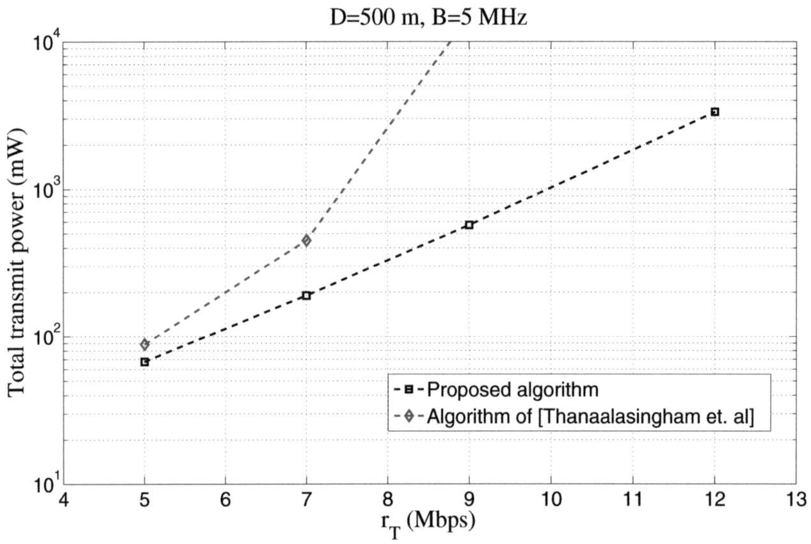

Fig. 9. Comparison between the proposed allocation algorithm and the all-reuse scheme of Thanabalasingham et al. (2006) in the case of 21 sectors (25 users per sector) vs the total rate requirement per sector

$$\max_{\{\mathcal{N}_k, P_{k,n}\}_{1 \le k \le K, n \in \mathcal{N}_k}} \sum_c \sum_{k \in c} R_k$$

$$\text{subject to the OFDMA orthogonality constraint and to (8) and } P_{O,k}(R_k) \le \epsilon_k . \qquad (21)$$

In M. Pischella & J.-C. Belfiore (2009), the problem is tackled in the context of MIMO-OFDMA systems where both the base stations and the users' terminals have multiple antennas. In the approach proposed by the authors to solve this problem, **the outage probability is replaced with an approximating function**. Moreover, subcarrier assignment is performed **independently** (and thus suboptimally) in each cell assuming equal power allocation and equal interference level on all subcarriers. Once the subcarrier assignment is determined, multicell power allocation *i.e.*, the determination of $P_{k,n}$ for each user k is done thanks to an iterative allocation algorithm. Each iteration of this algorithm consists in solving the power allocation problem separately in each cell based on the current level of multicell interference. The result of each iteration is then used to update the value of multicell interference for the next iteration of the algorithm. The convergence of this iterative algorithm is also studied by the authors. A solution to Problem (21) which performs **joint** optimization of subcarrier assignment and power allocation is yet to be provided.

In S. V. Hanly et al. (2009), a min-max outage-based multicell resource allocation problem is solved assuming that there exists a genie who can instantly return the outage probability of any user as a function of the power levels and subcarrier allocations in the network. When this restricting assumption is lifted, only a suboptimal solution is provided by the authors.

4.4 Resource allocation for real-world WiMAX networks: Practical considerations

- All the resource allocation schemes presented in this chapter assume that the transmit symbols are from Gaussian codebooks. This assumption is widely made in the literature, mainly for tractability reasons. In real-world WiMAX systems, Gaussian codebooks are not practical. Instead, discrete modulation (*e.g.* QPSK,16-QAM,64-QAM) is used. The adaptation of the presented resource allocation schemes to the case of dynamic Modulation and Coding Schemes (MCS) supported by WiMAX is still an open area of research that has been addressed, for example, in D. Hui & V. Lau (2009); G. Song & Y. Li (2005); J. Huany et al. (2005); R. Aggarwal et al. (2011).
- The WiMAX standard provides the necessary signalling channels (such as the CSI feedback messages (CQICH, REP-REQ and REP-RSP) and the control messages DL-MAP and DCD) that can be used for resource allocation, as explained in Byeong Gi Lee & Sunghyun Choi (2008), but does not oblige the use of any specific resource allocation scheme.
- The smallest unity of band allocation in WiMAX is **subchannels** (A subchannel is a group of subcarriers) **not** subcarriers. Moreover, WiMAX supports transmitting with different powers and different rates (MCS schemes) on different subchannels as explained in Byeong Gi Lee & Sunghyun Choi (2008). This implies that the per-subcarrier full-CSI schemes presented in Subsection 4.1 are not well adapted for WiMAX systems. They should thus be first modified to per-subchannel schemes before use in real-world WiMAX networks. However, the average-rate statistical-CSI schemes of Subsection 4.2 are compatible with the subchannel-based assignment capabilities of WiMAX.

5. Optimization of the reuse factor for WiMAX networks

The selection of the frequency reuse scheme is of crucial importance as far as cellular network design is concerned. Among the schemes mentioned in Section 3, fractional frequency reuse (FFR) has gained considerable interest in the literature and has been explicitly recommended for WiMAX in WiMAX Forum (2006), mostly for its simplicity and for its promising gains. For these reasons, we give special focus in this chapter to this reuse scheme.

Recall from Section 3 that the principal parameter characterizing FFR is the *frequency reuse factor* α. The determination of a relevant value α for the this factor is thus a key step in optimizing the network performance. The definition of an *optimal* reuse factor requires however some care. For instance, the reuse factor should be fixed in practice prior to the resource allocation process and its value should be independent of the particular network configuration (such as the changing users' locations, individual QoS requirements, etc).

A solution adopted by several works in the literature consists in performing system level *simulations* and choosing the corresponding value of α that results in the best average performance. In this context, we cite M. Maqbool et al. (2008), H. Jia et al. (2007) and F. Wang et al. (2007) without being exclusive. A more interesting option would be to provide *analytical* methods that permit to choose a relevant value of the reuse factor.

In this context, A promising analytical approach adopted in recent research works such as Gault et al. (2005); N. Ksairi & Ciblat (2011); N. Ksairi & Hachem (2010b) is to resort to **asymptotic analysis** of the network in the limit of **large number of users**. The aim of this approach is to obtain optimal values of the resuse factor that no longer depend on the particular configuration of the network *e.g.,* the exact positions of users, their single QoS requirements, etc, but rather on an asymptotic, or "average", state of the network *e.g.,* density of users' geographical distribution, average rate requirement of users, etc.

In order to illustrate this concept of *asymptotically optimal* values of the reuse factor, we give the following example that is taken from N. Ksairi & Ciblat (2011); N. Ksairi & Hachem (2010b). Consider the resource allocation problem presented in Section 4.2 and which consists in minimizing the total transmit power that should be spent in a 3-sector [9] WiMAX network using the FFR scheme with reuse factor α such that all users' *average* (i.e. *ergodic) rate* requirements r_k (nats/s) are satisfied (see Figure 10). Denote by $Q_T^{(K)}$ the total transmit power spent by the three base stations of the network when the **optimal** solution (see Subsection 4.2) to the above problem is applied. We want to study the behaviour of $Q_T^{(K)}$ as the number K of users tends to infinity [10]. As we already stated, the following holds under mild assumptions:

1. the asymptotic configuration of the network, as far as resource allocation is concerned, is completely characterized by i) the **average** (as opposed to **individual**) data rate requirements $\bar{r}$ and ii) a function $\lambda(x, y)$ that characterizes the asymptotic density of users' geographical positions in the coordination system (x, y) of their respective cells.

2. the optimal total transmit power $Q_T^{(K)}$ tends as $K \to \infty$ to a value Q_T that is given in closed form in N. Ksairi & Ciblat (2011):

$$\lim_{K\to\infty} Q_T^{(K)} \stackrel{\text{(def)}}{=} Q_T .$$

[9] The restriction of the model to a network composed of only 3 neighboring cells is for tractability reasons. This simplification is justified provided that multicell interference can be considered as mainly due to the two nearest neighboring base stations.

[10] As stated earlier, we also let the total bandwidth B (Hz) occupied by the system tend to infinity such that the ratio B/K remain constant

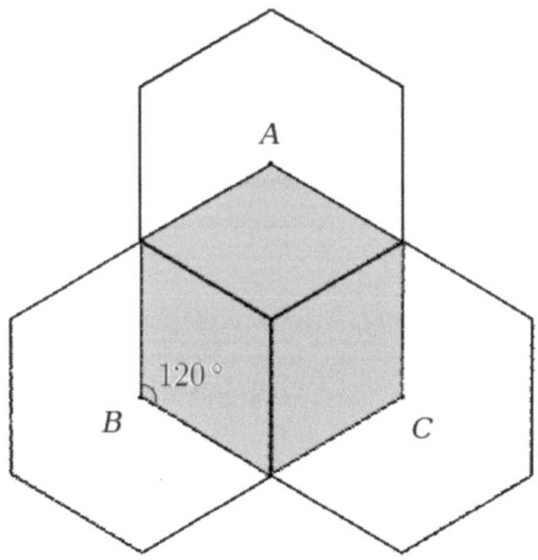

Fig. 10. 3-sectors system model

It is worth noting that the limit value Q_I **only** depends on i) the above-mentioned asymptotic state of the network *i.e.,* on the average rate $\bar{r}$ and on the asymptotic geographical density λ and ii) on the value of the reuse factor α.

It is thus reasonable to select the value α_{opt} of the reuse factor as

$$\alpha_{\text{opt}} = \arg\min_{\alpha} \lim_{K\to\infty} Q_T^{(K)}(\alpha)\,.$$

In practice, we propose to compute the value of $Q_T = Q_T(\alpha)$ for several values of α on a grid in the interval $[0,1]$. In Figure 11, α_{opt} is plotted as function of the average data rate requirement $\bar{r}$ for the case of a network composed of cells with radius $D = 500$m assuming uniformly distributed users' positions. Also note that complexity issues are of few importance, as

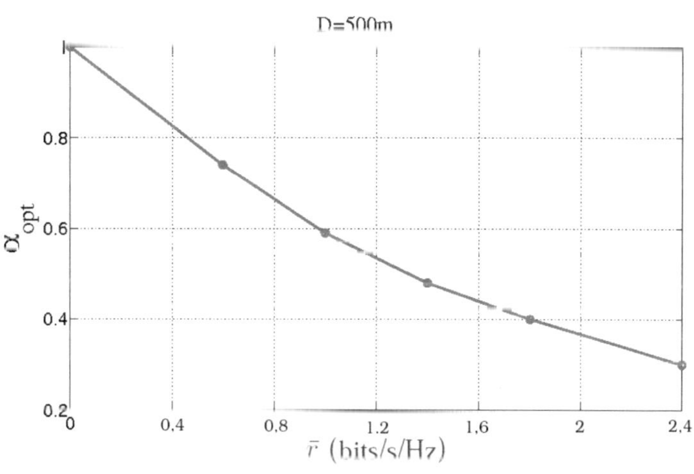

Fig. 11. Asymptotically optimal reuse factor vs average rate requirement. Source:N. Ksairi & Ciblat (2011)

the optimization is done prior to the resource allocation process. It does not affect the complexity of the global resource allocation procedure. It has been shown in N. Ksairi & Ciblat (2011); N. Ksairi & Hachem (2010b) that significant gains are obtained when using the asymptotically-optimal value of the reuse factor instead of an arbitrary value, even for moderate numbers of users.

6. References

Brah, F., Vandendorpe, L. & Louveaux, J. (2007). OFDMA constrained resource allocation with imperfect channel knowledge, *Proceedings of the 14th IEEE Symposium on Communications and Vehicular Technology*.

Brah, F., Vandendorpe, L. & Louveaux, J. (2008). Constrained resource allocation in OFDMA downlink systems with partial CSIT, *Proceedings of the IEEE International Conference on Communications (ICC '08)*.

Byeong Gi Lee & Sunghyun Choi (2008). *Broadband Wireless Access and Local Networks, Mobile WiMAX and WiFi*, Artech House.

D. Hui & V. Lau (2009). Design and analysis of delay-sensitive cross-layer OFDMA systems with outdated CSIT, *IEEE Transactions on Wireless Communications* 8: 3484–3491.

F. Wang, A. Ghosh, C. Sankaran & S. Benes (2007). WiMAX system performance with multiple transmit and multiple receive antennas, *IEEE Vehicular Technology Conference (VTC)*.

G. Song & Y. Li (2005). Cross-layer optimization for OFDM wireless networks - parts I and II, *IEEE Transactions on Wireless Communications* 3: 614–634.

Gault, S., Hachem, W. & Ciblat, P. (2005). Performance analysis of an OFDMA transmission system in a multi-cell environment, *IEEE Transactions on Communications* 12(55): 2143–2159.

Gesbert, D. & Kountouris, M. (2007). Joint power control and user scheduling in multi-cell wireless networks: Capacity scaling laws. submitted to IEEE Transactions On Information Theory.

Hoo, L. M. C., Halder, B., Tellado, J. & Cioffi, J. M. (2004). Multiuser transmit optimization for multicarrier broadcast channels: Asymptotic FDMA capacity region and algorithms, *IEEE Transactions on Communications* 52(6): 922–930.

I. E. Telatar (1999). Capacity of multi-antenna gaussian channels, *European Transactions on Telecommunications* 10: 585–595.

I.C. Wong & B.L. Evans (2009). Optimal resource allocation in OFDMA systems with imperfect channel knowledge, *IEEE Transactions on Communications* 57(1): 232–241.

J. Huang, R. Berry & M. L. Honig (2006). Distributed interference compensation for wireless networks, *IEEE Journal on Selected Areas in Telecommunications* 24(5): 1074–1084.

J. Huany, V. Subbramanian, R. Aggarwal & P. Berry (2005). downlink scheduling and resource allocation for OFDM systems, *IEEE Transactions on Wireless Communications* 8: 288–296.

Jang, J. & Lee, K. B. (2003). Transmit power adaptation for multiuser OFDM systems, *IEEE Journal on Selected Areas in Communications* 21(2): 171–178.

Kivanc, D., Li, G. & Liu, H. (2003). Computationally efficient bandwidth allocation and power control for OFDMA, *IEEE Transactions on Wireless Communications* 2(6): 1150–1158.

Lengoumbi, C., Godlewski, P. & Martins, P. (2006). Dynamic subcarrier reuse with rate guaranty in a downlink multicell OFDMA system, *Proceedings of IEEE 17th International Symposium on Personal, Indoor and Mobile Radio Communications*.

M. Maqbool, M. Coupechoux & Ph. Godlewski (2008). Comparison of various frequency reuse patterns for WiMAX networks with adaptive beamforming, *IEEE Vehicular Technology Conference (VTC)*, Singapore.

M. Pischella & J.-C. Belfiore (2008). Distributed weighted sum throughput maximization in multi-cell wireless networks, *Proceedings of the IEEE 19th International Symposium on Personal, Indoor and Mobile Radio Communications (PIMRC 2008)*.

M. Pischella & J.-C. Belfiore (2009). Distributed resource allocation in MIMO OFDMA networks with statistical CSIT, *Proceedings of the IEEE 10th Workshop on Signal Processing Advances in Wireless Communications*, pp. 126–130.

WiMAX Forum (2006). Mobile WiMAX - part II: A comparative analysis, *Technical report*. Available at http://www.wimaxforum.org/.

N. Ksairi, P. B. & Ciblat, P. (2011). Nearly optimal resource allocation for downlink ofdma in 2-d cellular networks, *accepted for publication in IEEE Transactions on Wireless Communications*.

N. Ksairi, P. Bianchi, P. C. & Hachem, W. (2010a). Resource allocation for downlink cellular ofdma systems: Part i—optimal allocation, *IEEE Transactions on Signal Processing* 58(2).

N. Ksairi, P. Bianchi, P. C. & Hachem, W. (2010b). Resource allocation for downlink cellular ofdma systems: Part ii—practical algorithms and optimal reuse factor, *IEEE Transactions on Signal Processing* 58(2).

H. Jia, Z. Zhang, G. Yu, P. Cheng & S. Li (2008). On the Performance of IEEE 802.16 OFDMA system under different frequency reuse and subcarrier permutation patterns, *IEEE International conference on Communications (ICC)*.

R. Aggarwal, M. Assaad, M. Koksal & C. Schniter (2011). Joint scheduling and resource allocation in the OFDMA downlink: Utility maximization under imperfect channel-state information, *IEEE Transactions on Signal Processing* (99).

S. Plass, A. Dammann & S. Kaiser (2004). Analysis of coded OFDMA in a downlink multi-cell scenario, *9th International OFDM Workshop (InOWo)*.

S. Plass, X. G. Doukopoulos & R. Legouable (2006). Investigations on link-level inter-cell interference in OFDMA systems, *IEEE Symposium on Communications and Vehicular Technology*

S. V. Hanly, L. .L H. Andrew & T. Thanabalasingham (2009). Dynamic allocation of subcarriers and transmit powers in an OFDMA cellular network, *IEEE Transactions on Information Theory* 55(12): 5445–5462.

Seong, K., Mohseni, M. & Cioffi, J. M. (2006). Optimal resource allocation for OFDMA downlink systems, *Proceedings of IEEE International Symposium on Information Theory (ISIT)*, pp. 1394–1398.

Staćzak, S., Wiczanowski, M. & Boche, H. (2009). *Fundamentals of resource allocation in wireless networks: Theory and algorithms*, Springer.

Thanabalasingham, T., Hanly, S. V., Andrew, L. L. H. & Papandriopoulos, J. (2006). Joint allocation of subcarriers and transmit powers in a multiuser OFDM cellular network, *Proceedings of the IEEE International Conference on Communications (ICC '06)*, Vol. 1.

Tse, D. & Visawanath, P. (2005). *Fundamentals of wireless communication*, Cambridge University Press.

Wong, C. Y., Cheng, R. S., Letaief, K. B. & Murch, R. D. (1999). Multiuser OFDM with adaptive subcarrier, bit and power allocation, *IEEE Journal on Selected Areas in Communications* 17(10): 1747–1758.

Wong, I. C. & Evans, B. L. (2007). Optimal OFDMA resource allocation with linear complexity to maximize ergodic weighted sum capacity, *Proceedings of the IEEE International Conference on Acoustics, Speech and Signal Processing (ICASSP)*.

Yu, W. & Lui, R. (2006). Dual methods for non-convex spectrum optimization for multi-carrier system, *IEEE Transactions on Communications* 54(6): 1310–1322.

6

Multi Radio Resource Management over WiMAX-WiFi Heterogeneous Networks: Performance Investigation*

Alessandro Bazzi[1] and Gianni Pasolini[2†]
[1]IEIIT-BO/CNR, Wilab
[2]DEIS-University of Bologna, Wilab
Italy

1. Introduction

In an early future communication services will be accessed by mobile devices through heterogeneous wireless networks given by the integration of the radio access technologies (RATs) covering the user area, including, for instance, WiMAX and WLANs. Today, except for rare cases, it is a choice of the user when and how using one RAT instead of the others: for example, WiFi based WLANs (IEEE-802.11, 2005) would be the favorite choice if available with no charge, while WiMAX or cellular technologies could be the only possibility in outdoor scenarios.

The automatic selection of the best RAT, taking into account measured signal levels and quality-of-service requirements, is the obvious next step, and it has somehow already begun with modern cellular phones, that are equipped with both 2G and 3G technologies: depending on the radio conditions they are able to seamlessly switch from one access technology to the other following some adaptation algorithms. Indeed, all standardization bodies forecast the interworking of heterogeneous technologies and thus put their efforts into this issue: IEEE 802.21 (*IEEE802.21*, 2010), for instance, is being developed by IEEE to provide a protocol layer for media independent handovers; IEEE 802.11u (IEEE802.11u, 2011) was introduced as an amendment to the IEEE 802.11 standard to add features for the interworking with other RATs, and the unlicensed mobile access (UMA) and its evolutions have been included as part of 3GPP specifications ((3GPP-TS-43.318, 2007) and (3GPP-TR-43.902, 2007)) to enable the integration of cellular technologies and other RATs.

It appears clear that the joint usage of available RATs will be a key feature in future wireless systems, although it poses a number of critical issues mainly related to the architecture of future heterogeneous networks, to security aspects, to the signalling protocols, and to the

†This chapter reflects the research activity made in this field at WiLAB (http://www. wilab.org/) over the years. Authors would like to acknowledge several collegues with which a fertile research environment has been created, including O. Andrisano, M. Chiani, A. Conti, D. Dardari, G. Leonardi, B.M. Masini, G. Mazzini, V. Tralli, R. Verdone, and A. Zanella.

multi radio resource management (MRRM) strategies to be adopted in order to take advantage of the multi-access capability.

Focusing on MRRM, the problem is how to effectively exploit the increased amount of resources in order to improve the overall quality-of-service provided to users, for instance reducing the blocking probability or increasing the perceived throughput.

Most studies on this topic assume that the generic user equipment (UE) is connected to one of the available RATs at a time, and focus the investigation on the detection of smart strategies for the optimum RAT choice, that is, for the optimum RAT selection and RAT modification (also called vertical handover); for example in (Fodor et al., 2004) the overall number of admitted connections is increased by taking into account the effectiveness of the various RATs to support specific services; the same result is also achieved in (Bazzi, 2010) by giving a higher priority to those RATs with smaller coverage, and in (Song et al., 2007) by using a load balancing approach.

Besides considering the different RATs as alternative solutions for the connection set-up, their parallel use is also envisioned, in order to take advantage of the multi-radio transmission diversity (MRTD) (Dimou et al., 2005) (Sachs et al., 2004), which consists in the splitting of the data flow over more than one RAT, according to somehow defined criteria. Different approaches have been proposed to this scope, having different layers of the protocol stack in mind: acting at higher layers, on the basis of the traffic characteristics, entails a lower capacity to promptly follow possible link level variations, whereas an approach at lower layers requires particular architectural solutions. In (Luo et al., 2003) the generation of different data flows at the application layer for video transmission or web-browsing (base video layer/enhanced video layer and main objects/in line objects, respectively) is proposed: the most important flow is then served through the most reliable link, such as a cellular connection, while the secondary flow is transmitted through a cheaper connection, such as a WiFi link. In (Hsieh & Sivakumar, 2005), separation is proposed at the transport layer, using one TCP connection per RAT. A transport layer solution is also proposed in (Iyengar et al., 2006), that introduces the concurrent multipath transfer (CMT) protocol based on the multihoming stream control transmission protocol (SCTP). A network level splitting is supposed in (Chebrolu & Rao, 2006; Dimou et al., 2005), and (Bazzi et al., 2008). Coming down through the protocol stack, a data-link frame distribution over two links (WiFi and UMTS) is proposed in (Koudouridis et al., 2005) and (Veronesi, 2005). At the physical layer, band aggregation is supposed for OFDM systems (for example, in (Batra et al., 2004) with reference to UWB) and other solutions that sense the available spectrum and use it opportunistically are envisioned in cognitive radios (Akyildiz et al., 2006).

Hereafter both the alternative and the parallel use of two RATs will be considered. In particular, a scenario with a point of access (PoA) providing both WiMAX and WiFi coverage will be investigated, and the performance level experienced by "dual-mode users" is assessed considering the three following MRRM strategies:

- *autonomous RAT switching*: the RAT to be used for transmission is selected on the basis of measurements (e.g, received power strength) carried out locally by the transmitter, hence with a partial knowledge of RATs'status;
- *assisted RAT switching*: the RAT to be used for transmission is selected not only on the basis of local measurements, but also on the basis of information exchanged with MRRM entities;
- *parallel transmission*: each UE connects at the same time to both RATs.

This chapter is organized as follows: the investigation outline, with assumptions, methodology and considered performance metric, is reported in Section 2; Section 3 and Section 4 introduce and discuss the MRRM strategies for multiple RATs integration; numerical results are shown in Section 5, with particular reference to the performance of an integrated WiFi-WiMAX network; the final conclusions are drawn in Section 6.

2. Investigation assumptions and methodology

Architectural issues

From the viewpoint of the network architecture, the simplest solution for heterogeneous networks integration is the so-called "loose coupling": different networks are connected through gateways, still maintaining their independence. This scenario, that is based on the mobile IP paradigm, is only a little step ahead the current situation of completely independent RATs; in this case guaranteeing seamless (to the end user) handovers between two RATs is very difficult, due to high latencies, and the use of multiple RATs at the same time is unrealistic.

At the opposite side there is the so-called "tight-coupling": in this case different RATs are connected to the same controller and each of them supports a different access modality to the same "core network". This solution requires new network entities and is thus significantly more complex; on the other hand it allows fast handovers and also parallel use of multiple RATs. For the sake of completeness, therefore, the scenario here considered consists of a tight-coupled heterogeneous network, that, for the scope of this chapter, integrates WiMAX and WiFi RATs.

Technologies. As already discussed, here the focus is on a WiMAX and WiFi heterogeneous network, where the following choices and assumptions have been made for the two RATs:

- *WiMAX*. We considered the IEEE802.16e WirelessMAN-OFDMA version (IEEE802.16e, 2006) operating with 2048 OFDM subcarriers and a channelization bandwidth of $7MHz$ in the $3.5GHz$ band; the time division duplexing (TDD) scheme was adopted as well as a frame duration of $10ms$ and a 2:1 downlink:uplink asymmetry rate of the TDD frame.
- *WiFi*. The IEEE802.11a technology (IEEE802.11a, 1999) has been considered at the physical level of the WLAN, thus a channelization bandwidth of 20 MHz in the 5 GHz band and a nominal transmission rate going from 6 Mb/s to 54 Mb/s have been assumed. At the MAC layer we considered the IEEE802.11e enhancement (IEEE802.16e, 2006), that allows the quality-of-service management.

Service and performance metric. The main objective of this chapter is to derive and compare the performance provided by a WiFi-WiMAX integrated network when users equipped with dual-mode terminals perform downlink best effort connections. The performance metric we adopted is the throughput provided by the integrated WiFi-WiMAX network. As we focused our attention, in particular, on best effort traffic, we assumed that the TCP protocol is adopted at the transport layer and we derived, as performance metric, the TCP layer throughput perceived by the final user performing a multiple RATs download.

Let us observe, now, that several TCP versions are available nowadays; it is worth noting, on this regard, that the choice of the particular TCP version working in the considered scenario is not irrelevant when the *parallel transmission* strategy is adopted. For this reason in Section

4 the issue of interactions between the TCP protocol and the *parallel transmission* strategy is faced, and the expected throughput is derived.

Investigation methodology. Results have been obtained partly analytically and partly through simulations, adopting the simulation platform SHINE that has been developed in the framework of several research projects at WiLab. The aim of SHINE is to reproduce the behavior of RATs, carefully considering all aspects related to each single layer of the protocol stack and all characteristics of a realistic environment. This simulation tool, described in (Bazzi et al., 2006), has been already adopted, for example, in (Andrisano et al., 2005) to investigate an UMTS-WLAN heterogeneous network with a *RAT switching* algorithm for voice calls.

MRRM Strategies. In this chapter the three previously introduced MRRM strategies are investigated:

- *autonomous RAT switching;*
- *assisted RAT switching;*
- *parallel transmission.*

In the case of the *autonomous RAT switching* strategy, the decision on the RAT to be adopted for data transmission is taken only considering signal-quality measurements carried out by the transmitter. This the simplest solution: the RAT providing the highest signal-quality is chosen, no matter the fact that, owing to different traffic loads, the other RAT could provide a higher throughout.

As far as the *assisted RAT switching* is concerned, we assumed that an entity performing MRRM at the access network side periodically informs the multi-mode UE about the throughput that can be provided by the different RATs, which is estimated by the knowledge of the signal-quality, the amount of users, the scheduling policy, etc. This entails that in the case of UE initiated connections, the UE has a complete knowledge of the expected uplink throughput over the different RATs. In the case of network initiated connections all information is available at the transmitter side, hence the expected downlink throughput is already known.

The *parallel transmission* strategy, at last, belongs to the class of MRTD strategy, that acts scheduling the transmission of data packets over multiple independent RATs. This task can be accomplished either duplicating each packet, in order to have redundant links carrying the same information, or splitting the packet flow into disjointed sub-flows transmitted by different RATs. In this chapter we considered the latter solution, that is, the parallel transmission "without data duplication" modality. We made the (realistic) assumption that the entity performing MRRM is periodically informed on the number of IP packets transmitted by each technology as well as on the number of IP packets still waiting (in the data-link layer queues) to be transmitted; by the knowledge of these parameters a decision on the traffic distribution over the two RATs is taken, as detailed later on. Let us observe that this assumption is not critical since the entity performing MRRM and the front-end of the jointly used RATs are on the same side of the radio link, thus no radio resource is wasted for signalling messages.

3. RAT switching strategies

The adoption of *RAT switching* strategies (both the *autonomous RAT switching* and the *assisted RAT switching*) does not require significant modifications in the in the PoA/UE behavior except for what concerns vertical handovers. When a dual-mode (or multi-mode) UE or the PoA somehow select the favorite RAT, then they act exactly as in a single RAT scenario. Most of the research effort is thus on the vertical handovers management, in order to optimize the resource usage, maintaining an adequate quality-of-service and acting seamlessly (i.e., automatically and without service interruptions).

Although the tight coupling architecture is with no doubt the best solution to allow prompt and efficient vertical handovers, also loose coupling can be used. In the latter case some advanced technique must be implemented in order to reduce the packet losses during handovers: for example, packets duplication over the two technologies during handovers is proposed for voice calls in (Ben Ali & Pierre, 2009), for video streaming applications in (Cunningham et al., 2004), and for TCP data transfers in (Naoe et al., 2007) and (Wang et al., 2007). (Rutagemwa et al., 2007) and (Huang & Cai, 2006) suggest to use the old connection in downlink until the base-station's queue is emptied while the new connection is already being used for the uplink. Since the issue of vertical handover is besides the scope of this work, hereafter vertical handovers are assumed to be possible and seamless to the end user.

Independently on how multiple RATs are connected and how the vertical handover is performed, there must be an entity in charge of the selection of the best RAT. A number of metrics can be used to this aim, such as, for instance, the perceived signal level or the traffic load of the various RATs. The easiest way to implement a RAT switching mechanism is that each transmitter (at the PoA and the UE) performs some measurements on its own and then selects what it thinks is the best RAT. This way, no information concerning MRRM is exchanged between the UE and the network.

Let us observe, however, that the PoA and the UE have a different knowledge on RAT's status: the PoA knows both link conditions (through measurements of the received signal levels, for instance) and traffic loads of each RAT; the UE, on the contrary, can only measure the link conditions. It follows that without an information exchange between the PoA and the UE, the RAT choice made by the UE (in case of UE initiated connection) could be wrong, owing to unbalanced traffic loads. We define this simple, yet not optimal, MRRM strategy as *autonomous RAT switching*.

A more efficient MRRM is possible if some signalling protocol is available for the exchange of information between the UE and the PoA; a possible implementation could relay, for example, on the already mentioned IEEE 802.21 standard, as done for example in (Lim et al., 2009). The MRRM strategy hereafter denoted as *assisted RAT switching*, assumes that the PoA informs the UE of the throughput that can be guaranteed by each RAT, taking into account also the actual load of the network. It is obviously expected that the increased complexity allows a better distribution of UEs over the various RATs.

4. Parallel transmission strategy

From the viewpoint of network requirements, the adoption of the *parallel transmission* strategy is more demanding with respect to the two *RAT switching* strategies above discussed.

In this case, in fact, the optimal traffic distribution between the different RATs must be continuously derived, on the basis of updated information on their status. It follows that interactions between the entity performing the MRRM and the front-end of the RATs should be as fast as possible, thus making the tight coupling architecture the only realistic architectural solution. Apart from the need of updated information, the loose coupling architecture would introduce relevant differences in the delivery time of packets transmitted over the different technologies, thus causing reordering/buffering problems at the receiver.

Indeed, also in the case of tight coupling architecture, the different delivery delays that the *parallel transmission* strategy causes on packets transmitted by different RATs conflict with the TCP behavior.

For this reason in the following subsections the issue of interactions between the TCP and the *parallel transmission* strategy is thoroughly discussed.

4.1 TCP issues

The most widespread versions of the TCP protocol (e.g., New Reno (NR) TCP (Floyd & Henderson, 1999)) work at best when packets are delivered in order or, at least, with a sporadic disordering. A frequent out-of-order delivery of TCP packets originates, in fact, useless duplicates of transport layer acknowledgments; after three duplicates a packet loss is supposed by the transport protocol and the Fast Recovery - Fast Retransmit phase is entered at the transmitter side.

This causes a significant reduction of the TCP congestion-window size and, as a consequence, a reduction of the throughput achievable at the transport layer.

This aspect of the TCP behavior has been deeply investigated in the literature (e.g. (Bennett et al., 1999) and (Mehta & Vaidya, 1997)) and modern communication systems, such as WiMAX, often include a re-ordering entity at the data-link layer of the receiver in order to prevent possible performance degradation.

Let us observe, now, that when the *parallel transmission* strategy is adopted, each RAT works autonomously at data-link and physical layers, with no knowledge of other active RATs. During the transmission phase, in fact, the packets flow coming from the upper layers is split into sub-flows that are passed to the different data-link layer queues of the active RATs and then transmitted independently one of the others.

It follows that the out-of-order delivery of packets and the consequent performance degradation are very likely, owing to possible differences of the queues occupation levels as well as of the medium access strategies and to the transmission rates of active RATs.

The independency of the different RATs makes very difficult, however, to perform a frame reordering at the data-link layer of the receiver and, at the same time, it would be preferable to avoid, for the sake of simplicity, the introduction of an entity that collects and reorders TCP packets coming from different RATs. For this reason, the adoption of particular versions of TCP, especially designed to solve this problem, is advisable in multiple RATs scenarios.

Here we considered the adoption of the Delayed Duplicates New Reno version of TCP (DD-TCP) (Mehta & Vaidya, 1997), which simply delays the transmission of TCP acknowledgments when an out-of-order packet is received, hoping that the missing packet is already on the fly.

The DD-TCP differs from the NR-TCP only at the receiving side of the transport layer peer-to-peer communication; this implies that the NR-TCP can be maintained at the transmitter side. Thus, this solution could be adopted, at least, on multi-mode user terminals, where the issue of out-of-order packet delivery is more critical owing to the higher traffic load that usually characterizes the downlink phase.

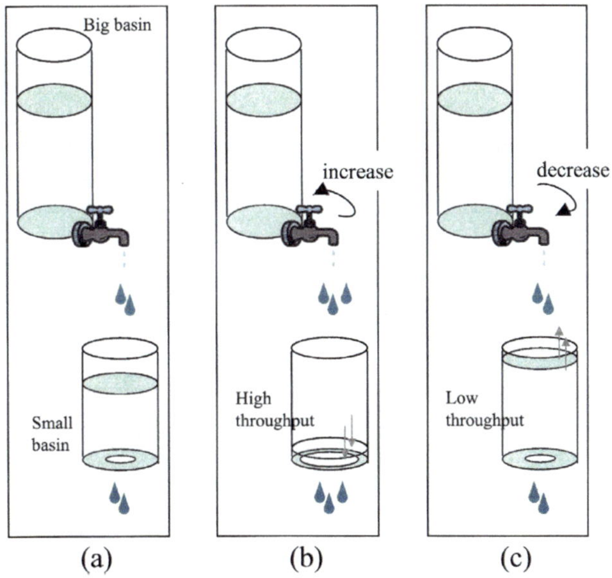

Fig. 1. Representation of the TCP mechanism.

The above introduced critical aspects of the TCP protocol and its interaction with the MRRM strategy in the case of *parallel transmission* are further investigated in the following, where an analytical model of the throughput experienced by the final user is derived.

Let us consider, to this aim, an heterogeneous network which is in general constituted by RATs whose characteristics could be very different in terms, for instance, of medium access strategies and transmission rates.

It is straightforward to understand that, in the case of *parallel transmission*, the random distribution of packets with uniform probability over the different RATs would hardly be the best solution. Indeed, to fully exploit the availability of multiple RATs and get the best from the integrated access network, an efficient MRRM strategy must be designed, able to properly balance the traffic distribution over the different access technologies.

In order to clarify this statement, a brief digression on the TCP protocol behavior is reported hereafter, starting from a simple metaphor.

Let us represent the application layer queue as a big basin (in the following, big basin) filled with water that represents the data to be transmitted (see Fig. 1-a). Another, smaller, basin (in the following, small basin) represents, instead, the data path from the source to the receiver: the size of the data-link layer queue can be represented by the small basin size and the transmission speed by the width of the hole at the small basin bottom.

In this representation the TCP protocol works like a tap controlling the amount of water to be passed to the small basin in order to prevent overflow events (a similar metaphor is used, for example, in (Tanenbaum, 1996)). It follows that the water flow exiting from the tap represents the TCP layer throughput and the water flow exiting from the small basin represents the data-link layer throughput.

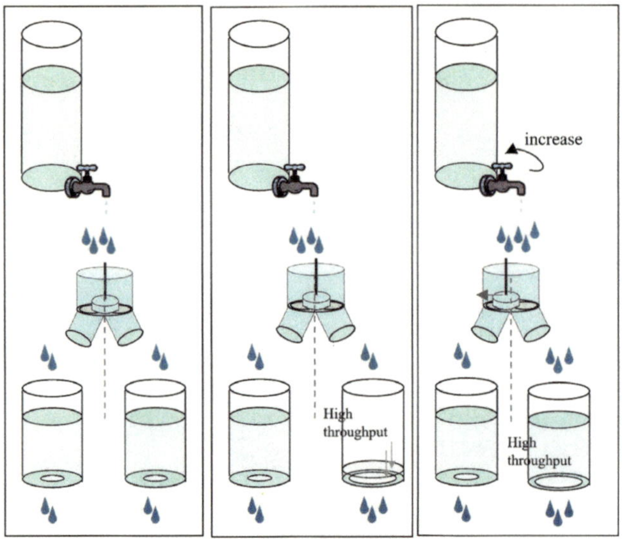

Fig. 2. Representation of the TCP mechanism with parallel transmission over two RATs.

As long as the small basin is characterized by a wide hole, as depicted in Fig. 1-b, the tap can increase the water flow, reflecting the fact that when a high data-link layer throughput is provided by the communication link, the TCP layer throughput can be correspondingly increased.

When, on the contrary, a small hole ($\rightarrow$ a low data-link layer throughput) is detected, the tap ($\rightarrow$ the TCP protocol) reduces the water flow ($\rightarrow$ the TCP layer throughput), as described in Fig. 1-c. This way, the congestion control is performed and the saturation of the data-link layer queues is avoided.

Now the question is: what happens when two basins (that is, two RATs) are available instead of one and the water flow is equally split between them?

Having in mind that the tap has to prevent the overflow of either of the two small basins, it is easy to understand that, in the presence of two small basins with the same hole widths, the tap could simply double the water flow, as depicted in Fig. 2-a.

In the presence of a small basin with a hole wider than the other (see Fig. 2-b), on the other hand, the tap behavior is influenced by the small basin characterized by the lower emptying rate (the leftmost one in Fig. 2-b), which is the most subject to overflow. This means that the availability of a further "wider holed" basin is not fully exploited in terms of water flow increase. Reasoning in terms of TCP protocol, in fact, the congestion window moves following the TCP layer acknowledgments related to packets received in the correct order. This means that, as long as a gap is present in the received packet sequence (one or more packets are missing because of a RAT slower than the other), the congestion window does not move at the transmitter side, thus reducing the throughput provided.

Coming back to the water flow metaphor, it is immediate to understand that, in order to fully exploit the availability of the further, "more performing", small basin, the water flow splitting modality must be modified in such a way that the water in the two small basins is kept at almost the same level (see Fig. 2-c). This consideration introduces in our metaphor the concept of resource management, which is represented in Fig. 2-c by the presence of a valve which dynamically changes the sub-flows discharge.

This concept, translated in the telecommunication-correspondent MRRM concept, will be thoroughly worked out in Section 4.4. To do this, however, an analytical formulation of the expected throughput in the case of multiple RATs adopting the *parallel transmission* strategy is needed, which is reported in the following subsection.

4.2 Throughput analytical derivation

Starting from the above reported considerations, we can derive a simple analytical framework to model the average throughput T perceived by the final user in the case of two heterogeneous RATs, denoted in the following as RAT_A and RAT_B, managed by an MRRM entity which splits the packet flow between RAT_A and RAT_B with probabilities P_A and $P_B = 1 - P_A$, respectively.

Focusing the attention on a generic user, let us denote with T_i the maximum data-link layer throughput supported by RAT_i in the direction of interest (uplink or downlink), given the particular conditions (signal quality, network load due to other users, ...) experienced by the user. Dealing with a dual mode user, we will denote with T_A and T_B the above introduced metric referred to RAT_A and RAT_B respectively.

Let us assume that a block of N transport layer packets of B bits has to be transmitted and let us denote with O the amount of overhead bits added by protocol layers from transport to data-link. After the MRRM operation the N packet flow is split into two sub-flows of, in average, $N \cdot P_A$ and $N \cdot P_B$ packets, which are addressed to RAT_A and RAT_B.

It follows that, in average, RAT_A and RAT_B empty their queues in $D_A = \frac{N \cdot (B+O) \cdot P_A}{T_A}$ and $D_B = \frac{N \cdot (B+O) \cdot P_B}{T_B}$ seconds, respectively.

Thus, the whole N packets block is delivered in a time interval that corresponds to the longest between D_A and D_B.

This means that the average TCP layer throughput provided by the integrated access network to the final user can be expressed as:

$$T = \begin{cases} \frac{N \cdot B}{D_A} = \frac{T_A}{P_A}\xi & \text{when } D_A > D_B \text{, that is when } \frac{T_A}{P_A} < \frac{T_B}{P_B}; \\ \frac{N \cdot B}{D_B} = \frac{T_B}{P_B}\xi & \text{in the opposite case, when } \frac{T_A}{P_A} \geq \frac{T_B}{P_B}, \end{cases} \tag{1}$$

or, in a more compact way, as:

$$T = min\left\{\frac{T_A\xi}{P_A}, \frac{T_B\xi}{P_B}\right\}, \tag{2}$$

where the factor $\xi = B/(B+O)$ takes into account the degradation due to the overhead introduced by protocol layers from transport to data-link.

Let us observe, now, that the term $T_A\xi/P_A$ of (2) is a monotonic increasing function of $P_B = 1 - P_A$, while the term $T_B\xi/P_B$ is monotonically decreasing with P_B.

Since $\frac{T_A}{P_A} < \frac{T_B}{P_B}$ when P_B tends to 0 and $\frac{T_A}{P_A} > \frac{T_B}{P_B}$ when P_B tends to 1, it follows that the maximum TCP layer throughput T_{max} is achieved when $\frac{T_A}{P_A} = \frac{T_B}{P_B}$, that is when:

$$P_A = P_A^{(max)} = \frac{T_A}{T_A + T_B}, \quad (3)$$

and consequently

$$P_B = P_B^{(max)} = 1 - P_A^{(max)} = \frac{T_B}{T_A + T_B}, \quad (4)$$

having denoted with $P_A^{(max)}$ and $P_B^{(max)}$ the values of P_A and P_B that maximize T.

Recalling (2), the maximum TCP layer throughput is immediately derived as:

$$T_{max} = min\left\{\frac{T_A\xi}{P_A}, \frac{T_B\xi}{P_B}\right\}\Bigg|_{P_A = P_A^{(max)}} = (T_A + T_B)\xi, \quad (5)$$

thus showing that a TCP layer throughput as high as the sum of the single TCP layer throughputs can be achieved.

Eqs. (3) and (4) show that an optimal choice of P_A and P_B is possible, in principle, on condition that accurate and updated values of the data-link layer throughput T_A and T_B are known (or, equivalently, accurate and updated values of the TCP layer throughput $T_A\xi$ and $T_B\xi$).

4.3 Parallel transmission strategy: Throughput model validation

In order to validate the above described analytical framework, a simulative investigation has been carried out considering the integration of a WiFi RAT and a WiMAX RAT, which interact according to the *parallel transmission* strategy.

The user is assumed located near the PoA that hosts both the WiMAX base station and the WiFi access point, thus perceiving a high signal-to-noise ratio.

Packets are probabilistically passed by the MRRM entity to the WiFi data-link/physical layers with probability P_{WiFi} (which corresponds to P_A in the general analytical framework) and to the WiMAX data-link/physical layers with probability $1 - P_{WiFi}$ (which corresponds to P_B in the general analytical framework), both in the uplink and in the downlink.

The simulations outcomes are reported in Fig. 3, where the average throughput perceived at the TCP layer is shown as a function of P_{WiFi} (see the curve marked with the circles).

In the same figure we also reported the average throughput predicted by (2), assuming $T_A\xi$ referred to the WiFi RAT and $T_B\xi$ to the WiMAX RAT.

The values of $T_A\xi$ and $T_B\xi$ adopted in (2) have been derived by means of simulations for each one of the considered technologies, obtaining $T_{WiFi} = T_A\xi = 18.53$ Mb/s and $T_{WiMAX} = T_B\xi = 12.76$ Mb/s.

With reference to Fig. 3, let us observe, first of all, the very good matching between the simulation results and the analytical curves derived from (2), which confirms the accuracy of the whole framework. Moreover, from (3) and (5) it is easy to derive $P_A^{(max)} = P_{WiFi} = 0.59$ and $T_{max} = 31.29$ Mb/s, in perfect agreement with the coordinates of the maximum that can be observed in the curve reported in Fig. 3.

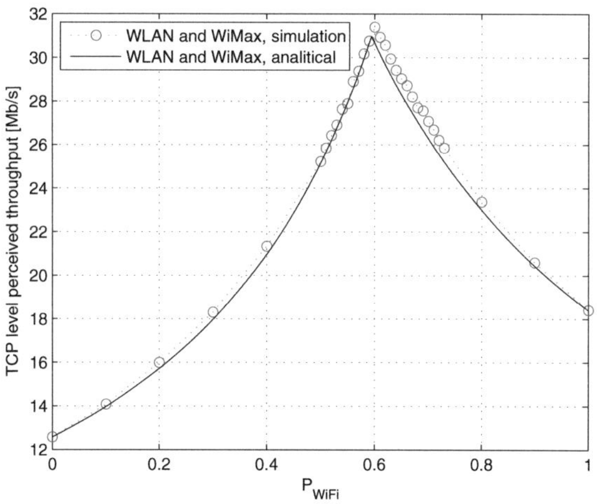

Fig. 3. TCP layer throughput provided by a WiFi-WiMAX heterogeneous network, as a function of the probability that the packet is transferred through the WiFi.

Let us observe, moreover, the rapid throughput degradation resulting from an uncorrect choice of P_{WiFi}. This means that the correct assessment of P_{WiFi} heavily impacts the system performance.

4.4 Traffic-management strategy

The results reported in Fig. 3 showed that in the case of *parallel transmission* the random distribution of packets with uniform probability over the two technologies is not the best solution. On the contrary, to fully exploit the availability of multiple RATs, an efficient traffic-management strategy must be designed, able to properly balance the traffic distribution over the different access technologies.

In order to derive the throughput realistically provided to the final user adopting the *parallel transmission* strategy, we must therefore check whether the optimum traffic balance can be actually achieved or not. In other words, we need to check whether a really effective traffic-management strategy, allowing the user terminal to automatically "tune and track" the optimal traffic distribution, exists or not.

For this reason we conceived an original traffic-management strategy, that we called *Smoothed Transmissions over Pending Packets (Smooth-Tx/Qu)*, that works as follows: packets are always passed to the technology with the higher ratio between the number of packets transmitted up to the present time and the number of packets waiting in the data-link queue; thus, system queues are kept filled proportionally to the transmission speed. The number of transmitted packets is halved every T_{half} seconds (in our simulations we adopted $T_{half} = 0.125$ s) in order to reduce the impact of old transmissions, thus improving the achieved performance in a scenario where transmission rates could change (due to users mobility, for instance).

The performance of such strategy have been investigated evaluating the throughput experienced by a single user in a scenario consisting of a heterogeneous access network with one IEEE802.11a-WiMAX PoA. Transmission *eirp* of 20 dBm and 40 dBm have been assumed for IEEE802.11a and WiMAX, respectively. The throughput provided by each technology

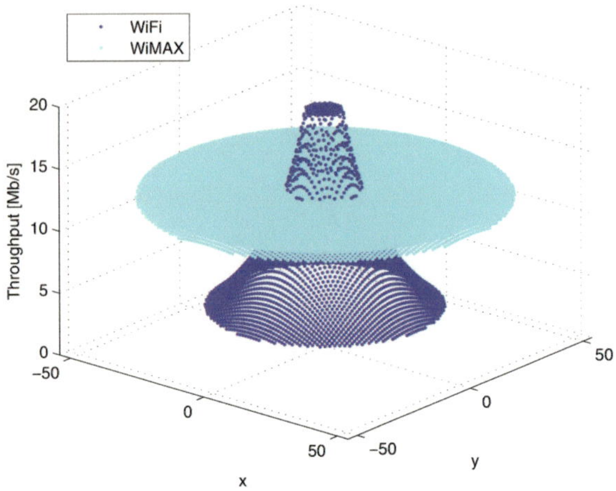

Fig. 4. The investigated scenario.

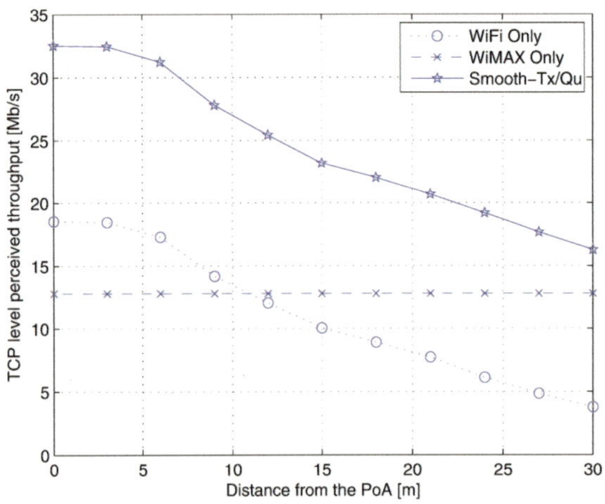

Fig. 5. WiFi-WiMAX heterogeneous networks. TCP layer throughput varying the distance of the user from the access point/base station, for different MRRM schemes. No mobility.

within the area of overlapped coverage is depicted in Fig. 4, where the couple $(x, \ y)$ represents the user's coordinates.

The user is performing an infinite file download and does not change its position. The outcomes of this investigation are reported in Fig. 5, that shows the average perceived TCP layer throughput as a function of the distance from the PoA.

Before discussing the results reported in Fig. 5, a preliminary note on the considered distance range $(0 - 30 \text{ m})$ is needed.

Let us observe, first of all, that WiMAX is a long range communications technology, with a coverage range in the order of kilometers. Nonetheless, since our focus is on the heterogeneous WiFi-WiMAX access network, we must consider coverage distances in the

User position/behavior	WiFi only	WiMAX only	Smooth Tx/Qu
1 Still, near the PoA	18.53	12.76	32.37
2 Still, 30 m far from the PoA	3.81	12.76	16.40
3 Moving away at 1 m/s, starting from the PoA	11.83	12.76	25.01
4 Near the PoA for half sim., then 30 m far away	10.04	12.61	21.03

Table 1. TCP layer average throughput. Single user, 1 WiFi access point and 1 WiMAX base station co-located. 10 seconds simulated.

order of few dozens of meters (i.e., the coverage range of a WiFi), where both RATS are available; for this reason the x-axis of Fig. 5 ranges from 0 to 30 meters.

The different curves of Fig. 5 refer, in particular, to the traffic-management strategy above described and, for comparison, to the cases of a single WiFi RAT and of a single WiMAX RAT.

Of course, when considering the case of a single WiMAX RAT, the throughput perceived by an user located in the region of interest is always at the maximum achievable level, as shown by the flat curve in Fig. 5. As expected, on the contrary, the throughput provided by WiFi in the same range of distances rapidly decreases for increasing distances.

The most important result reported in Fig. 5, however, is related to the upper curve, that refers to the previously described traffic-management strategy when applied in the considered heterogeneous WiFi-WiMAX network. As can be immediately observed, the throughput provided by this strategy is about the sum of those provided by each single RAT, which proves the effectiveness of the proposed traffic-management strategy.

The impact of the user's position and mobility has also been investigated: the results are reported in Table 1 and are related to four different conditions:

1. the user stands still near the PoA (optimal signal reception),
2. the user stands still at 30 m from the access PoA (optimal WiMAX signal, but medium quality WiFi signal),
3. the user moves away from the PoA at a speed of 1 m/s (low mobility),
4. the user stands still near the PoA for half the simulation time, then it moves instantaneously 30 m far away (reproducing the effect of a high speed mobility).

Results are shown for the above described traffic-management strategy as well as for the benchmark scenarios with a single WiFi RAT and a single WiMAX RAT and refer to the average (over the 10 s simulated time interval) throughput perceived in each considered case.

As can be observed the proposed strategy provide satisfying performance in all cases, thus showing that the optimum traffic balance between the different RATs can be achieved.

5. Performance comparison

In the previous section we derived the throughput provided to a single user when the *parallel transmission* strategy is adopted; in this section we also derive the performance of the *autonomous RAT switching* strategy and the *assisted RAT switching* strategy and we extend the investigation to the case of more than one user.

To this aim we considered the same scenario previously investigated, with co-located WLAN access point and WiMAX base station. The resource is assumed equally distributed among connections within each RAT; this assumption means that the same number of OFDMA-slots is given to UEs in WiMAX and that the same transmission opportunity is given to all UEs in WiFi (i.e., they transmit in average for the same time interval, as permitted by IEEE802.11e, that has been assumed at the MAC layer of the WiFi).

In Fig. 6, the complementary cumulative distribution function ($ccdf$) of the perceived throughput is shown when $N = 1, 2, 3, 5, 10$, and 20 users are randomly placed in the coverage area of both technologies: for a given value $\overline{T}$ of throughput (reported in the abscissa), the corresponding $ccdf$ provides the probability that the throughput experienced by an user is higher than $\overline{T}$.

For each value of N, 1000 random placements of the users were performed; the already discussed MRRM strategies are compared:

- *autonomous RAT switching;*
- *assisted RAT switching;*
- *parallel transmission.*

With reference to Fig. 6(a), that refers to the case of a single user, there is obviously no difference adopting the *autonomous RAT switching* strategy or the *assisted RAT switching* strategy. In the absence of other users the choice made by the two strategies is inevitably the same: WiFi is used at low distance from the PoA, while WiMAX is preferred in the opposite case.

The results reported in Fig. 6(a) also confirm that in the case of a single user the perceived throughput can significantly increase thanks to the use of the *parallel transmission* strategy, as discussed in Section 4.4. The significant improvement provided in this case by the *parallel transmission* strategy is not surprising: in the considered case of a single user, in fact, both the *autonomous RAT switching* strategy and the *assisted RAT switching* strategy leave one of the two RATs definitely unused, which is an inauspicious condition.

This consideration suggests that the number of users in the scenario plays a relevant role in the detection of the best MRRM strategy, thus the following investigations, whose outcomes are reported in figures from 6(b) to 6(f), refer to scenarios with $N = 2, 3, 5, 10$, and 20 users, respectively. As can be observed, when more than one user is considered the *dynamic RAT switching* always outperforms the *no RAT switching* and the advantage of using the *parallel transmission* strategy becomes less clear.

Let us focus our attention, now, on Fig. 6(b), that refers to the case of $N = 2$ users randomly placed within the scenario. When the *parallel transmission* strategy is adopted, the 100% of users perceive a throughput no lower than 7.9 Mb/s, whereas the *autonomous RAT switching* strategy and the *assisted RAT switching* strategies provides to the 100% of users a throughput no lower than 6.3 Mb/s. It follows that, at least in the case of $N = 2$ users, the *parallel transmission* strategy outperforms the other strategies in terms of minimum guaranteed throughput. Fig. 6(b) also shows that with the *parallel transmission* strategy the probability of perceiving a throughput higher than 9 Mb/s is reduced with respect to the case of the *assisted RAT switching* strategy. This should not be deemed necessarily as a negative aspect:

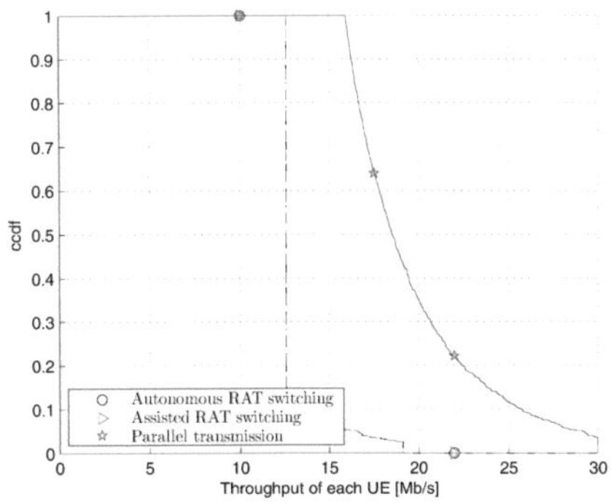

(a) One user.

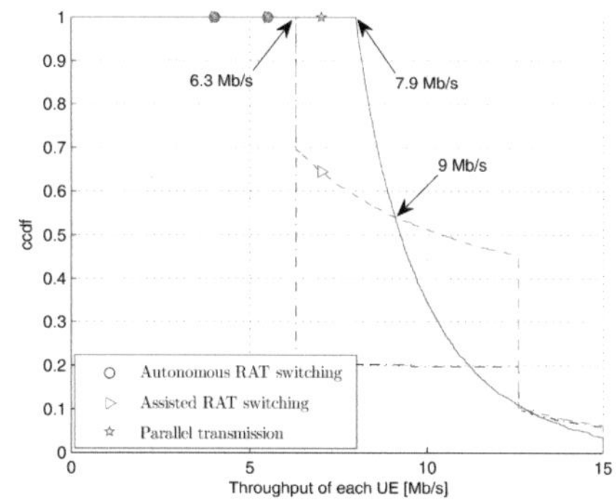

(b) Two users.

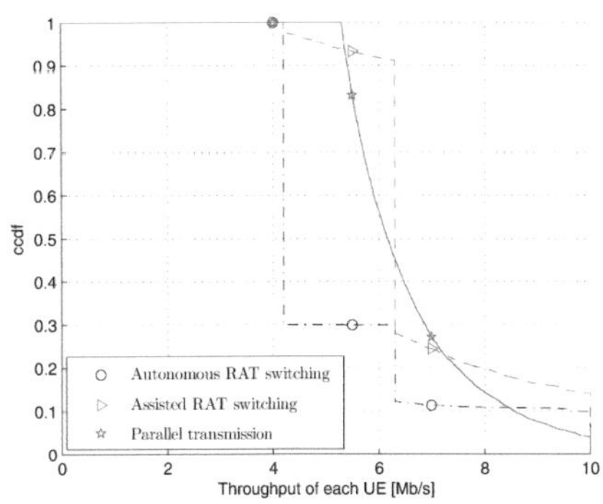

(c) Three users.

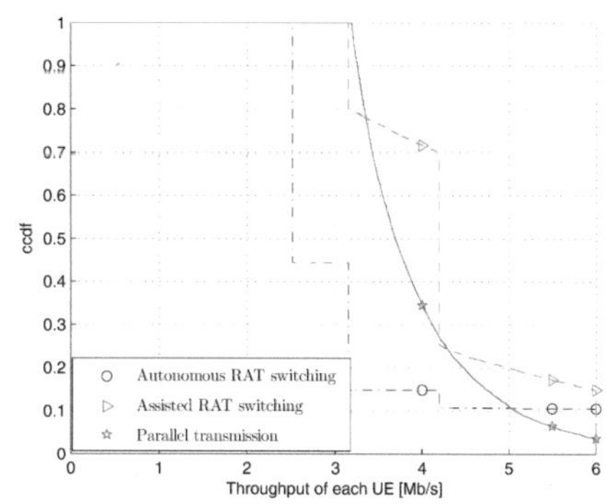

(d) Five users.

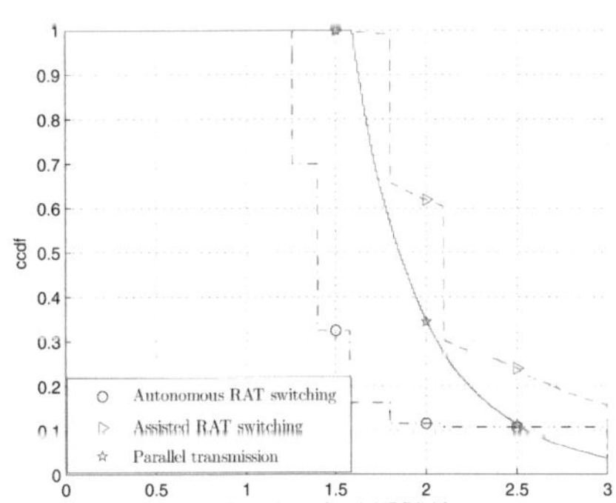

(e) Ten users.

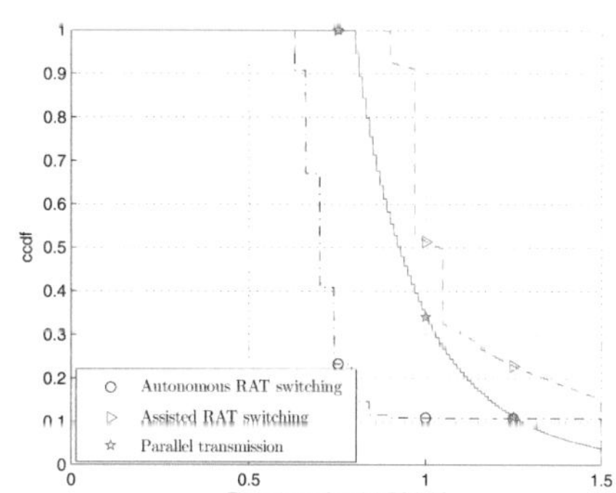

(f) Twenty users.

Fig. 6. *Ccdf* of the throughput perceived by N users randomly placed in the scenario.

everything considered we can state, in fact, that the parallel transmission strategy is fairer than the assisted RAT switching strategy (at least in the case of N = 2 users), since it penalizes lucky UEs (those closer to the PoA) providing a benefit to unlucky users.

Increasing the number of users to $N = 3, 5, 10$, and 20 (thus referring to Figs. 6(c), 6(d), 6(e), and 6(f), respectively), the *autonomous RAT switching* strategy confirms its poor performance with respect to both the other strategies, while the $ccdf$ curve related to the *assisted RAT switching* strategy moves rightwards with respect to the *parallel transmission* curve, thus making the *assisted RAT switching* strategy preferable as the number of users increases.

Let us observe, however, that passing from $N = 10$ to $N = 20$ users, the relative positions of the $ccdf$ curves related to the *parallel transmission* strategy and the *assisted RAT switching* strategy do not change significantly and the gap between the two curves is not so noticeable. It follows that in scenarios with a reasonable number of users the *parallel transmission* strategy could still be a good (yet suboptimal) choice, since, differently from the *assisted RAT switching* strategy, no signalling phase is needed.

6. Conclusions

In this chapter the integration of RATs with overlapped coverage has been investigated, with particular reference to the case of a heterogeneous WiFi-WiMAX network.

Three different MRRM strategies (*autonomous RAT switching*, *assisted RAT switching* and *parallel transmission*) have been discussed, aimed at effectively exploiting the joint pool of radio resources. Their performance have been derived, either analytically or by means of simulations, in order to assess the benefit provided to a "dual-mode" user. In the case of the *parallel transmission* over two technologies a traffic distribution strategy has been also proposed, in order to overcome critical interactions with the TCP protocol.

The main outcomes of our investigations can be summarized as follows:

- in no case the *autonomous RAT switching* strategy is the best solution;
- in the case of a single user the *parallel transmission* strategy provides a total throughput as high as the sum of throughputs of the single RATs;
- the *parallel transmission* strategy generates a disordering of upper layers packets at the receiver side; this issue should be carefully considered when the parallel transmission refers to a TCP connection;
- as the number of users increases the *assisted RAT switching* strategy outperforms the *parallel transmission* strategy.

7. References

3GPP-TR-43.902 (2007). Technical specification group gsm/edge radio access network; enhanced generic access networks (egan) study; v10.0.0; (release 10).

3GPP-TS-43.318 (2007). Technical specification group gsm/edge radio access network; generic access network (gan); v10.1.0; stage 2 (release 10).

Akyildiz, I. F., Lee, W.-Y., Vuran, M. C. & Mohanty, S. (2006). Next generation/dynamic spectrum access/cognitive radio wireless networks: a survey, *Computer Networks* 50(13): 2127 – 2159.

Andrisano, O., Bazzi, A., Diolaiti, M., Gambetti, C. & Pasolini, G. (2005). Umts and wlan integration: architectural solution and performance, *Personal, Indoor and Mobile Radio*

Communications, 2005. PIMRC 2005. IEEE 16th International Symposium on, Vol. 3, pp. 1769 –1775 Vol. 3.

Batra, A., Balakrishnan, J. & Dabak, A. (2004). Multi-band OFDM: a new approach for UWB, *Circuits and Systems, 2004. ISCAS '04. Proceedings of the 2004 International Symposium on*.

Bazzi, A. (2010). Wlan hot spots to increase umts capacity, *Personal Indoor and Mobile Radio Communications (PIMRC), 2010 IEEE 21st International Symposium on*, pp. 2488 –2493.

Bazzi, A., Pasolini, G. & Andrisano, O. (2008). Multiradio resource management: Parallel transmission for higher throughput?, *EURASIP Journal on Advances in Signal Processing* 2008. 9 pages.

Bazzi, A., Pasolini, G. & Gambetti, C. (2006). Shine: Simulation platform for heterogeneous interworking networks, *Communications, 2006. ICC '06. IEEE International Conference on*, Vol. 12, pp. 5534 –5539.

Ben Ali, R. & Pierre, S. (2009). On the impact of soft vertical handoff on optimal voice admission control in PCF-based WLANs loosely coupled to 3G networks, *IEEE Transactions on Wireless Communications* 8: 1356 – 1365.

Bennett, J., Partridge, C. & Shectman, N. (1999). Packet reordering is not pathological network behavior, *Networking, IEEE/ACM Transactions on* 7(6): 789 –798.

Chebrolu, K. & Rao, R. (2006). Bandwidth aggregation for real-time applications in heterogeneous wireless networks, *Mobile Computing, IEEE Transactions on* 5(4): 388 – 403.

Cunningham, G., Perry, P. & Murphy, L. (2004). Soft, vertical handover of streamed video, *5th IEE International Conference on Mobile Communication Technologies, 2004. (3G 2004).*, IEE, London, UK, pp. 432 – 436.

Dimou, K., Agero, R., Bortnik, M., Karimi, R., Koudouridis, G., Kaminski, S., Lederer, H. & Sachs, J. (2005). Generic link layer: a solution for multi-radio transmission diversity in communication networks beyond 3G, *Vehicular Technology Conference, 2005. VTC-2005-Fall. 2005 IEEE 62nd*.

Floyd, S. & Henderson, T. (1999). The newreno modification to tcp's fast recovery algorithm, *Request for Comments: 2582*, pp. 1 –11.

Fodor, G., Furuskär, A. & Lundsjö, J. (2004). On access selection techniques in always best connected networks, *In Proc. ITC Specialist Seminar on Performance Evaluation of Wireless and Mobile Systems*.

Hsieh, H.-Y. & Sivakumar, R. (2005). A transport layer approach for achieving aggregate bandwidths on multi-homed mobile hosts, *Wireless Networks* 11: 99–114.

Huang, H. & Cai, J. (2006). Adding network-layer intelligence to mobile receivers for solving spurious TCP timeout during vertical handoff., *IEEE Network* 20: 24 – 31.

IEEE-802.11 (2005). Iso/iec standard for information technology - telecommunications and information exchange between systems - local and metropolitan area networks - specific requirements part 11: Wireless lan medium access control (mac) and physical layer (phy) specifications (includes ieee std 802.11, 1999 edition; ieee std 802.11a.-1999; ieee std 802.11b.-1999; ieee std 802.11b.-1999/cor 1-2001; and ieee std 802.11d.-2001), *ISO/IEC 8802-11 IEEE Std 802.11 Second edition 2005-08-01 ISO/IEC 8802 11:2005(E) IEEE Std 802.11i-2003 Edition* pp. 1 –721.

IEEE802.11a (1999). Supplement to ieee standard for information technology - telecommunications and information exchange between systems - local and metropolitan area networks - specific requirements. part 11: Wireless lan medium access control (mac) and physical layer (phy) specifications: High-speed physical layer in the 5 ghz band, *IEEE Std 802.11a-1999* p. i.

IEEE802.11u (2011). Ieee standard for information technology-telecommunications and information exchange between systems-local and metropolitan networks-specific requirements-part ii: Wireless lan medium access control (mac) and physical layer (phy) specifications: Amendment 9: Interworking with external networks, *IEEE Std 802.11u* p. i.

IEEE802.16e (2006). Ieee standard for local and metropolitan area networks part 16: Air interface for fixed and mobile broadband wireless access systems amendment 2: Physical and medium access control layers for combined fixed and mobile operation in licensed bands and corrigendum 1, *IEEE Std 802.16e-2005 and IEEE Std 802.16-2004/Cor 1-2005 (Amendment and Corrigendum to IEEE Std 802.16-2004)* pp. 1 –822.

IEEE802.21 (2010).
URL: *http://www.ieee802.org/21/*

Iyengar, J., Amer, P. & Stewart, R. (2006). Concurrent multipath transfer using SCTP multihoming over independent end-to-end paths, *Networking, IEEE/ACM Transactions on* 14(5): 951 –964.

Koudouridis, G., Karimi, H. & Dimou, K. (2005). Switched multi-radio transmission diversity in future access networks, *Vehicular Technology Conference, 2005. VTC-2005-Fall. 2005 IEEE 62nd.*

Lim, W.-S., Kim, D.-W., Suh, Y.-J. & Won, J.-J. (2009). Implementation and performance study of ieee 802.21 in integrated ieee 802.11/802.16e networks, *Computer Communications* 32(1): 134 – 143.

Luo, J., Mukerjee, R., Dillinger, M., Mohyeldin, E. & Schulz, E. (2003). Investigation of radio resource scheduling in WLANs coupled with 3G cellular network, *Communications Magazine, IEEE* 41(6): 108 – 115.

Mehta, M. & Vaidya, N. (1997). Delayed duplicate acknowledgments: A proposal to improve performance of tcp on wireless links, *Tech. Rep., Texas A&M University* .

Naoe, H., Wetterwald, M. & Bonnet, C. (2007). IPv6 soft handover applied to network mobility over heterogeneous access networks., *IEEE 8th International Symposium on Personal, Indoor and Mobile Radio Communications, 2007. (PIMRC 2007).*, IEEE, Athens, Greece, pp. 1–5.

Rutagemwa, H., Pack, S., Shen, X. & Mark, J. (2007). Robust cross-layer design of wireless-profiled TCP mobile receiver for vertical handover., *IEEE Transactions on Vehicular Technology.* 56: 3899 – 3911.

Sachs, J., Wiemann, H., Lundsjo, J. & Magnusson, P. (2004). Integration of multi-radio access in a beyond 3g network, *Personal, Indoor and Mobile Radio Communications, 2004. PIMRC 2004. 15th IEEE International Symposium on*, Vol. 2, pp. 757 – 762 Vol.2.

Song, W., Zhuang, W. & Cheng, Y. (2007). Load balancing for cellular/wlan integrated networks, *IEEE Network* 21: 27–33.

Tanenbaum, A. S. (1996). *Computer Networks*, Prentice Hall, Upper Saddle River, NJ.

Veronesi, R. (2005). Multiuser scheduling with multi radio access selection, *Wireless Communication Systems, 2005. 2nd International Symposium on*, pp. 455 –459.

Wang, N.-C., Wang, Y.-Y. & Chang, S.-C. (2007). A fast adaptive congestion control scheme for improving TCP performance during soft vertical handoff., *IEEE Wireless Communications and Networking Conference, 2007. (WCNC 2007).*, IEEE, Hong Kong, pp. 3641–3646.

7

A Cross-Layer Radio Resource Management in WiMAX Systems

Sondes Khemiri Guy Pujolle[1] and Khaled Boussetta Nadjib Achir[2]
[1]*LIP6, University Paris 6, Paris*
[2]*L2TI, University Paris 13, Villetaneuse*
France

1. Introduction

This chapter addresses the issue of a cross layer radio resource management in IEEE 802.16 metropolitan network and focuses specially on IEEE 802.16e-2005 WiMAX network with Wireless MAN OFDMA physical layer. A wireless bandwidth allocation strategy for a mobile WiMAX network is very important since it determines the maximum average number of users accepted in the network and consequently the provider gain.

The purpose of the chapter is to give an overview of a cross-layer resource allocation mechanisms and describes optimization problems with an aim to fulfill three objectives: (i) to maximize the utilisation ratio of the wireless link, (ii) to guarantee that the system satisfies the QoS constraints of application carried by subscribers and (iii) to take into account the radio channel environment and the system specifications.

The chapter is organized as follows: Section 1 and 2 describe the most important concepts defined by IEEE 802.16e-2005 standard in physical and MAC layer, Section 3 presents an overview of QoS mechanisms described in the literature, Section 4 gives a guideline to compute a physical slot capacity needed in resource allocation problems, the cross-layer resource management problem formalization is detailed in section 5. Solutions are presented in section 6. Finally, section 7 summarizes the chapter.

2. Mobile WiMAX overview

This section presents an overview of the most important concepts defined by IEEE 802.16e-2005 standard in physical and MAC layer, that are needed in order to define a system capacity.

2.1 WiMAX PHY layer

We will give in this section details about PHY layer and we will focus specially on specified concepts that must be taken into account in allocation bandwidth problem namely, the specification of the PHY layer, the OFDMA multiplexing scheme and the permutation scheme for sub-channelization from which we deduce the bandwidth unit allocated to accepted calls in the system and the Adaptive Modulation and Coding scheme (AMC).

2.1.1 Generality

The IEEE 802.16 defines five PHY layers which can be used with a MAC layer to form a broadband wireless system.

These PHY layers provide a large flexibility in terms of bandwidth channel, duplexing scheme and channel condition. These layers are described as follows:

1. WirelessMAN SC: In this PHY layer single carriers are used to transmit information for frequencies beyond 11GHz in a Line of sight (LOS) condition.
2. WirelessMAN SCa: it also relies on a single carrier transmission scheme, but for frequencies between 2 GHz and 11GHz.
3. WirelessMAN OFDM (Orthogonal Frequency Division Multiplexing): it is based on a Fast Fourier Transform (FFT) with a size of 256 points. It is used for point multipoint link in a non-LOS condition for frequencies between 2 GHz and 11GHz.
4. WirelessMAN OFDMA (OFDM Access): Also referred as mobile WiMAX , it is also based on a FFT with a size of 2048 points. It is used in a non LOS condition for frequencies between 2 GHz and 11GHz.
5. Finally a WirelessMAN SOFDMA (SOFDM Access): OFDMA PHY layer has been extended in IEEE 802.16e to SOFDMA (scalable OFDMA) where the size is variable and can take different values: 128, 512, 1024, and 2048.

In this chapter we will focus only on the WirelessMAN OFDMA PHY layer. As we saw in previous paragraph many combination of configuration parameters like band frequencies, channel bandwidth and duplexing techniques are possible. To insure interoperability between terminals and base stations the WiMAX Forum has defined a set of WiMAX system profiles. The latter are basically a set of fixed configuration parameters.

2.1.2 OFDM, OFDMA and subchannelization

The WiMAX PHY layer has also the responsibility of resource allocation and framing over the radio channel. In follows, we will define this physical resource. In fact, the mobile WiMAX physical layer is based on Orthogonal Frequency Multiple Access (OFDMA), which is a multi-users extension of Orthogonal Frequency-Division Multiplexing (OFDM) technique. The latter principles consist of a simultaneous transmission of a bit stream over orthogonal frequencies, also called OFDM sub-carriers. Precisely, the total bandwidth is divided into a number of orthogonal sub-carriers. As described in mobile WiMAX (Jeffrey G. et al., 2007), the OFDMA sharing capabilities are augmented in multi-users context thanks to the flexible ability of the standard to divide the frequency/time resources between users. The minimum time-frequency resource that can be allocated by a WiMAX system to a given link is called a slot. Precisely, the basic unit of allocation in the time-frequency grid is named a slot. Broadly speaking, a slot is an $n \times m$ rectangle, where n is a number of sub-carriers called sub-channel in the frequency domain and m is a number of contiguous symbols in the time domain.

WiMAX defines several sub-channelization schemes. The sub-channelization could be adjacent i.e. sub-carriers are grouped in the same frequency range in each sub-channel or distributed i.e. sub-carriers are pseudo-randomly distributed across the frequency spectrum. So we can find:

- Full usage sub-carriers (FUSC): Each slot is 48 sub-carriers by one OFDM symbol.

- Down-link Partial Usage of Sub-Carrier (PUSC): Each slot is 24 sub-carriers by two OFDM symbols.
- Up-link PUSC and TUSC Tile Usage of Sub-Carrier: Each slot is 16 sub-carriers by three OFDM symbols.
- Band Adaptive Modulation and Coding (BAMC) : As we see in figure 1 each slot is 8, 16, or 24 sub-carriers by 6, 3, or 2 OFDM symbols.

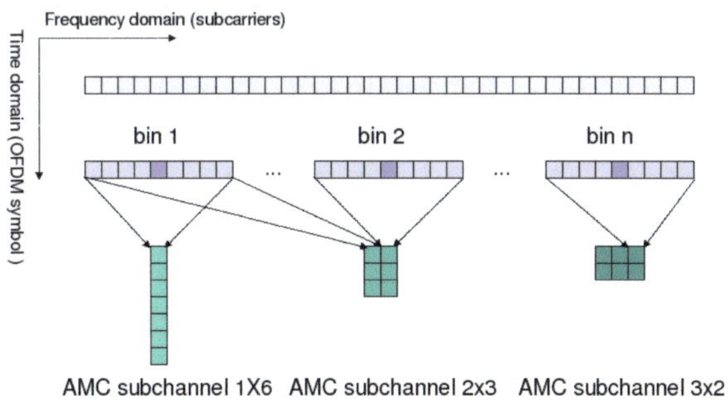

Fig. 1. BAMC slot format

In this chapter we will focus on the last permutation scheme i.e BAMC and we will explain how to compute the slot capacity.

2.1.3 The Adaptive Modulation and Coding scheme (AMC)

In order to adapt the transmission to the time varying channel conditions that depends on the radio link characteristics WiMAX presents the advantage of supporting the link adaptation called Adaptive Modulation and Coding scheme (AMC). It is an adaptive modification of the combination of modulation, channel coding types and coding rate also known as burst profile that takes place in the physical link depending on a new radio condition. The following table 1 shows examples of burst profiles in mobile WiMAX, among a total of 52 profiles defined in IEEE802.16e-2005 (IEEE Std 802.16e-2005, 2005): In fact when a subscriber station tries to

Profile	Modulation	Coding scheme	Rate
0	BPSK	(CC)	$\frac{1}{2}$
1	QPSK	$(RS+CC/CC)$	$\frac{1}{2}$
2	QPSK	$(RS+CC/CC)$	$\frac{3}{4}$
3	16 QAM	$(RS+CC/CC)$	$\frac{1}{2}$
6	64 QAM	$(RS+CC/CC)$	$\frac{3}{4}$

Table 1. Burst profile examples: (CC)Convolutional Code,(RS) Reed-Solomon

enter to the system, the WiMAX network undergoes various steps of signalization. First, the Down-link channel is scanned and synchronized. After the synchronization the SS obtains information about PHY and MAC parameters corresponding to the DL and UL transmission from control messages that follow the preamble of the DL frame. Based on this information negotiations are established between the SS and the BS about basic capabilities like maximum transmission power, FFT size, type of modulation, and sub-carrier permutation support.

In this negotiation the BS takes into account the time varying channel conditions by computing the signal to noise ratio (SNR) and then decides which burst profile must be used for the SS.

In fact, using the channel quality feedback indicator, the downlink SNR is provided by the mobile to the base station. For the uplink, the base station can estimate the channel quality, based on the received signal quality.

Based on these informations on signal quality, different modulation schemes will be employed in the same network in order to maximize throughput in a time-varying channel. Indeed,when the distance between the base station and the subscriber station increases the signal to the noise ratio decreases due to the path loss. Consequantely, modulation must be used depending on the station position starting from the lower efficiency modulation (for terminals near the BS) to the higher efficiency modulation (for terminals far away from the BS).

2.2 WiMAX MAC layer and QoS overview

The primary task of the WiMAX MAC layer is to provide an interface between the higher transport layers and the physical layer. The IEEE 802.16-2004 and IEEE 802.16e-2005 MAC design includes a convergence sublayer that can interface with a variety of higher-layer protocols, such as ATM,TDM Voice, Ethernet, IP, and any unknown future protocol.

Support for QoS is a fundamental part of the WiMAX MAC-layer design. QoS control is achieved by using a connection-oriented MAC architecture, where all downlink and uplink connections are controlled by the serving BS. Before any data transmission happens, the BS and the MS establish a unidirectional logical link, called a connection, between the two MAC-layer peers. Each connection is identified by a connection identifier (CID), which serves as a temporary address for data transmissions over the particular link. WiMAX also defines a concept of a service flow. A service flow is a unidirectional flow of packets with a particular set of QoS parameters and is identified by a service flow identifier (SFID). The QoS parameters could include traffic priority, maximum sustained traffic rate, maximum burst rate, minimum tolerable rate, scheduling type, ARQ type, maximum delay, tolerated jitter, service data unit type and size, bandwidth request mechanism to be used, transmission PDU formation rules, and so on. Service flows may be provisioned through a network management system or created dynamically through defined signaling mechanisms in the standard. The base station is responsible for issuing the SFID and mapping it to unique CIDs. In the following, we will present the service classes of mobile WiMAX characterized by these SFIDs.

2.2.1 WiMAX service classes

Mobile WiMAX is emerging as one of the most promising 4G technology. It has been developed keeping in view the stringent QoS requirements of multimedia applications. Indeed, the IEEE 802.16e 2005 standard defines five QoS scheduling services that should be treated appropriately by the base station MAC scheduler for data transport over a connection:

1. Unsolicited Grant Service (UGS) is dedicated to real-time services that generate CBR or CBR-like flows. A typical application would be Voice over IP, without silence suppression.
2. Real-Time Polling Service (rtPS) is designed to support real-time services that generate delay sensitive VBR flows, such as MPEG video or VoIP (with silence suppression).
3. Non-Real-Time Polling Service (nrtPS) is designed to support delay-tolerant data delivery with variable size packets, such as high bandwidth FTP.
4. Best Effort (BE) service is proposed to be used for all applications that do not require any QoS guarantees.

5. Extended Real-Time Polling Service (ErtPS) is expected to provide VoIP services with Voice Activation Detection (VAD).

Note that the standard defines 4 service classes for Fixed WiMAX: UGS, rtPS, nrtPS and BE.

In order to guarantee the QoS for these different service classes Call Admission Control (CAC) and resource reservation strategies are needed by the IEE 802.16e system.

2.2.2 QoS mechanisms in WiMAX

To satisfy the constraints of service classes, several QoS mechanisms should be used. Figure 2 shows the steps to be followed by the BS and SSs or MSSs to ensure a robust QoS management. To manage the QoS, we distinguish between the management in the UL and DL. For UL, at the

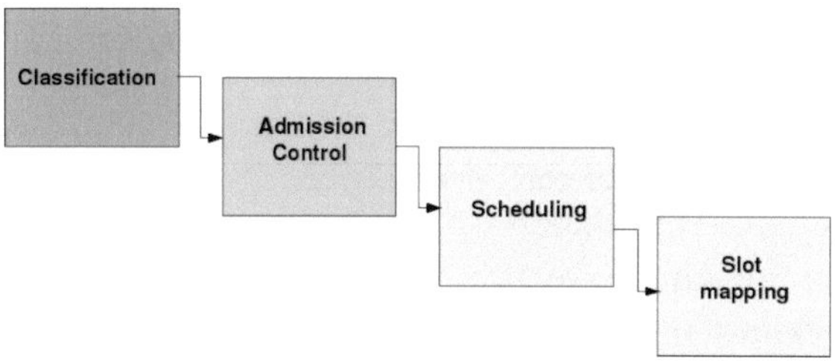

Fig. 2. QoS mechanisms

SS, the first step is the traffic classification that classifies the flow into several classes, followed by the bandwidth request step, which depends on service flow characteristics. Then the base station scheduler can place the packets in BS files, depending on the constraints of their services, which are indicated in the CID (Connexion IDentifier). The bandwidth allocation is based on requests that are sent by the SSs. The BS generates UL MAP messages to indicate whether it accepts or not to allocate the bandwidth required by the SSs. Then, the SS or MSS processes the UL MAP messages and sends the data according to these messages.

For the downlink, the base station gets the traffic, classifies it following the CID and generates the DL MAP messages in which it outlines the DCD messages that determine the burst profiles

The following section will describe each step. It should be noted that the standard does not define in detail each mechanism. But it is necessary to understand some methods that are used to satisfy the QoS for each mechanism.

1. **The classification** The classifier matches the MSDU to a particular connection characterized by an CID in order to transmit it. This is called CID mapping that corresponds to the mapping of fields in the MSDU (for example mapping the couple composed of the destination IP address and the TOS field) in the CID and the SFID.

   The mapping process associates an MSDU to a connection and creates an association between this connection and service flow characteristics. It is used to facilitate the transmission of MSDU within the QoS constraints.

   Thus, the packets processed by the classifier are classed into the diffrent WiMAX service classes and have the correspondant CID. The standard didn't define precisely the classification mechanism and many works in the literature have been developed in order to define the mapping in QoS cross layer framework. Once classified the connection requests are admitted or rejected following the call admission control mechanism decision.

2. **Call admission control (CAC) and Bandwidth Allocation** As in cellular networks, the IEEE 802.16 Base Station MAC layer is in charge to regulate and control bandwidth allocation. Therefore, incorporating a Call Admission Control (CAC) agent becomes the primary method to allocate network resources in such a way that the QoS user constraints could be satisfied. Before any connection establishment, each SS informs the BS about its QoS requirements. And the BS CAC agent have the responsability to determine whether a connection request can be accepted or should be rejected. The rejection of request happens if its QoS requirements cannot be satisfied or if its acceptance may violate the QoS guarantee of ongoing calls.

   To well manage the operation of this step, the WiMAX standard provides tools and mechanisms for bandwidth allocation and request that is described briefly as follows:

   (a) **Bandwidth request** At the entrance to the network, each SS or MSS is allocated up to 3 dedicated CID identifiers. These CIDs are used to send and receive control messages. Among these messages one can distinguish Up-link Channel Descriptor, Downlink Channel Descriptor, UL-MAP and DL-MAP messages, plus messages concerning the bandwidth request. The latter can be sent by the SS following one of these modes:
      - Implicit Requests: This mode corresponds to UGS traffic which requires a fixed bit rate and does not require any negotiation.
      - Bandwidth request message: This message type uses headers named *BW request*. It reaches a length of 32 KB per request by CID.
      - Piggybacked request: is integrated into useful messages and is used for all service classes, except for UGS.
      - Request by the bit *Poll-Me*: is used by the SS to request bandwidth for non-UGS services.

   (b) **Bandwidth Allocation modes**

      There are two modes of bandwidth allocation:
      - **The Grant Per Subscriber Station (GPSS):** In this mode, the BS guarantes the aggregated bandwidth per SS. Then the SS allocates the required bandwidth for each connection that it carries. This allocation must be performed by a scheduling algorithm. This method has the advantage of having multiple users by SS and therefore requires less overhead. However, it is more complex to implement because it requires sophisticated SSs that support a hierarchical distributed scheduler.
      - **The Grant Per Connection (GPC):** In this type of allocation the BS guarantes the bandwidth per connection, which is identified thanks to the individual CID (Connection IDentifier). This method has the advantage of being simpler to design than the GPSS mode but is adapted for a small number of users per SS and provides more overhead than the first mode.

      Thus, based on SS and MSS requests the base station can satisfy the other QoS application constraints by employing different allocation bandwidth strategies and call admission control policies. Recall that the latters have not been defined in the standard.

3. **Scheduling** In WiMAX, the scheduling mechanism consists of determinating the information element (IE) sent in the UL MAP message that indicates the amount of the allocated bandwidth, the allocated slots etc... A simplified diagram of the scheduler in the standard IEEE 802.16 is illustrated in the following figure:

   The scheduler in the WiMAX has been defined only for UGS traffic. Precisely for this class, the BS determines the IEs UL MAP message by allocating a fixed number of time slots in

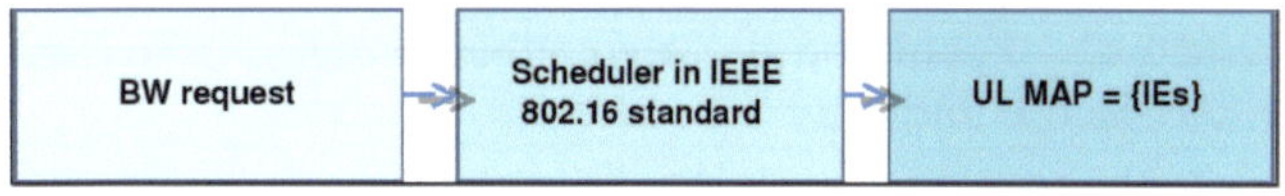

Fig. 3. Scheduler in IEEE 802.16 standard

each frame interval. The BS must take into account the state of queues associated to traffic and all queues among the SS, corresponding to UL traffic. For the remaining traffic classes the standard does not specify a particular scheduling algorithm, and left the choice to the operator to implement one of the algorithm that was described in the literature (Jianfeng C. et al., 2005) (Wongthavarawat K. et al, 2003).

4. **The mapping**
   This is the final step before sending user data in the radio channel. The idea is to assign sub-carriers in the most efficient possible way to scheduled MPDUs in order to satisfy QoS constraints of each connection. The mapping mechanism is left to the choice of the provider.

3. State of the art

3.1 Bandwidth sharing strategies: background

To maintain a quality of service required by the constraining and restricting services, there are different strategies of bandwidth allocation and admission control. Many bandwidth allocation policies have been developed in order to give for different classes a certain amount of resource. Among the classical strategies, one can citeComplete Sharing (CS), Upper Limit (UL), Complete Partitioning (CP), Guaranteed Minimum (GM) and Trunk Reservation (TR) policies. These policies are illustrated in figure 4 and will be introduced in the following sections. To this end, and in a seek of simplicity of the presentation, we will suppose in these sections that system defines only two service classes 1 and 2 (instead of the 5 classes defined in Mobile WiMAX). Moreover, we will also suppose that if a system accepts a call of class $i \in \{1,2\}$ it will allocate to this call a fixed amount of bandwidth denoted by d_i. Finally, let n_i denotes the number of class $i \in \{1,2\}$ calls in the system.

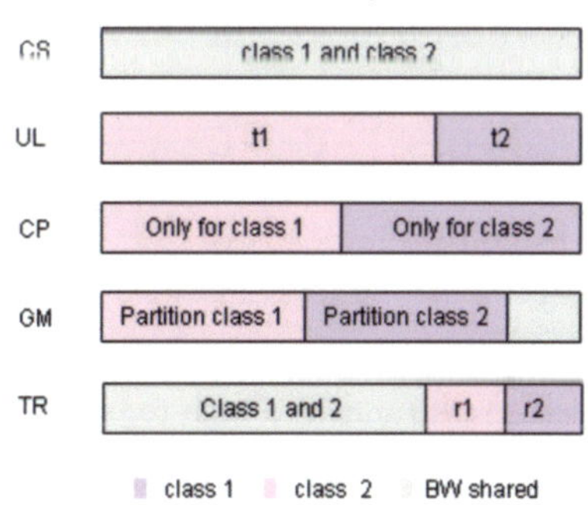

Fig. 4. Heuristic CAC policies

3.1.1 Complete Sharing (CS)

In this strategy, the bandwidth is fully shared among the different service classes. That is all classes are in competition. In other words, if we consider an offered capacity system equal to C and 2 types of service class (class 1 and 2). If class 1 (i.e. aggreagted calls) uses I units then the remained bandwidth $C - I$ could be allocated either to class 1 or to class 2. Formally, a call of class $i \in \{1,2\}$ is accepted if and only if:

$$d_i + \sum_{k=1}^{2} n_k d_k \leq C \tag{1}$$

3.1.2 Upper Limit (UL)

This policy is very similar to CS except that it aims to eliminate the case where one class can dominate the use of the resource, through the use of thresholds-based bandwidth occupation strategy. Precisely, thresholds t_1 and t_2 are associated to class1 and class 2, respectively. These thresholds represent the maximum numbers of bandwdith units that each class can occupy at a given time. So, a call of class $i \in \{1,2\}$ is accepted if and only if:

$$(1 + n_i)d_i \leq t_i \; and \; \sum_{k=1}^{2} n_k d_k \leq C \tag{2}$$

Note that this relation is not excluded :

$$\sum_{k=1}^{2} t_k > C$$

3.1.3 Complete Partitioning (CP)

This policy allocates a set of resources for every service class. These resources can only be used by that class. To this end the bandwidth is divided into partitions. Each partition is reserved to an associated service class. In this figure the capacity is divided into 2 partitions denoted by C_1 for class 1 and C_2 for class 2. Then, a call of class $i \in \{1,2\}$ is accepted if and only if:

$$(1 + n_i)d_i \leq C_i \tag{3}$$

Note that contrarily to the UL strategy the following relation must always be verified:

$$\sum_{k=1}^{2} C_k = C$$

3.1.4 Guaranteed Minimum (GM)

As illustrated in figure 4 the resource is divided into different partition. The policy gives each classes their associated partition of bandwidth, which we note M_1 for class 1 and M_2 for class 2. If this partition is fully occupied, each class can then use the remaining resource partition that is shared by all other classes. This is clearly an hybrid strategy between CP and CS. Formally, the CAC rule to follow in order to accept a call of class $i \in \{1,2\}$ is:

$$\sum_{k=1}^{2} max(d_k(n_k + 1_i(k)), M_k) \leq C, \; where \; 1_i(k) = 1 \; if \; k = i, 0 \; otherwise \tag{4}$$

Note that the following relation must always be verified:

$$\sum_{k=1}^{2} M_k \leq C.$$

3.1.5 Trunk Reservation (TR)

As illustrated in figure 4, there are not dedicated partitions per classes in this policy. In fact, class $i \in \{1,2\}$ may use resources in a system as long as the amount of remaining resources is equal to a certain threshold $r_i \in \{1,2\}$ bandwidth units. Thus each service class will protected thank to thresholds, which will avoid that any class occupies the totality of resource units. So a call of class $i \in \{1,2\}$ is accepted if and only if:

$$d_i + \sum_{k=1}^{2} n_k d_k \leq C - r_i \tag{5}$$

This rule guarantees that after applying this CAC policy and accepting the class i the remaining bandwidth is equal to r_i. Several comparison have been made between these policies and with optimal solution. One important challenge is to explain the method that thresholds imposed by GM, UL and CP strategies are computed or determined which is explained in (Khemiri S. et al., 2007).

So the main challenge is to setup these policy in an optimized way. This is could be done by choosing the optimal partition sizes or reservation thresholds in order to 1) guarantee the QoS constraints of the application provided by the system and in the other words to satisfy subscribers and 2) to provide a good system performance which satisfies the provider.

3.2 Scheduling and mapping in the literature

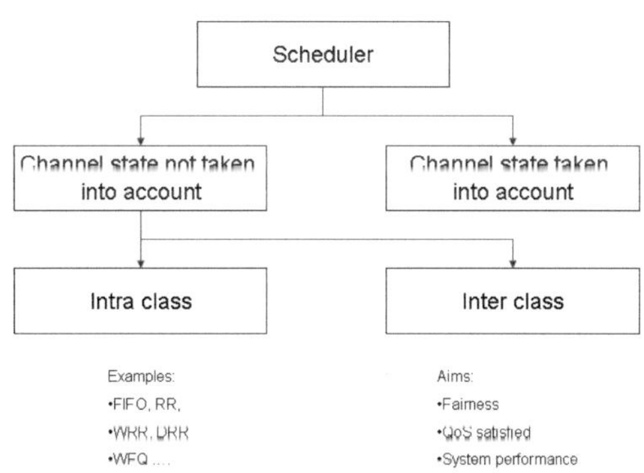

Fig. 5. Scheduler classification

In literature few studies have focused on both the scheduling and the selection of MPDUs and choice of OFDMA slots to be allocated (called mapping) to send the data in the frame.

Regarding scheduling, we can distinguish, as shown in Figure 5, two types of schedulers: a) the non-opportunistic schedulers are those who do not take into account the state of the channel we cite the best known, the RRs that ensure fairness and WRRs based on fixed weights and b) the opportunistic schedulers are those that take into account the channel state (Ball et al., 2005)(Rath H.K. et al., 2006)(Mukul, R et al.)(Qingwen Liu and Xin Wang and Giannakis,

G.B. et al.)(Mohammud Z. et al., 2010) an example is the MAXSNR which first selects the MSSs that have the maximum SIR. In (Ball et al., 2005), the authors present an algorithm called TRS that removes from queues MSSs with the SNR that is below a certain threshold. Further works (Rath H.K. et al., 2006) (Laias E. et al., 2008) improve conventional schedulers like DRR to make opportunistic one and this by introducing the SNRs threshold as a criterion for selecting MSSs to serve. Others are based on the prediction of the packets arrival like in (Mukul, R et al.).

Regarding the mapping, in (Einhaus, M. et al 2006), the authors propose an algorithm that uses a combined dynamic selection of sub-channels and their modulation with a power transmission allocation in an OFDMA packets but this proposal does not take into account the constraints of QoS packets. (Einhaus, M. et al) made a performance comparison between multiple resource allocation strategies based on fairness of transmission capacity in a multi-user scenario of a mobile WiMAX network that supports an OFDMA access technology. These compared policies are the MAXSNR, the maximum waiting time and the Round Robin strategies. The performance metrics analyzed are the delay and the rate. The evaluation was conducted using a WiMAX simulator based on OFDMA mechanism developed in NS2 simulator. The results presented indicate the significant impact of these policies on the tradeoff between rate and delay. Indeed, this work shows that a strategy based on taking into account to the radio channel conditions gives a better performance in term of capacity utilization than that of the delay. Thus the slot allocation strategies aiming to minimize the delay has resulted in reducing the efficiency of resource use. However, this work does not address the specifics in terms of QoS traffic and didn't provide any service differentiation between classes UGS, rtPS, and nrtPS Ertps. This work was improved in (Khemiri S. et al., 2010) by applying this strategy to a mobile WiMAX network: authors compared it to MAXSNR well known as a conventional mapping techniques. The results showed an improvement of a channel utilization.

In (Akaiwa, Y. et al 1993) and (katzela I. et al, 1996) Channel segregation performance has been examined by applying it to FDMA systems. This paper discusses its application to the multi-carrier TDMA system. Spectrum efficiency of the TDMA/FDMA cellular system deteriorates due to the problem of inaccessible channel: a call can be blocked in a cell even when there are idle channels because of the restriction on simultaneous use of different carrier frequencies in the cell. This solution shows that channel segregation can resolve this problem with a small modification of its algorithm. The performance of the system with channel segregation on the call blocking probability versus traffic density is analyzed with computer simulation experiments. The effect of losing the TDMA frame synchronization between cells on the performance is also discussed.

In (Wong et al., 2004) Orthogonal Frequency Division Multiple Access (OFDMA) base stations allow multiple users to transmit simultaneously on different subcarriers during the same symbol period. This paper considers base station allocation of subcarriers and power to each user to maximize the sum of user data rates, subject to constraints on total power, bit error rate, and proportionality among user data rates.

These works did not consider the double problem of MPDUs selection for transmission and the channel assignment technique.

4. Slot capacity

As we seen before, the PHY layer provides different parameter stettings which leads to interoperability problems. This is why *WiMAX forum* creates the WiMAX profiles which describes a set of parameters of an operational WiMAX system. These sets of parameters concerns: The System Bandwidth, the system frequency and the duplexing scheme. This section gives a computational method of slot capacity based on two WiMAX system profiles: 1) The Fixed WiMAX system profile and 2) The mobile WiMAX system profile.

This slot capacity, computed in term of bits, depends on permutation type and parameters which depends on the radio mobile environment like burst profile and defined by the SINR (Chahed T. et al, 2009) (Chahed T. et al, 2009). To compute this capacity its is needed to know system parameters, so we distinguish:

1. The OFDM slot capacity compute in case of Fixed WiMAX profile system.
2. The OFDMA slot capacity compute in case of Mobile WiMAX profile system.

The following table describes the parameters of each system profile:

Parameters	definition	Fixed	Mobile
B	System Bandwidth	3.5 MHz	10 MHZ
L_{FFT}	Subcarrier number or FFT size	256	1024
L_d	Data subcarrier number	192	720
G	Guard time	12.5%	12.5%
n_f	Oversampling rate	$8/7$	$28/25$
$(DL:UL)$	Duplexing rate	$3:1$	$3:1$
(c, M)	Modulation and coding scheme $c = coding\ rate$ $M = Constellation\ of\ the\ modulation$	depending on channel	depending on channel
TTG and RTG	transition Gap between UL and DL	$188\mu s$	$134.29\mu s$
T	Frame length ms	5 ms	5 ms
N	Number of user	N	N
Perm	Permutation mode	-	BAMC 1X6

Table 2. Mobile and fixed WiMAX system parameters

4.1 Fixed WiMAX case

Lets consider an SS n and one subcarrier f, we can determine the corresponding $SINR_{n,f}$ and then the modulation and coding scheme $(c_{n,f}, M_{n,f})$. One subcarrier can transmit the following number of bits (Wong et al., 2004) (Chung S. et al, 2000):

$$b_{n,f} = c_{n,f} log_2 \left(M_{n,f} \right) \tag{6}$$

An OFDM slot, denoted by s, is composed by L_d data subcarriers. The channel state of a user n described by $SINR_{n,s}$ can be deduced by computing the mean SINR of all data subcarriers. Once this SINR is determined we can deduce the MCS (c_n, M_n) and we can compute the SINR as follows:

$$SINR_{n,s} = \frac{1}{L_d} \sum_{f=1}^{L_d} SINR_{n,f} \tag{7}$$

So the number of bits that can transmit the minimum time-frequency resource or a the OFDM slot is defined as follows:

$$b_n = c_n \log_2 (M_n) L_d \tag{8}$$

Where $\frac{(1+G)L_{FFT}}{n_f B}$ corresponds to time duration of the OFDM symbol of L_{FFT} length, so the rate in bps provided by an OFDM frame for a modulation and coding scheme (c, M) is given by:

$$C = c \log_2 (M) L_d \frac{n_f B}{(L_{FFT}(1+G))} \tag{9}$$

In addition, the total number of OFDM symbols per frame is computed as follows:

$$nb_s = T \frac{n_f B}{(1+G)L_{FFT}} \tag{10}$$

We deduce the number of symbols dedicated to the UL noted nb_{UL} and the DL noted nb_{DL} using the ratio $(DL : UL)$:

$$nb_{DL} = \frac{D}{D+U} nb_s \tag{11}$$

$$nb_{UL} = \frac{U}{D+U} nb_s \tag{12}$$

The DL throughput is given by the following formula:

$$C_{DL} = \frac{C T_{useful} \frac{1}{T}}{nb_s} nb_{DL} \tag{13}$$

where $T_{useful} = T - (TTG + RTG)$ is the usable size of the frame by removing periods reserved for the UL and DL transmission gap and $\frac{1}{T}$ is the number of frames sent per second. The total number of OFDM slots in a mobile WiMAX frame corresponds to $S \times T$ where $S = L_d$ is the number of data subcarriers and $T_s = nb_s$ is the number of OFDM symbol in the frame, we obtain a frame with the format $((S = 192) \times (T_s = 69))$ OFDM slots.

4.2 Mobile WiMAX case

In mobile WiMAX, the slot format depends on the permutation scheme supported by the system. In the rest of this chapter, we chose to take an interest in the permutation BAMC 1×6. This choice is not limiting, but for reasons of clarity and simplification of the presentation.

Considering the permutation BAMC 1×6, the format of the OFDMA slot is 8 data subcarriers of 6 OFDM symbols. The total number of OFDMA slots in a mobile WiMAX frame corresponds to $S \times T_s$ where $S = \frac{L_d}{8}$ and T_s is the number of OFDM symbol in the frame which is equal to $T_s = \frac{nb_s}{6}$. So we get a frame whose size is $((S = 90) \times (T_s = 6))$ OFDMA slots.

To determine the capacity of this slot $s \in [1, S]$, it suffices to determine the burst profile $(c_{n,s}, M_{n,s})$ of OFDMA slot s for user n. To do this, simply determine the $SINR_{n,s}$ corresponding to:

$$SINR_{n,s} = \frac{1}{48} \sum_{f=1}^{8} \sum_{t \in 1}^{6} SINR_{n,f}(t) \tag{14}$$

Thus the number of bits provided by the OFDMA slot s is given by the following equation:

$$b_{n,s} = 6 * 8c_{n,s} \log_2 (M_{n,s}) \tag{15}$$

Finally, using the parameter presented in table 2 and the equations above we obtain the following table. It should be noted that the flow rates presented are calculated for the modulation and coding scheme $(64 - QAM, \frac{3}{4})$

Parameters	definition	Fixed	Mobile
(SxT_s)	Frame size (Total slot number)	(192×69)	(90×6)
C_{DL}	DL frame rate (Mbps)	8.51117	23.0905
C_{UL}	UL frame rate (Mbps)	2.83706	7.69682
C	Total frame rate (Mbps)	11.348	30.787
$b_{n,s}$	**Number of bit per slot (bits)**	869	219

Table 3. Mobile and Fixed WiMAX slot capacity

In the rest of this chapter we focus on the slot allocation problem combined with scheduling mechanism in mobile WiMAX OFDMA system which consists of how to assign PHY resource to a user in order to satisfy a QoS request in MAC layer.

5. Case study: System description and problem statement

5.1 System description

In this case study let's consider a WiMAX cell based on IEEE 802.16e 2005 technology supporting Wireless MAN OFDMA physical layer. The system offers a quadruple-play service to multiple mobile subscribers (MSS). These subscriber stations can have access anytime and anywhere to various application types like file downloading, video streaming, emails and VoIP. In this model let's suppose a typical downlink WiMAX OFDMA system and we consider that the system parameters corresponds to those of a mobile WiMAX profile, which is characterized by the second column of the table 3.

Recall that the minimum time-frequency resource that can be allocated by a WiMAX system to a given link is called a slot. Each slot consists of one sub-channel over one, two, or three OFDM symbols, depending on the particular sub-channelization scheme used. So a slot is an n x m rectangle, where n is a number of sub-channel in the frequency domain and m is a number of symbols in the time domain. The standard supports multiple *subchannelization* schemes (PUSC, BAMC, FUSC, TUSC, etc.), which define how an OFDMA slot is mapped over subcarriers. As we see in figure 6, the system frame is a matrix whose size is $((S = 90) * (T_s = 6))$ OFDMA slots, where S is the number of subchannels and T_s is the number of OFDMA symbols. So we can allocate up to $90 * 6 = 540$ OFDMA slots to a user n. Only the DL case will be studied. In order to model this system the physical and MAC layer characteristics will be presented in following.

5.1.1 QoS constraints

In order to guarantee the quality of service required by these applications, the service provider has to distinguish five service classes. Namely: UGS for VoIP, rtPS for video streaming, nrtPS for file downloading and ErtPS for voice without silence suppression. As BE for emails is not

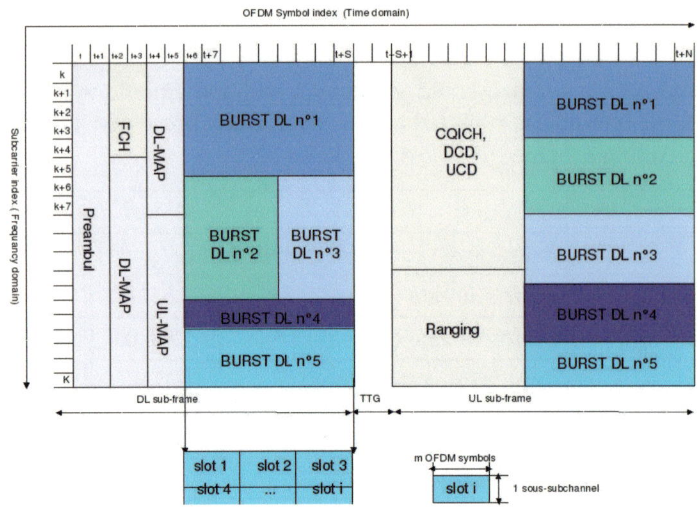

Fig. 6. OFDMA frame

constringent in term of QoS it will not be considered here. For notation simplicity, we will refer to UGS, rtPS, nrtPS and ErtPS as a class 1, 2, 3 and 4, respectively. Let $U = \{1,2,3,4\}$. To satisfy application QoS constraints provided by the system, we assume that there is a classifier implemented in the BS that associates each traffic users to a class $i \in U$ and we also suppose that there is a call admission control mechanism that ensures that the newly admitted calls do not degrade the QoS of the ongoing calls, and there is enough available system resources for the accepted call and if not the call is rejected. We suppose that to satisfy the QoS of each user n supporting a traffic class i, it suffices to have:

$$C_n \in [\underline{s}_i, \overline{s}_i], \forall i \in U \tag{16}$$

Where $\underline{s}_i$ and $\overline{s}_i$, are respectively the minimum and maximum class i data rate. Since we consider a mobile radio environment this system capacity vary with channel condition. This is why a scheduling mechanism must be used in order to select which MPDUs must be transmitted in addition to a physical resource assignment strategy in order to select the best slot (physical resource) that satisfies the QoS constraints of the selected MPDUs.

5.1.2 Cell division for AMC

In order to adapt the transmission to the time varying channel conditions that depend on the radio link characteristics WiMAX presents the advantage of supporting the link adaptation called adaptive modulation coding (AMC). AMC consist of an adaptive modification of the combination of modulation, channel coding types and coding rate also known as burst profile, that takes place in the physical link depending on a new radio condition.

The following table 4 shows examples of burst profiles in mobile WiMAX there are 52 in IEEE802.16e-2005 (Jeffrey G. et al., 2007)(IEEE Std 802.16e-2005, 2005):

Profile	Modulation L	Coding scheme	Rate
3	16 QAM	$(RS + CC/CC)$	$\frac{1}{2}$
5	64 QAM	$(RS + CC/CC)$	$\frac{1}{2}$
6	64 QAM	$(RS + CC/CC)$	$\frac{3}{4}$

Table 4. Burst profiles: (RS) Reed Solomon, (CC) Convolutional Code

We will demonstrate in this section that we can divide the WiMAX cell into several areas where each of them corresponds to one modulation scheme.

Lets consider our system as a WiMAX base station with a total bandwidth B operating at a frequency f. The BS and SS antenna height in meters is respectively given by h_{BS} and h_{SS}. The SS has a transmission power P_{SS}. If we model our system in presence of path loss defined by the COST-231 Hata radio propagation model (Jeffrey G. et al., 2007) (Roshni S. et al., 2007), we can deduce a variation of the SNR while varying the distance d between SSs and BS (Chadi T. et al., 2007) (Chadi T. et al., 2007)(Chadi T. et al., 2007). This model is chosen because it is recommended by the WiMAX Forum for mobility applications in urban areas which is the case of our system.

In order to know the variation of the SNR with distance, the path loss for the urban system environment is needed. According to the COST-231 Hata model, the pathloss is given by:

$$P_{loss}\left[dB\right] = 46.3 + 33.9log_{10}\left(f\right) - 13.82log_{10}\left(h_{BS}\right) + \\ (44.9 - 6.55log_{10}\left(h_{BS}\right))log_{10}\left(d\right) - F_a\left(h_{SS}\right) + C_F \quad (17)$$

Where P_{loss} is the path loss, and $F_a\left(h_{SS}\right)$ is the station antenna correction factor, C_F is a correction factor.

$$F_a\left(h_{SS}\right) = (1.11log_{10}\left(f\right) - 0.7)h_{SS} - (1.56log_{10}\left(f\right) - 0.8) \quad (18)$$

For illustration lets consider an example of a WiMAX system with total bandwidth $B = 20MHz$, operating at a frequency $f = 2Ghz$, with an SS transmission power $P_{SS} = 10Watt = 10dBm$, $h_{BS} = 30m$, $h_{SS} = 1m.d = 0$ to 20 Km,$C_F = 3dB$. The path loss is defined as:

$$P_{loss}\left[dB\right] = 41.17 + 35.26log_{10}\left(d\right) \quad (19)$$

By considering the following link budget :

$$SNR = P_{SS} - \left[P_{loss} + N\right] \quad (20)$$

Where N is the thermal noise equal to : $N\left[dBm\right] = 10log\left(\tau TB\right)$ here $\tau = 1.38 \cdot 10^{-23} W/KHz$ is the Boltzmann constant and T is the temperature in Kelvin ($T = 290$) as defined in (Chadi T. et al., 2007) $N\left[dBm\right] = -100.97dBm$. we can deduce the SNR as follows:

$$SNR = P_{SS} + 59.8 - 35.26log_{10}\left(d\right) \quad (21)$$

Using Matlab tool the variation of the SNR while varying the distance between SSs and BS from 0 to 20 Km is given by the figure 7 This figure shows that we can distinguish areas corresponding to the modulation region. We assume that our system supports only

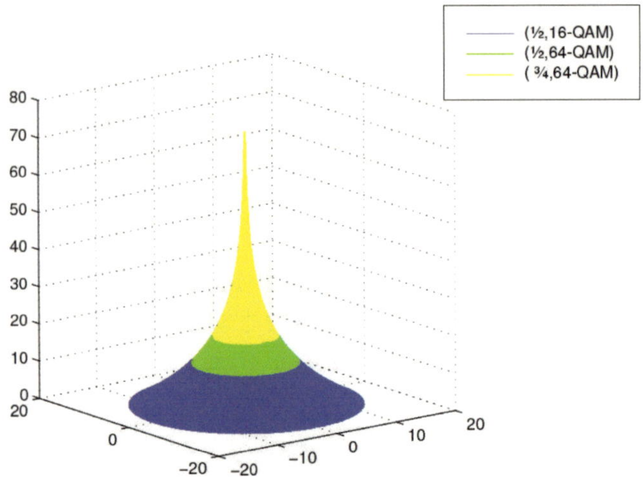

Fig. 7. SNR variation versus distance BS-SS

3 modulation schemes, so following SNR thresholds described in table 4 we obtain three modulation regions.

We assume that the cell's bandwidth is totally partitioned, so that each partition is adapted to a specific modulation scheme. According to the adaptive modulation and coding scheme, we can divide this cell into 3 uniform areas in which we suppose that only one modulation scheme is used. As figure 8 shows we choose 3 modulation and coding schemes as following:

1. $(\frac{1}{2}, 16QAM)$ corresponds to the SNR interval $I_1 = [0, 11.2[$ dB.
2. $(\frac{1}{2}, 64QAM)$ corresponds to the SNR interval $I_2 = [11.2, 22.7]$ dB.
3. $(\frac{3}{4}, 64QAM)$ corresponds to the SNR interval $I_3 =]22.7, +\infty[$ dB.

Note that the $(\frac{3}{4}, 64QAM)$ modulation (burst profile number 6) is used in the nearest area of the BS, then $(\frac{1}{2}, 64QAM)$ modulation (burst profile number 5) in the second area, finally $(\frac{1}{2}, 16QAM)$ (burst profile number 3) is employed in the third area.

Fig. 8. The system partition areas

Thus at the BS transmitter, the station must select for each user $n \in [1, N]$ the MCS for each selected slot $s \in [1, S]$ using the signal to noise level $SNR_{n,s}$.

In figure 8, we designed three zones illustrated by three concentric perfect circles corresponding to the three types of modulation. It is just an example, because this obviously does not square with reality since the channel undergoes disturbances other than the path Loss that vary the channel between two stations even they are at the same distance from the BS.

5.1.3 Mobility

In order to be close to a realistic WiMAX network, we take into account some assumptions. We assume that N users are MSSs whose trajectory is a perfect concentric circle with radius $n \in [1, N]$ km. The velocity of the MSS n corresponds to $V_n = n * V$ where n is the user index and V is a velocity expressed by m/s. Each signal will be transmitted through a slowly time-varying, frequency-selective Rayleigh channel with a bandwidth B. Each OFDMA slot s allocated to a user n will be sent with a power denoted by $p_{n,s}$. We will discuss here the choice of this power.

In this case study, let's consider that we allocate a fixed power $p_{k,s} = \frac{P}{S}$ for each subcarrier since we didn't focus on a power allocation problem. We assume that each user experiences an independent fading and the channel gain of user k in subcarrier s is denoted as $g_{k,s}$ We can easily deduce that the nth user's received signal-to-noise ratio (SNR) for the slot s which corresponds to the average signal to noise ratios of all sub-carriers that form this slot, is written as follows:

$$SNR_{n,s} = p_{n,s} \frac{g_{n,s}^2}{\sigma^2} \tag{22}$$

Where, $\sigma^2 = N_0 \frac{B}{L_{FFT}}$ and N_0 is power spectrum density of the Additive white Gaussian noise (AWGN). The slowly time-varying assumption is crucial since it is also assumed that each user is able to estimate the channel perfectly and these estimates are made known to the transmitter via a dedicated feedback channel. Specifically, the SNR will be sent periodically (once per frame) in control messages. Then they are used as input to the resource allocation algorithms. We suppose that the channel condition didn't change during the frame duration, i.e 5 ms.

5.2 Parameters and problem statement

As we consider a mobile WiMAX system supporting Adaptive Modulation and Coding we can deduce from (Wong et al., 2004) and (Chung S. et al, 2000) the OFDMA slot capacity denoted by $b_{n,s}$ corresponding to the number of bits that a given subcarrier s can transmit if we know channel condition for a given user n, so we have:

$$b_{n,s} = 48 c_{n,s} \log_2 (M_{n,s}) \tag{23}$$

Where $(c_{n,s}, M_{n,s})$ is the modulation and coding scheme of a slot s allocated to the MSS n defined as follows: $(c_{n,s}, M_{n,s})$=$(\frac{1}{2}, 16QAM)$ if $SNR_{n,s} \in I_1$, $(c_{n,s}, M_{n,s})$=$(\frac{1}{2}, 64QAM)$ if $SNR_{n,s} \in I_2$ and $(c_{n,s}, M_{n,s})$=$(\frac{3}{4}, 64QAM)$ if $SNR_{n,s} \in I_3$. As we see in 6 the OFDMA frame is a matrix with dimension $S \times T_S$. Let's have an allocation matrix of a nth user denoted by A_n, this matrix is expressed as following:

$$A_n = \left[a_{s,t}^n\right]_{(s,t) \in \{1,S\} \times \{1,T_s\}} \tag{24}$$

Where, $a_{s,t}^n = \mathbb{1}_{\{\mathbb{1}_{(s,t)}=n\}}$, ie, $a_{s,t}^n = 1$ if and only if $\mathbb{1}_{(s,t)}(i,j) = n$, 0 otherwise. By using equations 23 and 24, we can deduce the total capacity B_n which corresponds to the total bit number provided to the user n after a slot allocation following the allocation matrix A_n:

$$B_n = \sum_{s=1}^{S} \sum_{t=1}^{T_s} a_{s,t}^n b_{n,s} \tag{25}$$

The total system capacity if the call admission controll mechanism accept N MSSs is:

$$C = \sum_{n=1}^{N} C_n = \frac{n_f B}{(1+G)L_{FFT}} \sum_{n=1}^{N} \sum_{s=1}^{S} \sum_{t=1}^{T_s} a_{s,t}^n c_{n,s} \log_2 (M_{n,s}) \tag{26}$$

It is clear that the choice of the matrix allocation is crucial for the optimal use of resources. The aim of this case study is to present an efficient cross-layer resource assignment strategy that takes into account two aspects: 1)the varying channel condition and 2) the QoS constraints of user's MPDUs scheduled to be transmitted into the physical frame.

Problems related to resource allocation and power assignment aim to solve the following mutli-constraints optimization problem (Wong et al., 2004) (Cheong et al., 1999):

Problem 1 Slot allocation problem

maximize: $\max_{p_{n,s}, a_{t,s}} C$

subject to:

$$C1: \sum_{n=1}^{N} \sum_{s=1}^{S} \sum_{t=1}^{T_s} a_{t,s} p_{n,s} \leq P_{total}$$

$$C2: C_n \in [\underline{s}_i, \overline{s}_i], \quad \forall i \in U$$

$$C3: p_{n,s} \geq 0, \quad \forall (n,s) \in [1,N]X[1,S]$$

$$C4: a_{t,s} \in 0,1, \quad \forall (s,t) \in [1,S]X[1,T_s]$$

Where C1 corresponds to the power constraint, C2 corresponds to the QoS constraint discribed in 16 , C3 and C4 ensure the correct values for the power and the subcarrier allocation matrix element, respectively.

This problem is NP-hard problem (Mathias et al, 2007) and was often treated by taking into account only the physical layer without respecting constraints related to quality of service. Generally, this problem is split into two subproblems: subproblem (1) consists on power assignment problem, where only the power will be considered as the variable of the problem, and subproblem (2) consists on maximizing the instantaneous system capacity C once the power is allocated. In our case study, we will not consider power allocation issues and we will assume that all subcarriers have the same transmit power, i.e, $p_{n,s} = p\forall(n,s) \in [1,N]X[1,S]$. The SNR variation is only related to the channel variation. So our problem statement is the following, if we consider the OFDMA frame is like a puzzle game with slots as game pieces, where the game rule is that these slots must be allocated to each MSSs according to their demand. The difficulty of this game is that of the slot capacity is variable and depends on the channel state. In the next we answer the two questions: Which MPDUs to serve? and which slot to assign to satisfy the bandwidth request of the selected MPDUs? In the next section, we propose solutions to both questions.

6. Solutions

In order to answer to questions asked in the previous section, one solution is to combine scheduling mechanism with a slots mapping while taking into account three aspects: 1) The QoS constraints of each traffic class, 2) the specific features of the system like Permutation scheme and 3) OFDMA access technology and the radio channel variation which results in the choice of modulation and therefore the variation of the allocated slots capacity.

To treat this problem five steps, as described in figure 9, are needed: step 1 for call admission control, step 2 for scheduling, step 3 for user selection, step 4 for the selection of the traffic granularity and step 5 for slots selection.

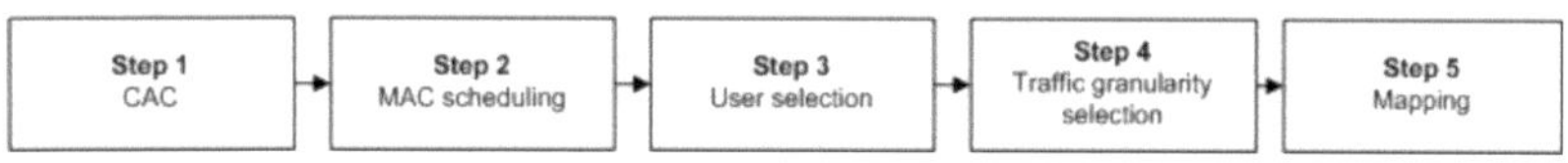

Fig. 9. The 5 steps solution

The main objective of these steps is to find a compromise between QoS constraints of service classes and the bandwidth utilization. We will describe in the following all these steps and we will present several proposals for step three, four and five.

6.1 Step 1: Call admission controll

One solution is to use a CAC block presented in (Khemiri S. et al., 2008) based on Complete Partitioning (CP) between service classes and we assume that all connections accepted in the system are the result of applying this CAC strategy. We also suppose that at the MAC layer all MPDUs of the traffic transported by the MSSs are fragmented so that a single frame can carry the largest MPDU in the traffic.

6.2 Step 2: Scheduling

Before presenting step 3, 4 and 5, it is important to choose the scheduler that guarantee the QoS constraints of applications provided to subscribers at the MAC layer. Several works have been proposed to efficiently schedule traffic in WiMAX (Jianfeng C. et al., 2005) (Wongthavarawat K. et al, 2003), one solution is to use a hybrid two-stage scheduler presented in figure 10.

Here the idea is to use two Round Robin (RR) schedulers in a first stair to provide fair distribution of bandwidth especially between ErTPS, UGS and rtPS classes since they are real time traffic. In the second stair we propose to use a Priority queuing scheduler in order to give a high priority for VoIP applications and real time traffic and a lower priority for video streaming and web browsing applications.

As we see in figure 10, we use two types of scheduler:

- **Priority Queuing (PQ)**: In this scheduler, each queue has a priority. A queue can be served only if all higher priority queues are empty.
- **Weighted Round Robin (WRR)**: In this discipline, each queue has a weight which defines the maximum number of packets that can be served during each scheduler round.

This hybrid scheduler handles differently real time and non real time traffic: In the first stage, each traffic class is associated to a queue. The classifier stores the packets in the queue that corresponds to the appropriate packet service class. Queues associated with real time flows

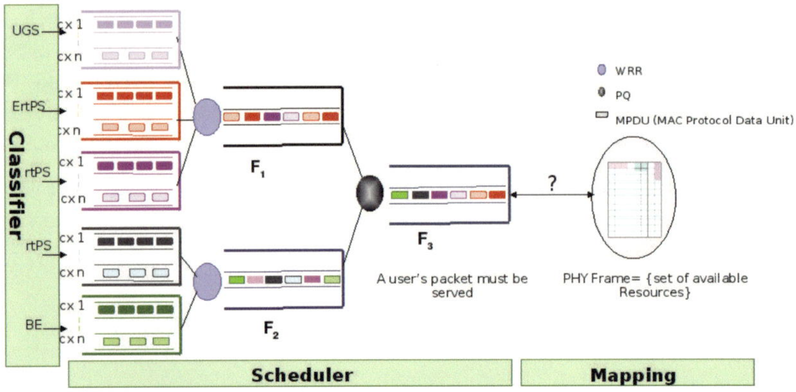

Fig. 10. DL hybrid scheduling block

(UGS, rtPS and ErtPS) are managed by the WRR scheduler and queues corresponding to non real time flows (nrtPS and BE) are managed by the same WRR discipline. This stage guarantees a fixed bandwidth for UGS and ErtPS classes and a minimum bandwidth for rtPS while ensuring fairness between flows because the rtPS packets have variable size and this flow could monopolize the server if the traffic is composed by packets with larger size than those of Class 1 and 2.

In the second stage, output of the two WRR schedulers are enqueued in two queues F1 and F2, packets of these queues are managed by a priority PQ scheduler which gives higher priority to real time stream (stored in F1) which are more constringent in term of throughput and delay than the non-real time traffic (stored in F2) which are less time sensitive.

Once scheduled the MPDUs are placed in a FIFO queue of infinite size. The next step is to choose the users and therefore MPDUs that must be served in this queue, it is also necessary to determine how much MPDUs will be served and what are the slots allocated to them?

6.3 Step 3: The users selection

We consider that for each source that transmitting a traffic class i a system have to allocate an $\underline{s_i}$ minimum required bandwidth to satisfy its QoS constraints. If we consider that this source has traffic with k service classes to send, the BS has to allocate a minimum required bandwidth denoted by $\underline{S_n}$ for each user n to satisfy its QoS constraints. If we assume that this user carries traffic with the five service classes $i \in U$, so this bandwidth $\underline{S_n}$ corresponds to:

$$\underline{S_n} = \sum_{i=1}^{5} \underline{s_i} \tag{27}$$

Where $\underline{s_i}$ is the required bandwidth to satisfy QoS constraints of class i. Note that these parameters varies periodical in time. Without loss of generality let's suppose that each user has only one type of traffic class to receive. So either it should be noted $\underline{S_n} = \underline{s_i}$. let's consider that for every user n in the system we can obtain the cumulative rate $S_n = \underline{s_i}$ which corresponds to the number of bits per seconds that the system has to allocate to this user. As before the mapping, all traffic are processed by a described scheduling mechanism, a weight ϕ_i that corresponds to the priority of a class i is assigned to each traffic class. Let's denote by

Q_i the following satisfaction parameter:

$$Q_i = \phi_i \frac{s_i}{\underline{s}_i} \quad (28)$$

This parameter will serve to select users that are not satisfied in order to serve them first. The user satisfaction is defined as follows: All users that verifying the condition $s_i \leq \underline{s_i}$, that we call QoS satisfaction condition (QSC), are called not satisfied users . To determine what user to choose, the algorithm selects the user that is least satisfied i.e the one that checks the least satisfaction condition QSC and thus satisfies the equation 29:

$$n = \arg\min_{u \in N} Q_u \quad (29)$$

If there are many that corresponds to the minimum several solutions are used: one solution is to choose randomly one of them or the user that request the maximum of bandwidth $(\underline{s}_(i))$ or the user that corresponds to the maximum of the value $\left(s_i - \underline{s_i}\right)$ otherwise select the user that it has the prior service class $(UGS > ErtPS > rtPS > nrtPS > BE)$.

In what follows, for simplicity the first option is used.

6.4 Step 4: The selection of the traffic granularity

Once the user is selected to be served, the next step is to know how much user MPDUs it will be served? Three solutions to choose the amount of MPDUs to be served are presented as follows:

1. All user MPDUs: All MPDUs belonging to the selected user that are in the queue will be served. The disadvantage is that a user could monopolize physical resources. We denote this method a **TP** strategy for Total user packets.
2. MPDUs by MPDUs: In this proposal, we process only one MPDUs by selected user. Once slots are allocated to it, we move to the next user. This avoids the disadvantage of the first proposal. We denote this method **PP** for Packet Per Packet.
3. Only the number of bits needed is treated in order to reduce the user delay: In this case, each user has a credit we will denote $Credit_n(t)$ which corresponds to the amount of bandwidth allocated until time t, (t is a multiple of the duration of the frame $(t = xT,\ T = Frame\ duration)$). This credit will be updated whenever the system allocates one or more slots by adding the amount of bits provided by each allocated slot. At time t, to guarantee the QoS constraints of the user n that receiving a traffic class i, the user will be allocated at least $B_n = x\underline{s}_i$. B_n is the number of bits that should be served to ensure the user's request. We can then define the delay or retard as follows:

   $$Retard_n(t) = B_n - Credit_n(t) \quad (30)$$

   Two cases arise:

   - If $Retard_n(t) > 0$, i.e what we need to allocate to the user, is more than what we have allowed him, in this case the user is in retard and we must serve more than the $Retard_n(t)$ to retrieve the user n retard .
   - If $Retard_n(t) \leq 0$, in this case the user is not in retard and we serve only one MPDU of this user.

Lets note this strategy as **RR** for Retrieve Retard.

6.5 Step 5: Slots selection

The last step is the selection of slots to be allocated to MPDUs to be served by system. Two solutions are presented in this section:

1. Iterative solution: It is an instinctive idea. The BS allocates randomly the available slots in order to satisfy the selected user request in term of bits. We can call this solution as a FIFO strategy since the first user selected will be the first served.
2. MAXSNR solution: The basic idea is to select with a selfish behavior, so the BS choose the best slots in term of SNR for selected users and didn't care if the set of the allocated slots could be the best for other users. To determine if a slot is better or not, we proceed as follows: When we allocate a slot s to a given user n, that corresponds in term of bits to $b_{n,s}$. This parameter is easily deduced from the SNR of the allocated slot s to the user n and expressed by equation 23. Lets denote by $F_{n,s} = \frac{b_{n,s}}{b_n^{max}}$ the factor which indicates if a given slot s is the best one to be allocated to the user n. Here $b_n^{max} = \max_{l \in S_n} \left[b_{n,l} \right]$, where S_n is the set of free slots to be allocated to user n. More this factor is close to 1 more the slot is better.

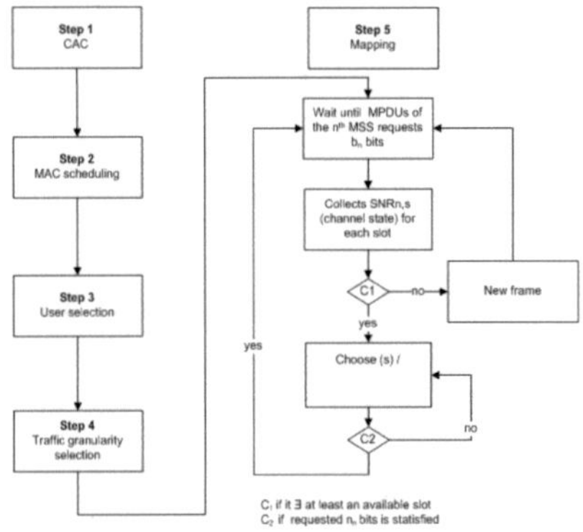

Fig. 11. Slot selection

7. Evaluation and discussion

7.1 Simulation parameters

This solution can be evaluated by using the following tools:

1. Opnet (Laias E. et al., 2008), (Shivkumar et al, 2000): This simulator is used to generate the traffic carried by the MSS and to implement the two stages of the scheduler block in step 2 9 that we described below.
2. Matlab: This mathematical tool is used to generate the MSSs signal at the physical layer and introduce the channel perturbation due to mobility and signal attenuation.

We then implement the steps 3, 4 and 5 of proposed block 9, using the programming language C++. These tools interact according to the following:

To evaluate the performance of the methods described above, we define three types of flows. Each flow models a service class: UGS, rtPS and nrTPS. This choice is justified by the fact

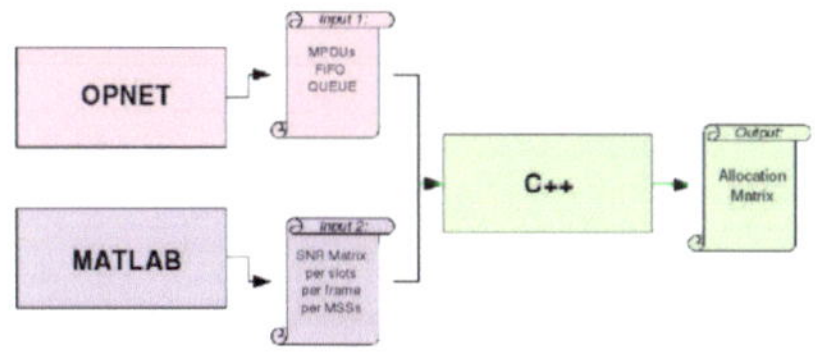

Fig. 12. Simulation tools

that classes UGS and ErtPS have same behavior and that the BE is a traffic which has no significant influence on the capacity as the BS allocate the rest of the remaining bandwidth. To characterize these streams, we set two parameters: the MPDUs size and the packet inter-arrival time. The following table shows the parameters used for the studied traffic :

Class	Application	Mean rate (Kbps)	Arrival time (s)	Distribution and packet size(bits)
UGS	VoIP(G711)	64	Constant: 0.02	Constant: 1280
rtPS	Video streaming (25 pictures/s)	$3.5\ 10^3$	Constant: 2.287510^{-4}	Geometric:mean=12.510^{-4}
nrtPS	FTP	$3.5\ 10^3$	Constant: 2.287510^{-4}	Geometric: mean=12.510^{-4}

Table 5. Traffic parameters

Note that we could easily introduce the packet loss due to the physical channel perturbation and assume that all the slots with $SNR_{n,s} \in I_0 = [0, 6.4[dB$ are considered as lost and no data will be sent in these slots. In fact, $6.4dB$ corresponds to the sensitivity threshold of all MSSs receiving antennas, and therefore below this threshold, the received data will not be noticeable by these antennas. However, as we do not introduce retransmission mechanisms, we assume that the BS affects the least efficient modulation in terms of spectral efficiency to the user whose SNR is in I_0 which corresponds to MCS $(\frac{1}{2}, QPSK)$.

The topology of the simulated network consists of a BS with system capacity equal to 7.4 Mbps which serves for the first scenario 3 MSSs with 3 traffics classes UGS, rtPS and nrtPS and for the second scenario 6 MSSs where 2 MSSs receives UGS traffic, 2 other receives rtPS traffic and the rest receives nrtPS traffic.

These SS are randomly distributed around the BS, and they turn around a BS. The mobile SS velocity vary from 0.1 to 20 m/s and the trajectory is a perfect circle with radius varying from 1m to 2 km. The duration time of our simulation is 20s.We choose system parameters corresponding to the mobile WiMAX profile, with 10 MHz bandwidth and an FFT size of 1024. The mobile WiMAX frame with 5ms duration provides 69*4 units of physical resource or OFDMA slots. The base station provides the following applications to MSS: We apply a slowly time-varying, frequency-selective Rayleigh channel that we described in 5.1.3. Each MSS n moves with velocity $V_n = n * V$ where n is the user index and $V = 10m/s$. Thus the MSS $n = 6$ will move with speed $V_6 = 60m/s = 216Km/h$ and the MSS $n = 1$ will move with a velocity $V_1 = 36Km/h$.

We then varied the SNR channel for only one MSS and we kept the SNR fixed and equal to 11 dB, then we varried the channel for all MSSs, we studied a total of 5 scenarios which we summarized in the following table:

The channel variation is given by the figure 13 which corresponds to Cumulative Distribution Function CDF of the modulation schemes.

We then apply the different methods of choosing the granularity of traffic TP, RR and PP to which we added the FIFO method which corresponding to serve MPDUs as they arrive in

scenario: 6 MSSs						
Channel state	UGS(1)	UGS(2)	rtPS(1)	rtPS(2)	nrtPS(1)	nrtPS(2)
1	F	F	F	F	F	F
2	P	P	P	P	P	P
3	P	F	F	F	F	F
4	F	F	P	F	F	F
5	F	F	F	F	P	F

Table 6. Studied scenarios, F: SNR fixed 11 dB, V: SNR varied, (1): MSS_1, (2): MSS_2

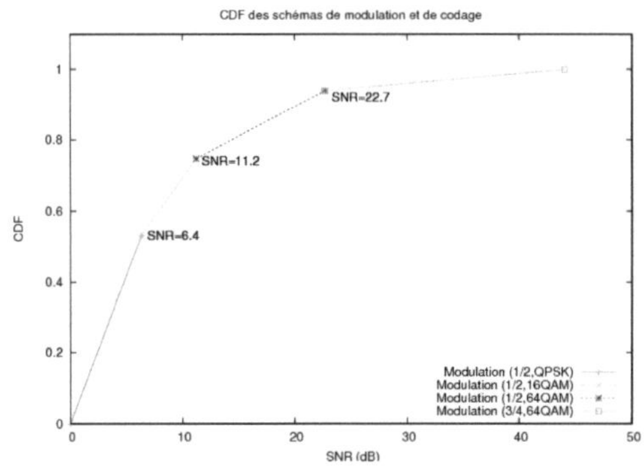

Fig. 13. Modulation scheme distribution (CDF) when the channel is varrying

the queue. We have combined these methods with the ITERATIV and MAXSNR mapping solutions explained above.

The simulation duration is 10s which is equivalent to 2000 frames sent and 5 hours time machine and we chose the following weights $\phi_i = 1$ for UGS class, $\phi_i = 2$ for rtPS class and $\phi_i = 3$ for nrtPS class. Simulation results are presented in the next section.

7.2 Performance parameters

In this evaluation we focused on several evaluation parameters such as the average data rate of each MSS, the average delay of each service class, the utilization ratio and packet loss. In what follows we give the results for the second scenario with 6 MSSs, the first scenario with 3 MSSs shows the same results. To facilitate understanding of our analysis and results we follow the following notations:

1. State F: all users channel SNR are set to 11dB.
2. State P: all users channel SNR are perturbed.
3. State UGS-P: only users receiving UGS traffic have a perturbed channel.
4. State rtPS-P: only users receiving rtPS traffic have a perturbed channel.
5. State nrtPS-P: only users receiving nrtPS traffic have a perturbed channel.

The first parameter that we evaluate is the utilization ratio which corresponds to the ratio between the average number of slots used and the total number of slots $(90 * 6 = 540)$. This ratio is expressed with the following equation:

$$U = \frac{E[\sum_{n=1}^{N} \sum_{s=1}^{S} \sum_{t=1}^{T_s} a_{s,t}^{n}]}{540} \tag{31}$$

We are also interested in the average delay per class i per user expressed as follows:

$$D_i = E[T_{s,i} - T_{g,i}] \tag{32}$$

Where $T_{s,i}$ is the service time and $T_{g,i}$ is the MPDUs generation time for class i. Finally, it is also important to estimate the MPDUs loss which corresponds to those that they could not be served on time, this loss is expressed as the mean number of lost packets per user per frame, denoted $Loss_i(t)$. We assume that a UGS or rtPS packet is lost only if it waits longer than 40 ms in the queue before to be served.

$$Loss_i(t) = \frac{\sum_{d_i >= 40} n_{MPDUS,d_i}(t)}{2000} \tag{33}$$

$n_{MPDUS,d_i}(t)$ is the number of MPDUs of class i that should b served at time t and the waiting time is $d_i = T_{s,i} - T_{g,i}$.

7.3 Analysis

As we have several combinations of channel perturbations and mapping and user selection strategies in 5 blocks we obtain about sixty curves. Here are results that we obtained for the performance parameters that we described before: For the utilization ratio in figure 14 we have a heavy traffic load, between 96% and 100%. The required average rate of all classes are

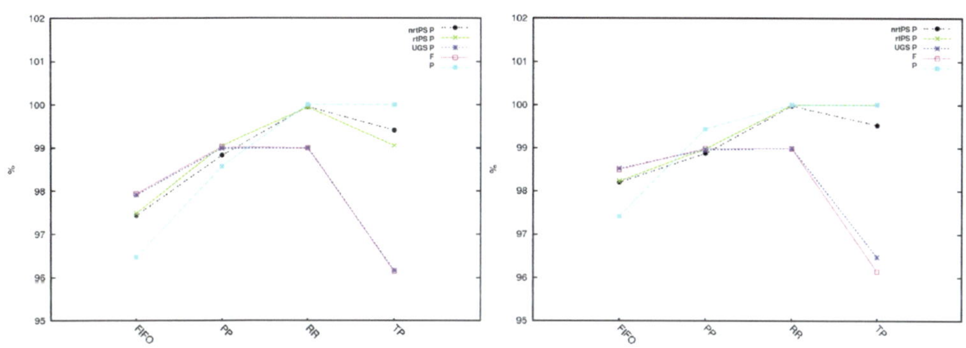

fig(a) MAXSNR fig(b) ITERATIVE

Fig. 14. Frame average utilization ratio

satisfied with all strategies, TP ensures exactly the requested rate without bandwidth waste and therefore it optimizes the use of the system capacity, an example for rtPS is given in figure 15.

As we see in figure 16 TP strategy shows also a best performance regarding delays since there is no delay for rtPS which is a real time constringent application. We observed loss for the rtPS traffic for FIFO, RR and PP strategies and we can deduce that MAXSNR mapping solution is better than the ITERATIVE one. The block user selection is efficient since in its absence (ie when we use FIFO method), rtPS delay is greater than 40 ms which is equivalent to rtPS packet loss. As a conclusion the combination that it is recommended is to use TP as a selection traffic granularity method with MAXSNR as a mapping slot strategy after processing traffic by our proposed hybrid scheduling block.

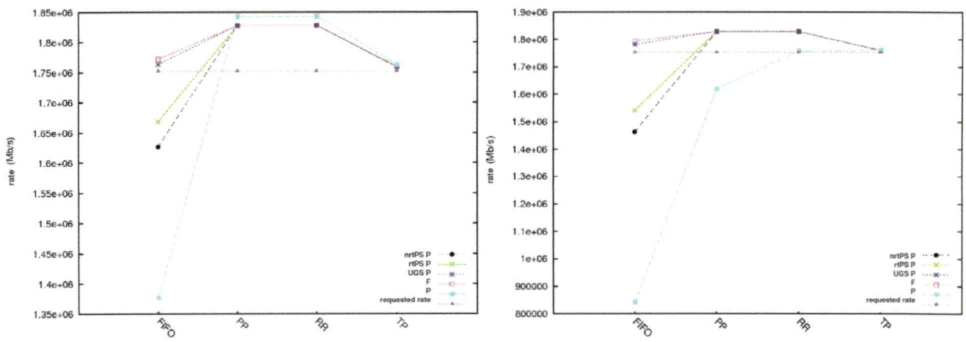

fig(a) MAXSNR fig(b) ITERATIVE

Fig. 15. rtPS average rate

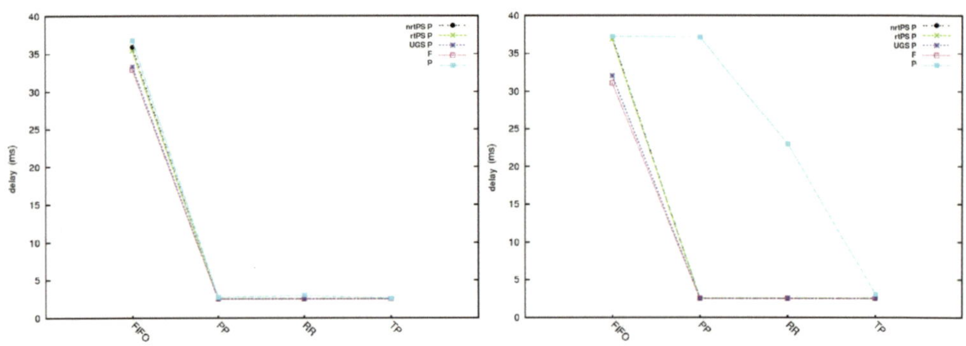

fig(a) (MAXSNR) fig(b) (ITERATIVE)

Fig. 16. rtPS average delay

8. Conclusion

This chapter presents one of the fundamental requirements of next generation OFDMA based wireless mobile communication systems which consist on the cross-layer scheduling and resource allocation mechanism.

The purpose of the first part of the chapter was to give an overview of QoS mechanisms in WiMAX systems and to explain the optimization problems related with these features. The rest of this chapter presents case study in order to analyse and discuss several solution developed to guarantee QoS management of a mobile WiMAX system.

Nevertheless, the growth of network access technologies in the mobile environment has raised several new issues due to the interference between the available accesses. This is why the novel resource allocation solution must integer a new concepts like SON (Self-Organizing network) features in a framework of general policy management. The next generation wireless communications standard (i.e., IEEE 802.16e/m, 3GPP-LTE and LTE-Advanced ...) has to include smart QoS management systems in order to obtain an optimal ubiquitous operating system any time and any where.

9. References

Akaiwa, Y. & Andoh, H. (1993). Channel segregation-a self-organized dynamic channel allocationmethod: application to TDMA/FDMA microcellular system. *Name of Journal in Italics*, Vol.11, No.6, (august 1993) page numbers (949 - 954)

Ball, C.F. Treml, F. Gaube, X. & Klein, A. (1996). Performance analysis of temporary removal scheduling applied to mobile WiMax scenarios in tight frequency reuse, *Proceedings of Personal, Indoor and Mobile Radio Communications*, pp. 888-894, Germany, september 2005, Berlin

Chadi T. & Tijani C. (2001). AMC-Aware QoS Proposal for OFDMA-Based IEEE802.16 WiMAX Systems. *proceeding of GLOBECOM*,page numbers (4780-4784), ISSN

Chadi T. & Tijani C. (2001). Modeling of streaming and elastic flow integration in OFDMA-based IEEE802.16 WiMAX. *Computer Communications*

Chadi T. & Tijani C. (2001). On capacity of OFDMA-based IEEE802.16 WiMAX including Adaptive Modulation and Coding (AMC) and inter-cell interference. *LANMAN'2007, Princeton NJ*

Chahed T., Cottatelluci L., Elazouzi R., Gault S. & He G.. (2009). *Information theoretic capacity of WIMAX, Radio resources management in WiMAX: from theoretical capacity to system simulations*, Wiley-Iste/Hermes Science Publisher.

Chahed T., Chammakhi Msaada I., Elazouzi R., Filali F.& Elayoubi S-E. (2009). *WiMAX network capacity and radio resource management, Radio resources management in WiMAX: from theoretical capacity to system simulations*, Wiley-Iste/Hermes Science Publisher.

Cheong Yui W., Roger S. C., Khaled L. & Ross M. (1999). Multiuser OFDM with Adaptive Subcarrier, Bit, and Power Allocation. *IEEE Journal on Selected Areas of Communications*, Vol.17, No.1, (month and year of the edition) page numbers (1747-1758)

Chung S. & Goldsmith A. (2000). Degrees of freedom in adaptive modulation: A unified view. *IEEE Trans. Commun*,(January 2000)

Einhaus, M. & Klein, O. (2006). Performance Evaluation of a Basic OFDMA Scheduling Algorithm for Packet Data Transmissions, *Proceedings of the IEEE Symposium on Computers and Communications*, pp. 7, 0-7695-2588-1, Cagliari, Italy, 2006

Einhaus, M. and Klein, O. & Walke, B. (2008). Comparison of OFDMA Resource Scheduling Strategies with Fair Allocation of Capacity. *Proceedings of 5th IEEE Consumer Communications & Networking Conference 2008*, pp. 5, 1-4244-1457-1, Las Vegas, USA, january and 2008, Las Vegas, USA

IEEE802.16eStd05. (2005). *IEEE Standard 802.16e-2005, Amendment to IEEE Standard for Local and Metropolitan Area Networks - Part 16: Air Interface for Fixed Broadband Wireless Access Systems- Physical and Medium Access Control Layers for Combined Fixed and Mobile Operation in Licensed Bands*

Jeffrey G. et al (2007). *Fundamentals of WiMAX Understanding Broadband Wireless Networking*, Prentice Hall Communications Engineering and Emerging Technologies Series

Jianfeng C. et al (2005). An Integrated QoS Control Architecture for IEEE 802.16 Broadband Wireless Access Systems

katzela I. & Naghshineh M. (1996). Channel assignement schemes for cellular mobile telecommunication systems: a comprehensive survey. *IEEE personal communications*,

S. Khemiri and K. Boussetta and N. Achir & G. Pujolle (2007). Optimal Call Admission Control for an IEEE 802.16 Wireless Metropolitan Area Network, *International Conference on Network Control and Optimization (NETCOOP*, pp. 105-114, Avignon, France, June, 2007, Lectures Notes in Computer Sciences, Springer-Verlag

S. Khemiri, K. Boussetta and N. Achir & G. Pujolle (2007). Bandwidth partitioning for mobile WiMAX service provisioning, *Proceedings of IFFIP Wireless Days*, pp. 20-26, Dubai, November 2008, IEEExplorer

S. Khemiri and K. Boussetta and N. Achir & G. Pujolle (2007). A combined MAC and Physical resource allocation mechanism in IEEE 802.16e networks, *Proceedings of VTC'2010*, pp. 6-10, 15502252, Taipei, may 2010, IEEExplorer

Laias E., Awan I. & Chan P.M. (2008). An Integrated Uplink Scheduler in IEEE 802.16, *Proceedings of Second UKSIM European Symposium on Computer Modeling and Simulation, 2008. EMS '08*, pp. 518-523, sep 2008

Mathias Bohge and James Gross and Adam Wolisz and Tkn Group and Tu Berlin and Michael Meyer and Ericsson Gmbh (2007). Dynamic Resource Allocation in OFDM Systems: An Overview of Cross-Layer Optimization Principles and Techniques. *IEEE Network*, Vol.21, (2007) page numbers (53-59)

Mohammud Z. ,P. Coon, C. Nishan , Joseph P.McGeehan, Simon M. D. Armour & Angela Doufexi(2010). Joint Call Admission Control and Resource Allocation for H.264 SVC Transmission Over OFDMA Networks, *Proceedings of VTC'2010*, pp. 1-5, 15502252, Taipei, may 2010, IEEExplorer

Mukul, R. and Singh, P. and Jayaram, D. and Das, D. and Sreenivasulu, N. and Vinay, K. & Ramamoorthy, A. (2006). An adaptive bandwidth request mechanism for QoS enhancement in WiMax real time communication, *Proceedings of Wireless and Optical Communications Networks, 2006 IFIP International Conference*, pp. 1-5, , Bangalore, April 2006

Qingwen L., Xin W. & Giannakis G.B. (2006). A cross-layer scheduling algorithm with QoS support in wireless networks, *IEEE Transactions Vehicular Technology*, Vol.55, No.3, (2006) page numbers (839 -847), 0018-954

Rath H.K.,Bhorkar, A. & Vishal S. (2006). Title of conference paper, *Proceedings of Global Telecommunications Conference*, pp. 1-5, San Francisco, november 2006

Shivkumar K., Raj J., Sonia F., Rohit G. & Bobby V. (2000). The ERICA switch algorithm for ABR traffic management in ATM networks. *IEEE/ACM Trans. Netw.*, Vol.8, No.1, (2000) page numbers (87-98)

Roshni S., Shailender T., Alexei D. & Apostolos P. (2007). Downlink Spectral Efficiency of Mobile WiMAX, *Proceedings of VTC'2007*, pp. 2786-2790

Wongthavarawat, K. and Ganz A. (2003). Packet Scheduling for QoS Support in IEEE 802.16 Broadband Wireless Access Systems. *International Journal of Communication Systems*, vol.16, (February 2003) page numbers (81-96)

Wong, I.C. Zukang Shen Evans, B.L. & Andrews, J.G. (2004). A low complexity algorithm for proportional resource allocation in OFDMA systems, *Proceedings of IEEE Signal Processing Systems, 2004. SIPS 2004*, pp. 1-6, october 2004

Part 2

Quality of Service Models and Evaluation

8

A Unified Performance Model for Best-Effort Services in WiMAX Networks

Jianqing Liu[1], Sammy Chan[1] and Hai L. Vu[2]
[1]*City University of Hong Kong*
[2]*Swinburne University of Technology*
[1]*Hong Kong S.A.R.*
[2]*Australia*

1. Introduction

Based on the work from the IEEE Working Group 802.16 and ETSI HiperMAN Working Group, the WiMAX (Worldwide Interoperability for Microwave Access) technology is defined by the WiMAX Forum to support fixed and mobile broadband wireless access. In the standard (IEEE 802.16 standard, 2009), it defines several air interface variants, including WirelessMAN-SC, WirelessMAN-OFDM, WirelessMAN-OFDMA and WirelessMAN-HUMAN. WiMAX networks can be operated in two different modes: point to multi-point (PMP) mode and mesh mode. Under the PMP mode, all traffics from subscriber stations (SSs) are controlled by the base station. Mesh mode is a distributed architecture where traffics are allowed to route not only between SSs and the base station but also between SSs. In this chapter, we focus on the WirelessMAN-SC air interface operating in the PMP mode.

In WiMAX networks, quality of service (QoS) is provided through five different services classes in the MAC layer (Andrews et al., 2007):

1. Unsolicited grant service (UGS) is designed for real time applications with constant data rate. These applications always have stringent delay requirement, such as T1/E1.
2. Real-time polling service (rtPS) is designed for real-time applications with variable data rate. These applications have less stringent delay requirement, such as MPEG and VoIP without silence suppression.
3. Extended real-time polling service (ertPS) builds on the efficiency of both UGS and rtPS. It is designed for the applications with variable data rate such as VoIP with silence suppression.
4. Non-real-time polling service (nrtPS) is designed to support variable bit rate non-real-time applications with certain bandwidth guarantee, such as high bandwidth FTP.
5. Best effort service (BE) is designed for best effort applications such as HTTP.

To meet the requirements of different service classes, several bandwidth request mechanisms have been defined, namely, unsolicited granting, unicast polling, broadcast polling and piggybacking. In this chapter, we present a performance model for services, such as BE service, based on the broadcast polling mechanism which is contention based and requires the SSs to use the truncated binary exponential backoff (TBEB) algorithm (Kwak et al., 2005) to resolve contention. There is some previous research work on the contention free and contention based bandwidth request mechanisms. Delay analysis of contention free unicast polling request mechanism is proposed in (Iyengar et al., 2005). In (Vinel et al., 2005), average delay of random access with broadcast polling in saturation IEEE 802.16 networks is studied. An analytical model of contention based bandwidth request for IEEE 802.16 networks is proposed in (He et al., 2007), in which bandwidth efficiency and channel access delay are obtained. In (Vu et al., 2010), the throughput and delay performances of best-effort services in IEEE 802.16 networks is analysed. Both (He et al., 2007; Vu et al., 2010) consider the saturated case that each SS always has traffics to send. In (Ni & Hu, 2010), the authors propose a model for the unsaturated case of the request mechanisms in WiMAX. Fallah *et al.* propose a 2-dimensional Markov chain (MC) model to evaluate the average access delay and the capacity of the contention slots in delivering bandwidth request (Fallah et al., 2008). Fattah *et al.* extend (Fallah et al., 2008) to analyze the IEEE 802.16 networks with subchannelization (Fattah & Alnuweiri, 2009). Chuck *et al.* also use the 2-dimensional MC model to obtain the performance of bandwidth utilization and delay (Chuck et al., 2010). However, (Fallah et al., 2008; Fattah & Alnuweiri, 2009; Chuck et al., 2010) assume that the probability of an SS sending a request is an input parameter of their models, instead being a function of the backoff process. Moreover, all existing works only explicitly model mean packet delay, but not the complete distribution.

This chapter significantly extends our work in (Vu et al., 2010) by proposing a unified model for the performance of the best-effort service of WiMAX networks. This model can capture the performances of both unsaturated and saturated cases, and derives the expressions for network throughput and packet delay distribution, rather than just mean packet delay. Each SS will be modeled as a $M/G/1$ queueing system, where the bandwidth request arrival follows a Poisson process, and the service time is determined by the broadcast polling mechanism. Since our model explicitly models the broadcast polling mechanism, it provides a more accurate estimate of the service time of bandwidth request and packet delay than (Fallah et al., 2008; Fattah & Alnuweiri, 2009; Chuck et al., 2010). The validity of our model will be evaluated by extensive simulations. Our model can be used by operators to configure the parameter settings at the MAC layer for performance optimization.

The rest of this chapter is organized as follows. In Section 2, we first briefly introduce the contention based broadcast polling mechanism. Section 3 proposes fixed point equations to analyze the system. Section 4 derives the expressions of some performance measures. Section 5 verifies the analytical results by simulations. Section 6 degenerates the unsaturated model to saturated networks. Finally, Section 7 concludes the chapter.

2. Broadcast polling

We consider an IEEE 802.16 network consisting of N SSs operating in the PMP mode through WirelessMAN-SC air interface. The SSs access the network through the time division multiple access technology. The MAC frame structure defined in the IEEE 802.16 standard for TDD in PMP mode is shown in Fig. 1. Each frame has a duration of Δ and is divided into uplink and downlink subframes. At the beginning of a downlink subframe, which has a duration T_{DL}, there are two important messages called downlink map (DL-Map) and uplink map (UL-Map) messages. They specify the control information for the downlink and uplink subframes respectively. In the UL-Map, there is data or information element indicating whether there are transmission opportunities for bandwidth requests (REQs) and data packets. The uplink subframe is composed of bandwidth request bursts with duration T_{RE} and data bursts with duration T_{DA}, respectively. At frame i, when an SS has a data packet to send, it first sends a bandwidth request for transmitting its data in one of the transmission opportunities within the request interval of the uplink subframe. Upon receiving the bandwidth requests, the BS then allocates bandwidth and data slots for data transmission in the uplink data interval of frame $i+1$ based on its scheduler.

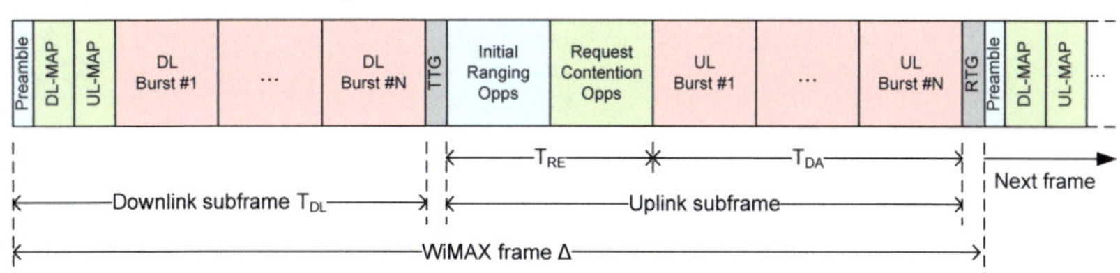

Fig. 1. IEEE 802.16 MAC frame structure with times division duplexing (TDD).

Let us consider a scenario where broadcast polling is used by the BS with m (fixed) transmission opportunities for bandwidth requests which are referred to as *request slots*. In this case, if there is only one request submitted to a request slot, the request is successful. On the other hand, if there are two or more SSs sending their requests in the same request slot, collision will happen and TBEB is used to solve this contention problem. Let W_i be the contention window for backoff state i, and each SS randomly selects a backoff time in the range $[0, W_i - 1]$. With TBEB, W_i is given by:

$$W_i = \begin{cases} 2^i W, & 0 \leq i \leq r, \\ 2^r W, & r < i < R, \end{cases}$$

where r is referred to as the truncation value, W is the initial contention window and R is the maximum allowable number of attempts. If the request still fails after R attempts, the packet will be discarded. Then if there are other packets queueing in the buffer, the packet at the head of the queue will send bandwidth request in the next frame.

In this chapter, the SSs are only allowed to request bandwidth to transmit one packet per request, and all packets are assumed to have the same length. Let t_{RE} be the length of a request (or backoff) slot. Furthermore, we assume that the BS always allocates the same amount of uplink capacity consisting of $d \leq m$ data slots in every uplink subframe for uplink traffic. Each data slot is of length T ($T \gg t_{RE}$) which is the transmission time of a packet. As the standard does not define scheduling algorithms for both BS and SSs, we assume here that the BS uplink scheduler will uniformly allocate bandwidth to SSs whose bandwidth request is successful in the previous frame. Let j be the number of requests that do not collide. If $j < d$ then in the next frame there will be $(d-j) > 0$ unused data slots, which are wasted. However, if $j > d$ then $(j-d) > 0$ requests must be declined because there are only d slots available in the next frame; those $(j-d)$ requests are also considered unsuccessful.

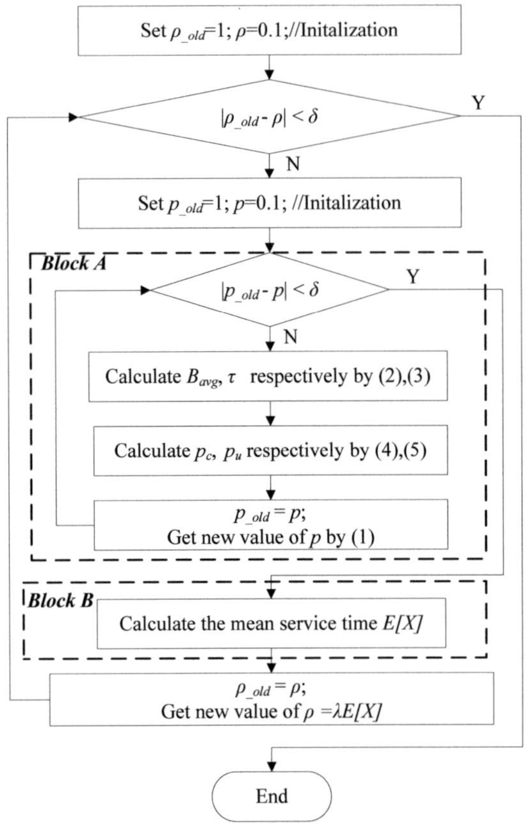

Fig. 2. An overview of the nested fixed point equations.

3. Fixed point equations

In this section, we will use the fixed-point method (Agarwal et al., 2001) to analyze the queueing behaviour at an SS. We assume that packets arrive at an SS according to a Poisson process with rate λ and each SS has an infinite buffer. An SS can therefore be modelled as a $M/G/1$ queueing system. We develop two sets of fixed point equations, one nested by the other, to calculate the failure probability p of an REQ and the offered load to the queue ρ, respectively. The relationship between these two sets of fixed point equations is illustrated by the flow chart shown in Fig. 2. The inner set, labelled as ***Block A***, calculates the p for a given ρ. The outer set includes one more block, labelled as ***Block B***, and calculates ρ which is relevant to the mean service time of an REQ.

3.1 Failure probability of an REQ

As in (He et al., 2007), a request is regarded as unsuccessful either when the request experiences collision during transmission (with probability p_c) or when the request is successfully transmitted but the BS could not allocate bandwidth to it due to insufficient data slots (with probability p_u). For simplicity, these two events are assumed to be independent. Then p can be expressed as

$$p = 1 - (1 - p_c)(1 - p_u). \tag{1}$$

Based on TBEB, we can derive the average number of backoff slots B_{avg} an SS has to wait before sending requests as

$$B_{avg} = \frac{m}{2} + \eta \sum_{i=0}^{r-1} p^i (\frac{2^i W - 1}{2}) + \eta (\frac{2^r W - 1}{2}) \sum_{i=r}^{R-1} p^i, \tag{2}$$

where $\eta = (1-p)(1-p^R)^{-1}$, and $(1-p^R)$ is a normalization factor.

Knowing B_{avg}, the probability that an SS attempts to send the requests in a slot can be written as

$$\tau = \rho / (B_{avg} + 1), \tag{3}$$

where $\rho = \lambda E[X]$ and $E[X]$ is the average REQ service time, which will be derived in next subsection.

Given that there are N SSs in the system, the probability p_c can be expressed as

$$p_c = 1 - (1-\tau)^{N-1}. \tag{4}$$

Let ξ be the probability that a collision-free request is made in a given slot, given that there are N SSs, each attempting to send requests with probability τ. Under the assumption that requests are independent, we have

$$\xi = N\tau(1-\tau)^{N-1}.$$

The probability that there are j collision-free requests among m request slots, $0 \le j \le n = min(m, N)$, is then given by a truncated binomial distribution

$$Q(j) = \frac{\binom{m}{j} \xi^j (1-\xi)^{m-j}}{\sum_{i=0}^{n} \binom{m}{i} \xi^i (1-\xi)^{m-i}}.$$

The probability that a collision-free request is unsuccessful due to lack of bandwidth in the subsequent frame can be expressed as

$$p_u = \frac{\sum_{j=d+1}^{n} (j-d) Q(j)}{\sum_{j=0}^{n} j Q(j)}. \tag{5}$$

Equations (1) to (5) form the inner set of fixed point formulations for p. As shown in ***Block A*** of Fig. 2, for a given ρ, p can be obtained by repeatedly solving these equations until p converges. The resultant p obtained is subsequently used in the outer set of fixed point equations evolving around the traffic load of an SS, ρ. In the following, we will develop the outer set of fixed point equations for ρ.

3.2 Mean service time of an REQ

This subsection presents the details of ***Block B*** of Fig. 2, which calculates the mean service time of REQs.

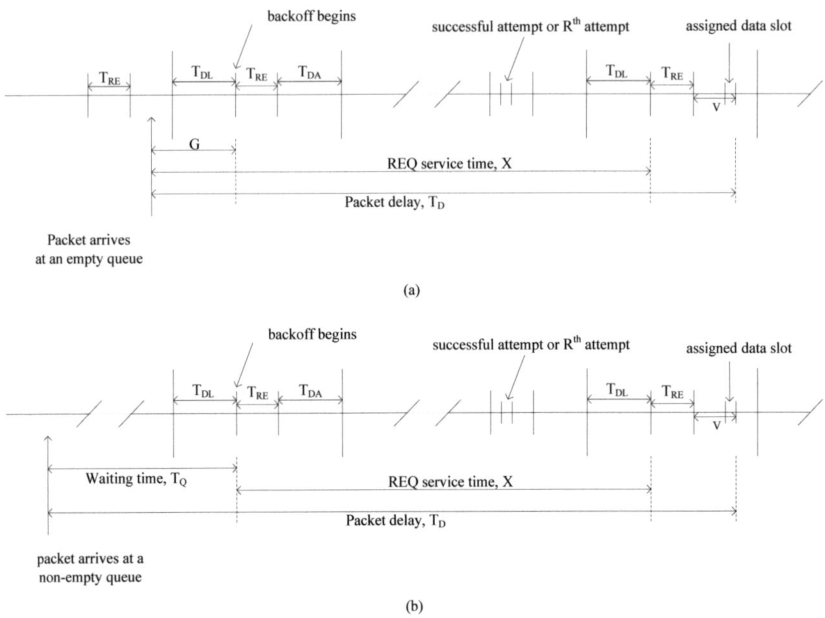

Fig. 3. The service time of an REQ when (a) its packet arrives at an empty queue, (b) its packet arrives at a non-empty queue

Referring to Fig. 3, the definition of REQ's service time depends on whether the queue is empty or not upon the arrival of a new packet at an SS. We specify below separately these two cases:

1. S0: The queue is empty (with probability $1 - \rho$, Fig. 3(a)). If a packet arrives at an empty queue, its REQ's service time will include the time period from its arrival until the start of the request interval where the backoff of the first attempt is initiated, and its backoff process from the beginning of the first request interval until the beginning of the request interval prior to which a successful request or the Rth request attempt is made.

2. S1: The queue is non-empty (with probability ρ, Fig. 3(b)). If a packet arrivals at a non-empty queue, it will be placed in the buffer until it becomes the head-of-the-line (HOL) packet. The REQ service time of this packet is defined as the time duration from the beginning of the request interval where the backoff of the first attempt is initiated until the beginning of the request interval prior to which a successful request or the Rth request attempt is made.

Consider case **S0**, let G be a random variable representing the time period from packet's arrival until the start of the request interval where the backoff of the first request for that packet is initiated. The cumulative distribution function of G is written as

$$F_G(g) = \begin{cases} \frac{e^{-\lambda\Delta}(e^{\lambda g}-1)}{1-e^{-\lambda\Delta}} & 0 \le g \le \Delta, \\ 1 & g \ge \Delta. \end{cases} \quad (6)$$

The probability density function (pdf) of G is written as

$$f_G(g) = \begin{cases} \frac{\lambda e^{\lambda g}}{e^{\lambda\Delta}-1} & 0 \le g \le \Delta, \\ 0 & g > \Delta. \end{cases}$$

Based on (6), the average of G can be obtained as

$$E[G] = \frac{\Delta}{1 - e^{-\lambda\Delta}} - \frac{1}{\lambda}, \tag{7}$$

where $E[\cdot]$ is the average operator. And, we can obtain the Laplace-Stieltjes transform of $f_G(g)$

$$\mathcal{L}_G(s) = \frac{\lambda(e^{-s\Delta} - e^{-\lambda\Delta})}{(\lambda - s)(1 - e^{-\lambda\Delta})}.$$

Next, we need to analyze the collision resolution process by TBEB. Let $H^{(i)}, 0 \le i < R$, be a discrete random variable representing the number of backoff frames incurred by the ith attempt of an REQ. Since the backoff period is uniformly chosen from $[0, W_i - 1]$ in the ith attempt, the probability mass function (pmf) of $H^{(i)}$ is given by

$$H^{(i)} = \begin{cases} j & \text{w.p.} \quad m/W_i, j = 1, 2, \ldots, A_i - 1 \\ A_i & \text{w.p.} \quad 1 - \frac{(A_i - 1)m}{W_i} \end{cases}$$

where w.p. stands for "with probability" and $A_i = \lceil W_i/m \rceil$, which is the smallest integer greater than or equal to W_i/m. Hence, the average number of backoff frames incurred by the ith attempt of an REQ can be expressed as

$$E[H^{(i)}] = A_i - A_i(A_i - 1)\frac{m}{2W_i} \quad i = 0, 1, \ldots, R - 1.$$

Then, the Laplace-Stieltjes transform of $H^{(i)}$ can be obtained as follows

$$\mathcal{L}_{H^{(i)}}(s) = \sum_{j=1}^{A_i - 1} \frac{m}{W_i} e^{-js} + \left(1 - \frac{(A_i - 1)m}{W_i}\right) e^{-A_i s}. \tag{8}$$

Let $Y^{(i)}, 0 \le i < R$, be a discrete random variable representing the accumulated backoff time that an SS has spent from backoff state 0 to backoff state i,

$$Y^{(i)} = \sum_{j=0}^{i} H^{(j)}\Delta.$$

So, the Laplace-Stieltjes transform of $Y^{(i)}$ can be given as

$$\mathcal{L}_{Y^{(i)}}(s) = \prod_{j=0}^{i} \mathcal{L}_{H^{(j)}}(\Delta s). \tag{9}$$

Therefore, the accumulated backoff time Y for an arbitrary REQ is given as

$$Y = \begin{cases} Y^{(i)} & \text{w.p.} \quad (1-p)p^i, i = 0, 1, \ldots, R-2 \\ Y^{(R-1)} & \text{w.p.} \quad p^{R-1}. \end{cases} \tag{10}$$

From (10), the pdf of Y, denoted by $f_Y(y)$, can be obtained, and $E[Y]$ can be written as

$$E[Y] = (1-p) \sum_{i=0}^{R-2} p^i E[Y^{(i)}] + p^{R-1} E[Y^{(R-1)}], \tag{11}$$

where

$$E[Y^{(i)}] = \Delta \sum_{j=0}^{i} E[H^{(j)}].$$

And the Laplace-Stieltjes transform of Y can be written as

$$\mathcal{L}_Y(s) = \sum_{i=0}^{R-2} (1-p)p^i \mathcal{L}_{Y^{(i)}}(s) + p^{R-1} \mathcal{L}_{Y^{(R-1)}}(s). \tag{12}$$

Note that $\mathcal{L}_G(s), \mathcal{L}_{H^{(i)}}(s), \mathcal{L}_{Y^{(i)}}(s) and \mathcal{L}_Y(s)$ will be used in Section 4.2 where the distribution of packet delay is derived.

At the instant of packet arrival, the queue at the SS may be in one of two cases: **S0** *or* **S1**.

For case **S0**, the service time of an REQ is $X_0 = G + Y$, noting that G and Y are independent, so the pdf of X_0 can be written as

$$f_{X_0}(x) = \int_{-\infty}^{\infty} f_G(x-y) f_Y(y) dy.$$

So, $E[X_0] = E[G] + E[Y]$, and the Laplace-Stieltjes transform of X_0 can be written as

$$\mathcal{L}_{X0}(s) = \mathcal{L}_G(s) \mathcal{L}_Y(s). \tag{13}$$

For case **S1**, the service time of an REQ is $X_1 = Y$, and the Laplace-Stieltjes transform of X_1 is therefore given by that of Y.

Thus the service time of an REQ is given by

$$X = \begin{cases} X_0 & \text{w.p.} \quad 1-\rho \\ Y & \text{w.p.} \quad \rho \end{cases} \tag{14}$$

and the mean service time can be written as

$$\begin{aligned} E[X] &= (1-\rho)(E[G] + E[Y]) + \rho E[Y] \\ &= E[Y] + (1-\rho)E[G]. \end{aligned} \tag{15}$$

Hence, the outer set of fixed point equations is completed by updating ρ as in Fig. 2.

4. Performance metrics

4.1 Throughput

Recall that a packet is discarded after its request has failed R attempts, the throughput of each SS is given by $\lambda(1-p^R)$. Since the network provides a capacity of d data slots in each frame with duration Δ, the normalized network throughput Γ is thus given by

$$\Gamma = \frac{N\lambda(1-p^R)}{d/\Delta}. \tag{16}$$

4.2 Distribution of packet delay

Recall that X is a random variable representing the service time experienced by an REQ, irrespective whether the REQ will be successful or unsuccessful. Let us define a related random variable X', which represents the service time experienced by a successful REQ. In addition, referring to Fig. 3, we define another random variable V which represents the time from the beginning of a data subframe to the end of a packet transmission. Hence, for a successful REQ, the corresponding packet delay $D(t)$ is comprised of the waiting time of the REQ in the queue $W_q(t)$, X' of the REQ, T_{RE} and V, which can be written as

$$D(t) = W_q(t) + X' + T_{RE} + V. \tag{17}$$

So, the Laplace-Stieltjes transform of $D(t)$ can be written as

$$\mathcal{L}_D(s) = \mathcal{L}_{W_q}(s)\mathcal{L}_{X'}(s)\mathcal{L}_V(s)e^{-sT_{RE}}. \tag{18}$$

To calculate $\mathcal{L}_D(s)$, we first need to derive $\mathcal{L}_{W_q}(s)$. In (Welch, 1964), the waiting time distribution has been derived for the generalized $M/G/1$ queueing process. Hence, we can apply this result for our model. The waiting time distribution for our model can be rewritten as

$$\mathcal{L}_{W_q}(s) = \frac{(1-\lambda E[Y])\{\lambda[\mathcal{L}_{X0}(s) - \mathcal{L}_Y(s)] - s\}}{[1-\lambda(E[Y]-E[X_0])][\lambda - s - \lambda\mathcal{L}_Y(s)]}. \tag{19}$$

The service time experienced by successful REQs X' is given by

$$X' = \begin{cases} Y' & \text{w.p.} \quad \rho \\ Y' + G & \text{w.p.} \quad 1-\rho \end{cases} \tag{20}$$

where the random variable Y' is the accumulated backoff time for successful REQs only, and is given by

$$Y' = Y^{(i)} \quad \text{w.p.} \quad \eta p^i. \tag{21}$$

Hence, the Laplace-Stieltjes transform of Y' is expressed as

$$\mathcal{L}_{Y'}(s) = \sum_{i=0}^{R-1} \eta p^i \mathcal{L}_{Y^{(i)}}(s) = \eta \sum_{i=0}^{R-1} p^i \prod_{j=0}^{i} \mathcal{L}_{H^{(j)}}(\Delta s) \tag{22}$$

Therefore, the Laplace-Stieltjes transform of X' can be written as

$$\mathcal{L}_{X'}(s) = \rho\mathcal{L}_{Y'}(s) + (1-\rho)\mathcal{L}_{Y'}(s)\mathcal{L}_G(s). \tag{23}$$

Using (19) and (23), the remaining term in (18) that needs to be determined is $\mathcal{L}_V(s)$. Let $q(j)$ be the probability that there are j successful requests other than the tagged SS in a frame. The probability $q(j)$ follows a truncated binomial distribution

$$q(j) = \frac{Q(j+1)}{1-Q(0)}, \quad 0 \le j \le n-1. \tag{24}$$

Using the assumption that the BS randomly allocates data slots to successful requests, the pmf of V can be expressed as

$$V = iT \quad \text{w.p.} \quad \sum_{j=i}^{n'} \frac{q(j-1)}{j}, i = 1,2,...,n', \tag{25}$$

where $n' = min(n,d)$. Then, the Laplace-Stieltjes transform of V can be written as

$$\mathcal{L}_V(s) = \sum_{i=1}^{n'} e^{-iTs} \sum_{j=i}^{n'} \frac{q(j-1)}{j}. \tag{26}$$

From (19), (23) and (26), $\mathcal{L}_D(s)$ can be determined. Hence, by the properties of Laplace-Stieltjes transform, any moments of the delay distribution can be derived straightforwardly. In particular, the mean packet delay $\overline{D}$ is given by

$$\overline{D} = -\frac{d\mathcal{L}_D(s)}{ds}\Big|_{s=0},$$

and the variance of packet delay is given by

$$\sigma_D^2 = \frac{d^2\mathcal{L}_D(s)}{ds^2}\Big|_{s=0} - \overline{D}^2.$$

5. Model validation and numerical results

In this section, we verify our analytical model using computer simulation and investigate the performances under various configurations of N, W and λ. To this end, we have developed an event-driven simulation program to simulate the broadcast polling mechanism of IEEE 802.16. The simulator was written in C++. In the simulation model, the channel is operated in TDD mode, in which a frame is divided into a downlink and uplink subframe. The MAC and physical layer parameters were configured in accordance with default parameters taken from the standard (IEEE 802.16 standard, 2009). In particular, the frame duration is 1 $m$sec consisting of 2500 mini slots each of 0.4 μsec length. Each bandwidth request consists of 6 mini slots including 3 mini slots for subscriber station transition gap (SSTG), 2 mini slots for preamble and one mini slot for a bandwidth request message of 48 bits. The length of a data slot including the preamble and transition gap is 37.6 μsec (i.e. 94 mini slots). Each SS has an infinite buffer fed by a Poisson traffic source with mean arrival rate λ packet per msec. The head-of-queue packet of each SS makes bandwidth request and follows the TBEB mechanism. Based on the contention result, the processes of bandwidth allocation and packet transmission are then carried out. The duration of each simulation is 5000 seconds long, with an initial transient period of 300 seconds. For the analytical results, we set δ of Fig. 2 equal to 10^{-8}. As shown in the following figures, the numerical results match well with values obtained from simulation.

Therefore, our model is suitable for studying the impact of different parameters on the performance of contention-based services of IEEE 802.16.

We evaluate the impact of the number of SSs (N) and the initial backoff window (W) on various performance metrics. We set $r = 4, R = 8, m = 10, d = 8, \lambda = 0.1$. The results are shown in Fig. 4(a) to Fig. 4(f). The failure probability of REQ (p) under different N with $W = 8, 16, 32$ are plotted in Fig. 4(a). As expected, larger N leads to more request contentions and thus larger p. On the other hand, p decreases as W increases. This is because when W increases, there are more choices of a request slot in each backoff stage. As a result, the probability that an SS transmits a request in a request slot (τ) becomes smaller. So, p_c and p decrease.

Fig. 4(b) plots the mean service time of REQs against N with $W = 8, 16, 32$, respectively. Since p increases with N, it means that larger N increases the average number of attempts of a successful REQ. This results in a larger mean service time. Similarly, larger W leads to larger backoff time which constitutes the service time of REQs. Therefore, the mean service time also increases with W.

Fig. 4(c) and Fig. 4(d) plot the mean and variance of packet delay against N for various W, respectively. Since the mean service time contributes part of the mean packet delay, as expected from Fig. 4(b), the mean packet delay also increases with both N and W.

Fig. 4(e) also indicates that larger W results in higher traffic load for a given N. However, increasing W does not increase the net throughput when N is fixed. Therefore, it is actually better to choose small W and tolerate a slightly higher REQ unsuccessful probability.

Next, we evaluate the impact of the packet arrival rates (λ) on the performance metrics. We set $r = 4, R = 8, m = 10, d = 8, N = 30, W = 8, 16, 32$. The results are shown in Fig. 5(a) to Fig. 5(f). Essentially, increase in λ means increasing the offered traffic load ρ. Therefore, this set of results would resemble to that of varying N. The failure probability of REQ under different λ and W are plotted in Fig. 5(a). As packet arrival rate increases, each node is more likely to make requests and hence p also increases.

At last, we also consider how d influences the performance of the mean packet delay and normalized network throughput. As shown in Fig. 6(a), mean packet delay does not change too much against d for a given N. On the other hand, the normalized network throughput varies greatly, so it is important to choose suitable values of m and d.

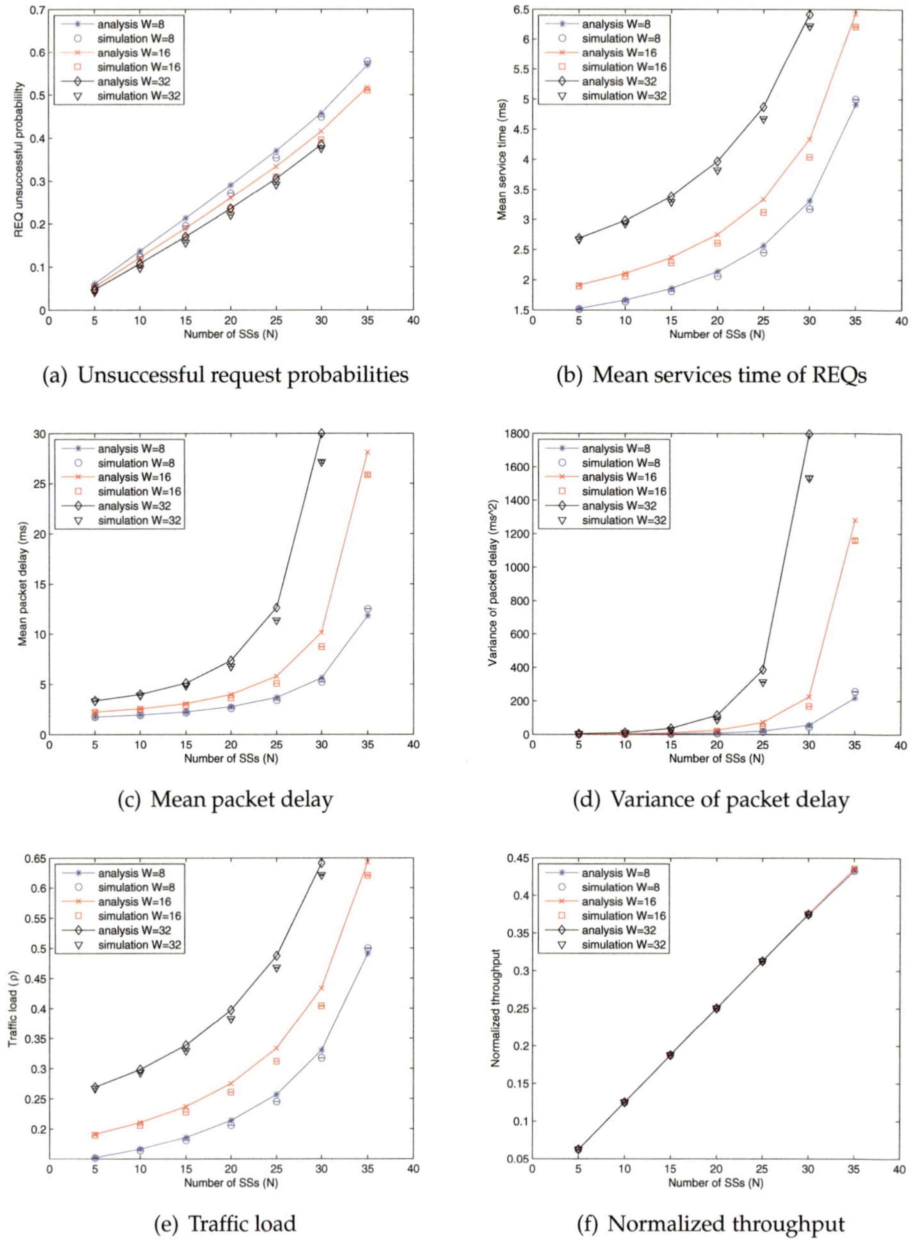

(a) Unsuccessful request probabilities

(b) Mean services time of REQs

(c) Mean packet delay

(d) Variance of packet delay

(e) Traffic load

(f) Normalized throughput

Fig. 4. Results for varying N and W, when $r = 4, R = 8, m = 10, d = 8, \lambda = 0.1$.

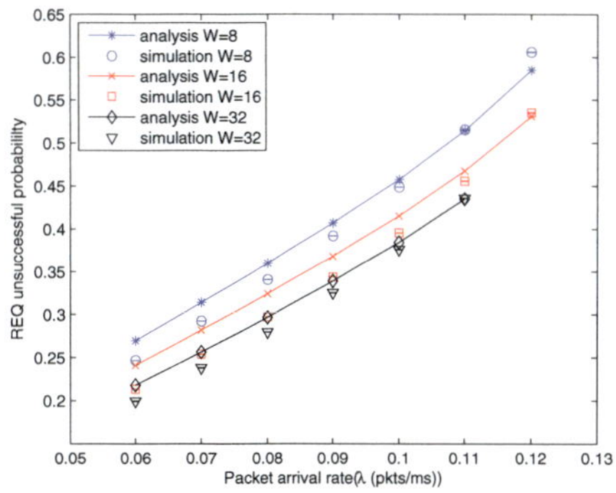

(a) Unsuccessful request probabilities

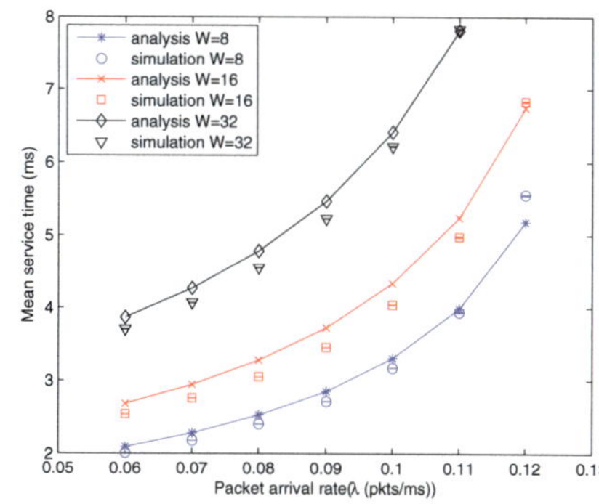

(b) Mean services time of REQs

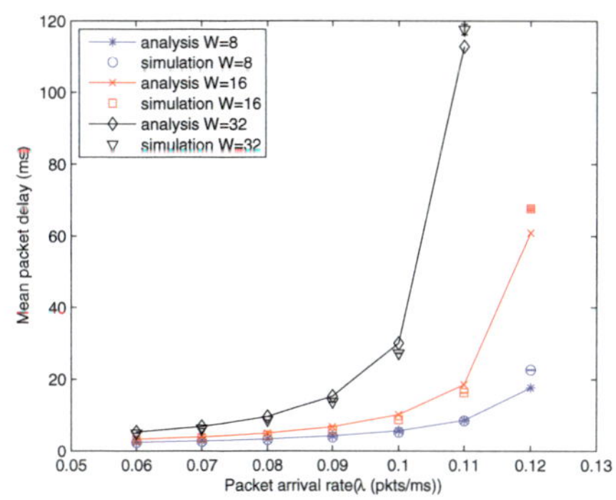

(c) Mean packet delay

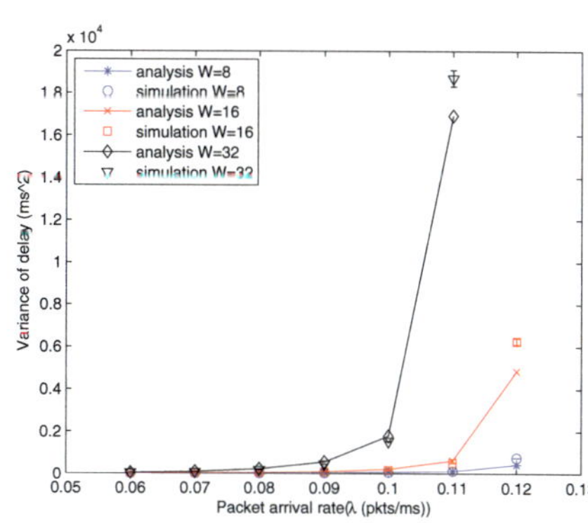

(d) Variance of packet delay

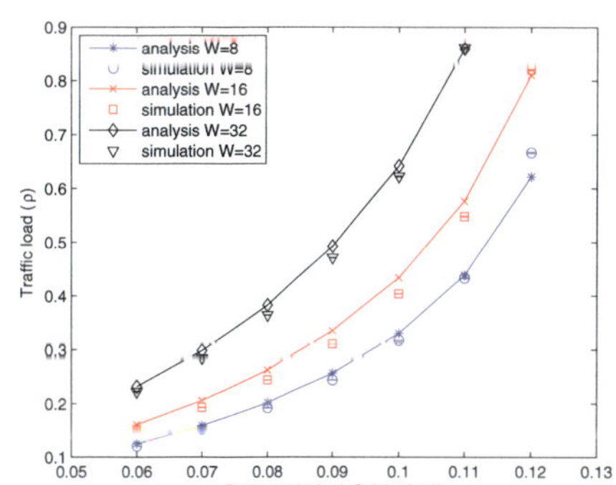

(e) Traffic load

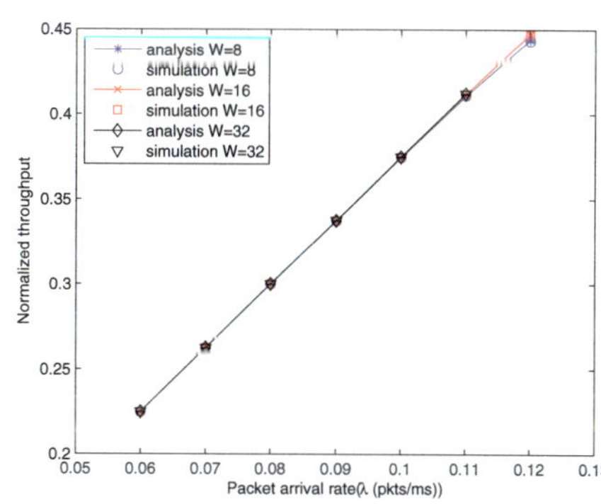

(f) Normalized throughput

Fig. 5. Results for varying λ and W, when $r = 4, R = 8, m = 10, d = 8, N = 30$.

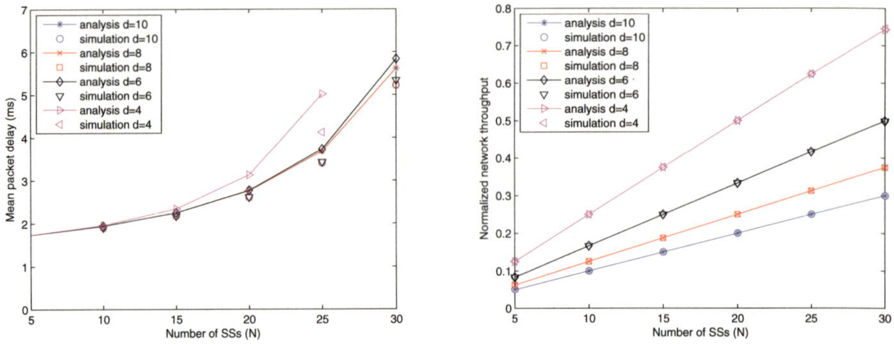

(a) Mean packet delay (b) Normalized throughput

Fig. 6. Results for varying N and d, when $r = 4, R = 8, m = 10, W = 8, \lambda = 0.1$.

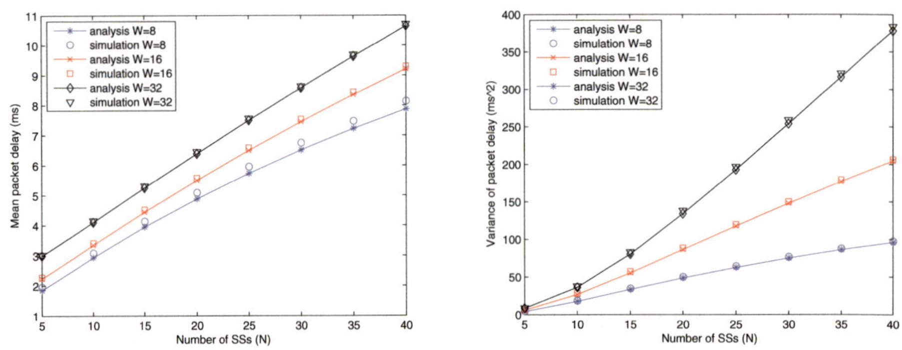

(a) Mean packet delay (b) Variance of packet delay

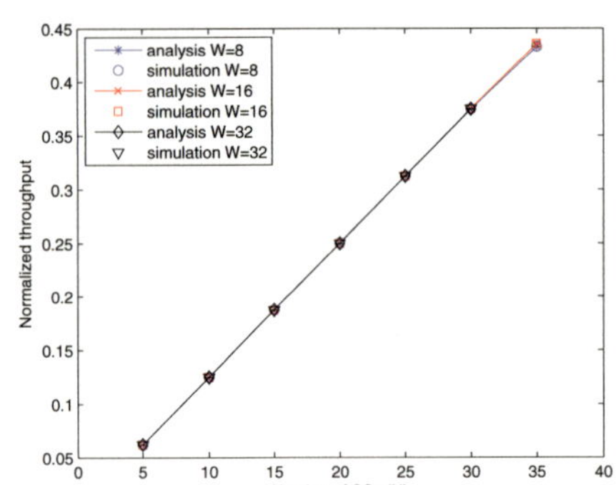

(c) Normalized throughput

Fig. 7. Results for saturated networks, when $r = 4, R = 8, m = 10, d = 8, \lambda = 0.1$.

6. Saturated networks

As defined in Section 1, saturated networks mean that each SS always has a packet to send. In other words, $\rho = 1$. Hence, the outer set in Fig. 2 is not required for the saturation case and (3) becomes

$$\tau = 1/(B_{avg} + 1) \tag{27}$$

Meanwhile, the case **S0** in Section 3 does not exist. Therefore, the service time of an REQ X is equal to Y. For the same reason, the service time of an successful REQ X' is equal to Y'. Obviously, there is no need to calculate the waiting time in the queue of an REQ for saturated networks. So the delay of a packet can be changed to packet access delay as the time duration from the beginning of the request interval in which a request initiates the TBEB process till the end of the transmission of the packet, which is given by

$$D_{sat} = Y' + T_{RE} + V. \tag{28}$$

So, the Laplace-Stieltjes transform of D_{sat} can be written as

$$\mathcal{L}_{D_{sat}}(s) = \mathcal{L}_{Y'}(s)\mathcal{L}_V(s)e^{-sT_{RE}}. \tag{29}$$

And the normalized network throughput for saturated works is given by

$$\Gamma_{sat} = \frac{\sum_{j=1}^{d} jQ(j) + \sum_{j=d+1}^{k} dQ(j)}{d}. \tag{30}$$

In order to verify this degenerated model for the saturated network, the mean and variance of packet access delay and throughput against N with different W are plotted as Fig. 7(a) to Fig. 7(c). It can be seen that the analytical and simulation results again match very well.

7. Conclusion

In this chapter, we have developed a unified performance model to evaluate the performances of the contention-based services in both saturated and unsaturated IEEE 802.16 networks. Different from some related works which assume that the probability of an SS sending a bandwidth request is an input parameter, our model takes into account the details of the backoff process to evaluate this probability. By solving two nested sets of fixed point equations, we have obtained the failure probability of a bandwidth request and the probability that a subscriber station has at least one REQ to transmit. Based on these two probabilities, the network throughput and the distribution of packet delay are derived. The model has been validated by simulations and shown to be accurate. Using the model, we have been able to investigate the impact of various parameters on the performance metrics of the 802.16 network.

8. References

IEEE 802.16-2009. IEEE Standard for Local and Metropolitan Area Networks. Part 16: Air Interface for Fixed Broadband Wireless Access Systems, *IEEE*, May 2009.

J. G. Andrews; A. Ghosh & R. Muhamed (2007). *Fundamentals of WiMAX: Understanding Broadband Wireless Networking*, Prentice Hall, ISBN 0-13-222552-2.

B. Kwak; N. Song & L. E. Miller. Performance Analysis of Exponential Backoff. *IEEE/ACM Trans. on Networking*, vol. 13, no. 2, 2005, pp. 343-355.

R. Iyengar; P. Iyer & B. Sikdar. Delay Analysis of 802.16 based Last Mile Wireless Networks. *Proceedings, IEEE Globecom'05*, 2005, pp. 3123-3127.

A. Vinel; Y. Zhang; M. Lott & A. Tiurlikov. Performance Analysis of the random access in IEEE 802.16. *Proceedings, IEEE International Symposium on Persoal, Indoor and Mobile Radio Communications*, Berlin, September 2005.

J. He; K. Guild; K. Yang & H. H. Chen. Modeling Contention Based Bandwidth Request Scheme for IEEE 802.16 Networks. *IEEE Communications Letters*, vol. 11, no. 8, August 2007, pp. 698-700.

H. L. Vu; S. Chan & L. Andrew. Performance Analysis of Best-Effort Service in Saturated IEEE 802.16 Networks. *IEEE Trans. on Vehicular Thechnology*, vol. 59, no. 1, 2010, pp. 460-472.

Q. Ni & L. Hu. An Unsaturated Model for Request Mechanisms in WiMAX. *IEEE Communications Letters*, vol. 14, no. 1, Jan. 2010, pp. 45-47.

Y. P. Fallah; F. Agharebparast; M. R. Minhas; H. M. Alnuweiri & V. C. M. Leung. Analytical Modeling of Contention-Based bandwidth Request Mechanism in IEEE 802.16 Wireless Networks. *IEEE Trans. on Vehicular Technology*, vol. 5, no. 5, 2008, pp. 3094-3107.

H. Fattah & H. Alnuweiri. Performance Evaluation of Contention-Based Access in IEEE 802.16 Networks with Subchannelizaion. *IEEE ICC on Communications*, 2009, pp. 1-6.

D. Chuck; K. Chen & J. M. Chang. A Comprehensive Analysis of Bandwidth Request Mechanisms in IEEE 802.16 Networks. *IEEE Trans. on Vehicular Technology*, vol. 59, no. 4, 2010, pp. 2046-2056.

R. P. Agarwal; M. Meehan & D. O'Regan. Fixed point theory and applications. Cambridge University Press, New Yourk, ISBN 0-52-180250-4, 2001.

Peter D. Welch. On a Generalized M/G/1 Queueing Process in Which the First Customer of Each Busy Period Receives Exceptional Service. *Operations Research*, vol. 12, no. 5, 1964, pp. 736-752.

9

A Mobile WiMAX Architecture with QoE Support for Future Multimedia Networks

José Jailton[1], Tássio Carvalho[1], Warley Valente[1], Renato Frânces[1], Antônio Abelém[1], Eduardo Cerqueira[1] and Kelvin Dias[2]
[1]*Federal University of Pará,*
[2]*Federal University of Pernambuco,*
Brazil

1. Introduction

The permanent evolution of future wireless network technologies together with demand for new multimedia applications, has driven a need to create new wireless, mobile and multimedia-awareness systems. In this context, the IEEE 802.16 Standard (IEEE 802.16e, 2005), also known as WiMAX (WorldWide Interoperability for Microwave Access) is an attractive solution for last mile Future Multimedia Internet (Sollner, 2008) , particularly because of its wide coverage range and throughput support.

The IEEE 802.16e extension, also known as Mobile WiMAX, supports mobility management with the Mobile Internet Protocol version 6 (MIPv6). This provides service connectivity in handover scenarios, by coordinating layer 2 (MAC layer) and layer 3 (IP layer) mobility mechanisms (Neves, 2009) . In addition to mobility control issues, an end-to-end quality level support for multimedia applications is required to satisfy the growing demands of fixed and mobile users, while increasing the profits of the content providers.

With regard to Quality of Service (QoS) control, the WiMAX system provides service differentiation based on the combination of a set of communication service classes supported by both wired IP-based and wireless IEEE 802.16-based links. In the case of the former, network elements with IP standard QoS models, such as Differentiated Services (DiffServ) and Integrated Services (IntServ), Multiprotocol Label Switching (MPLS) can be configured to guarantee QoS support for applications crossing wired links. In the latter, several IEEE 802.16 QoS services can be defined to provide service differentiation in the wireless interface (IEEE 802.16e, 2005).

Four services designed to support different type of data flows can be defined as follows: (i) Unsolicited Granted Service (UGS) for Constant Bit Rate (CBR) traffic, such as Voice over IP (VoIP). (ii) The Real Time Polling Service (rtPS) for video-alike traffic. (iii) The Non-Real Time Polling Service for an application with minimum bandwidth guarantees, such as File Transfer Protocol (FTP). Finally, (iv) the Best Effort (BE) service which does not have QoS guarantees (e.g., web and e-mail traffic) (Neves, 2009) (Ahmet et Al, 2009).

Existing QoS metrics, such as packet loss rate, packet delay rate and throughput, are generally used to measure the impact on the quality level of multimedia streaming from the perspective of the network , but do not reflect the user's experience. As a result, these QoS parameters fail to reflect subjective factors associated with human perception. In order to overcome the limitations of current QoS-aware multimedia networking schemes with respect to human perception and subjective factors,, recent advances in multimedia-aware systems, called Quality of Experience (QoE) approaches, have been introduced. Hence, new challenges in emerging networks involve the study, creation and the validation of QoE measurements and optimization mechanisms to improve the overall quality level of multimedia streaming content, while relying on limited wireless network resources (Winkler, 2005).

In this chapter, there will be an overview of the most recent advances and challenges in WiMAX and multimedia systems, which will address the key issues of seamless mobility, heterogeneity, QoS and QoE. . Simulation experiments were carried out to demonstrate the benefits and efficiency of a Mobile WiMAX environment in controlling the quality level of ongoing multimedia applications during handovers. These were conducted, by using the Network Simulator 2 (ns-2, 2010) and the Video Quality Evaluation Tool-set Evalvid. Moreover, well- known QoE metrics, including Peak Signal-to-Noise Ratio (PSNR), Video Quality Metric (VQM), Structural Similarity Index (SSIM) and Mean Option Score (MOS), are used to analyze the quality level of real video sequences in a wireless system and offer support for our proposed mechanisms.

2. WiMAX network infrastructure

A number of WiMAX schemes, such as mobility management for the handover and user authentication, require the coordination of a wide range of elements in a networking system. The implementation of these features is far beyond the definition] of IEEE 802.16, since this only adds to the physical layer components that are needed for modulation settings and the air interface between the base stations and customer, together with the definitions of what comprises the Medium Access Control (MAC) layer.

With the WiMAX Forum, it was possible to standardize all the main elements of a WiMAX network, including mobile devices and network infrastructure components. In this way, interoperability between the networks was ensured even when they had different manufacturers. However, there are several outstanding issues related to QoS, QoE, seamless handover and multimedia approaches that must be addressed before the overall performance of the Multimedia Mobile WiMAX system can be improved.

2.1 General architecture

The development of a WiMAX architecture follows several principles, most of which are applicable to general issues in IP networks. Figure 1 illustrates a generic Heterogeneous Mobile WiMAX scenario.

The WiMAX architecture should provide connectivity support, QoS, QoE and seamless mobility, independently of the underlying network technologies, QoS models and available service classes. The system should also enable the network resources to be shared, by allowing a clear distinction to be drawn between the Network Access Provider (NAP), an organization that provides access to the network and the Network Service Provider (NSP), an entity that deals with customer service and offers access to broadband applications and large Service Providers (ASP).

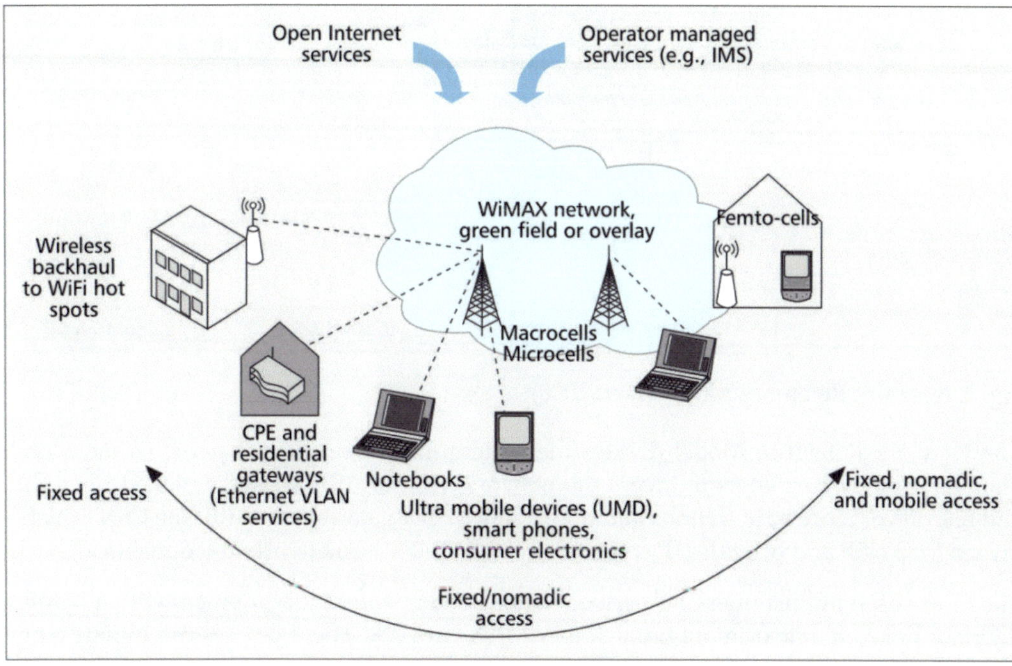

Fig. 1. Heterogeneous Mobile WiMAX System (Eteamed ,2008).

This section addresses the end-to-end network system architecture of WiMAX, based on the WiMAX Forum's Network Working Group (NWG), which includes issues related to and beyond the scope of (IEEE 802.16-2009). The Network Reference Model (NRM) with the WiMAX Architecture will also be introduced and various functional entities and their respective connections and responsibilities explained.

2.2 Network architecture

The WiMAX network architecture is usually represented by a NRM in most modern research papers and technical reports. This model describes the functional entities and reference points for an interoperable system based on the WiMAX Forum. The NRM usually has some Subscriber Stations (MS) (clients, customers, subscriber stations, etc), Access Service Network (ASN) and Connectivity Service Network (CSN) with their interactions which are expected to continue through the reference points. Figure 2 shows the defined reference points R1 to R8 which represent the communications between the network elements.

The WiMAX NRM differentiates between NAPs and NSPs, where the former are business entities that provide the infrastructure and access to the WiMAX network that contains one or more ASNs. At a high level, these NAPs are the service providers and their infrastructure with a shared wireless access. The NSPs are business entities that provide IP connectivity and WiMAX services to the subscriber stations in accordance with service level agreements or other agreements. The NSP can have control over the CSN (Iyer, 2008).

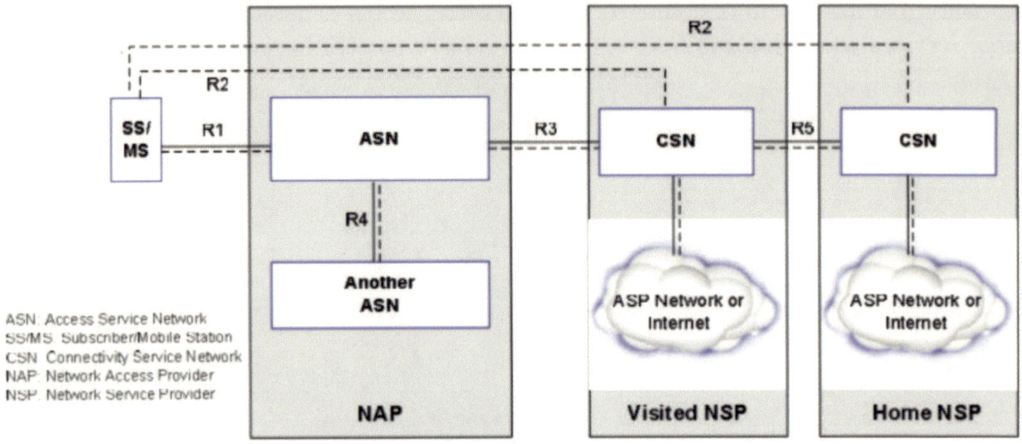

Fig. 2. Network Reference Model (Iyer, 2008)

The Network Reference Model divides the system into three distinct parts: (i) the Mobile Stations used by customers to access the network, (ii) the ASN which is owned by a NAP and has one or more base stations and one or more ASN gateways and (iii) the CSN which is owned by a NSP and provides IP connectivity and all IP core network functionalities.

The SS are used by customers, subscriber stations and any mobile equipment with a wireless interface linked to one or more hosts of a WiMAX network. These devices can initiate a new connection once the presence of a new base in an ASN has been verified.

The ASN is the ingress point of a WiMAX network, where the MS must be connected. Hence, the MS has to follow a set of steps and corresponding functions for authentication and boot process to request and receive access to the network and, thus establish , the connectivity (Ahmadi, 2009) (Vaidehi & Poorani, 2010). The ASN can have one or more Base Stations (BS) and one or more ASN-GW (Access Service Network - Gateway). All the ASNs have the following mandatory functions:

- IEEE 802.16-2009 layer 2 connectivity with the Mobile Station;
- AAA (Authentication, Authorization and Accounting) Proxy: messages to client's home network with authentication, authorization and accouting to the mobile station;
- Radio Resource Management and the QoS policy;
- Network discovery and selection;
- Relay functionality for establishing IP connectivity with WiMAX MS;
- Mobile functions such as handover (support for mobile IP), location control, etc.

The CSN supports a set of network functions that provide IP connectivity to the WiMAX clients and customers. A CSN usually has many network elements such as routers, database, AAA servers, DHCP servers, gateways, providers, etc. The CSN can provide the following functions:

- IP address allocation to the mobile station;
- Policy, admission control and QoS managements based on service level agreements (SLA)/a contract with the user;

- Support for roaming between NSPs;
- Mobility management and mobile IP home agent functionality;
- Connectivity, infrastructure and policy control;
- Interoperability and billing solution;
- AAA proxy for devices, clients and services such as IP multimedia services (IMS).

The combination of these three elements form the WiMAX network reference model defined by the WiMAX Forum, together with the IEEE Standard 802.16-2009. Each function requires interaction between two or more functional entities and may operate one or more physical devices.

2.3 QoS architecture

WiMAX is one of the most recent broadband technologies for Wireless Metropolitan Area Networks (WMANs). To allow users to access, share and create multimedia content with different QoS requirements, WiMAX implements a set of QoS Class of Services (CoS) at the MAC layer as discussed earlier, (UGS, rtPS, ertPS, nrtPS and BE).

The UGS is designed to support real-time and delay/loss sensitive applications, such as voice. It is characterized by fixed-size data packets, requiring fixed bandwidth allocation and a low delay rate. The rtPS is similar to UGS regarding real-time requirements, but it is suitable for delay-tolerant with variable packet sizes, such as Moving Pictures Experts Group (MPEG) video transmission and interactive gaming.

The ertPS was recently defined by the IEEE 802.16 standard to support real-time content with a QoS/QoE requirement between UGS and rtPS. The BS provides grants in an unsolicited manner (as in UGS), with dynamic bandwidth allocation which is needed for some voice applications with silence suppression.

The nrtPS is associated with non real-time traffic with high throughput requirements, such as FTP transmission. The BS performs individual polling for SSs bandwidth requests. The BE is designed for applications without guarantees in terms of delay, loss or bit-rate. An example is web browsing and e mail (Chrost & Brachman, 2010) (Alison & Ilyas, 2007).

Each CoS has a mandatory set of QoS parameters that must be included in the service flow definition when the class of service is adapted to a service flow. The main parameters are the following: traffic priority, maximum latency, jitter, maximum and minimum data rate and maximum delay. Table 1 provides an overview of the five WiMAX class of services, typical applications and corresponding QoS parameters.

The MAC layer of the IEEE 802.16 standard is connection-oriented. Signaling messages between BS and SS must be exchanged so that a service flow can be established between them. A Service Flow (SF) is a MAC transport service that provides unidirectional transport of packets on the uplink or on the downlink. Each service flow is characterized by a set of QoS parameters that indicate the latency and jitter that is necessary and ensures throughput. In addition, each service flow receives a unique Service Flow Identifier (SFID) from the BS, a long integer of 32 bits, to allow each individual service flow to be identified. For any active service flow, a connection is discovered by a Connection Identifier (CID), a piece of information coded in 16 bits. A connection is a unidirectional mapping between a BS and a

SS MAC peers for the purpose of transporting the traffic of a service flow. Thus, a CID will be assigned for each connection between BS and SS associated with a service flow.

Scheduling service	Corresponding data delivery service	Typical applications	QoS specifications
Unsolicited Grant Service (UGS)	Unsolicited grant service (UGS)	Voice (VoIP) without silence suppression	Maximum sustained rate Maximum latency tolerance Jitter tolerance
Extended Real-Time Polling Service (ertPS)	Extended realtime variable-rate service (ERT-VR)	VoIP with silence suppression	Maximum sustained rate Minimum reserved rate Maximum latency tolerance Jitter tolerance Traffic priority
Real-Time Polling Service (rtPS)	Real-time variable-rate service (RT-VR)	Streaming audio or video	Maximum sustained rate Minimum reserved rate Maximum latency tolerance Traffic priority
Non-Real-Time Polling Service (nrtPS)	Non-real-time variable rate service (NRT-VR)	File Transfers Protocol (FTP)	Maximum sustained rate Minimum reserved rate Traffic priority
Best-Effort Service (BE)	Best-effort service (BE)	Web browsing, e-mail	Maximum sustained rate Traffic priority

Table 1. WiMAX scheduling and data delivery service classes, including applications and QoS parameters.

Figure 3 outlines the WiMAX QoS architecture as defined by the IEEE 802.16 standard. It can be observed that schedulers, QoS parameters and classifiers are present in the MAC layer of both the Base Station (BS) and Subscriber Station (SS). The BS is responsible for managing and maintaining the QoS for all of the packet transmissions. The BS manages this by actively distributing usage time to subscriber stations through information embedded in the transmitted management frames, as illustrated in Figure 4.

Communication between BS and SS can be initiated by the BS (mandatory condition) or by the SS (optional condition). In both cases, it is necessary for there to be a connection request to the Connection Admission Control (CAC) located in the BS. The CAC is responsible for accepting or rejecting a connectivity request. Its decisions are based on the QoS parameters contained in the request messages - Dynamic Service Addition Request (DSA-REQ). If the QoS parameters are within the limits of the available resources, and this is the case, the BS then replies with an acceptance message - Dynamic Service Addition Response (DSA-RSP) - and assigns a unique SFID for the new service flow.

The service flow is then classified and mapped into a particular connection for transmission between the MAC peers. The mapping process associates a data packet with a connection, which also creates a link with the service flow characteristics of this connection.

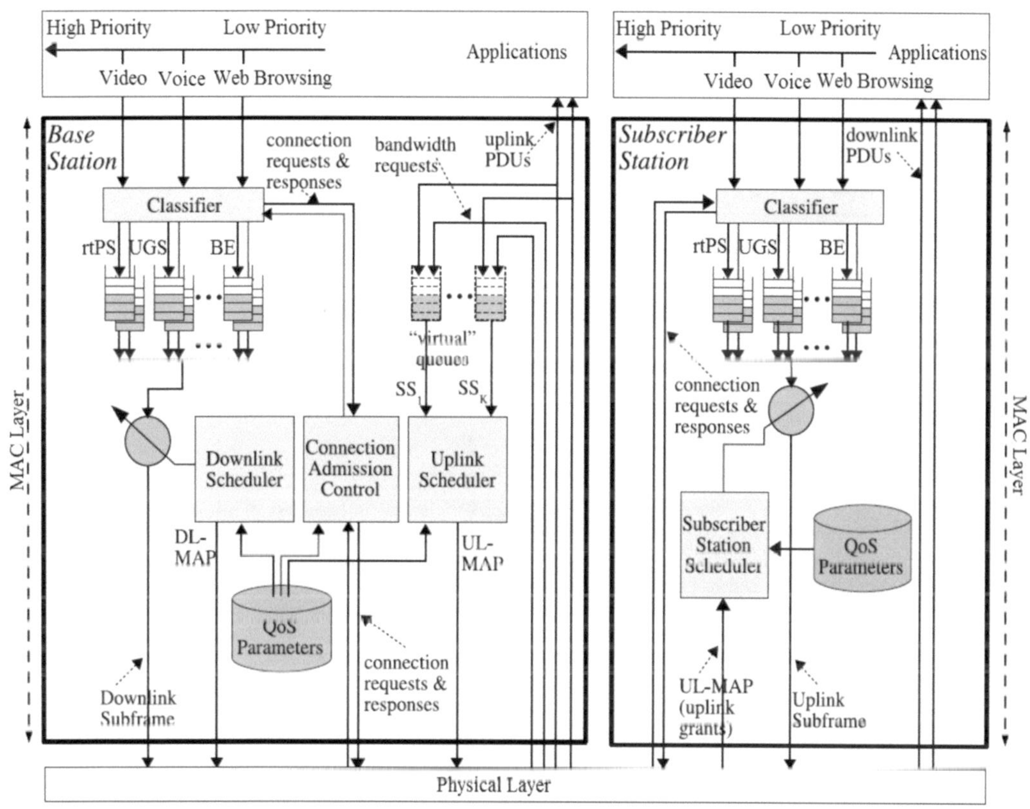

Fig. 3. Overall Architecture of WiMAX QoS.

After the process of classification has been completed,, the most complex aspect of the provision of QoS to individual packets is performed by the three schedulers: downlink and uplink schedulers located at BS, and responsible for managing the flows in the downlink and uplink respectively, and subscriber station schedulers, which together manage flows in the uplink or the SS-to-BS flows.

The aim of a scheduler is generally to determine the burst profile and the transmission periods for each connection, while taking into account the QoS parameters associated with the service flow, the bandwidth requirements of the subscriber stations and the parameters for coding and modulation.

The Downlink Scheduler's task is relatively simple compared to that of the Uplink Scheduler, since all the downlink queues reside in the BS and their state is locally accessible to the scheduler. The decisions regarding the time allocation of bandwidth usage are transmitted to the SSs through the DL-MAP (Downlink Bandwidth Allocation Map) MAC management message, located in the downlink sub-frame, as shown in Figure 4. This field notifies the SSs of the timetable and physical layer properties for transmitting subsequent bursts of packets.

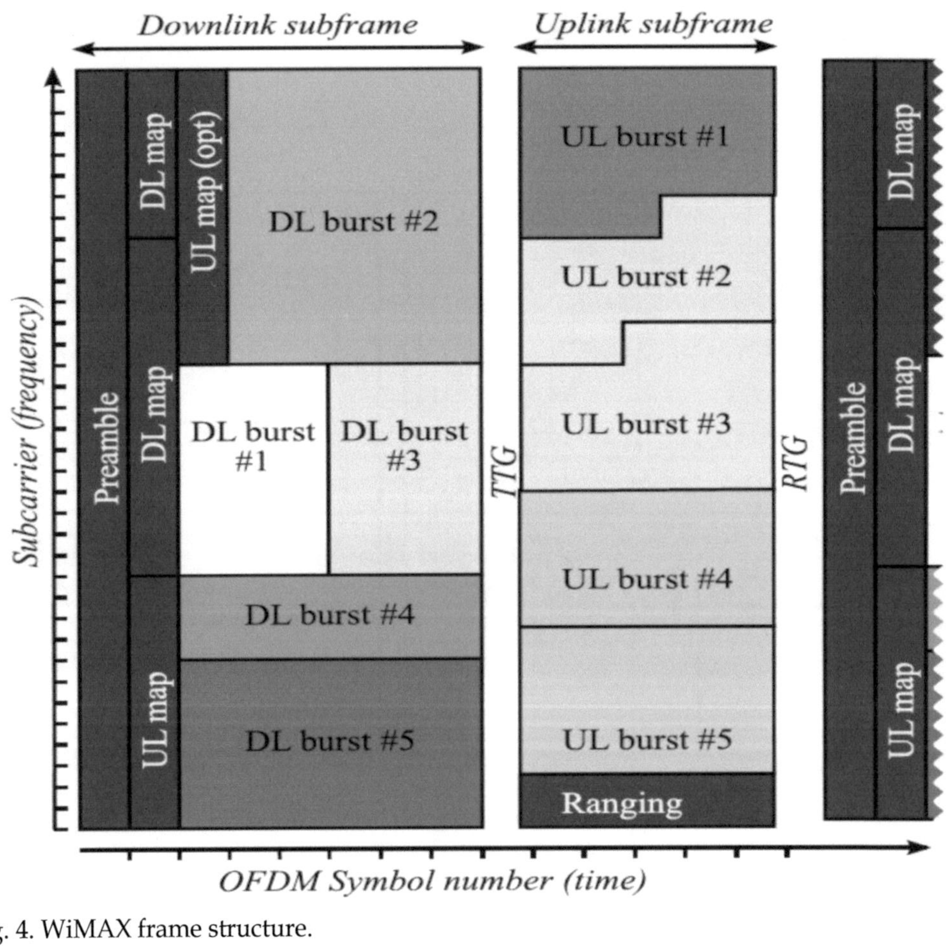

Fig. 4. WiMAX frame structure.

The task of the Uplink Scheduler is much more complex. Since queues of uplink packet flows are distributed among the SSs, their states and QoS requirements have to be obtained through bandwidth requests. The information gathered from the remote queues, forms the operational basis of the uplink scheduler and is displayed as "virtual queues", as can be seen in Figure 1. The uplink scheduler will select uplink allocations based on the bandwidth requests, QoS parameters and priorities of the service classes. These decisions are transmitted to the SSs through the UL-MAP (Uplink Bandwidth Allocation Map) which is the MAC management message for regulating the uplink transmission rights of each SS. Thus, , the UL-MAP controls the amount of time that each SS is provided with access to the channel in the immediately following or the next uplink sub-frame(s) (Sekercioglu, 2009).

The uplink sub-frame of the WiMAX management frame should also be mentioned.. This sub-frame basically contains three fields: initial ranging (Ranging), bandwidth requests (BW-REQ) and specific slots.

Initial ranging is used by SSs to discover the optimum transmission power, as well as the timing and frequency offset needed to communicate with the BS. The bandwidth requests contention slot is used by the SSs for transmitting bandwidth request MAC messages. These are the slots that are specifically allocated to the individual SSs for transmitting data.

The scheduler of an SS visits the queues and selects packets for transmission. The selected packets are transmitted to the BS in the allocated time slots as defined in the UL-MAP, which is constructed by the BS Uplink Scheduler and broadcast by the BS to the SSs (Nuaymi, 2007).

The WiMAX does not define the scheduling algorithm that must be implemented. Any of the known scheduling algorithms can be used: Round Robin (RR) (Ball et Al, 2006), Weighted Round Robin (WRR), Weighted Fair Queuing (WFQ), maximum Signal-to-Interference Ratio (mSIR) (Chen et Al, 2005), and Temporary Removal Scheduler (TRS) (Ball et Al, 2005).

3. WiMAX mobility

The IEEE 802.16e controls the handover, when an SS changes its current BS to a new BS within a continuous ongoing session. There are two types of handover. When the SS moves to a new BS, it stops the connection with the current BS before establishing the connection with the new BS; this procedure is also known as hard handover or break - before - make. When the SS establishes the connection with the new BS, before it stops the connection with the current BS, this procedure is called seamless handover or make - before - break (Manner, 2004).

When the SS enters the coverage area of a BS, the association process begins by obtaining the downlink parameters. The BS sends two messages to the SS (when it is inside the cell): the DL-MAP (Downlink MAP) and DCD (Downlink Channel Description). The DL-MAP message contains three elements, the physical specifications, the DCD value and the id BS. The DCD message describes the physical characteristics of the downlink channel. The next step corresponds to obtaining the uplink UCD (Uplink Channel Description) messages and UL-MAP (Uplink MAP). The UCD describes the physical characteristics of the uplink channel and the UL-MAP contains the physical specifications and also the time allocation of resources. After the downlink and uplink parameters, the SS sends the Ranging Request (RNG-REQ) to BS to discover the link quality (signal strength, modulation), and the BS replies with the Ranging Response (RNG -RSP). Finally, the last step is the registration between SS and BS to acquire an IP address. The SS sends a Registration Request (REG-REQ) and BS replies with a Registration Response (REG-RSP).

Another important feature of the IEEE 802.16e standard is the exchange of information between neighboring BSs. The BS sends the same information to another BS in the UCD / DCD messages transmitted. The Information is exchanged on the backbone through the Mobility Neighbor Advertisement (MOB_NBH_ADV) message.

Figure 5 illustrates the handover signaling for a WiMAX network. In this scenario, the SS is initially served by/connected to the WiMAX network, but periodically the SS listens and tries out other connectivity opportunities.

1. The SS detects a new link connectivity to the WiMAX Network.
2. The Current BS sends the downlink and uplink parameter messages to the SS.
3. The SS requests information about the network by Ranging Messages
4. The SS registers in current BS by means of Registrations Messages .
5. The current BS supports the QoS flow Services.
6. The Current BS communicates with the Target BS about network information by means of Mobility Neighbor Advertisement (MOB_NBR_ADV)
7. A new link connectivity is detected and the current link goes down. The SS iniates the handover to Target BS.
8. The SS repeats steps 2, 3, 4, 5 and 6 with the Target BS

3.1 Handover policy

It is necessary to create seamless mobility schemes for Mobile WiMAX Systems to improve the handover process, while ensuring QoS and QoE support for ongoing applications. . To achieve this, an algorithm for handover policy should use two metrics: WiMAX Link failure probability and SS speed. The link failure probability means the possibility of a "break" SS connection with current BS; this value represents the signal strength obtained from the physical layer. The link failure probability P is shown in Equation 1.

$$P = \frac{(Factor \times Rxthreshold) - Avg}{(Factor \times Rxthreshold) - Rxthreshold} \tag{1}$$

Where:

Avg = average signal strength

$Factor$ = connectivity factor

$Rxthreshold$ = clear signal strength

A GPS module installed at mobile nodes is required to improve the accuracy of the system with regard to the position and speed of the mobile users, as was the case with current smart phones and laptops. As a result, it will be possible to inform the BS about position and speed issues affecting the mobile user. This involves defining three mobility profiles: high, medium and low. Each mobility profile will be associated with the precise period of time

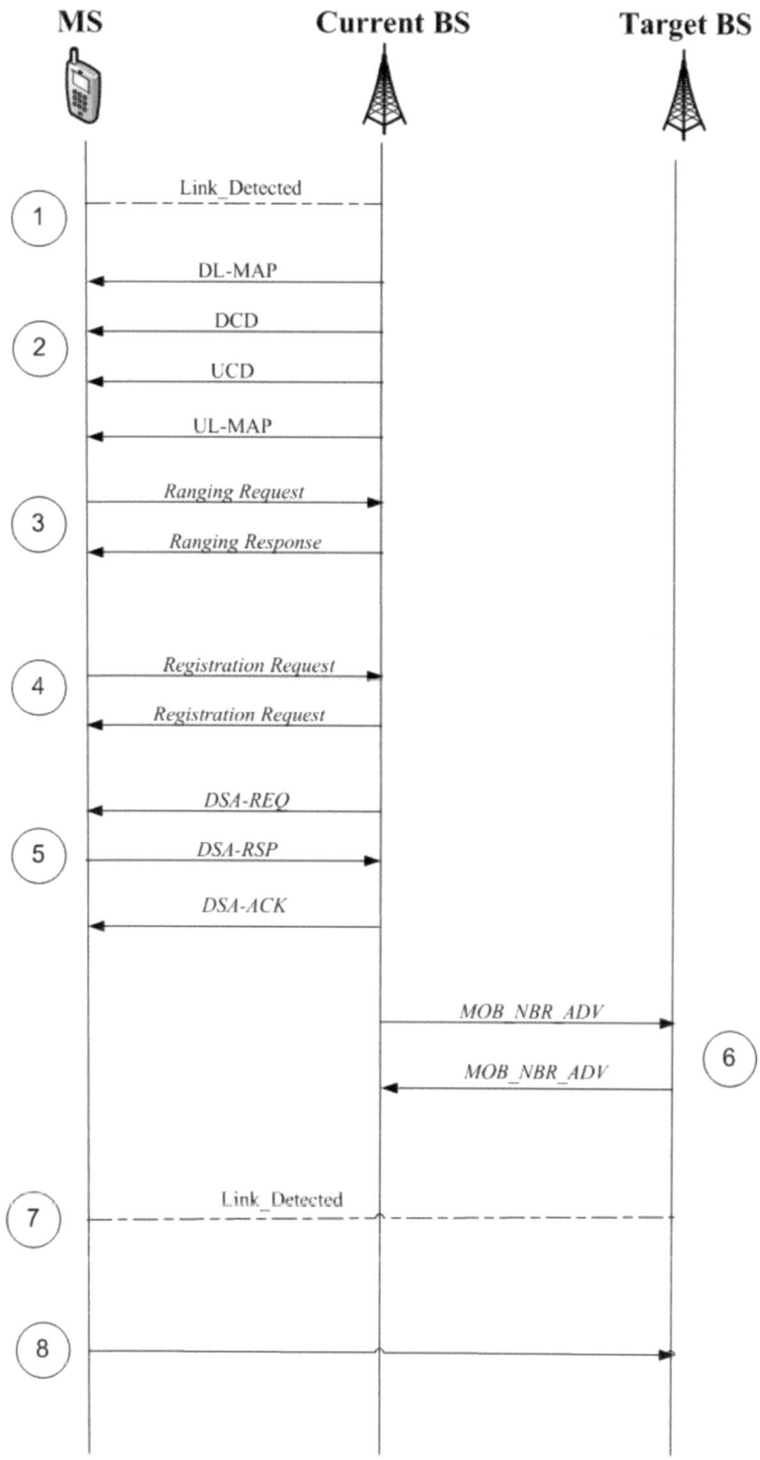

Fig. 5. The handover signaling for a WiMAX network

needed to initiate the handover. The high mobile node will remain the shortest time inside the cell, in this situation, and the handover process will be triggered before the other mobile nodes. The mobility information and link failure probability are the two components used as metrics to start the process of making a handover decision in the Mobile WiMAX architecture (Dial et Al, 2008).

1. Low mobility users (down to 7 m/s) - the handover process is initiated when the link failure probability is equal to 90%.
2. Medium mobility user (from 7 m/s and equal to 15 m/s) - the handover process is initiated when the link failure probability is equal to 70%.
3. High mobility users (from 15m/s) - the handover process is initiated when the link failure probability is equal to 50%.

When the handover process is triggered (Figure 6), the new BS sends the uplink and downlink (DL-MAP, DCD, UL-MAP, and UCD) messages to the SS. Then the SS receives a notification of the new BS with "better physical conditions" than the current BS. When the SS is in the intersection coverage (in the current and new BS), the SS can still receive packets from the current BS once it has carried out the connection process with the new BS. The SS establishes a connection with the new BS before it breaks the connection with the current BS.

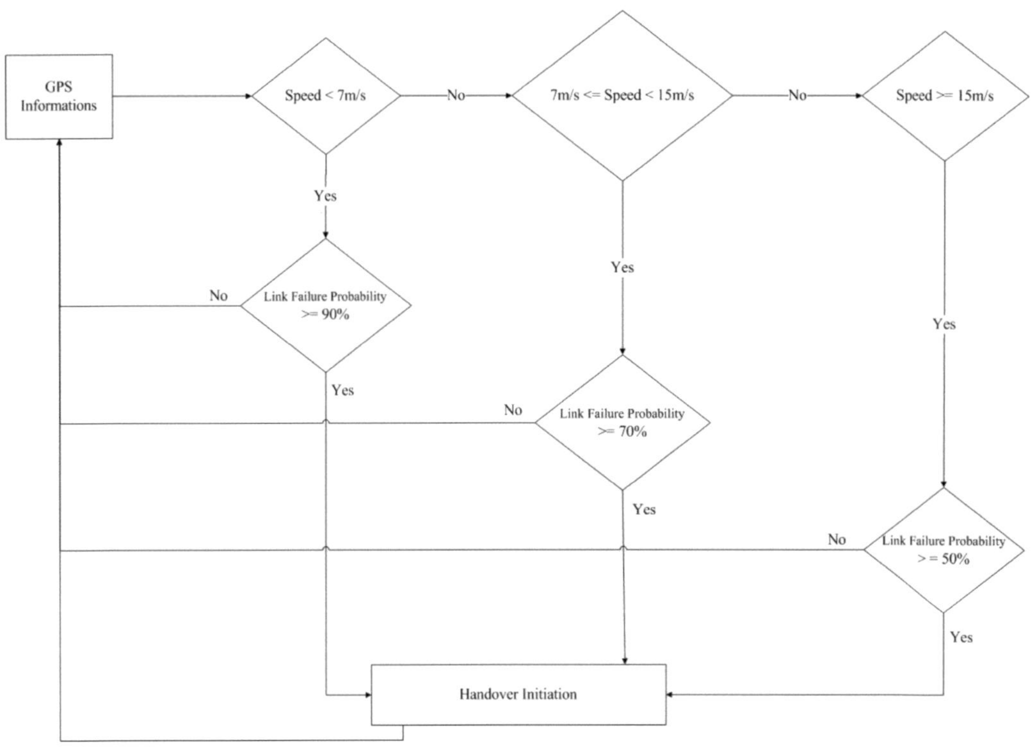

Fig. 6. Handover Policy Scheme

4. Evaluation of performance

The Simulations experiments were carried out with the aid of Network Simulator 2 to show the benefits and impact of the proposed Mobile WiMAX system in a simulated environment with all the handover policies. For the WiMAX simulations it was used a module developed by The National Institute of Standards and Technology (NIST, 2007), the module was based on the IEEE 802.16e with mobility support (Nist WiMAX, 2007). The results demonstrate the effectiveness of the architecture in supporting a seamless handover, QoS and QoE assurance. Figure 7 and Table 2 below show the topology used for the tests.

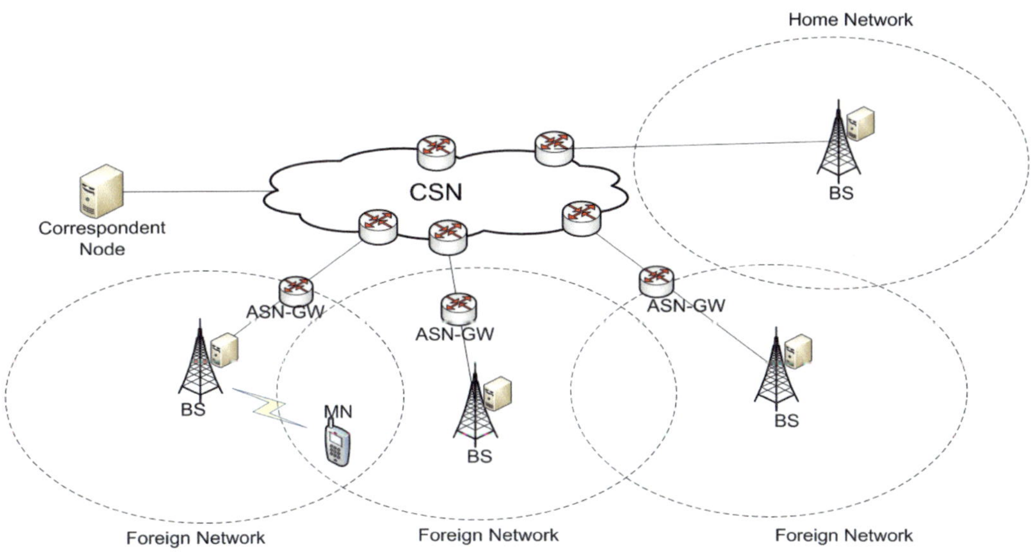

Fig. 7. Simulated Topology

Parameters		Value
Wired	Link Capacity	4 Mbps
	Link Delay	50 ms
	Buffer	50
	Queue	CBQ
WiMAX	Cover Area	1km
	Frequency	3,5GHz
	Standard	IEEE 802.16e
	Modulation	OFDM

Table 2. Simulated Parameters

4.1 CBR traffic

In the first experiment, the simulations were conducted with three mobile nodes with different mobility (low, medium and high). Due to the high mobility, the SS remains a short time inside the cell and will make three handovers. The SS with medium mobility will make two handovers and the SS with low mobility will make just one handover. The simulations were performed with CBR applications. In these simulations, the network/packet information that was measured, comprised the throughput and sequence number of packets received by each SS. Although the CBR application uses UDP as a transport protocol, we include a sequence number field to determine the losses during the handover process. For each mobility partner, a different CBR rate is used (Table 3)

Mobility	CBR application
Low	200Kbps
Medium	400Kbps
High	600Kbps

Table 3. CBR Traffic

In the first case, the simulations were performed without a handover policy. All the mobile nodes are disconnected when they change their BSs; in other words, during the handover process they break the connection with the current BSs, and after taking this step, they (re)connect with the new BS (Break - Before - Make). When the mobile nodes change their BSs, they do not receive a CBR packet application. Figure 8 and 9 below confirm this information.

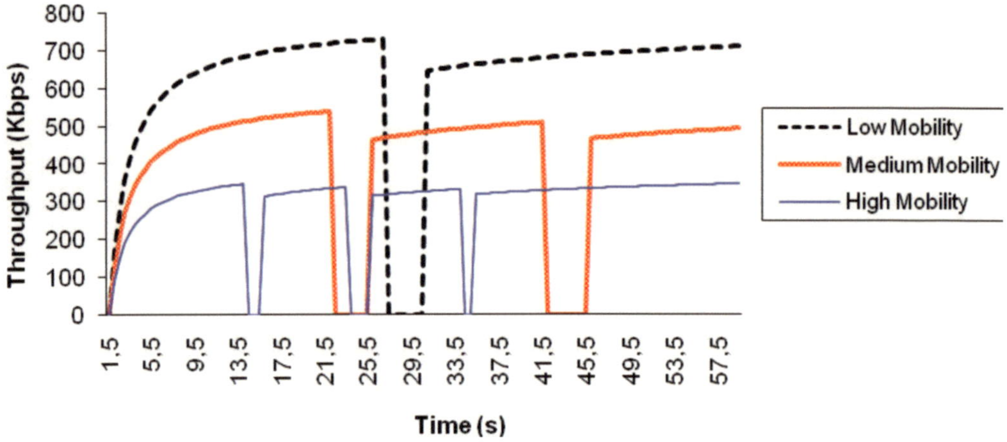

Fig. 8. Throughput without a handover policy

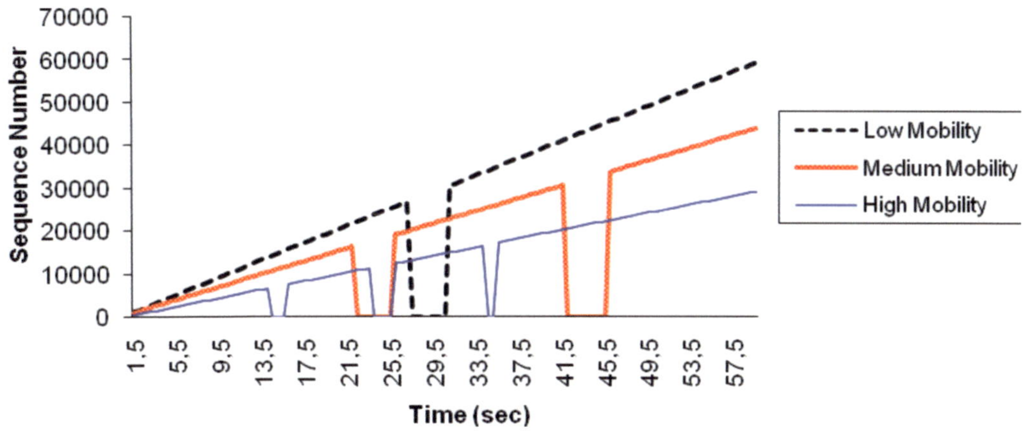

Fig. 9. Sequence Number without a handover policy

In the second case, the simulations were accomplished with the proposed handover policies. All the mobile nodes still continuously connected when they changed their BSs; in other words, during the handover process they did not break the connection with their current BSs so that they could connect with the new BS (Make - Before - Break). This meant that, the mobile nodes still received CBR packets applications during the handover. Graphs 10 and 11 below confirm this information.

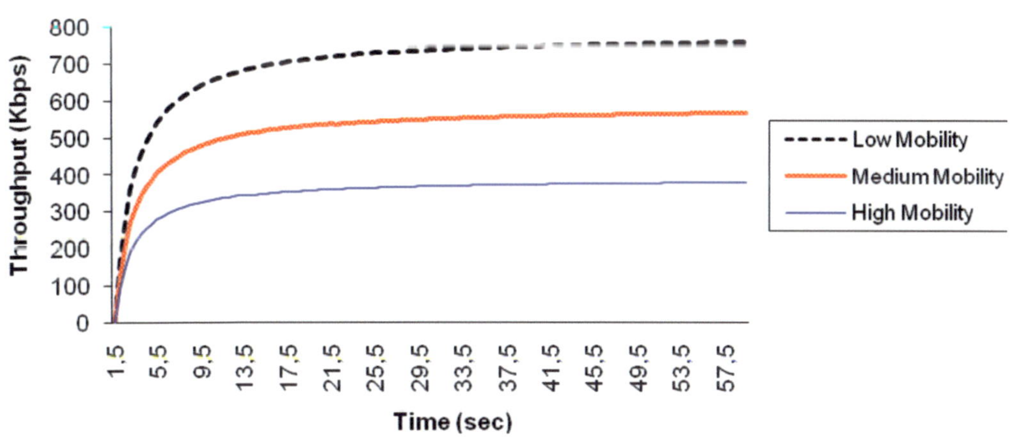

Fig. 10. Throughput with a handover policy

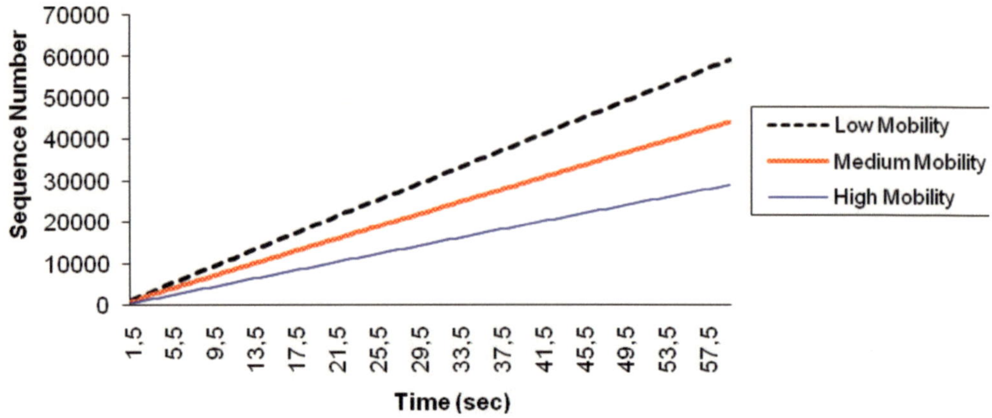

Fig. 11. Sequence Number with a handover policy

In the same scenario, by means of the Random Waypoint Mobility Model, 90 simulations were performed with the CBR application with 600kbps rate for different mobility and positions. Figure 12 shows the average throughput for each specific situation with and without a handover policy. The SSs with high speed did more handovers than others, and thus, more time should be spent without connection during the handover. In other SSs, the handover process damages the CBR application. With a handover policy, the throughput is almost constant, because the mobile nodes make a seamless handover.

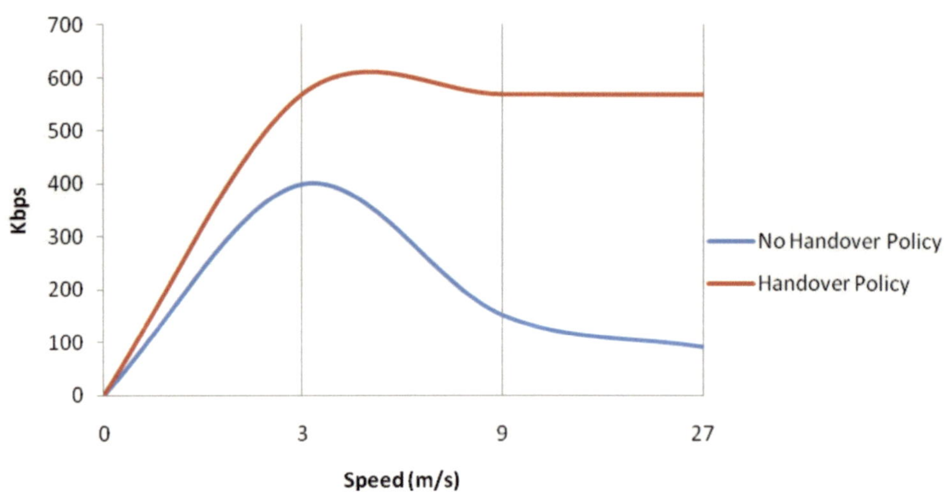

Fig. 12. Throughput x Mobility

In the simulations without the proposed handover policies, the average throughput for low, medium and high mobility were equal to 400kbps, 151kpbs and 91kbps, respectively. In simulations using the handover policies, the average throughput for low, medium and high mobility were equal to 569kbps, 567kbps and 568kbps, respectively. The growth in throughput for the low mobility of the SS was 49.25 %, for its medium mobility the growth was 250% and for its high mobility was 517%. Table 4 shows the comparative values of the throughput between simulations with and without the handover policy.

Mobility	No Handover Policies	With Handover Policies	Growth
Low	400,17	569,96	49,25%
Medium	151,44	567,91	250%
High	91,7	568,04	517%

Table 4. Average Throughput

4.2 Video traffic

The simulations with video have durations of 70 seconds and during this period, the video traffic was generated by the CN and sent to the SSs in an uninterrupted form. Table 5 shows the parameters set for the video simulations.

Parameters	**Value**
Resolution	352 x 288
Frame Rate	30 Frame/sec
Color Scale	Y, U, V
Packet Length	1052
Packet Fragmentation	1024

Table 5. Simulation of Video Parameters

First, the simulations were performed without the handover policy. In the simulations conducted in this way, suggest that SSs are not connected during the corresponding time of the handover process and resulted in lost packets. Following this, the simulations were performed with the handover policy in the same scenarios and in the same circumstances as those of previous simulations. The SSs experienced a seamless handover, when the video quality was maintained during the change of BS. The SS that experienced a hard handover did not receive 5% of the packets, and as a result, there was, a reduction in the quality of the video. Figure 13 compares the number of frames received for each situation.

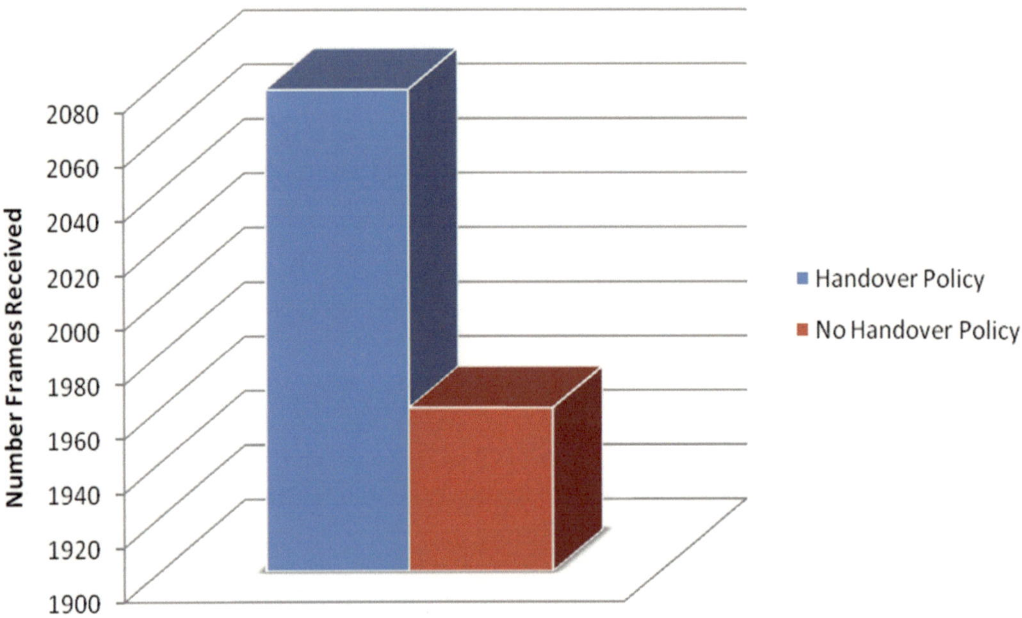

Fig. 13. Number of Decoded Frames

As well as the QoS analysis of the handover in the network architecture, we also investigated the impact of the handover on user perceptions. This was carried out by using the Evalvid tool (Evalvid, 2011) that allows control of real video quality called "Bridge (far)" or (Bridge (far) in simulations.

The benefits of the proposed solution are clear when we look at the frames in Figures 14, 15, 16 and 17. Figures 15 e 17 show frames of video received by the SS during the seamless handover. It was possible to ensure the highest video quality throughout the transmission. However, when the hard handover is experienced, the video quality is noticeably degraded. In addition, some objects in the picture are not received, as shown in Figure 14. Due to user mobility, the object containing the "bird" was not received and, thus has, not been decoded.

Fig. 14. Frame without a handover policy

Fig. 15. Frame with a handover policy

When Figure 16 and 17 are compared, it is clear that there is degradation in the quality of the frame without a handover policy.

Fig. 16. Frame degraded without a handover policy

Fig. 17. Frame with a handover policy

The QoE metrics confirm the previous statement; the video with a handover policy has 32dB PSNR. This value describes the video as "good", while the video without a handover policy has 29dB PSNR. This value describes the video as "acceptable." Figures 18 below show the similarities between the videos.

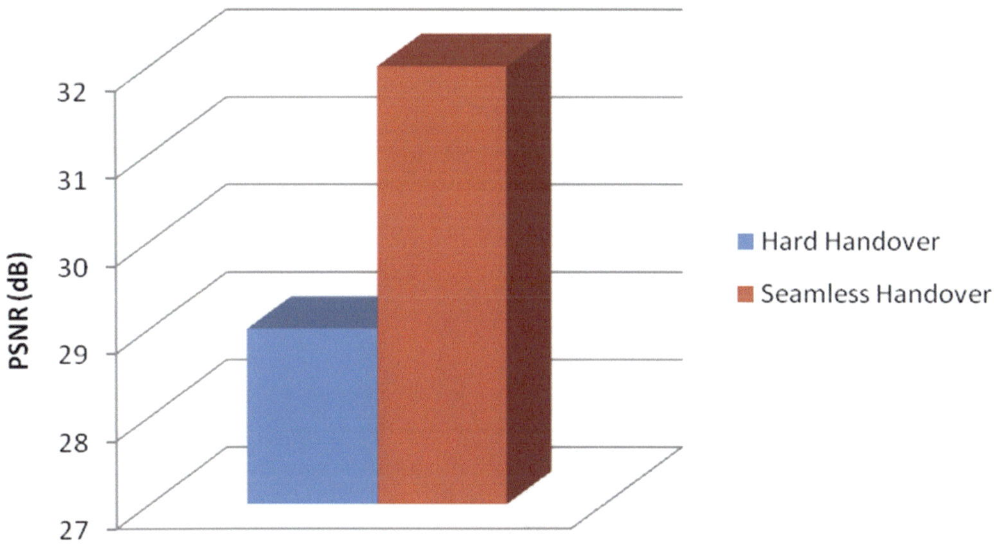

Fig. 18. Video PSRN without a handover policy x Video PSNR with a handover policy

Apart from PSNR, another metric that confirms the superiority of the video with a handover policy over the video without it, is SSIM. The value 1 means the exact same video. The SSIM for the video with seamless handover was 0.9. For the video with hard handover, the SSIM was equal to 0.7. Figures 19 below display the SSIM video.

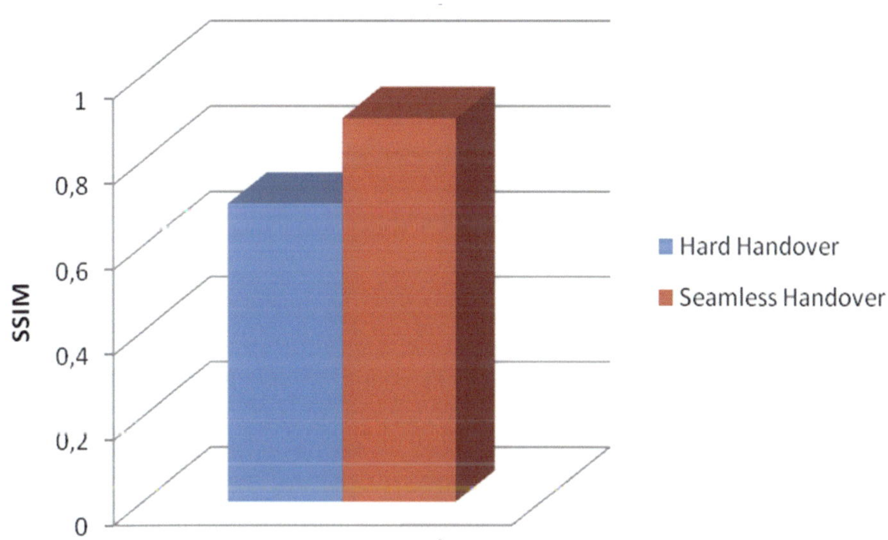

Fig. 19. Video SSIM without a handover policy x Video SSIM with a handover policy

The Video Quality Metrics are considered the most complete metrics because compare the following aspects: noise, distortion and color. In this situation, the value 0 means the exact same video. The VQM for the video with seamless handover was 1.4. For the video with hard handover, the VQM was equal to 2.6. Figures 20 below display the VQM video.

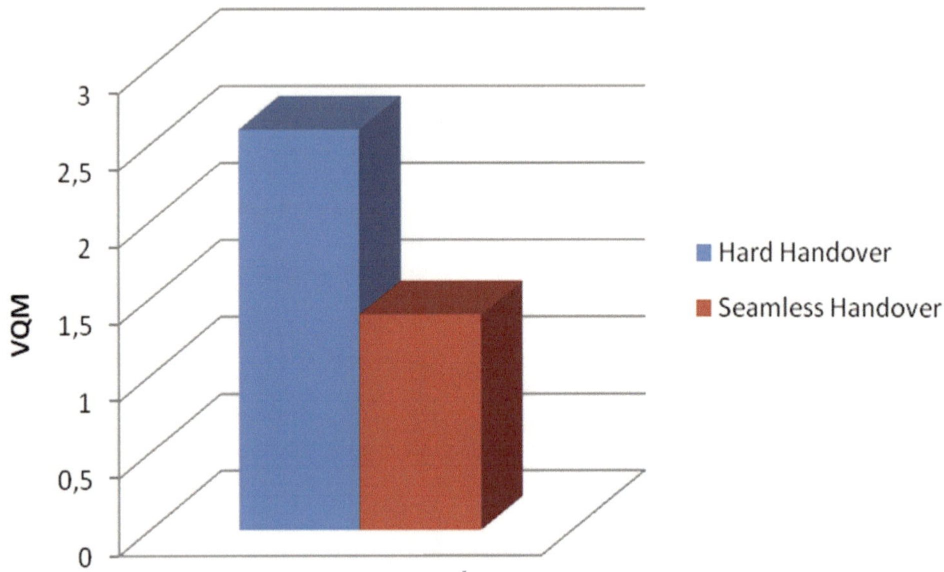

Fig. 20. Video VQM without a handover policy x Video VQM with a handover policy

5. Conclusion

In this chapter, a new architecture has been outlined that integrates the IEEE 802.16e, or as it is popularly known, the mobile WiMAX. This architecture draws on new technology and helps the handover process to provide the maximum QoS and QoE for the SS. It also includes a mobility prediction algorithm to avoid losses during the exchange of the BS. The algorithm takes account of the link quality between a mobile user and the current BS and the information about the SS received by GPS, which determines the moment when the handover should be triggered. Future work is recommended including new metrics in the algorithm, the performance of load balancing and and a plan to integrate other wireless technologies and thus form a heterogeneous architecture (e.g. the integration of WiMAX with UMTS and / or IEEE 802.11).

6. References

IEEE 802.16e (2005) Part 16: Air Interface for Fixed and Mobile Broadband Wireless Access Systems Amendment 2: Physical and Medium Access Control Layers for Combined Fixed and Mobile Operation in Licensed Bands.

M. Sollner, C. Gorg, K. Pentikousis, J. M. Cabero Lopez, M. Ponce de Leon, and P. Bertin (2008), "Mobility scenarios for the Future Internet: The 4WARD approach", in Proc. 11th International Symposium on Wireless Personal Multimedia Communications (WPMC), Saariselkä, Finland, September.

Neves, P.; Matos, R.; Sousa, B.; Landi, G.; Sargento, S.; Pentikousis, K.; Curado, M.; Piri, E.; (2009), "Mobility management for NGN WiMAX: Specification and implementation," *World of Wireless, Mobile and Multimedia Networks & Workshops,*

2009. WoWMoM 2009. IEEE International Symposium on a , vol., no., pp.1-10, 15-19 June 2009

Winkler S. (2005) , Perceptual video quality metrics - a review, Digital Video Image Quality and Perceptual Coding, eds. H. R. Wu, K. R. Rao, chap. 5, CRC Press.

ns-2 (2010) The Network Simulator (2009) http://www.isi.edu/nsnam/ns/

Ahmet et Al (2009), " A Survey f MAC based QoS implementations for WiMAX Networks".

Eteamed, Kamran (2008). "Overview of Mobile WiMAX Technology and Evolution". WiMAX, A Technology Update. Intel Corporation. IEEE Wireless Communications Magazine, Volume 46, Pages 31-40.

Iyer, Prakash; Natarajan, Nat; Venkatachalam, Muthaiah; Bedekar, Anand; Gonen, Eren; Etemad, Kamran; Taaghol, Pouya; (2007) "All-IP Network Architecture for Mobile WiMAX". IEEE Mobile WiMAX Symposium, Digital Object Identifier: 10.1109/WiMAX.2007.348700, Pages 54-59.

Ahmadi, Sassan. (2009) "An overview of next-generation mobile WiMAX technology". Intel Corporation. IEEE Wireless Communications Magazine, Volume 47, Digital Object Identifier: 10.1109/MCOM.2009.5116805, Pages 84-98, 2009.

Vaidehi, V. Poorani, M. (2010) "Study of Handoff Performance in a Mobile WiMAX Network". 11th International

Conference on Control, Automation Robotics & Vision (ICARCV), Digital Object Identifier: 10.1109/ICARCV.2010.5707316, Pages 2319-2324.

L. Chrost and A. Brachman, (2010) "Towards a common benchmark in WiMAX environment", Elsevier Computer Communications, July.

S.A. Ahson, M. Ilyas, (2007) "WiMAX: Standards and Security", CRC Press, Inc., Boca Raton, FL, USA.

Y. A. Sekercioglu, M. Ivanovich and A. Yegin, (2009) "A Survey of MAC based QoS implementation for WiMAX networks", Elsevier Computer Networks, May.

L. Nuaymi, (2007) "WiMAX Technology for Broadband wireless Access" John Wiley & Sons Ltd, the atrium, Southern gate, Chichester, West Sussex PO198SQ, England.

C. F. Ball, F. Treml, K. Ivanov, and E. Humburg. (2006) "Perfermance evaluation of ieee802.16 wimax with fixed and mobile subscribers in tight reuse". European Transactions on Telecommunications, March.

J. Chen, W. Jiao, and Q. Guo. (2005) "An integrated qos control architecture for ieee 802.16 broadband wireless access systems". Global Telecommunications Conference, November.

C. F. Ball, F. Treml, X. Gaube, and A. Klein. (2005) Performance analysis of temporary removal scheduling applied to mobile wimax scenarios in tight frequency reuse. European Transactions on Telecommunications, September.

Manner. K; (2004) " RFC 3753 – Mobility Related Terminology". June.

Dias et. al., (2008) "Approaches to Resource Reservation for Migrating Real-Time Sessions in Future Mobile Wireless Networks". Accepted for Publication in Springer Wireless Networks (WINET) Journal.

Evalvid. (2011) "http://www.tkn.tu-berlin.de/research/evalvid/"

NIST (2007) "http://www.nist.gov"
NIST WiMAX (2007) "http://www.nist.gov/itl/antd/emntg/ssm_tools.cfm"

10

Evaluation of QoS and QoE in Mobile WIMAX – Systematic Approach

Adam Flizikowski[1], Marcin Przybyszewski[2], Mateusz Majewski[2] and Witold Hołubowicz[3]
[1]University of Technology and Life Sciences, Bydgoszcz,
[2]ITTI Ltd., Poznań,
[3]University of Adam Mickiewicz, Poznań,
Poland

1. Introduction

International standardization organizations, responsible for preparing specifications (such as IMT-Advanced) for emerging 4G networks, define requirements for system level simulations for the candidate technologies [1], [2]. The goal behind those documents is to facilitate System-Level-Simulations by providing common methodology to perform such simulations (i.e. for WiMAX). According to [1] cell-level simulations can be an intermediate step between Link and System-level simulation where the capacity of a single cell and a single Base Station, providing service for multiple users, is evaluated by means of comprehensive tests. Still the IEEE standardized simulation methodology [1] does not specify how to evaluate (WiMAX) system capacity with various connection admission control mechanisms. Therefore as a first step we focus on the problem of adjusting simulation methodology to facilitate simulations covering CAC with Time Division Multiplexing Access scheme (TDMA), OFDM and uplink traffic. The applied evaluation methodology is derived from the best-practices in IEEE 802.16m Evaluation Methodology Document and WiMAX Forum's System-Level-Simulation (SLS) methodology. Afterward the introduced methodology is utilized to find answers to the following problems:

- To what extent does the capacity change when different FEC codes are deployed (Convolutional Turbo Coding - CTC, non binary Low-Density Parity-Check - nbLDPC)
- What is the user perception of the service quality (Quality of Experience - QoE) and what are the differences in the system performance when different FEC codes are deployed?
- How to improve resource estimation, especially when considering connection requests arriving in large batches?
- How the performance of traffic - aware admission control algorithms changes, when some users follow VoIP traffic pattern with silence-suppression enabled?
- Does the performance of measurement based CAC change, if the system experiences situations, in which connection requests arrive in large batches?

Since QoS support is an important part of WiMAX network, the system under test (SUT) controls resources using admission control (AC) mechanism. Arrival Rate aided Admission Control (ARAC) and its predecessor EMA - based Admission Control (EMAC) [41] are designed for controlling the VBR traffic. Moreover ARAC can cope with the problem of connections arriving in large batches. EMAC relies on calculating simple exponential weighted moving average (EWMA) of the overall resource consumption. ARAC differentiates between new and ongoing connections thus providing more accurate resource estimations.

To improve the fidelity level of the simulator and introduce mobile channels, method called Link-To-System interface (L2S) has been implemented. This approach removes constraints that arise when AWGN channel is being used. In particular a method based on mutual information (MI) called RBIR (Mutual Information Per Received Bit | Received coded Bit Information Rate) was selected. It is important, since attempting to simulate scenarios close to reality requires combining admission control and user mobility. The mobility model used is based on traces following the Leavy-walk distribution. Users' movements have been captured for a given geographical area and combined with maps generated by the Radio Mobile radio coverage planning tool [4]. Thus we are able to present results of assessing quality of VoIP (Voice Over IP) conversations also in the case of novel non-binary Low Density Parity-Check (nb-LPDC) coded WiMAX networks. The corresponding work is described within this chapter.

Finally, using L2S technique allows comparing SUT's performance using either nbLDPC or well-recognised CTC codes. Thus we eventually provide a comparison of CTC and nbLDPC codes in terms of resulting system capacity and quality of experience (QoE) as perceived by VoIP flows - it is shown that DaVINCI codes perform slightly better than CTC in the total cell utilization and decreased dropping probability. The QoE metrics measured show slightly more users are satisfied in a single cell with DaVINCI codes than when CTC is used.

The rest of the chapter is organized as follows: in Section 2 authors describe the related work and provide background information on previous work dealing with CAC and QoE in WiMAX networks. In Section 3 the authors provide information on how to evaluate WiMAX with CAC and compare this methodology with standardized SLS simulation approaches. In Section 4 a description of ns-2 and Matlab integration using Link-To-System (L2S) mapping can be found. Additionally information on simulator configuration is given. In Section 5 authors present the results collected for nbLDPC and CTC codes in QoS-aware WiMAX system. Discussion on QoE results is provided in Section 6. The authors conclude with Section 7.

2. Related work

The concept of QoS in broadband wireless networks has evolved during the past decade. More and more resource consuming applications emerge and by the time IPv6 protocol has been fully deployed, QoS capable systems will play an important role in IP-based wireless broadband networks. The importance of how QoS-aware networks can influence future wireless traffic is presented in [7] where authors compare existing QoS framework for WiMAX and LTE. The emphasis is put on the main differences in handling QoS in both 4G systems. Even though the underling technologies differ in many aspects, it is important to note that future 4G candidate networks are designed to provide services with guaranteed quality. Therefore QoS-aware mechanisms like Connection Admission Control or Packet Scheduling are to be deployed in order to align network capabilities with user needs and expectations when using a service [8].

Admission control algorithms can be classified according to method used to assess current system load. In parameter - based admission control (PBAC or DBAC) information about current state of the system's available resources is based solely on declarations made by applications. Therefore the performance of this kind of admission control is highly dependent on accuracy of the declarations, availability and types (depending on the system) of descriptors. Another approach is to use traffic measurements to estimate the current system load. This technique is used by MBAC (measurement - based admission control) algorithms.

One of the challenges is to estimate the incoming traffic characteristics using only provided descriptors. Especially it can prove hard to estimate required resources in a system utilizing Adaptive Coding and Modulation (ACM). Applications usually express their bandwidth requirements in bits (bytes) per second. In OFDM/ OFDMA systems utilizing ACM each user can use coding and modulation scheme most appropriate to his channel conditions. Therefore even a an application generating constant amount of traffic can require different number of OFDM symbols (/OFDMA slots). Therefore achieved transfer rates of a wireless link can vary significantly over short period of time. This adds a "second dimension" to the problem of estimating resources required by an application, since it is hard to predict how particular channel conditions will vary over time. This is in contrast to classic approach to admission control, where capacity of a link in terms of a maximum throughput / number of calls is considered constant. As a consequence, in such an ACM-enabled system, OFDM symbols (or slots for OFDMA) should be considered a scarce resource, since number of symbols available for a given system remains constant. PBAC algorithms seem more suited for systems where it is easy to properly describe flow characteristics (e.g. CBR traffic is usually easily described) and the required slots / symbols of a given flow do not fluctuate significantly over time (due to e.g. variations channel conditions).

The problem of estimating free resources can be mitigated (to some extent) by focusing on MBAC algorithms coupled with appropriate congestion control algorithms. MBAC algorithms are appropriate for systems where flow characteristics are not easily defined (or available traffic descriptors are not sufficient) and the required slots / symbols of a given flow can fluctuate significantly over time (due to e.g. variations in channel conditions). Although new connections requirements still have to be obtained through declarations, the percentage of bandwidth being used in reality by ongoing connections is known (usually at a base station level) thanks to measurements of traffic. If channel conditions of multiple users have became worse and the system approaches congestion, congestion control algorithm tries to minimize system load. This can be achieved in many ways, e.g. by signalling AC algorithm to block a part (or all) of the new connections requests, changing downlink / uplink scheduling priorities, or even by dropping some of the ongoing connections. Still it needs to be discussed, if e.g. dropping previously accepted connection is an acceptable congestion control policy. Still, few articles exist that are dedicated to this problem in admission control.

Nevertheless CAC in cellular networks has been a hot research topic for a few past years, since users' demand for mobile applications is constantly rising. A technique called Complete Sharing (CS) assumes that all connections are accepted as long as the system has sufficient resources to serve the new call / connection. This technique is the least complicated CAC algorithm and at the same time it is easy to implement. Another classic approach to admission control in cellular networks assumes allocation of dedicated resources for higher priority calls / connections (so called Guard Channel - GC) [9]. Guard Channel approach has been originally proposed in [10] for cellular networks. In this technique part of resources always remains reserved for higher priority connections (so called Fixed Guard Channel). This technique is adapted to WiMAX in [11] - [13] in order to prioritize handoff connections over arriving connection requests, thus ensuring required QoS for handoff connections. In Fixed Guard Channel, if there are multiple service classes present (as in e.g. WiMAX), an optimal value of guard channel is calculated usually using multidimensional Markov chains. However this process is relatively computationally intensive and may prove difficult to conduct in real-time for changing radio environment. This problem can be minimized by using a vector / table containing pre - defined, GC values optimal for a given traffic conditions [14]. Defining appropriate configurations for such a vector / table may prove hard / inefficient for systems with multiple classes of services, systems with ACM etc.

In [14] authors use reinforcement learning (Q-learning) algorithm to construct dynamic call admission control policies – TQ-CAC and NQ-CAC. TQ-CAC utilizes predefined tables, whereas NQ-CAC takes advantages of neural networks. This solution is evaluated for a cellular network with two classes of traffic. Both presented algorithms achieve lower blocking probabilities of handoff calls and higher rewards than simple greedy CAC scheme. Still, presented algorithms offer similar (NQ-CAC) or worse (TQ-CAC) performance - in terms of blocking probability - than simple guard channel approach.

Admission Control performance in LTE is described in [15]. Authors assume a single cell configuration to assess Uplink Admission Control where the admission criterion of the new user depends on the difference between the total and requested number of Physical Resource Blocks. Other results considering multi cell deployment scenarios are presented in [16] where authors describe and compare static and dynamic CAC in LTE. Additionally a delay-aware connection admission control algorithm is proposed and evaluated. Other approaches for ensuring QoS in LTE networks can be found in papers [17], [18].

On the other hand there are approaches aiming not only at assuring network service quality but also consider the quality as perceived by the end user. Perceived QoS (or Quality of Experience – QoE) is often considered as the "ultimate measure" of system performance. According to ITU-T one can describe QoS as the 'degree of objective service performance' and QoE as the 'overall acceptability of an application or service, as perceived subjectively by the end user' [19]. While QoS evaluation is only a matter of measuring vital network parameters, QoE measurements are much more complicated as they usually involve modelling the human component in the measurement process (in a direct or indirect manner). The user-centric QoE measurement process has been already conducted by ITU-T and captured in Recommendation P.800 [20]. The leading QoE evaluation method for voice is the Mean Opinion Score (MOS). This approach facilitates users' QoE assessment. When conducting subjective tests the MOS scale is used by users to rate the quality of the perceived audio signal. This makes such QoE measurement impractical as it requires time, resources and equipment. Therefore objective measurement approaches are used to estimate user QoE without the direct involvement of the user itself. A number of QoE measuring methods has been proposed during past years, each of them designed to capture perception relevant measurements (voice, audio). During the DAVINCI project authors have tackled the problem of voice quality measurements for VoIP in wireless IP systems.

Different approaches are proposed and a variety of solutions are investigated on how to evaluate VoIP quality over a wireless link - but only a fraction of them considers WiMAX networks. Some articles focus on the subjective measurement approach as a method for evaluating quality of experience [21] [22] and some try to correlate the subjective measurements with objective approach [23]. Objective approach measurements usually use PESQ (Perceptual Evaluation of Speech Quality) or PSQA (Perceptual Speech Quality Assessment) [24], [25], [26]. Both methods are suitable for single device (telephone) quality assessment but require expensive hardware and laboratory. Due to the constraints present in PESQ and PSQA other objective measurement approaches are proposed. The e-model approach was described in several publications [27], [28], [29] as a method for evaluating QoE over a wireless link using VoIP applications. Variations of the e-model implementation [30] as well as new approaches [31], [32] are investigated to evaluate QoE under QoS-aware mobility mechanisms [33]. In this paper authors focus on QoE solutions designed for wireless environments, especially WiMAX systems [19] [34]. The following section reviews the System-Level Simulation methodology and introduces Cell-Level simulation in WiMAX.

3. Cell-level versus system-level simulations

Link-level simulations are typically performed at the first stage of evaluation of a radio technology to provide results and fundamental knowledge of the behaviour of the air interface. Key performance indicators include spectral efficiency, robustness of the codes and modulations, influence of the HPA non linearity and so on. Usually such analysis is accomplished by performing simulations in an environment limited to transmitter and receiver circuitry. The role of PHY Layer simulation is to capture the relevant factors which influence the transmitted signal and to provide basic understanding of radio link-level performance. Real-world WiMAX network deployments are by definition attached to particular geographical area where multiple base stations provide service to hundreds of moving users in an environment characterized by path loss, signal distraction and fading. To evaluate performance of such system with novel FEC codes the standardized system-level simulation methodology has to be considered [1]. The extension of the link-level simulation towards system-level simulation may start by adding multiple users in one cell as defined in [1] and [2]. Numerous studies were conducted towards development of System-Level Simulations methodology and the mandatory recommendations to perform them are given in [1] and [2]. However the above documents do not state how to asses performance of WiMAX with Call Admission Control algorithms. To perform simulations with CAC algorithms authors narrow the scope to a Cell-Level Based approach as presented in Fig. 1.

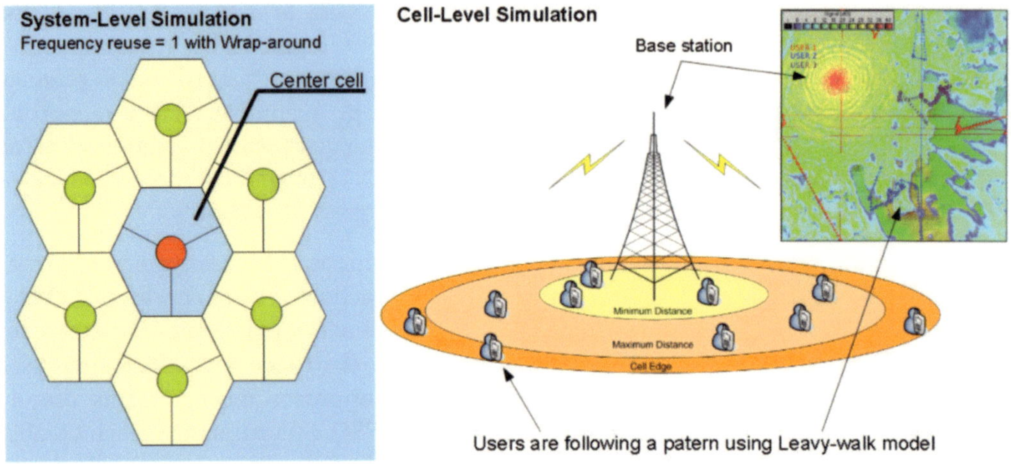

Fig. 1. System-Level Simulations versus Cell-Level Simulations own elaboration based on [1], [5]

As opposed to the approach described in [1] authors deploy one cell with single base station with no cell sectorization (as presented in Fig. 1). This straightforward approach is more suitable for simulations with CAC as it can produce results closer to reality by providing the control of the user movement patterns (conforming to Leavy-Walk model [35]) and apply them in a real-life scenario by generating maps with SNR distribution using the Radio Mobile application. In a limited geographical area the movement of mobile users is usually predictable. People are driving or walking to work/school each day taking the same path. In the end they follow a specific pattern on a day-by-day basis [49]. The SNR conditions of each user's channel may vary and depend also on the exact user location at a given moment. This observation is the underlying assumption for our methodology. We first assign a specific mobility pattern to each user. After aligning this pattern with the underlying map, we pick particular SNR values which correspond to the signal strength distribution on the map. Finally this procedure provides us with SNR trace files for our simulator. Each scenario can be repeated numerous times to increase reliability of results. Thus, even though users will take the same path each time, SNR distribution may change due to fading and path loss. The SNR matrices were prepared using the Radio Mobile application. The matrices represent two distinct geographical areas - rural and hilly terrains, both limited to 16 square kilometres. Mobility models are generated using Matlab source files provided by [35]. Radio mobile uses the ITS (Irregular Terrain Model) radio propagation model, developed by Longley & Rice. All calculations in this model are based on the distance of a terminal and the variation of the signal. Signal frequency can vary from 20 MHz to 20 GHz. This general purpose model is used in many fields of science, and can be utilized for WiMAX based network simulations. In the following section the simulation environment based on concept of L2S interface is described.

4. Link to System (L2S) interface

In a real cell-deployment user traffic flows are influenced by various transmission impairments of the air interface. Thus it is important to provide an accurate channel model which captures the channel characteristics to provide conditions closer to reality. As a preliminary work on WiMAX system performance authors have investigated the capabilities of the NS2 NIST patch and implemented (literature based) Guard-channel based CAC algorithms to measure the performance with nbLDPC codes. The outcome was the development of VIMACCS patch which includes mechanisms for Connection Admission Control deployed for cell level simulation. Implemented and evaluated CAC algorithms for nbLDPC codes included Complete Sharing CAC (CSCAC), Dynamic Hierarchical CAC (DHCAC) and Fair CAC (FCAC) [3][6].

The evident challenges in acquiring reliable simulation environment arise from numerous facts related to physical layer with nbLDPC FEC codes: computational complexity of nbLDPC decoder, the need of adapting decoder implementation to external cell level simulator requirements, requirement for facilitating multiple OFDM subcarriers experiencing different channel conditions.

In the first stage of development it was clear that the (FEC decoder) integration process would be computationally demanding [36]. At that time the available implementation of nbLDPC codes was not optimized for real-time transmission. Thus the decoding process took too much time to be executed on a standard PC with event based simulator in the loop. To reduce the excessive simulation times a method based on effective Signal-to-Noise-and-Interference computation has been evaluated and integrated into Matlab. This method is used to produce a PHY Layer abstraction which in turn can be deployed with different realizations of the decoder. By using eSINR computation we can omit the need for implementing the decoder and in result decrease the computation time. This method is described in the evaluation methodology documents [1] [2] and referred to as the Link-To-System mapping interface. First we compute the AWGN vs. CWER curves for every Modulation Coding Scheme (MCS) using the nbLDPC decoder. The results are not only useful for the PHY Layer abstraction but also provide basic information about the link-level performance. Once the AWGN vs. CWER lookup tables have been generated they can be used to predict the CWER value in mobile non-linear channels. In result we obtain AWGN Lookup Tables (LUTs) which, when used together with a L2S interface, can be used instead of the decoder itself and provide accurate CWER prediction in mobile channels. For more information about performing effective SINR computation the reader is referred to [37] and [38]. Authors decided to use a method based on Mutual Information [1] [37]. In particular the Mutual Information Per Received Bit (RBIR) method was implemented. The Mutual Information is calculated according to formula:

$$SI(SINR_n, m(n)) = \log_2 M - \frac{1}{M}\sum_{m=1}^{M} E_U\left\{\log_2\left(1 + \sum_{k=1,k\neq m}^{M} \exp\left[-\frac{|X_k - X_m + U|^2 - |U|^2}{1/SINR_n}\right]\right)\right\} \quad (1)$$

In the above equation we take U as the zero mean complex Gaussian with variance ½($SINR_n$) per OFDM symbol, where $SINR_n$ is the post-equalizer SINR at the n-th symbol or sub-carrier; *m(n)* is the number of bits at the n-th symbol (or sub-carrier) and *X* is the constellation alphabet. Now assuming that a number of *N* subcarriers was used to transmit a codeword (in case FFT-256 is used *N* is equal to 192) then the normalized mutual information per received bit (RBIR) is given by:

$$RBIR = \frac{\sum_{n=1}^{N} SI(SINR_n, m(n))}{\sum_{n=1}^{N} m(n)} \tag{2}$$

Eventually the above mentioned equations are used to model the behaviour of a mobile radio channel and to generate LUT tables with ESINR values. The LUT tables follow the behaviour of physical layer with a decoder implementation, but without the complexity trade-off. In turn L2S can be used within NS2 simulator to provide more realistic results for simulations with CAC in mobile channels. Since we want to compare system capacity/performance for given FEC schemes, two distinct LUT tables were generated - one for nbLDPC codes and one for CTC.

Network configuration parameters	**Value**
Carrier frequency	3.5 GHz
Bandwidth	3.5 MHz
Number of sub-carriers	256
Number of data sub-carriers	192
Cyclic prefix	1/8
Modulation	QPSK, 16-QAM, 64-QAM
Coding scheme	nbLDPC, CTC
Codeword length	48, 96, 144, 288
Rates	1/2, 2/3,3/4, 5/6
Velocity	0.83 m/s
Scheduler	Priority scheduler
Traffic type	UDP CBR or VBR
Transmission direction	Uplink

Table 1. Configuration parameters for integrated simulator

The LUTs were calculated with the assumption that Adaptive Modulation and Coding (AMC) mechanism is enabled thus the target CWER value of ca. 10^{-3} is selected – see Table 5 for details. A more detailed simulator configuration is provided in Table 1.

5. Connection admission control performance assessment - WiMAX networks

This section presents the results of test methodology that focused on the three major questions:

- What is the system capacity and performance when different FEC codes are deployed (CTC, nbLDPC) under declaration based admission control and varying system load?
- To what extent does the capacity change if some users follow the VoIP traffic pattern with silence-suppression enabled – depending on the admission control algorithm used (EMAC, ARAC)?

The above questions have been assessed by applying the testing methodology that assumes worst case user mobility [39]. In simulations with admission control we decided to follow an approach similar to the one presented in [40]. This approach assumes that admission control could be triggered not only by the arrival of a new connection request. Such an approach seems logical in a system utilizing adaptive coding and modulation, since resource requirements of a given connection can change over time. Therefore admission control is triggered in situations when:

- new connection request arrives
- peer's MCS (Modulation and Coding Scheme) changes
- parameters of a given service flow have been changed.

Since admission control is triggered also when parameters of a given flow have been changed, admission control algorithms are functioning also as Congestion Control algorithms. In this chapter we have evaluated the three following admission control algorithms:

- Complete Sharing Admission Control (CSCAC)
- EMA - based Admission Control (EMAC)
- Arrival Rate aided Admission Control (ARAC) - modified version of the algorithm proposed in [41].

Complete Sharing Admission Control is a parameter based admission control making admission decision based on the declarations provided by arriving connections requests. Connections are accepted as long as there are free resources available at the base station. CSCAC is used in simulations with nbLDPC and CTC codes (section 5.2)

Moreover two measurement-based admission control algorithms (MBAC) have been compared (section 5.3). First we propose a measurement based connection admission control algorithm for the CAC module, which is aware of the current network state and is able to cope with the problem of batch arrivals. It is called Arrival Rate aided Admission Control (ARCAC or ARAC) and represents another approach to Measurement-Aided Admission Control (MAAC) algorithm presented in [41]. We then compare the proposed ARAC algorithm with algorithm utilizing exponentially moving average (EMA-MBAC) this algorithm has also been presented in [41] and in this chapter is referred to as EMAC. Since EMAC does not provide protection against problem of estimating resources when connections start arriving in large batches (EMAC underestimates number of used symbols - Fig. 2), in [41] authors propose a threshold – based solution.

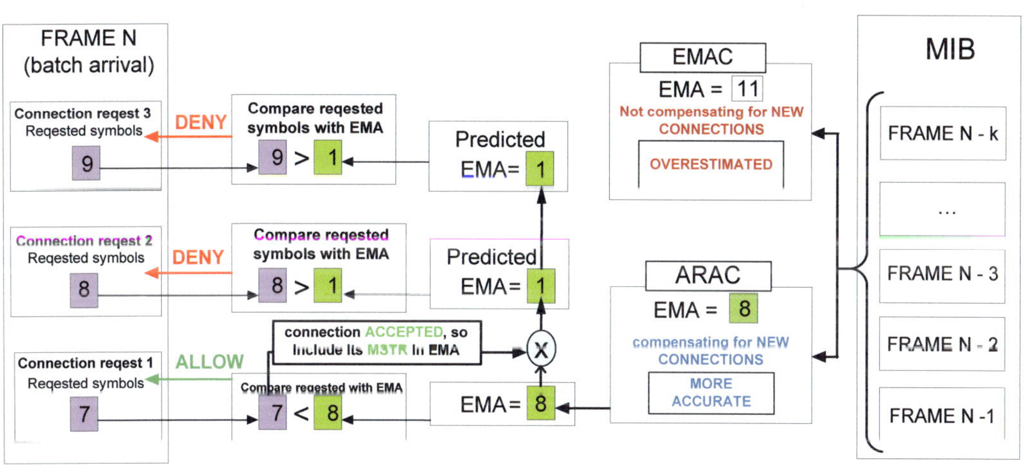

Fig. 2. EMAC vs. ARAC – example of the process of estimating resources for four frames

Value of guard channel (threshold) is adjusted based on the value of the declared Minimum Reserved Traffic Rate (MRTR) of existing connections and recent bandwidth utilization. Instead of using predefined thresholds, the proposed ARCAC takes an advantage of the fact that Base Station (BS) has the ability to monitor information about current arrival rate. Based

```
EMAC:

for each frame Ni in the current measurement window Twindow:
{
      S_all = number of all symbols of frame Ni
      //S_used = sum of symbols used by ongoing connections during frame Ni
      for each connection Cj:
      {
        S_Cj = symbols used by Cjth connection during frame Ni
        S_used = S_used + S_Cj
      }

      S_req = number of symbols required by incoming new connection (based on connection's MSTR)


      //compute predicted free symbols during Nith frame
      S_pred_i = (S_all - S_used - S_req ) / S_all
      treat S_pred_i as the ith sample for EMA calculations
}

when finished compute EMA of free resources

are there enough free resources to accept connection?
YES -> accept connection
NO -> reject connection
```

Fig. 3. Pseudo code for EMAC algorithm

```
for each frame Ni in the current measurement window Twindow:
{
      S_all = number of all symbols of frame Ni
      // S_used = sum of symbols used during frame Ni taking into consideration:
        // a) connections ongoing during Nith frame
        // b) freshly accepted connections that did not exist during Nith frame

      for each connection Cj:
      {
        Tconn = time the Cj connection exists
        // ? does the connection Cj exist longer than the measurement window?
        // and ? did the connection Cj exist already in the current Nith frame?
        if (Tconn > Twindow and Tconn > TcurrentFrame) {
              // the connection did exist during current frame
              S_Cj = symbols used by Cjth connection during frame Ni
        } else {
              // the connection did not exist during current frame
              S_Cj = predict symbols that the Cjth connection would have used using e.g. MSTR
        }
        S_used = S_used + S_Cj;
      }
      S_req = number of symbols required by incoming new connection (based on connection's MSTR)

      //compute predicted free symbols during Nith frame
      S_pred_i = (S_all - S_used - S_req ) / S_all
      treat S_pred_i as the ith sample for EMA calculations
}

when finished compute EMA of free resources

are there enough free resources to accept connection?
YES -> accept connection
NO -> reject connection
```

Fig. 4. Pseudo code for ARAC algorithm

on this value BS calculates, if the measured EMA of resources used does take into consideration recently accepted connections. If connection requests start arriving in large batches, in order to predict future value of average free symbols ARCAC also takes into consideration QoS parameters (e.g. MSTR) of connections that have already been accepted, but do not exist long enough to influence average symbols utilization (Fig. 2). Below we present the pseudo - code of both EMAC (Fig. 3) and ARAC (Fig. 4).

5.1 Traffic characteristics for simulations with CAC

All simulated nodes are generating VoIP traffic which is widely used for its suitability for evaluating QoS performance (stringent QoS requirements) although large number of streams is needed to shift the system under test towards it's capacity limits. There are two types of traffic characteristics used throughout the simulation - namely CBR (Constant Bit Rate) and VBR (Variable Bit Rate) streams. The contributing nodes include thirty WiMAX nodes for each simulation, although intensity of the requests for connections sent by each one is governed by generator that fulfils requirements of a given arrival rate.

The VBR flows are represented by VoIP traffic streams conforming to the ON-OFF distribution typical for voice codecs with silence-suppression. Thus depending on the type of codec used user packets are classified as the UGS traffic class (CBR) or rtPS (when silence-suppression is used). The UGS connections are transmitting packets with CBR and 64 kbps. For VBR rtPS VoIP we use two codecs - namely G.711 and G.729 with "one-to-one" voice detection model. In order to use realistic VoIP traffic models, the NS2 VoIP traffic generators developed as part of EuQoS European project [42] were integrated into our simulator (ViMACCS).

All simulated users are assumed to be mobile. Their mobility path follows the well-known mobility pattern - namely Leavy-walk distribution. To increase the reality of the simulated environment a COTS tool for coverage planning was used to provide SNR distribution in a given geographic area. Since the first aim of early stages of measurements (section 5.2) was to evaluate system capacity, it was essential to overload the base station. This condition can be achieved sooner if large (1000B) packets are being used. On the other hand, in order to fulfil the requirements of the ITU G.107 QoE method, packets should be small (64B). Thus the results in section 6.3 are following similar configuration but with smaller packets. The following section shows the results obtained during cell-level simulations with CAC.

5.2 Parameter based admission and congestion control with nbLDPC and CTC FEC schemes

In this scenario we compare results obtained for the two aforementioned FEC schemes - nbLDPC and CTC. We assume "worst - case" scenario where all users are moving in a dynamically changing SNR environment.

As mentioned before, user mobility patterns are generated according to the Leavy-Walk model [35]. SNR map has been generated for two villages - one near the city of Warsaw (Poland) and one near the city of Katowice (Poland). The *Map 1* represents good SNR conditions (on average) whereas *Map 2* mimics a bad SNR environment. The arrival rate of user requests follows Poisson process. The CSCAC is configured to handle both admission and congestion control algorithm. Simulation parameters have been presented in Table 2. The code word error rate (CWER) for both FEC schemes in presence of ACM is assumed to be 1%. All simulations have been repeated 20 times in order to increase statistical reliability of results. All figures present average values together with 95% confidence interval.

For simulations with nbLDPC we can observe lower Dropping Probabilities (Fig. 5) than for simulations with CTC. This is due to less MCS transitions (Fig. 6) as for similar simulation conditions there is less MCS changes for the nbLDPC codes. This results in average system throughput being slightly higher (by 5-10%) for simulations with the nbLDPC FEC (Fig. 7) as less resources are freed prematurely due to connections being dropped. This also finds reflection in BW utilization, which is slightly higher for nbLDPC (Fig. 8), and Blocking Probability (Fig. 9), which is higher for nbLDPC (fewer resources freed prematurely means higher probability that new connection requests will be rejected due to insufficient resources). It has to be noted that data connection's MCS change triggers admission control - thus in high mobility scenarios the offered traffic arrival rate should be adjusted by the average number of instantaneous MCS changes to make it realistic from a resource point of view.

Network configuration parameters	**Value**
Arrival rate	20 to 140 conn/minute (Poisson)
SF class	UGS
Average Connection Time	20 s (exponential)
Traffic characteristics	UDP CBR (1000 B at 20 ms)
FEC	CTC | nbLDPC
L2S	Enabled
MAP	Enabled - MAP 1; MAP 2
Simulation time	200 s
CAC	CSCAC (parameter - based)
Congestion Control	Enabled
Scenario Repetitions	20
CWER	0.01

Table 2. Configuration for CAC simulation with two FEC schemes

Network configuration parameters	**Value**
Arrival rate	25 to 250 conn/minute (Poisson)
SF class	UGS | rtPS
Average Connection Time	20 s (exponential)
Traffic characteristics (Codecs)	G.711 G.729
Voice Detection Model	One-to-one
L2S	Enabled
MAP	MAP 1
Simulation time	200 s
CAC	MBCAC | ARCAC
Congestion Control	Enabled
Scenario Repetitions	8
FEC	nbLDPC
CWER	0.01

Table 3. Configuration for simulations with the two MBCAC algorithms

In case of a environment with lower average SNR values, the nbLDPC gain observed for Map 1 is still present for Map 2, but becomes almost negligible (e.g. in terms of average system throughput - Fig. 10). This is due to nature of nbLDPC codes, as nbLDPC gain is most visible for high order modulations. In case of low SNR, when more robust modulations are being used (e.g. QPSK), nbLDPC gain becomes insignificant. It is worth noticing, that results obtained in this section are similar to the results obtained by authors in [43] where DAVINCI/nbLDPC gain in average sector throughput has been found to be approximately 5% higher compared to that achieved with CTC codes.

5.3 Measurement based admission and congestion control with nbLDPC FEC scheme

In this section we compare two measurement based admission control (MBCAC or MBAC) algorithms. Approach to simulation environment remains the same as for section 5.2 although within the set of mobile nodes there are now 60% of users that use VoIP codecs with silence suppression enabled. For all VoIP sources we assumed one - to - one conversation model.

Simulations are conducted only for *Map1*. In order to be able to measure performance of MBCAC algorithms alongside CBR VoIP traffic we introduce VBR VoIP traffic with silence suppression, which is marked as rtPS traffic. The amount of nodes using each type of VoIP traffic is equal (eg. 10 users with G.711, 10 with G.729 and 10 with CBR). The nbLDPC (DAVINCI) FEC scheme is used for all simulations. As in previous section admission control algorithm is used also as a congestion control algorithm.

All the figures below present average values together with 95% confidence interval (outliers in the charts). Simulation parameters can be found in Table 3. Figures Fig. 11 to Fig. 13 present average delays for VoIP for both tested Admission Control algorithms. It can be observed, that all VoIP connections experience lower delays when ARAC is used as admission control algorithm.

The reason is that if multiple connection requests arrive in a short period of time, ARAC can estimate remaining resources more accurately than EMAC. This becomes more evident for high arrival rates. For G.711 codec and high arrival rates difference in delay reaches approximately 25% and for G.729 approximately 23%. These findings are in compliance with the results obtained by researchers in [41], where using EMAC algorithm also caused increase in delay. The highest sensitivity to increased arrival rate can be observed for VoIP connections with silence suppression. These streams are scheduled as rtPS service class (G.711 and G.729).

UGS always takes priority over rtPS, thus its delay remains virtually constant. At the same time Blocking Probability for ARAC is similar to EMAC (approx. 2% difference for high arrival rates - Fig. 14). If we assume, that each MCS change should trigger CAC algorithm (working as a congestion control), EMAC is characterized by moderately lower Dropping Probabilities (ap. 14% for high arrival rates - Fig. 15). Although delay observed for both CAC algorithms is still acceptable for VoIP conversation it should be noted, that tests have been conducted assuming end application is located in the local network adjacent to the BS serving the VoIP source, therefore assuming the core network delay to be "zero" between the caller and callee. Therefore it should be noted that depending on the core network delay (especially when it exceeds 80ms) the ARAC should be considered a more robust choice.

Results obtained in this section show that ARAC provides means to cope with batch arrivals. As it utilizes data available at BS rather than incrementally adjusts values of guard channel, it can be considered as an alternative choice to threshold - based solutions like MAAC presented in [41].

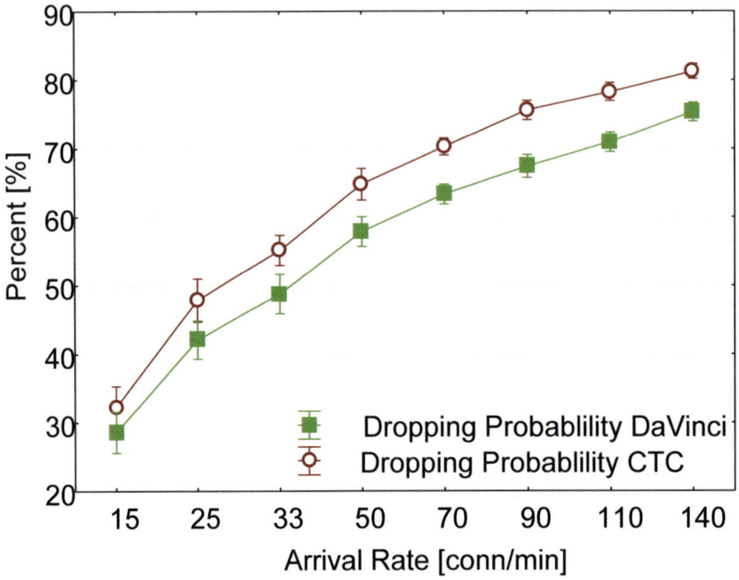

Fig. 5. Dropping Probabilities for DV and CTC *Map1*

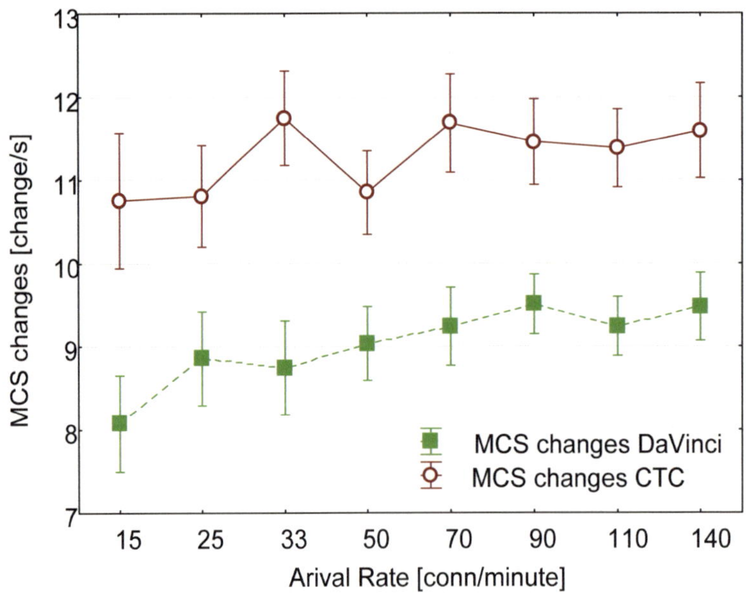

Fig. 6. MCS changes for DV and CTC *Map1*

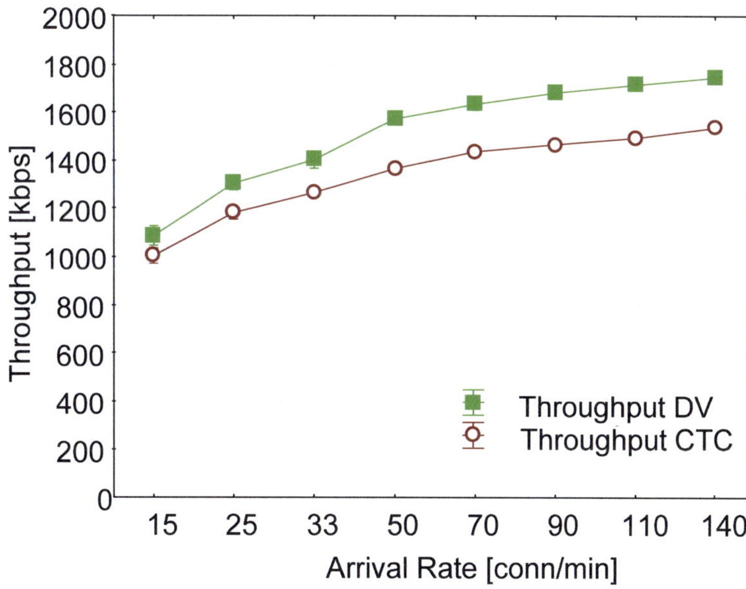

Fig. 7. System throughput for DV and CTC – *Map1*

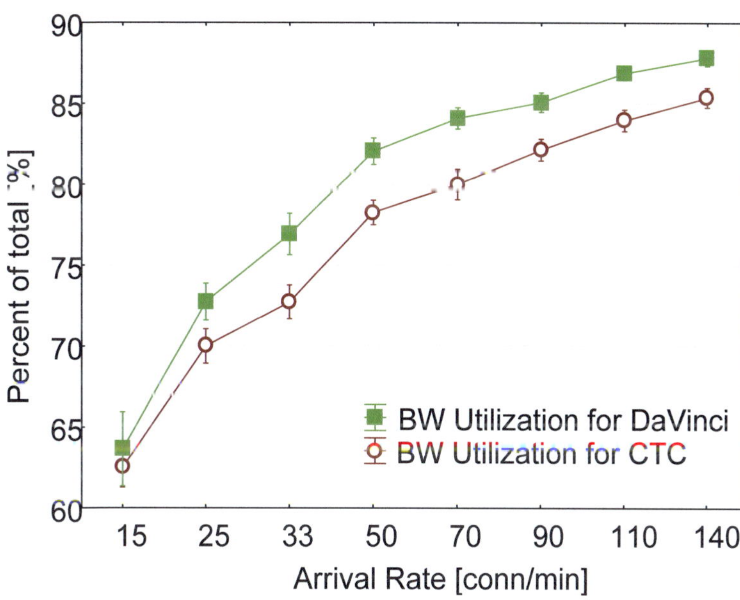

Fig. 8. Bandwidth utilization for DV and CTC – *Map 1*

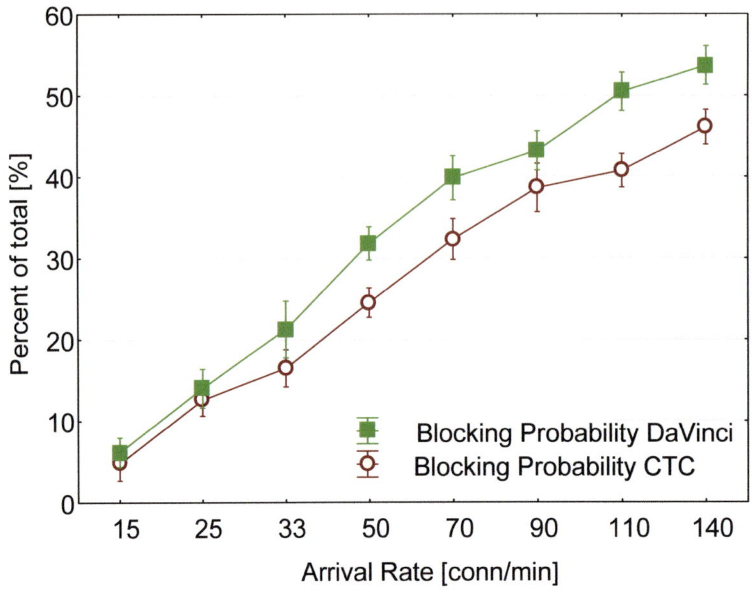

Fig. 9. Blocking probabilities for DV and CTC - *Map1*

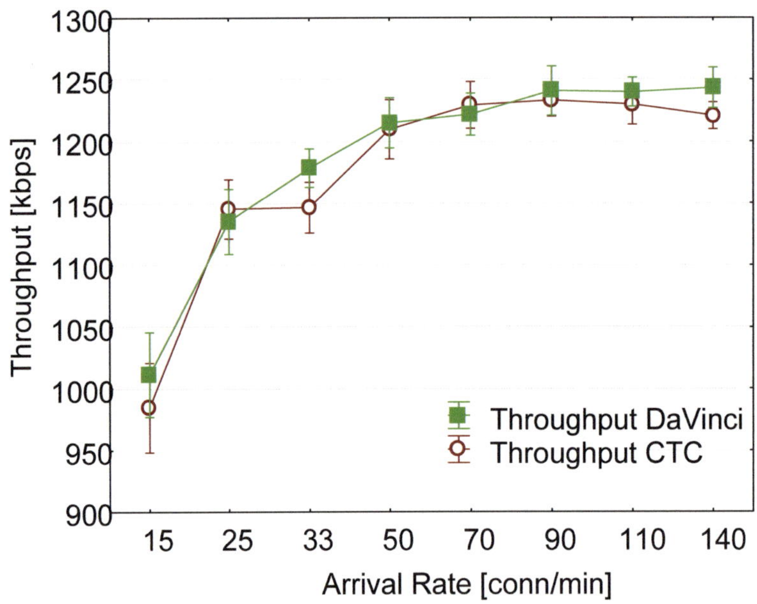

Fig. 10. System throughput for DV and CTC - *Map2*

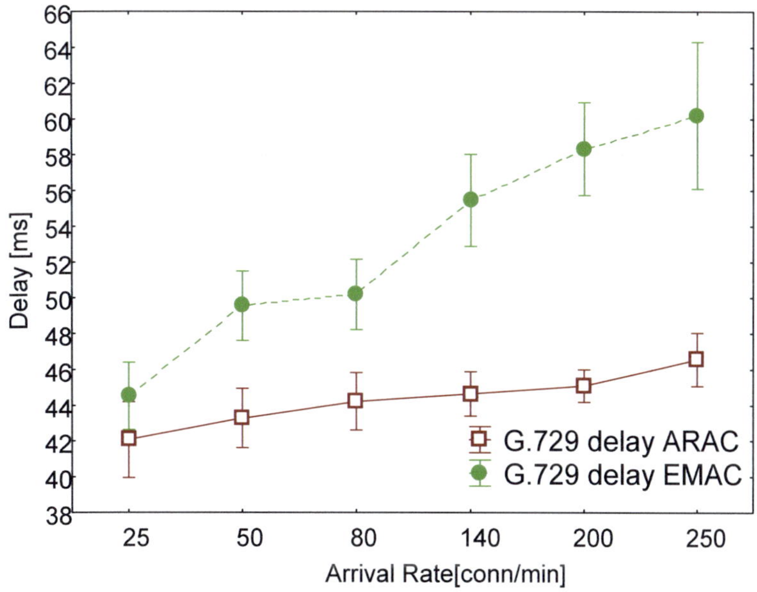

Fig. 11. G.729 VoIP delay for ARAC and EMAC (rtPS)

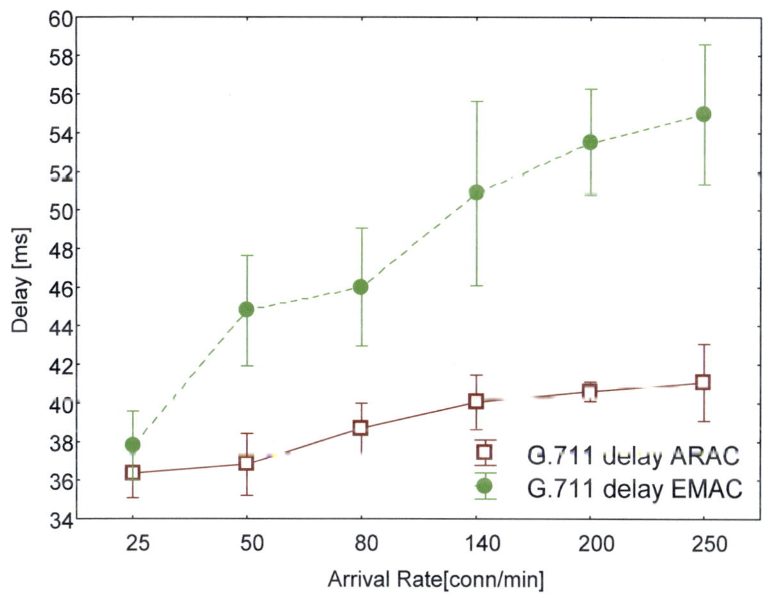

Fig. 12. G.711 VoIP delay for ARAC and EMAC (rtPS)

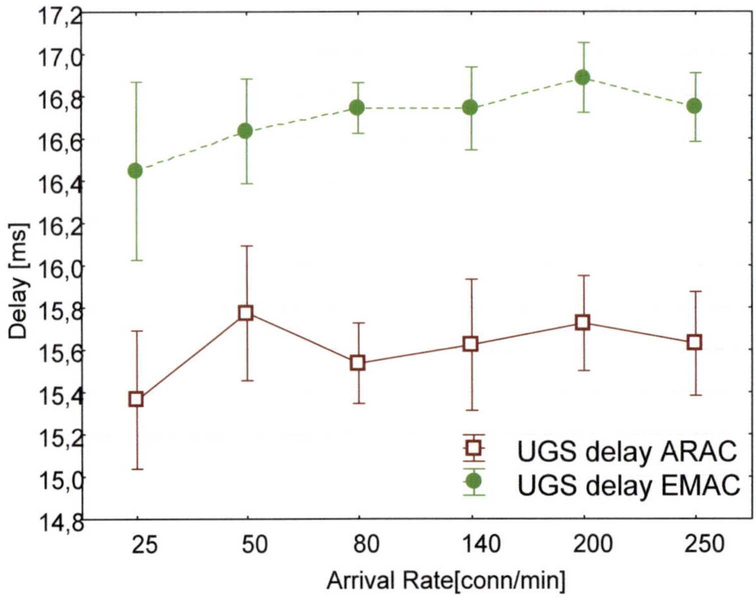

Fig. 13. CBR VoIP delay for ARAC and EMAC (UGS)

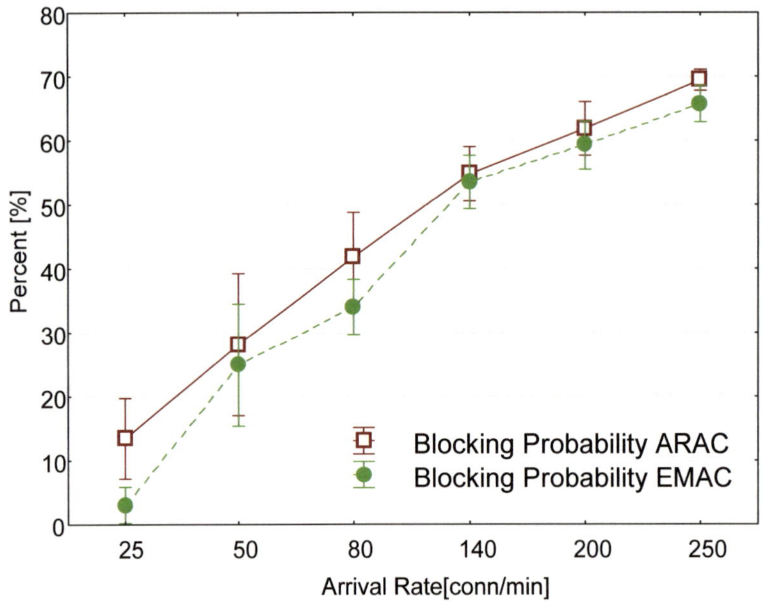

Fig. 14. Blocking probabilities for ARAC and EMAC

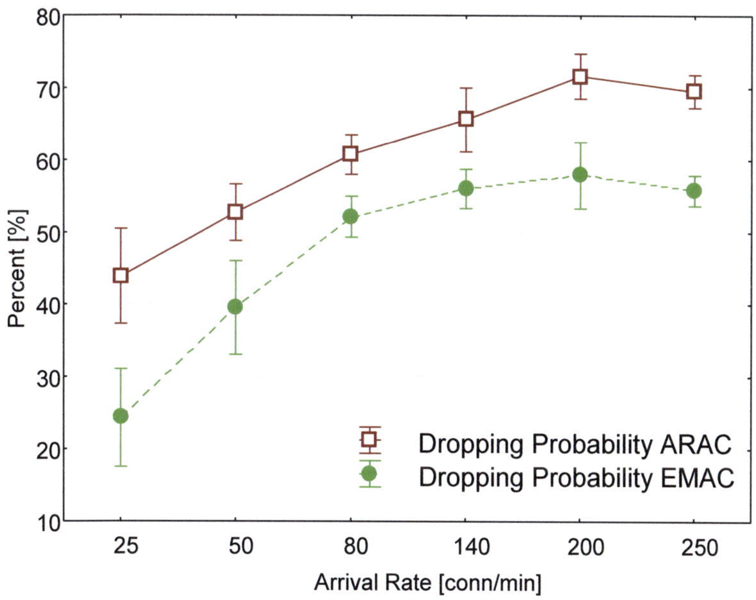

Fig. 15. Dropping probabilities for ARAC and EMAC

6. QoE VoIP performance assessment in WiMAX networks

Among key goals of our research was to assess the degree to which the new coding scheme affect the voice quality as perceived by the VoIP user using conversational service (VoIP). Since measurements using COTS HW implementing LDPC are not feasible (at the moment of writing) with nbLDPC codes authors implemented Matlab based E-model to estimate an appropriate grade of the signal quality in form of R-factor.

6.1 E-model for QoE calculation

The E-model (ITU G.107) was originally used to help PSTN network planners and telephone service providers to perform basic evaluation test for voice quality to determine the system requirements for telephone line [44]. However there are several publications which prove that a consistent and reliable approach towards the adoption of the E-model in an IP wireless environment for VoIP quality assessment is possible [45], [46]. The authors are using the simplified model that adjusts the equations defined by ITU-T for PSTN E-model to assess VoIP connection quality as proposed in [45]. The output of the E-model is calculated as follows:

$$R = 93.35 - I_d - I_e + A \tag{3}$$

Where I_d is the delay impairment and Ie the packet loss impairment. The calculated R-factor can be further used to map the objective measurement to subjective MOS scale resulting in an approximation of the user perceived quality. This allows overcoming the disadvantages of the subjective approach and achieve reliable results as shown in[45]. This approach has also been employed by authors in article [47].

6.2 Simulation parameters

In this subsection authors describe the simulation scenario used to perform QoE assessment. User's application is sending 200B voice packets in a 20 ms time interval. Each simulation run requires a number of repetitions and for each set of repetitions the number of users in the system increases as specified in Table 4. The number of users was chosen to show the point where the perceived quality falls bellow acceptable limits (from the point with most users satisfied to dissatisfied). Additionally in the scenarios users are moving at a constant speed of 3 km/h (pedestrian speed). They follow Leavy-walk mobility pattern on a map generated by radio planning tool [5]. In this scenario it is assumed that both CAC and congestion control algorithms are turned off. Simulations are performed for ACM with nbLDPC and CTC codes. The simulation parameters are gathered in Table 4.

The next section presents the results obtained during simulations with NS2 and the L2S physical layer abstraction interface (described above). The results include the delay, packet loss impairments and show how this parameters influence the perceived quality (in R-factor scale).

Parameter	Value
Nodes	30 to 33 (for MAP 1), 23 to 26 (for MAP 2)
SF class	UGS (no rtPS)
Traffic	UDP CBR (200 B at 20 ms)
FEC	CTC | nbLDPC
L2S	Enabled
MAP	Enabled (Map1, Map2)
Mobility	All users are mobile
Velocity	3 km/h
Simulation time	200 s (for MAP 1), 100 s (for MAP 2)
CAC	Disabled
Congestion Control	Disabled
Scenario Repetitions	6 (for MAP1), 3 (for MAP2)

Table 4. Parameters for simulation scenario

6.3 Results for VoIP QoE

In this subsection authors present the results of evaluating the QoE of a VoIP connection in WiMAX network. Authors measured latency (Fig. 16) and packet loss (Fig. 17) as a function of the number of active users in the system. The measurements were conducted for both maps. The captured parameter values were fed into the E-model equations for computing the R-factor Fig. 18.

The resulting R-factor represents the estimated degradation of QoE. The results depicted in Fig. 18 show that the R-factor is within acceptable limits for up to 32 (MAP1) and 25 (MAP2) users respectively. As more users are being served in a cell the quality drops instantly. A small performance gain of nbLDPC codes over CTC was achieved in terms of QoE. For simulations with worse SNR conditions (*Map2*) the nbLDPC gain further increases. Additionally when comparing the results for *Map1* and *Map2* it can be seen that QoE drops very fast when the channel conditions are bad (low SNR values).

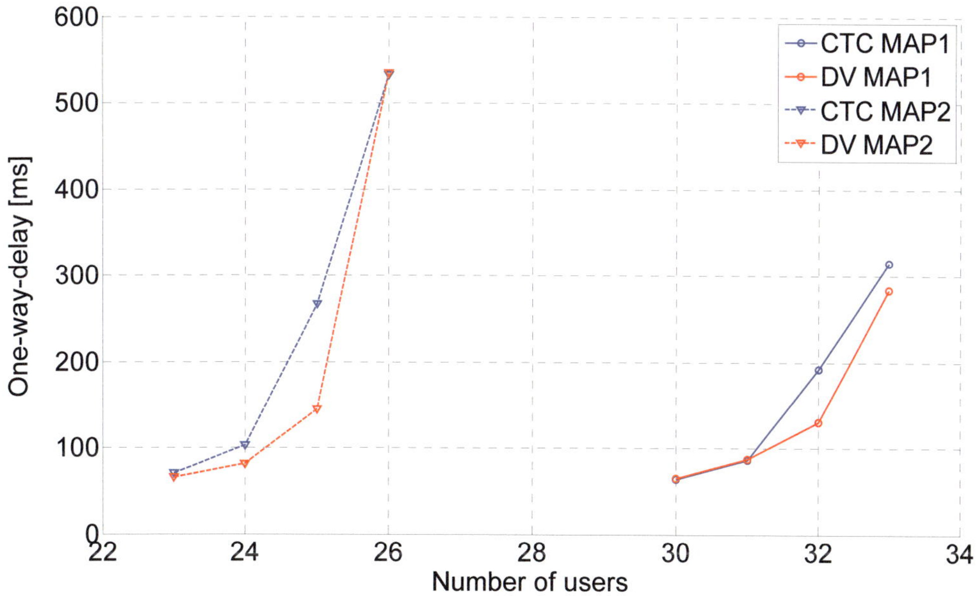

Fig. 16. Average delay for DV (nbLDPC) and CTC

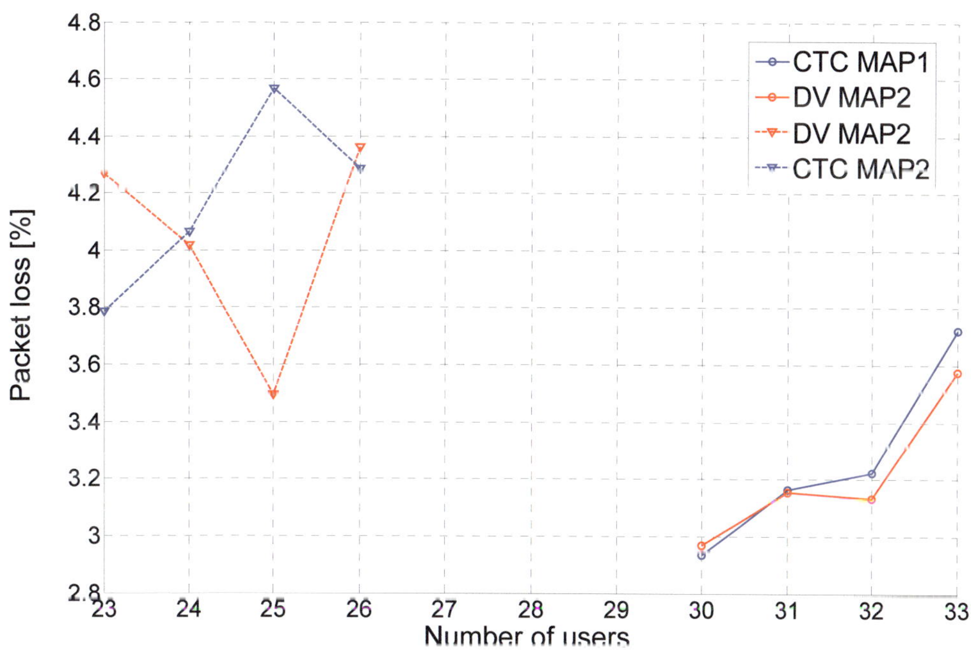

Fig. 17. Average packet loss for DV (nbLDPC) and CTC

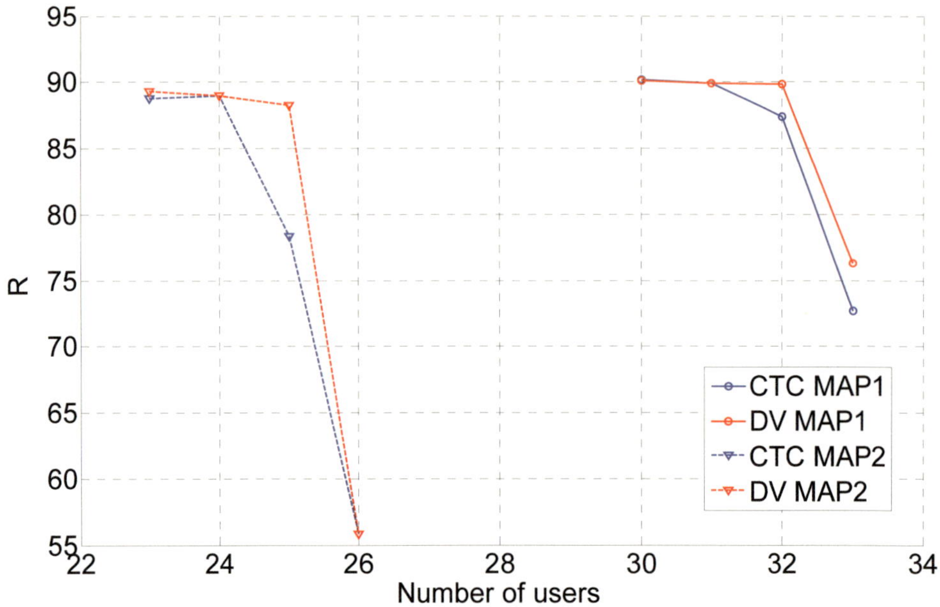

Fig. 18. R-factor for DV (nbLDPC) and CTC

7. Conclusions

The main focus of this chapter was to apply simulation methodology to facilitate cell - level simulations covering QoE measurements and CAC in WiMAX network with Time Division Multiplexing Access scheme (TDMA), OFDM and uplink traffic. The research addresses also the topic of what impact the dynamics of the system (such as resource optimization techniques e.g. AMC) has on admission control and quality of service. In order to evaluate the performance of envisaged algorithms and assess their impact on the system, authors have developed a cell-level simulation environment that relies on the proposed methodology. Previous work in the field is enhanced by improving the fidelity level of the proposed IEEE 802.16 simulator. In order to compare SUT's performance using either nbLDPC or legacy CTC (Convolutional Turbo Coding) codes in a mobile channel, a method called Link-To-System interface (L2S) has been implemented. In particular a method based on mutual information (MI) called RBIR (Mutual Information Per Received Bit | Received coded Bit Information Rate) was selected.The simulation environment relies on Network Simulator 2 integrated with Matlab software.

For admission control simulations with nbLDPC and CTC codes we come to conclusion, that achievable gain of nbLDPC can only be observed if users experience relatively good channel conditions. For higher modulations we observe less MCS transitions for nbLDPC codes, which results in lowering dropping probability and slightly increasing average system throughput. Nevertheles if users experience moderate or bad channel conditions, gain achieved thanks to nbLDPC codes becomes insignificant.

System under test (SUT) controls resources using either novel admission control mechanism ARAC (adopted by authors) or its predecessor EMAC, introduced in [41]. The algorithms are both traffic - aware and designed for controlling the VBR traffic with burst arrivals but one of them relies on calculating simple exponential weighted moving average (EWMA) of the overall resource consumption, whereas the other in the process of resource estimation differentiates between the new and the ongoing connections, thus providing more accurate resource estimations. Simulation results show that both of presented algorithms can provide appropriate QoS levels in the tested configuration. However ARAC provides protection against connections arriving in large batches. Therefore average delays of ARAC are generally lower than that of EMAC and reach the difference of approximately 23 - 25ms at maximum (depending on the codec used). These differences could prove crucial in a system with non - negligible core network delays. Results of CAC comparison prove that proposed ARAC algorithm decreases the delay experienced by VoIP connections the more the higher the arrival rate for the cost of increased blocking probability

Eventually authors provide results of assessing quality of VoIP (Voice Over IP) conversations. CTC and nbLDPC codes are compared in terms of system capacity and resulting quality of experience (QoE) performance of VoIP flows. It is shown that DaVINCI/nbLDPC codes outperform CTC in the total cell utilization and decreased dropping probability. The QoE metrics measured show slightly more users are satisfied in a cell with DaVINCI codes than when CTC is used. Therefore the nbLDPC FEC codes have proven to be a reliable coding scheme.

8. Attachments

Below in table (Table 5) the thresholds for the AMC mechanism are given. Code rate, codeword sizes and SNR thresholds are given for the codes being compared (CTC, nbLDPC).

Mod	BPSK		QPSK				16-QAM				64-QAM		
Rate	1/2	2/3	1/2	2/3	3/4	5/6	1/2	2/3	¾	5/6	2/3	3/4	5/6
Codeword length	48	48	96	96	96	96	144	144	144	144	288	288	288
SNR CTC	-0,50	1,20	1,78	3,90	4,97	6,30	7,09	9,69	11,06	12,43	14,35	15,97	17,64
SNR DAVINCI	-0,12	1,37	1,77	4,04	5,17	6,69	7,43	10,03	11,50	12,87	14,52	16,16	17,89
DAVINCI gain	0,38	0,17	-0,01	0,14	0,20	0,39	0,34	0,34	0,44	0,44	0,17	0,19	0,25

Table 5. SNR threshold for DAVINCI and CTC [48]

9. References

[1] IEEE 802.16m Evaluation Methodology Document (EMD), 2008
[2] WiMAX System Evaluation Methodology, 2008
[3] IEEE 802.16-2009: Part 16: Air Interface for Broadband Wireless Access Systems, 2009
[4] Radio Mobile http://www.cplus.org/rmw/english1.html
[5] A. Flizikowski, W. Hołubowicz, M. Przybyszewski, Sławomir Grzegorzewski, "Admission control and system capacity assessment of WiMAX with ACM and

nb-LDPC codes - simulation study with ViMACCS NS2 patch", The International Conference on Advanced Information Networking and Applications, Perth, Australia

[6] Adam Flizikowski, Marcin Przybyszewski, Mateusz Majewski, Rafał Kozik, "Evaluation of guard channel admission control schemes for IEEE 802.16 with integrated nb-LDPC codes", International Conference on Ultra Modern Telecommunications, ICUMT St.-Petersburg, Russia 2009

[7] Alasti M., Neekzad B.,Jie Hui, Vannithamby R.,"Quality of service in WiMAX and LTE networks [Topics in Wireless Communications]", IEEE Communications Magazine, vol.48, no.5, p.104-111, May 2010

[8] Anas M., Rosa C., Calabrese, F.D. ,Pedersen K.I. ,Mogensen P.E., "Combined Admission Control and Scheduling for QoS Differentiation in LTE Uplink", IEEE 86th Vehicular Technology Conference, 2008. VTC 2008-Fall.

[9] Kalikivayi, S.; Misra, I.S.; Saha, K., "Bandwidth and Delay Guaranteed Call Admission Control Scheme for QOS Provisioning in IEEE 802.16e Mobile WiMAX" Global Telecommunications Conference, 2008. IEEE GLOBECOM 2008. IEEE Nov. 30 2008-Dec. 4 2008

[10] D. Hong and S. S. Rappaport, Traffic model and performance analysis for cellular mobile radio telephone systems with prioritized and nonprioritized handoff procedures, IEEE Trans. Veh. Technol., 1986, vol. VT-35,pp. 77–92.(9)

[11] Liping Wang, Fuqiang Liu, Yusheng Ji, and Nararat Ruangchaijatupon. Admission Control for Non-preprovisioned Service Flow in Wireless Metropolitan Area Networks, ECUMN2007, 14-16, 2007, Toulouse, France.

[12] Kuo G. S., Yao H. J., A QoS-Adaptive Admission Control for IEEE 02.16e-based Mobile BWA Networks Consumer Communications and Networking Conference, 2007. CCNC 2007. 4th IEEE

[13] Xin Guo, Wenchao Ma, Zihua Guo, and Zifeng Hou , Dynamic Bandwidth Reservation Admission Control Scheme for the IEEE 802.16e Broadband Wireless Access Systems WCNC 2007, 11-15 March 2007(9)

[14] Sidi-Mohammed Senouci, Andre-Luc Beylot, and Guy Pujolle. 2004. "Call admission control in cellular networks: a reinforcement learning solution." Int. J. Netw. Manag. 14, 2 (March 2004), 89-103.

[15] Siomina I., Wanstedt S., "The impact of QoS support on the end user satisfaction in LTE networks with mixed traffic", IEEE 19th International Symposium on Personal, Indoor and Mobile Radio Communications, 2008. PIMRC 2008.

[16] Sueng Jae Bae, Bum-Gon Choi, Min Young Chung, Jin Ju Lee, Sungoh Kwon, "Delay-aware call admission control algorithm in 3GPP LTE system", TENCON 2009 - 2009 IEEE Region 10 Conference.

[17] Anas M., Rosa C., Calabrese F.D., Michaelsen P.H., Pedersen K.I., Mogensen, P.E., "QoS-Aware Single Cell Admission Control for UTRAN LTE Uplink", IEEE Vehicular Technology Conference, 2008. VTC Spring 2008.

[18] Manli Qian, Yi Huang , Jinglin Shi, Yao Yuan, Lin Tian, Dutkiewicz E., "A Novel Radio Admission Control Scheme for Multiclass Service in LTE Systems", IEEE Global Telecommunications Conference, Issued: Nov. 30 2009-Dec. 4 2009, p.1-4, 2009

[19] M.D. Katz, F.H.P. Fitzek , "WiMAX Evolution. Emerging Technologies and Applications", Wiley 2009

[20] ITU-T Recommmendation P.800, Series P: Telephone Transmission Quality, Methods for objective and subjective assessment of quality - Methods for subjective determination of transmission quality, 1996
[21] R.G. Garroppo, S. Giordano, D. Gacono, A. Cignoni, M. Falzarano, "Wimax testbed for interconnection of mobile navy units in operation scenarios", University of Pisa in Italy, May 2008
[22] T. Deryckere, W. Joseph, L. Martens, "A software tool to relate technical performance to user experience in a mobile context", Department of Information Technology, August 2008
[23] A.A. Webster, C.T. Jones, M.H. Pinson, S.D. Voran, S. Wolf, "An objective video quality assessment system based on human perception", The Institute for Telecommunication Sciences, February 20
[24] Perceptual evaluation of speech quality (PESQ): An objective method for end-to-end speech quality assessment of narrow-band telephone networks and speech codecs, Telecommunications Standarization Sector of ITU
[25] K. Piamratm A. Ksentini, C. Viho, J. Bonnin, "Perceptual evaluation of speech quality (PESQ) : An objective method for end-to-end speech quality assessment of narrow-band telephone networks and speech codecs", University of Rennes, September 2009
[26] A. Paulo Couta de Silva, M. Varela, E. de Souza Silva, Rosa M.M. Lea, G. Rubino, "Quality assessment of interactive voice applications", Federal University of Rio De Janeiro, January 2008
[27] R.G. Cole, J.H. Rosenbluth, "Voice over IP performance monitoring", Computer Communications Review, 2001
[28] L. Carvalho, E. Mota, R. Auiar, A.F. Lima, J.N. de Souza, A. Baretto, "An E-model Implementation for Speech Quality Evaluation in VoIP Systems", 2004
[29] J.Q. Walker, "Assessing VoIP call quality using the E-model", netIQ
[30] A. Meddahi, H. Afifi, D. Zeghlache, "Packet-E-Model: E-model for Wireless VoIP Quality Evaluation", IEEE Symposium on Personal, Indoor and Mobile Radio Communication Proceedings
[31] Hyun-Jong Kim, Dong-Hyeon Lee, Jong-Min Lee, Kyoung-Hee Lee, Won Lyu, Seong-Gon Choi, "The QoE Evaluation Method through the QoS-QoE correlation model", Fourth International Conference on Networked Computing and Advanced Information Management, 2008
[32] A. Raja, R. Muhammad Atif Azad, C. Flanagan, C. Ryan, Real-Tiem, "Non-intrusive Evaluation of VoIP", Springer 2007
[33] F. Bernardo, N. Vucevic, A. Umbert, M. Lopez-Benitez, "Quality of Experience Evaluation under QoS-aware Mobility Mechanisms",
[34] S. Sengupta, M. Chatterjee, S. Ganguly, R. Izmailov, "Improving R-score of VoIP Streams over w WiMAX", University of Central Florida
[35] North Carolina State Univeristy, Human Mobility Model and DTN Group http://netsrv.csc.ncsu.edu/twiki/bin/view/Main/MobilityModels
[36] A.Flizikowski. R.Kozik, H.Gierszal, M.Przybyszewski, W.Hołubowicz, "WiMAX system level simulation platform based on NS2 and DSP integration", Broadbandcom2009, Wrocław, Poland

[37] J. Zhang, H. Zheng, Z. Tan, Y. Chen, „Principle of Link Evaluation", Communications and Network, 2009 06-19 p.6-19

[38] A. Kliks, A. Zalonis, I. Dagres, A. Polydoros, H. Bogucka, „PHY Abstraction Methods for OFDM and NOFDM Systems", Journal of telecommunications and information technology

[39] INFSCO-ICT-216203 DA VINCI D5.4.1 v1.0: Evaluation criteria for multimedia services, issue 1

[40] Olli Alanen; "Quality of Service for Triple Play Services in Heterogeneous Networks"; p. 38-39; University of Jyväskylä 2007; Finland

[41] J. Lakkakorpi, A.Sayenko *Measurement-Based Connection Admission Control Methods for Real-Time Services in IEEE 802.16e* 2009 Second International Conference on Communication Theory, Reliability, and Quality of Service

[42] EuQoS project http://www.euqos.eu/

[43] Alain Mourad, Ismael Gutierrez; "System level evaluation, Issue 2"; Jan. 2010, Samsung Electronics UK Ltd., United Kingdom

[44] ITU-T Recommendation G.107, The E-model, a computational model for use in transmission planning

[45] R.G. Cole, J.H. Rosenbluth, *Voice over IP performance monitoring,* Computer Communications Review, 2001

[46] J.Q. Walker, *Assessing VoIP call quality using the E-model,* netI

[47] A. Flizikowski, M. Majewski, M. Przybyszewski, W. Hołubowicz, "QoE assessment of VoIP over IEEE 802.16 networks with DaVinci codes using E-model", Future Network & Mobile Summit, Italy, June 2010

[48] INFSO-ICT-216203 DAVINCI D2.2.1 v1.0: Link level evaluation, Issue 1

[49] Chaoming Song, Zehui Qu, Nicholas Blumm, and Albert-László Barabási "Limits of Predictability in Human Mobility" Science 19 February 2010: 327 (5968), 1018-1021.Lima, P.; Bonarini, A. & Mataric, M. (2004). *Application of Machine Learning,* InTech, ISBN 978-953-7619-34-3, Vienna, Austria

Part 3

WiMAX Applications and Multi-Hop Architectures

11

Efficient Video Distribution over WiMAX-Enabled Networks for Healthcare and Video Surveillance Applications

Dmitry V. Tsitserov and Dmitry K. Zvikhachevsky
Lancaster University,
School of Computing and Communications,
UK

1. Introduction

In this chapter we present an efficient video distribution technique which is equally applicable to both E-health and surveillance applications running over IEEE802.16/WiMAX technology platform. The developed scheme contributes to resolving of ever-struggling challenge of optimal bandwidth allocation between competitive data-consuming applications in wireless communications. The introduced approach for combined utilization of WIMAX QoS guarantee mechanism with object/quality-segmented video streams enables to achieve an improved level of system performance when compared with conventional distribution algorithms. The test scenarios were verified through NS-2 computer simulations, whereas the obtained results report better model system behavior estimated in QoS metrics, such as per flow, summary throughputs, an average end-to-end delay, particularly evaluated as bandwidth utility gain.

The whole chapter consists of two sections which are structurally common, but focused on the specific application area. The first section is devoted to WiMAX consideration for E-health applications, while the second one addresses the same issues regarded video surveillance. Each section highlights important technical aspects of the communication technology which is well-suited for the relevant applications. There is also a brief review of up-to-date related research initiatives that are built on the existent standards like IEEE802.16/WiMAX and IEEE802.11/WiFi in each section. The detailed description of the experimental models, covering the suggested distribution technique, the case-study scenarios with simulation settings and appropriate results are separately accommodated in the according sub-sections. Finally, the chapter ends with the consequent conclusions.

2. Efficient video distribution in E-health systems via WiMAX technology

2.1 E-health environment and diligent communication platform

Recent technological breakthrough in wireless communications have extended the boundaries and enlarged the scope of the application fields that vividly contribute to human safety and healthcare.

E (electronic)-health terminology lumps a variety of medicine and communication services associated with rendering of healthcare practice and delivering it to patients. The existent range of e-health definitions, including health care providers, consumer health informatics, health knowledge management, electronic health records, first response service e.t.c only discover how broad and purpose-specific the e-health sphere turns out to be. With development of new technologies E-health have been following and implementing these state-of the arts for advanced care services, such as from conventional PC archive records to the video conferencing suggested for online surgery monitoring. The obvious commonplace of the outlined contemporary innovations is to enhance efficiency of healthcare, improve reliability and facilitate service acceptability throughout a patient-GP/medical specialist-hospital communication chain (Zvikhachevskaya, 2010). In order to support efficient delivery of healthcare and neighboring services to the consumers, a profound and cutting-edge telecommunication technology has to be opted for. Proper selection of the desired transport technology should be based on aggregation of the application-driven factors that conform to the advanced information systems applied in E-health, user-accessibility and comfort, flexible scalability and to be upgrade-appreciated. There are some healthcare services and its relevant technical applications that are presented in Table 1.1. (Zvikhachevskaya, 2010).

Technical application	Healthcare services Example(s)
Video conferencing	• Virtual multi-disciplinary team meeting in Cancer Care • Support for Minor Injury Units • Training and supervision • Prison to hospital
Remote monitoring of physiological or daily living signs (real time or asynchronously)	• Falls monitoring • Physiological monitoring of chronic COPD and Heart Failure (CHF) at home
Virtual visiting	• Remote supervision of home dialysis • Nurse visits to terminally ill patients
Store and forward referrals (for example sending history plus images for expert opinion)	• Teledermatology
Web access to own health records and guidance	• HealthSpace
Telephone and Call-centres	• Tele consultation • Reminders for medication and appointments

Table 1.1. Examples of the e-Health Technologies (Zvikhachevskaya, 2010)

As it follows from the examples, provided in the Table 1.1, an adequate E-health infrastructure with a diversified service range should rely on telecommunication technology which accommodates a number of dominant properties not limited to:

- **Resource availability**. In healthcare-related services, the timely and errorless data distribution is crucial since human life and safety might be at stake. Due to the complicated nature of application-dependant traffic, such as multimedia for video conferencing, emergency video from first response ambulance and call-centre voice transfer, fair and sufficient resource allocation is inherently challenging, in particular, when system bandwidth is shared by multiple services within the same network.

- **High date rate.** Interactive sessions like on-line consultancy together with video conference facilities require high data rate support.
- **Flexible QoS support.** An effective QoS provisioning in E-health networks is expected to classify traffic and delegate relevant system budget in line with given priority. (Zvikhachevskaya, 2010). Priority might be set for specific categories of patients, data flows, medical services. For example in (Zvikhachevskaya, 2010; Skinner et al., 2006; Bobadilla et al., 2007), the 2 priority-level approach is introduced for on-line and off-line clinical activities. On-line application type includes multimedia connections of audio and video exchange, biomedical signals and vital parameters (such as ECG signal, blood pressure, oxygen saturation, etc.) transmission. Of-line type specifies clinical routine accesses to databases, queries to medical report database. Triple urgency model is presented in (Hu & Kumar, 2003), in which the patients calls, that sensor-based telemedicine network covers, are referred to one out of 3 levels of urgency. The first level involves ambulance and emergency calls and is given the highest priority with rate-guaranteed and delay-bounded service parameters. The second level faces calls from seriously ill patients in needs of urgent information exchange. Finally, calls from wrist-worn sensors, detecting regular body conditions of the observed patients are treated with Best-Effort service provision. In addition to prioritized treatment, relevant QoS parameters of delay, rate variations, packet dropping rate and others are to be sturdily considered while performing resource allocation between demanding medical applications.
- **Wireless and portability support.** Wireless connectivity allows to cover rural destinations and remote WLANs (wireless local area network) frequently employed in small offices and medical departments. This also targets patients unable to regular visit clinics and conduct medical consultancy in hospitals located distantly. Wireless technology enables comfortable accessibility of on-line medical communication through active usage of portable mobile devices like smart-phones, I-pods, laptops that are in use by almost everyone. With progressive growth of portable wireless communication gadgets flooding the wireless market, these devices may potentially serve as a first-aid mini point which is able to rapidly connect you to your GP and get you adequately advised on medicine prescription regardless of your destination and activity. Moreover, based on GPS data support, integrated in most mobile phones, the immediate ambulance help may be delivered, if required.
- **Mobility Support**. Mobile communications bring forward important benefits for both the e-health end-users and the medical services and staff. Ambulance, equipped with a required mobile communication unit, is capable of immediate data transfer for an urgent call initiating with a basic response center, while moving along. The patients under observation with a mobile device in use are again in state of fast 2-way communication to prevent hazardous effects (Zvikhachevskaya, 2010). In healthcare services the failure to timely react might yield distressing results. Mobility factor enhances efficiency of treatment decision-making, patients care and makes e-health services more comfortable and accessible.

- **IP-compatible platform**. IP supported transport technology allows to be successfully interfaced with multitude of information systems and properly integrated into the hybrid network architecture with easy access to Web domains and public LANs whatever data path medium they counts on.

Therefore, a justified healthcare service delivery may by based on the broadband wireless standards, such as WiFi, LTE-Advanced, WiMAX, 3G/GPRS that present broadband wireless connectivity with WLANs as well as can act as fast-speed wireless transport communication platform (WiMAX, LTE-Adv, 3G, GPRS). Having observed the outlined above, it is important to note that an utmost wireless technology is not consistent to completely substitute wired communications and technologies yet, due to the restricted coverage, limited channel capacity and the available wired global infrastructure, the E-health network is a part of. The wireless segments of the global E-health network, however, can be on par with alternative wired paths, scaling from backhaul transmission to last-mile and broadband WLAN access solutions.

An example of how possible E-health services can be delivered across wireless broadband connection nodes is presented in Fig.1.1. (Zvikhachevskaya, 2010)

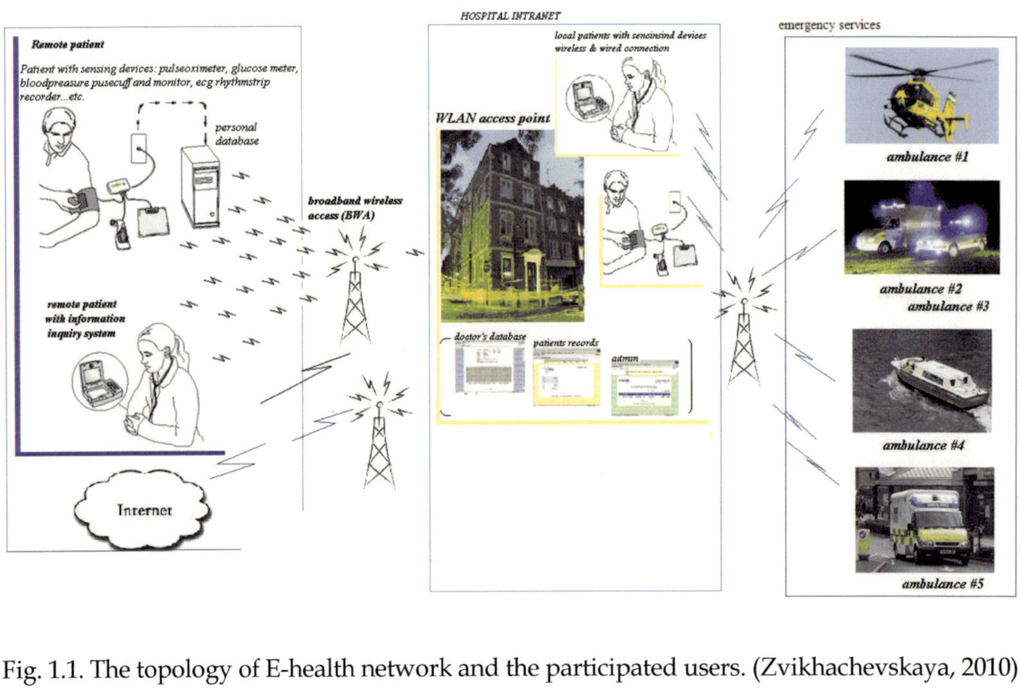

Fig. 1.1. The topology of E-health network and the participated users. (Zvikhachevskaya, 2010)

In this figure emergency services from multiple ambulances together with ordinary healthcare data of remote patients enter a hospital LAN through WBA (wirelses broadband access). Two-way communication is organized between the hospital centre and the involved users. The variety of core factors, such as a user remote distance, required traffic consumed, channel capacity, user moving speed, QoS guarantee and others will dominate the decision behind a suitable wireless system or combination of those.

2.2 Research advantages in E-health wireless communications

There have recently been exposed research initiatives aimed at investigating of E-health system models within a wireless-supported framework. In (Y. Lin, et al., 2004) a mobile monitoring system is introduced to regular record patient`s medical parameters like heart-rate, three-lead electrocardiography through accommodation of PDA (personal digital assistant) at a patient side and hospital WLAN technologies respectively. The objective of research carried out by Kutar and outlined in (Hu & Kumar, 2006) is to assess telemedicine wireless sensor network behavior on the ground of 3G technology. An energy–efficient query resolution tool is examined when a guaranteed QoS mechanism for arriving multimedia calls is required in a large-scale network topology. Mobile WAP phone communication is proposed in (Maglaveras, et al., 2002) to maintain interactive data exchange among a generic contact centre and remote patients. The promising outcome justifies such an implementation, specifically siutable for applications of the chronic disease type. Much research efforts were focused on exploration of reliable and feasible QoS means to support quality-distinctive traffic distribution in the context of versatile telemedicine services. Due to multiple telemedicine scenarios, the involved services are aggregated into a single healthcare network that should secure a certain level of performance to data streams, the particular users, associated with the relevant applications. For example, real-time IPTV, VoIP data are delay-sensitive and data rate-guarantee considered and it is always a QoS-related issue when network capacity is bounded with insufficient resources. Handy traffic management, therefore, is of great importance for E health service provisioning. Addressing this problem, (Hu & Kumar, 2003) have examined the use of energy-efficient query resolution mechanism for QoS-relied handling of arriving multimedia calls within a mobile wireless sensor network proposed for 3-G telemedicine applications. QoS consideration for wireless video transfer over ATM connections in medical environment was observed in (Dudzik, et al., 2009). In this review, ATM-based architecture allows ensure low delay and high bandwidth demands in mobile video services which positively impact on treatment efficiency of distant patients. IEEE 802.11 standard for WLAN connectivity was thoroughly explored for the purpose of its utilization across e-health mobile applications. Although, the standard is incapable of suiting real-time video and voice traffic demands on account of no priority provision and lack of service differentiation between various data flows, there is a great deal of research activity targeting QoS-accumulated techniques to maintain a certain level of QoS assurance in healthcare services. (Vergados et. al, 2006) Vergados pushes forward a challenge by proposing (Differentiated Services) wireless network architecture to support some e-Health applications with different QoS constraints. The developed DiffServ architecture is designed for emergency e-Health service and incorporates QoS mechanism that gets medical data transmission appropriately linked to different classes of service. The used resource allocation scheme considers urgency hierarchy of each application and its service-oriented QoS boundaries. The performance evaluation proves the obvious advantages of the proposed architecture in mobile telemedicine.

Yi Liu in (Y. Liu, et al., 2006) studies the emerging IEEE 802.11e standard for Wireless Local Area Networks (WLANs) with emphasis made on incoming data admission policy. In this QoS strategy, channel access parameters (CAPs) are assigned to different access categories (ACs). An admission control scheme is exploited to get the wireless system resources ultimately consumed in such a way, that let the upcoming real time traffic enter the network whilst leaving the existent data connections within the agreed QoS characteristics. The novel admission and congestion control scheme, introduced in the paper, performs regular analysis of traffic QoS requirements to assess admission control parameters for further updating the CAPs with help of adaptive channel conditions feedback. The extensive simulation of the proposed scheme demonstrates viability of guaranteed QoS mechanisms for real-time traffic in terms of guaranteed throughput indications, restricted delay and maximum dropping rate under efficient resource utilization.

D.Gao and J.Cai in (Gao &Cai, 2005) have given a broad overview of the cutting edge admission control techniques for QoS-supported traffic management across the evolving IEEE 802.11e-enabled WLANs. This survey faces the research outcomes that have highlighted both EDCA and HCCA admission control schemes. It has been shown in this manuscript how utilization of the novel MAC QoS-related elaborations in EDCA and HCCA allow for telemedicine multimedia applications to be well considered in the quality and data admission control context of WLANs.

IEEE 802.16 or WiMAX standard also provides a great deal of efficient properties which make its utilization attractive across telemedicine application scenarios.

In contrast to IEEE802.11 standards suite, IEEE 802.16 is able to cover more spacious areas (over 50 km in radius against 150-200m achieved with WiFI) with higher data rates of up to 72 MBpsec in optimal conditions. In addition, the diversified and powerful QoS-supported platform adopted in WiMAX allows handling numerous data types in conformance with specific telemedicine applications service demands, what is relatively limited for wireless E-health networks with WiFi-enabled assess technology (Noimanee, 2010).

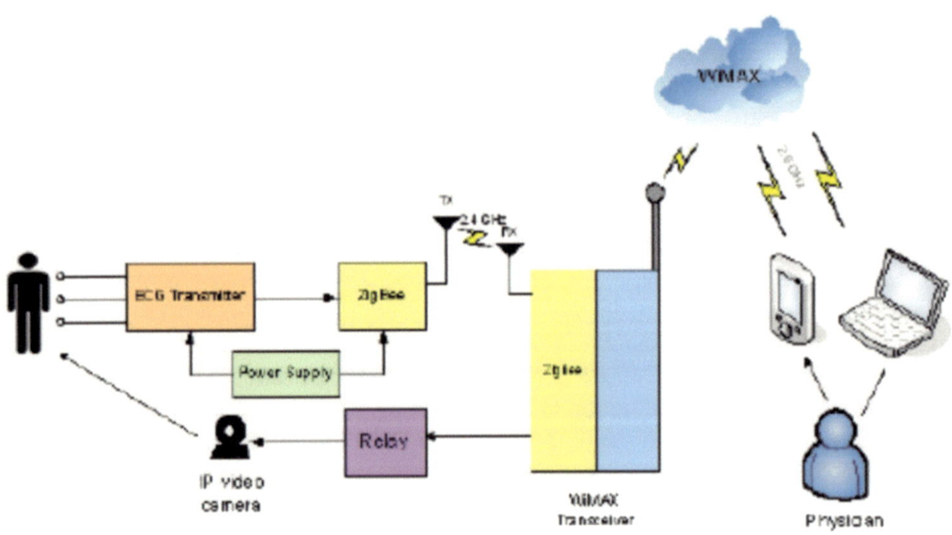

Fig. 1.2. The structure of on-line consult-based medical WiMAX system.

Considering that, WiMAX attracts intensive E-health practical and computer system modeling tailored to a particular telemedicine scenario. Thus, in (Noimanee, 2010) the authors designed and tested the global architecture for on-line monitoring and consultancy of remote patients with heart-related abnormal functionality detected through ECG signal measurement. The proposed solution enables for remote patients to regular send ECG signals measured on portable wrist-worn devices through the ZigBee/IEEE RF module to the responsible physicians through a WiMAX transceiver. In case of abnormal symptoms, the medical staff is able to remotely monitor relevant patients on application-run PDA or a wireless Laptop by activating the nearby IP surveillance camera via WiMAX connections.

The structure of the proposed system is shown on Figure 1.2 and encompasses 4 main sub-segments, in particular:

1. ECG transceiver equipped with ZigBee module for sending ECG signals.
2. IP camera for panning patient video.
3. WiMAX access point allows delivering patient video and ECG signals to physicians.
4. Physicians personal equipment to view panning video and perform data analysis which supports medical consult-based services.

The highlighted WiMAX-based telemedicine system have demonstrated much satisfaction on delivering monitoring and consultancy services through wireless communication channels in the course of real experiments with engaged factory equipment. WiMAX technology proves to be efficient means for fast and easy data transfer, video monitoring and effective patient-physician collaboration.

In our investigation we also adhere to IEEE802.16 technology for its fine suitability to the general E-health network essentials, namely: high data rate together with long distance coverage, IP compatibility with co-existing neighboring network paths, prioritized treatment of different traffic types and QoS management, mobility support. In many examples of E-Health services local area connections are not sufficient. IEEE 802.16/WiMAX technology can eliminate these drawbacks by providing broadband connectivity over existing networks for m-Health both fixed and mobile m-Health users in a wireless metropolitan area network environment. In addition, IEEE802.16 standard is one of the emerging candidates for the next generation of International Mobile Telecommunications (IMT) - advanced systems. This facilitates further modernization and scalable integration of previously installed WiMAX systems into on-going AMT-Advanced network framework. Therefore, we select IEEE 802.16e standard as a baseline specification for our simulations.

We propose a novel algorithm for video distribution over IEEE 802.16 networks for m-Health applications. We assume that the proposed technique will operate over existing wireless broadband systems installed in hospitals or any of m-Health dedicated environments. Therefore, there is a need for accommodating additional m-Health related traffic over existing networks. The proposed technique also allows utilization of the value-added services with intensive bandwidth requirements.

This work is based on our previous research (Tsitserov et. al, 2008; Markarian et. al, 2010) which is concerned with the distribution of object-oriented MPEG streams over WiMAX network with exploitation of service flows embedded in WiMAX specifications. In this paper we analyze bandwidth resource allocation depending on a scheduling algorithm and apply splitting of video traffic to evaluate system critical states. Based on the developed software model we optimize the process of video data segmentation and verify the developed technique through case study scenarios, such as E-Health applications.

In case studies, various QoS-dependant streams were emulated to quantify the achievable improvement in the overall network throughput and identify the critical issues that influence the performance. As it follows from the experimental results, the proposed segmentation of real-time data flows provides both quantitative and qualitative system resources utilization. In the next subsection the developed performance model for segmented distribution of medical video data and discussions on advantages and issues of using WiMAX technology for E-Health applications are described. Further on, the developed scheduling algorithm together with simulation parameters and results are presented. In conclusion, test results and open problems are summarized and discussed.

2.3 Distribution framework and simulation model

A *Service mapping*

The QoS concept incorporated in the IEEE802.16 standard assumes the ability to manage incoming traffic based on application requirements. Although the set of functionalities and recommendations specified for QoS support in WiMAX are conceptually approved, the scheduling design and explicit structure is left up to vendors and research bodies for further development and implementation (J. G. Andrews, 2007). In the rest of this paper we will explore these areas and apply our results for efficient video distribution over WiMAX networks, ensuring full compatibility with existing and emerging standard specifications.

Users of fixed and mobile E-Health applications can access services via IEEE 802.16/WiMAX technology. Hence, owing to the guaranteed large bandwidth available, it can help to considerably reduce the transmission delay, for e.g. of video and high resolution ultrasound and radiology images. High bandwidth according to (J. G. Andrews, 2007; , Niyato et. al, 2007, Istepanian et. al, 2006, Zvikhachevskaya et. al, 2010) can as well help to support simultaneous transmission of various types of E-Health traffic. IEEE 802.16/WiMAX standard also allows application of encrypted functionalities via the MAC layer security features for healthcare data transmission.

One of the main issues related to the application of IEEE 802.16BWA (broadband wireless access) based technology for E-Health applications is service **mapping.** Recently, a number of publications have addressed this issue (Istepanian et. al, 2006; Philip, 2008). Each of the proposed solutions has their own respective advantages and drawbacks. Although, there is a room for further optimization of this technique, the following mapping scheme is universally accepted for transferring E-Health data over WiMAX network (Philip, 2008):

- Allocate Unsolicited Grant Service (UGS) type of QoS to the biosignal traffic and voice conversation;
- Real-time Priority Service (rtPS) service for the video transmission;
- Non-real-time Priority Service (nrtPS) - to the file transfer, such as x-ray images and ultrasound results;
- Best Effort (BE) service class is to be allocated for the database access, e-mail exchange and web.

In the following research we utilize the above service mapping approach for the efficient E-Health related video streaming over IEEE 802.16 networks.

B *Distribution framework*

A novel concept is proposed to utilize object orientation of MPEG video streams for segmented distribution over IEEE 802.16 QoS-supported MAC infrastructure. We utilize a coded representation of media objects where each object is a part of complex audiovisual scene and can be perceived and processed separately. Most video distribution techniques aim at delivering MPEG streams with defined recommendation for protocol stack exploited within the communication procedures. QoS - supported network transmission technologies provide mechanisms for MPEG video distribution over its infrastructure inherently dictated by the dynamic nature of video traffic. In WiMAX networks service categories like rtPS, extended-real-time Priority Service (ertPS) and nrtPS are used for video-application data delivery depending on QoS needs for a certain video flow. Each Elementary Steam (ES) belonging to MPEG audiovisual flow can be characterized by stringent QoS requirements which are generally referred to one out of five service categories exploited in WiMAX.

Therefore, MPEG video can be transmitted through a defined MAC service connection of WiMAX system or, alternatively, many service connections of different service classes can be assigned to incoming MAC SDUs (service data units) of elementary streams segmented from the basic MPEG audiovisual scene.

The structural framework of traffic distribution in WiMAX simple topology is illustrated in Fig.1.3. In this figure, a Base Station (BS) is fully responsible for Up Link (UL) and Down Link (DL) traffic scheduling. Virtual UL scheduling process is integrated into the BS MAC architecture. The diagram schematically demonstrates data and signalling flows for UL communication between Service Station (SS) and BS. UL traffic from upper layer of MAC SDU units will be classified on the basis of QoS demands inherently allocated between

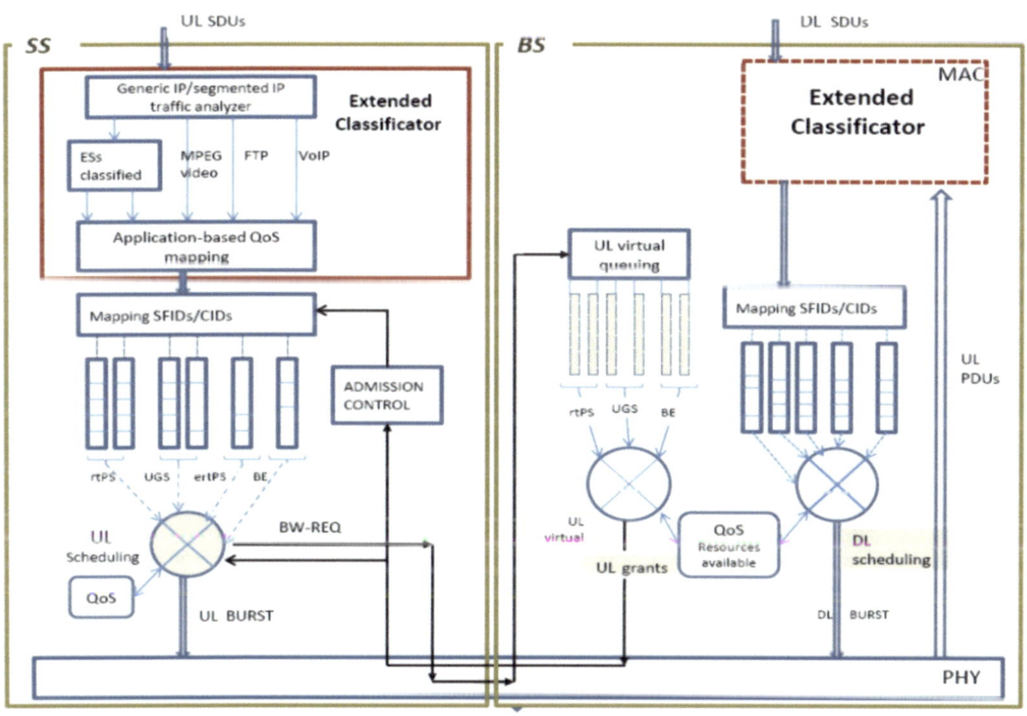

Fig. 1.3. Novel distribution framework to support object-based MPEG-4 video streams in WiMAX.

already existed service connections or put in a buffer for further connection established in line with grant/rejection generated by a BS. In order to set up a new CID (connection identifier), initiated by incoming traffic, the Mobile Station (MS) utilizes a well-known handshaking procedure to request bandwidth resources from the BS (Andrews, 2007).With appearing needs of bandwidth increase for existing service connections due to the dynamic behaviour of real time video data, for example, it is the responsibility of the MS UL scheduler either to re-allocate available resources between established connections or address the BS for additional provisional QoS set. As shown in Fig.1.3, this request opportunity is realized by BW-REG signalling message, outgoing over the WiMAX control channel. When the BS grants the necessary bandwidth, the MS UL scheduler decides whether to delegate this allocation to the maintained CID connection or set up a new one. The scheduling policy and design are beyond of WiMAX standard scope and equipment vendors are encouraged for proprietary solutions, complied with general standard specifications (Andrews, (2007).

Each service connection with packets waiting in the queue has a CID and service flow identifier (SFID) mapping to deliver packets with certain QoS guarantees to a destination address. The scheduling algorithm plays a major role in assigning burst profiles to awaited packets, and will be re-allocating the available resources, implement dropping and a connection admission policy corresponding to the distribution function and a mechanism presented in its design.

We extended the functionality of the conventional classifier/analyzer module integrated in WiMAX MAC layer to a number of specific tasks required to support the proposed algorithm. The upper SDU units will be analyzed with the purpose of determining IP packets belonging to segmented ESs or generic packets with MPEG payload. Furthermore, those from ESs are to be classified on the basis of QoS needs and then sent to the mapping block for correlating packets with QoS categories offered by WiMAX. Classified elementary streamed (ES) packets will be finally marked as application-based traffic in the category with similar QoS application needs. After that, the mapping module distributes traffic between unique service connections for supported QoS queues.

C *Descripton of the Extended Clasificator*

In order to support the proposed modifications, we introduce *Extended Classificator* as shown in Fig. 1.3. The significant value of the integrated *Extended Classificator* is to simultaneously treat packets from both conventional MPEG-structured and segmented ES video streams to provides freedom to end-users for optional use of either one or another, or both, video transmission schemes. This separation could face a quality difference and be beneficially applied by service operators in commercial implementations.

For the purpose of data identification we introduce a traffic analyzer module, which is capable of determining incoming IP packets. These packets can belong to certain MPEG-generic, MPEG segmented or conventional application payload types. In this architecture we add functions, such as handling and identification of MPEG and MPEG-ES related traffic. The classification of IP packets, such as from a Hypertext Transfer Protocol (HTP), Voice over IP (VoIP), and other services is specified by the standard. WiMAX MAC convergence sublayer is dedicated to manage the upper layer generated packets, as specified in packet header suppression (PHS) technology of the IEEE 802.16 standard (Andrews,

2007). The proposed technique is fully compatible with the WiMAX specification and does not require any alterations in the standard. MPEG ES segmentation process is to be performed at upper layers. IP packets, incoming to WiMAX system elements, contain signalling information about its segmented parameters and initial audiovisual source.

In our previous research (Markarian et. al, 2010) we were focused on mapping ES packets to specific categories of traffic applications, such as MPEG-4 video, as presented in (Markarian et. al, 2010). The mapping rules proposed in this paper introduce modification for WiMAX-enabled cross-layer data forwarding and are shown in Fig.1.4. In this diagram, each group of ES refers to a certain application type with the following classification for related IEEE802.16 service classes.

The header of the each layer bears significant information about associated links between ES data and further QoS treatment of incoming packets. We propose a cross-layer entity operating as a mapping/classification table to set up matching rules between communicating layers for delivering packets through the protocol suite. The operating layer is able to classify the incoming SDU by addressing to cross-layer table for inserting the correct information in the defined header field to inform lower layer of the requested services. However, the design and development of detailed protocol suite for ES-IP packet correlation mechanism and synchronization is beyond the scope of this paper and a topic of further detailed research. Meanwhile, it should be noted, that synchronization signalling data should be integrated into the single ES with premium QoS to provide guaranteed resources for delivery, as just the case with UGS service class.

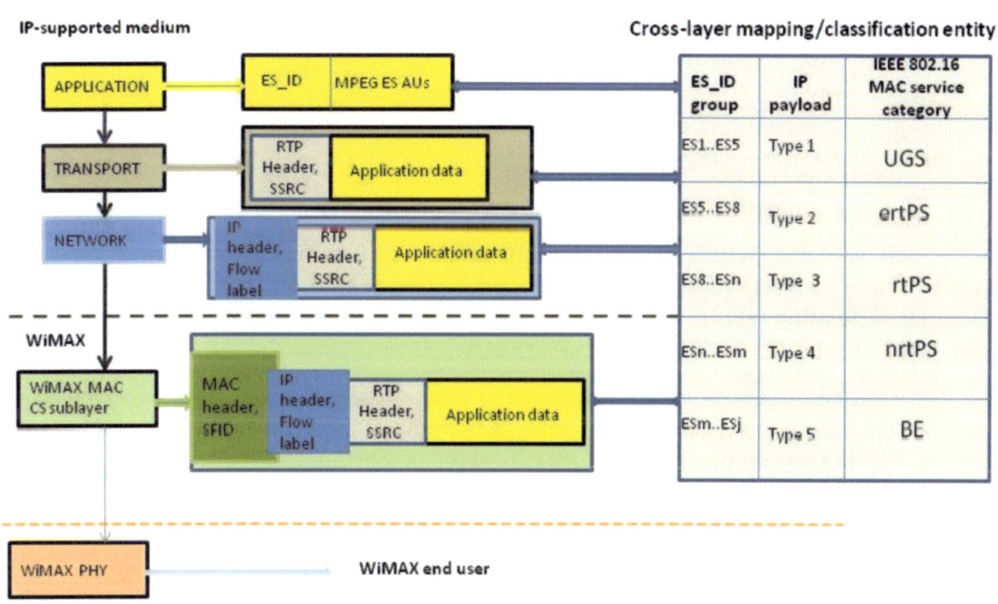

Fig. 1.4. Modified protocol-based cross-layer architecture.

One of the key aspects of video distribution over WiMAX is selecting of the right service class which will not be affected by the performance of the physical (PHY) layer. For example, Automatic Repeat Request (ARQ) of PHY layer dramatically improves the bit error ratio performance in pure (niseless) channel conditions. However, this mechanism introduces delays to the transmission of the video packets (Andrews, 2007; G.Markarian, 2010 a).

For our study we have chosen conventional Weighted Round Robin (WRR) algorithm for UL scheduling and developed a software model on the basis of the proposed program WiMAX module, elaborated for cooperative modelling within NS-2 simulator environment (Chen et. al, 2006). The results of this investigation could also be used in the future research related to optimum scheduling design.

2.4 Simulation scenarios and experimental results

A Tests of a Novel Segmented Distribution Scheme with the Stress to E-Health Applications

In the developed simulation model we implemented the direct functional correlation between the ESs and QoS scheduling categories offered in WiMAX. We assume that every ES with its QoS set can refer to a certain IEEE 802.16 MAC connection identified for the related service class UGS, rtPS, nrtPS, etc. which is associated with the specific healthcare application. Thus, this simulation approach means that the ES required for delivery of a data flow generated from a defined object with specific behaviour would get appropriate scheduling service as an individual stream with QoS-based application requirements.

Service Class	Type of E-Health Transmitted Data	Packet size, Byte	Data rate, Mbps
UGS	Live Teleconference (video)	200	2
rtPS	Medical Video Transmission (surgery, tutorial, presentation, video consultation)	150	1
BE	Request to the Database	40	0,02

Table 1.2. Test Parameters For The First Scenario.

In the first scenario, which represents the conventional approach (Andrews, 2007), we establish three connections with different service classes, as indicated in Table 1.2. Fig. 1.5 illustrates simulation results for the conventional transmission of the first scenario, which is described in Table 1.2.

The next simulation set (as given in Table 1.3) presents simulation settings for the second scenario sets, where the developed technique is applied. The aim of this simulation is not only to test the technique but also to compare its performance over the conventional transmission scenario 1 and demonstrate advantages of the developed technique.

As shown in Table 1.3 both UGS and rtPS streams were split according to the proposed video distribution algorithm. For example, in scenario 2.1 of Table 1.3, the total UGS load of 2Mbps is divided into two UGS streams of 1Mbps each. Furthermore, the original 1Mbps connection referred to rtPS service is separated into two streams. These streams are ertPS and BE with data rates 0.6Mbps and 0.4 Mbps respectively. In scenarios 2.2 and 2.3 (Table 1.3) the original UGS traffic rate is unchanged and the BE rate is constant through the whole simulation set.

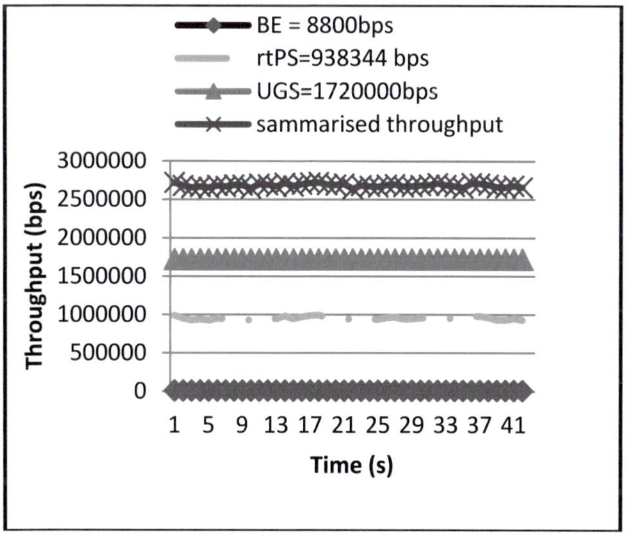

Fig. 1.5. Throughput comparison for the first (conventional) Scenario.

Fig. 1.6 shows comparative results in terms of summary throughput gain (system capacity gain), achieved for the second scenario in agreement with the parameters presented in Table 1.3.

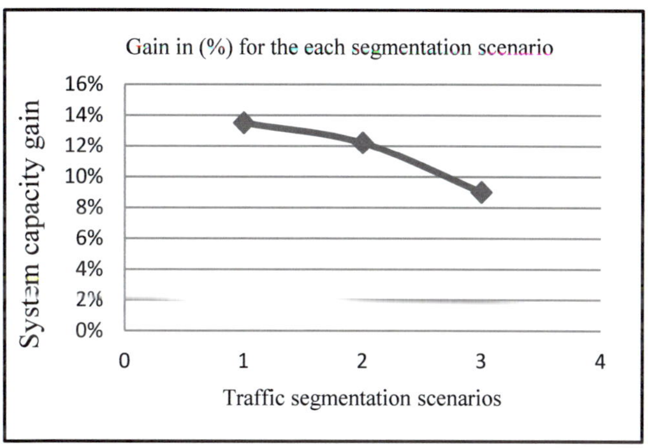

Fig. 1.6. System bandwidth gain for the Second Scenario.

The percentage gain is calculated on the basis of the comparison of the average summary throughput of conventional scenario with summary throughput results, obtained for the presented segmentation set scenarios:

$$T = \frac{T_{initial}}{T_{segmented}}, (\%) \quad (1)$$

where Tinitial - is the summarised throughput for the initial video stream; Tsegmented – is the summarised throughput for segmented scenarios.

$$\mathrm{T_{initial}} = \sum_{i=1}^{n} \mathrm{T_i} = \mathrm{T_{UGS}} + \mathrm{T_{rtPS}} + \mathrm{T_{BE}}; \quad (2)$$

where T_{UGS}, T_{rtPS}, T_{BE}– throughput results for UGS, rtPS and BE connections respectively.

$$T_{segmented} = \sum_{i=1}^{n} \sum_{k=1}^{m} T_{ij} ; \quad (3)$$

where i - number of service groups, k - number of segmented streams within each service group.

Scenario number	UGS1, load, Mbps	UGS2 load, Mbps	ertPS load, Mbps	rtPS load, Mbps	BE1 load, Mbps	BE2 load, Mbps	Summary load, Mbps	Total bandwidth, Mb
№ 2.1	1	1	0,6	0	0,4	0,02	3,02	3.5
№ 2.2	2	0	0,4	0,5	0,1	0,02	3,02	3.5
№ 2.3	2	0	0,3	0,5	0,2	0,02	3,02	3.5

Table 1.3. Simulation Parameters for the Second Scenario Set

As illustrated in Fig. 1.6, the best gain ratio approximately 14% was obtained when most data are forwarded via connections that were served by rtPS and UGS services. In addition, this best indication is explained by exploiting of separation of the initial UGS stream of 2 Mbps load on two UGS connections accounted for 1Mb load per each.

This fact supports our assumption that the segmented approach would lead to better performance in the comparison with traditional IEEE 802.16 MAC delivery. Moreover, as expected, the WRR scheduler first serves packets with a higher priority service connection. Hence, the least successful indications with about 9% capacity gain are provided for the scenario № 2.3.

Based on our evaluated results we conclude, that two sub-connected segmentation models might be a trade-off solution for delivery video data with 2-enchanced quality layers, with rtPS service reserved for E-health video conference transmission. Observing the performance of the described scenario, different video distribution models can be effectively exploited taking into account the scheduling design. Scheduling can evenly improve the performance, as our theoretical concept was experimentally approved with the simple WRR algorithm to which no specific properties were added for a selected service class-oriented priority provision.

The third scenario set is presented in Table 1.4. It is dedicated to study the variation in the overall network throughput when the segmentation scheme is applied. For example, in scenario 3.1 the initial 1 Mbps rtPS stream was separated on 0.5Mbps rtPS, 0,4 ertPS and 0.1 BE connections; while in scenario 3.2 the same 1Mbps rtPS video was simulated as 0.1Mbps rtPS, 0.4 ertPS and 0.5Mbps BE separate streams. The throughput for the each connection was analyzed. System capacity gain results for the 3 set are presented in Fig.1.7. As it can be seen from this figure, the summarized throughputs for each splitting scheme (Table 1.4) are compared to the conventional simulation model which is presented in Table 1.2.

This proves our expectation that the variation of the video stream splitting has an impact on the overall system throughput. Knowing this fact, for each type of transmitted video (surgery, tutorial, presentation, video consultation, etc) it is possible to predict the throughput gain and hence envisage the gain for the other type of transmitted data: ertPS (VoIP) or/and BE (web, database access) services, as it was shown for the specific scenario.

Scenario number	UGS load, Mbps	rtPS load, Mbps	ertPS load, Mbps	BE1 load, Mbps	BE2 load, Mbps	Summary load, Mbps	Total Bandwidth, Mb
№ 3.1	2	0,5	0,4	0,1	0,02	3,02	3,5
№ 3.2	2	0,25	0,25	0,25	0,02	3,02	3,5
№ 3.3	2	0.1	0,4	0,5	0,02	3,02	3,5

Table 1.4. Simulation Parameters for Third Scenario Set

Fig. 1.7 illustrates the gain in percentage among test sets in the third scenario. In this figure numbers 1, 2 and 3 of the traffic segmentation scenarios indicate scenarios 3.1, 3.2 and 3.3, respectively. The maximum bandwidth gain is obtained in the third set (scenario 3.3) and raise up to 16% above the conventional scenario.

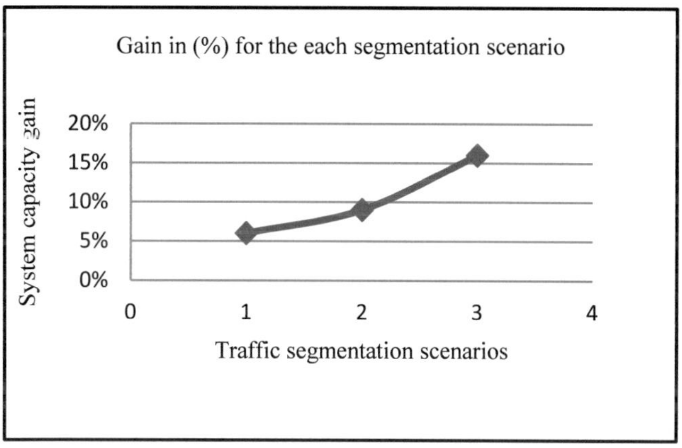

Fig. 1.7. System bandwidth gain for the third scenario.

It should be noted that we set the same values of the total system load and system bandwidth for all of the experiments. All the streams were re-allocated among the varied numbers of transport connections of defined QoS classes. It was made to model the variations of quality-selected video streams to compare network performance for the considered test scenarios. Our feasibility study demonstrates complete compatibility with the IEEE 802.16 standard.

2.5 Results overview

In this chapter section we described a novel video distribution approach designated for E-Health applications over IEEE 802.16 networks. The technique incorporates resource distribution, scheduling and content-aware video streaming taking advantage of a flexible QoS functionality offered by IEEE 802.16 technology. The proposed technique was thoroughly investigated under various scenarios, which included streaming video over MAC layer service connections. It is shown that the technique allows 9-16% increase in overall network bandwidth while maintaining full compatibility with IEEE802.16/WiMAX specifications. The exact gain is dependent upon initial system configuration and selection of WiMAX user parameters. In addition, simulation results shows that WiMAX-enabled E-health infrastructure is able to selectively handle numerous telemedicine application-driven traffic with required quality parameters within the available link budget.

3. WiMAX-supported video distribution in surveillance applications

3.1 IEEE 802.16/WiMAX practical benefits in video surveillance

Video surveillance technology has been exponentially increasing its presence among most public and private premises since its first introduction in the 1940s as a security tool for banking industry (Lalwani & Kulasekare, 2011). Current demand for cost-effective and reliable video surveillance system is spread over most public places, like schools, universities, shopping malls, including specific security aspects, such as public transport and street traffic monitoring with aim of crime prevention and fast lawful response. To address the full range of technical issues associated with deployment, maintenance and target-oriented behavior, a contemporary video surveillance system has to employ mobility support, IP-complied platform, scalable and cost-effective installation. Since, there is multitude of high resolution video flows, simultaneously transported over any video security systems, an efficient QoS and resource allocation mechanisms are to be present in this system to optimally utilize available bandwidth.

Meanwhile, most IP wireless video surveillance systems adhere to WiFi/IEEE802.11 standard suite and therefore show essential drawbacks, related to limited distance performance of up to 100 meters. It also does not allow to operate across large areas and assumes indoor applications with inherently small outdoor surrounding coverage (Lalwani & Kulasekare, 2011). In addition, WiFi is unable to provide strict security transmission standards and flexible QoS-based prioritized treatment of video flows what makes WiMAX/IEEE802.16 more favorable for video surveillance applications because of its PHY and MAC adopted properties.

Having designed, as a wireless backhaul of broadband data, WiMAX can efficiently manage video surveillance and adjacent voice, data traffic with deterministic QoS tool to ensure reliable and secure video transmission (Henshaw, 2008). Overlooking a deep insight into the standard, in overall, WiMAX-based video surveillance solution provides numerous value-added features, in particular (Henshaw, 2008):

- **Cost-economy and accessible system deployment** (fiber trenching and optic adjustment efforts of similar wired network cost 5-10 times more than its WiMAX system equivalent implementations)
- **Rapid deployment and system configuration** (as compared with fiber and copper wires WiMAX equipment can be virtually mounted anywhere and exploited under severe weather conditions with an ability of fast unit removing or location update, while making that economically inefficient for wired infrastructure. WiMAX system configuration can last up to a few hours when wired system installation requires months to accomplish preliminary trenching works.
- **Flexibility and scalability.** (Small and portable end-system WiMAX-enabled cameras are not permanently attached to a fixed location and can be removed to a new location.

System expansion is not limited and can be easily upgraded with new subscriber units, quickly re-configured by a central BS/video-server terminal owing to flexible resource management facilities which contribute to the IEEE802.16 specifications.

- **Reliable and secure communication** (OFDMa-based transmission opportunities together with embedded error correction and packet restore mechanisms provide high security standards for video signal propagation)
- **High system capacity** (Multi-stream video data from surveillance cameras strive for bandwidth-consuming and delay-sensitive QoS demands, whereas timely bandwidth fair allocation is of primary importance. MAC QoS support of WiMAX enables to handle multiple video traffic in case of gradual system extension and suits higher video quality needs if image resolution is varied upon request).
- **Mobility and IP support**. Mobility support in WiMAX enables to use surveillance equipment on public transport and transfer on-line video monitoring traffic to police vehicles. IP support makes the use of any WiMAX network segments feasible within any IP-compatible MAN or LAN infrastructure.

Research initiative on fruitful utilization of WiMAX technology for video surveillance applications has involved as the real test-bed investigation scenarios, so as computer modeling also. The chief commonplace of the conducted experiments is to demonstrate the suitable opportunities and practical benefits of the standard employment within the broad scope of specific video security applications.

In WEIRD project (Ciochina & Condrachi, 2008) video surveillance application was integrated into the functional network operator, Romania Orange, with WiMAX technology used as a broadband access solution. The actual surveyed area embraces a local Buharest test-bed and some other test-beds performed in Portugal and Italy (Ciochina & Condrachi, 2008). The key aspect of the test-bed scenarios was the use of actually working base-stations, engaged in service of the live ORANGE WiMAX network customers. Besides handling the traffic from real subscriber abonents, the base station manages video streams from the surveillance cameras installed across University campus. Throughout the experiments, video streams with different rates, resolution and quality were created and transported over WiMAX links with various QoS categories to elaborate a trade-off solution. The throughput, delay and jitter QoS metrics evaluated in course of multiple test scenarios show that WiMAX technology enables high quality video streaming for the set of video surveillance applications. The simulation provides more relevance to result analysis in terms of both research and business needs taking into consideration the real market performance environment.

Computer modeling provided in (Lalwani, S. Kulasekare, 2011) was aimed at estimation of WiMAX practicality for video surveillance application. QoS parameters like throughput, end-to-end delay, jitter and packet loss were selected for performance assessment basis and verified through sets of case-study scenarios with help of OPNET14 program simulation. Experimental scenarios diverge by number of users, its localization against the BS, various uplink coding and modulation schemes. The rtPS (real time polling service) service attribute was selected for video surveillance traffic, since this service class supports variable data-rate and packet-size parameters and is considered as a relevant category for video streaming by default in WiMAX recommendations. The conclusions presented in the manuscripts, prove to be theoretically-expected and define the backward correlation between the number of users, its distant localization from the Base Station and end to end delay. The higher order of modulation scheme results in packet loss increase as well as a longer distance between mobile nodes and the Base Station considerably affects uplink packet loss probability. Altogether, for all scenarios throughput and delay indications still remain within acceptable constraints, such as up to 5 Mbps and less than 0.5ms, respectively, that is quite suitable for video surveillance application.

The issues of live video surveillance on public transport were investigated in (Ahmad & Habibi, 2011). The real-time video communication from moving vehicles faces a significant technological challenge that is caused by multipath fading and consequent low throughput at high vehicle speeds due to technical constraints of the existing communication technologies. Despite the WiMAX/IEEE 802.16 ability to offer a guaranteed minimal date rate, it fails to cope with high packets error rate and maintain video traffic throughput sufficient for acceptable video quality in wireless mobile and speedy conditions. Due to ineffectiveness of lost packets retransmission recovery schemes, associated with considerable data overheads that get jitter and yet-low date rate application-unsuitable, error-control mechanisms, such as forward error control (FEC), are well-fitted for high speed wireless communication (Ahmad & Habibi, 2011). These recovery mechanisms therefore have no corrupted packets retransmission involved. However, FEC schemes use variable number of parity bits, a FEC code size consists of. The FEC code size is completely relied on feedback data which bear actual information about current communication environment. In real mobile wireless conditions, fluctuating noise level creates untrue channel characteristics for adjusting an optimal FEC code size with resulting data missing or overhead. In (Ahmad & Habibi, 2011) a novel FEC scheme was proposed to adaptively compute FEC code size in WiMAX video communications. The presented scheme is based on Reed-Solomon error correction code and includes 3 integral parts (Ahmad & Habibi, 2011):

1. Assessment of bit error probability at different vehicular speeds in WiMAX
2. Utilization of these estimates for proactive adjusting FEC code size in live video communication.
3. Use of de-activation/offline camera mode when the WiMAX resources are considered to be insufficient for maintaining all video flows.

Simulation results, a computer performed, demonstrate that the proposed scheme makes WiMAX technology an efficient means for real-time video delivery at high vehicular speeds with the developed technique in use (Ahmad & Habibi, 2011).

In the following we present and explore an efficient method for delivery of real-time video in multi-camera surveillance system which incorporates quality differentiation approach based on object tracking detection and QoS categorized policy brought forward by WiMAX technology. The aim of the conducted experiment is to verify the abilities of the proposed scheme to ensure more efficient management of WIMAX-based network capacity. Issues of optimal utilization of the saved bandwidth for transmission of additional traffic from active surveillance system elements were as well under exploration through NS-2 software computer simulation.

In (Tsitserov et. al, 2008) MPEG-based video distribution of object-oriented elementary streams over IEEE 802.16 networks is proposed. Further on, we expand this technique and suggest a selective quality control of the outgoing video streams, depending on a nature of objects detected or identified within a span. With the WiMAX flexible tools involving adjustment of service parameters for data transportation, we can ensure system resources for superior (high definition) HD video, whereas, (standard definition) SD or (low definition) LD video traffic will be assigned to available bandwidth in accordance with defined priorities. Such a control of the video quality allows solving the key surveillance challenge : detailed identification of a selected object for further recording and later analysis. Thus, dynamic allocation of available bandwidth in accordance with the proposed criteria enables optimization of system bandwidth. The simulation results show that the proposed technique is able to enlarge the covered surveillance area at the expense of saved bandwidth or allocate the released resources for additional data distribution upon selected case-study scenarios.

The rest of this sub-chapter is organized as follows: in the next section the proposed solution is described in details. In section 2.3 we present case-study scenarios together with simulation parameters. Experimental results analysis is given in section 2.4, and finally, in section 2.5, conclusions are provided for consideration and further research potential.

3.2 The basic attributes of the verified method

For the purpose of optimal utilization of the available system budget, we admit that dynamic regulation of outgoing video traffic will totally result in economy of bandwidth consumed and enable for extension of the surveillance area coverage. In most scenarios superior video quality allows detecting criminal identity, or details of negative factors. In conventional monitoring and surveillance systems, motion detection combined with object detection sensors are used for activation of monitoring functions of video cameras (Emilio Maggio & Andrea Cavallaro, 2011). The schematic illustration of the WiMAX-based surveillance network is presented in the Fig.2.1. Each camera has an embedded motion sensor (or any alarm event sensor) to react to some supposed actions within viewing ability of a particular camera. As depicted on this Figure, №1, №2 and №3 cameras keep following a moving object until it is tracked by №4 and №5 cameras. According to our approach, HD video is transmitted from first 3 cameras, but №1 and №5 deliver low or standard definition video. Once image is caught by cameras № 4 and № 5, these cameras will switch to HD while first 3 cameras will turn back to SD.

All video flows are received by the BS and then transported globally to the monitoring center for recording and archiving. Moreover, mobile or fast response teams are aware of the controlled sector in case of the total surveillance area is divided between fast response groups.

The BS will multicast total traffic to all groups or forward specific video streams to a dedicated user. To realize that, the BS should involve an Operation Server for video stream processing and re-distribution of data flow upon user request. In our solution, we also provide various QoS boundaries for quality segmented video flows based on service class categories, introduced in WiMAX. Therefore, HD streams will be given higher priority and served first as UGS data, then SD video corresponds to rtPS class, and LD flows are classified as BE data and served in the last turn together with additional control data. Therefore, the best service and most resources are allocated to streams with HD quality to

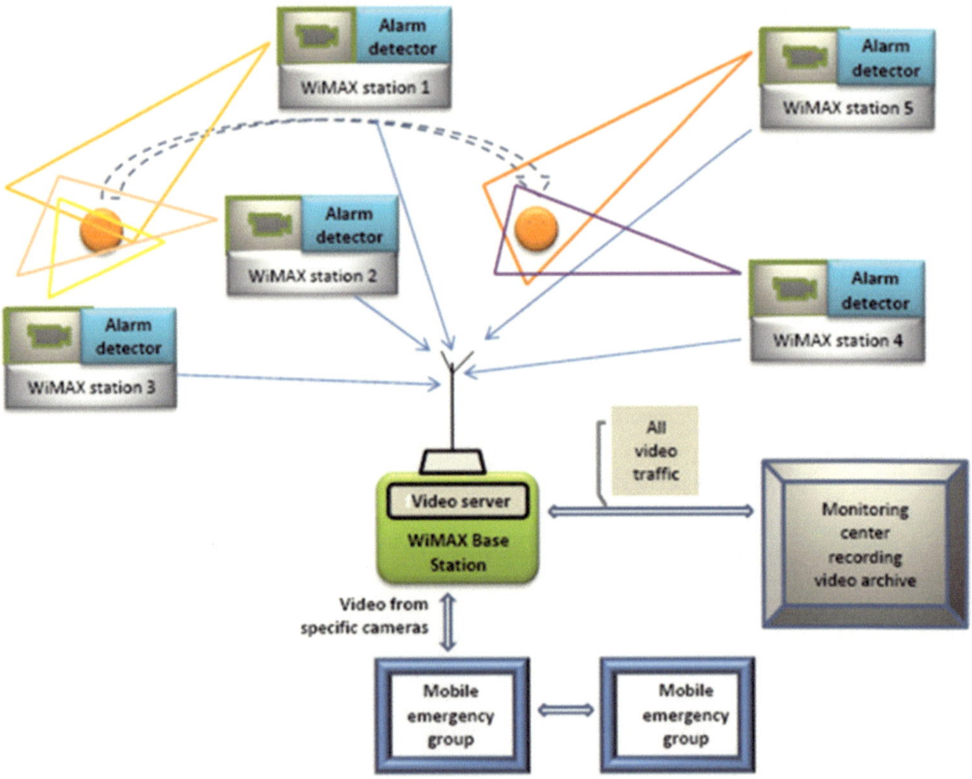

Fig. 2.1. WiMAX-based video surveillance system.

support high throughput for intensive traffic, but the rest of the bandwidth is delegated to video flows with less stringent boundaries for latency and throughput. In case of data drops or video artifacts in SD and LD video, most important information will be reflected in HD, triggered by event alarm, and can be re-produced with upper quality at the expense of better service treatment of HD video data. With introduction of categorized treatment of quality-selected video flows the surveillance network can dynamically re-allocate WiMAX resources between stations with cameras in such a way, that the whole controlled sector will be constantly covered and monitored, whereas any suspicious event is to be immediately fixed and recorded with a high resolution at premium quality. In comparison with a frequently-used video monitoring and wireless transport technology, like WiFi (IEEE 802.11) or direct PMP (point-to-multipoint) digital communication, no service guarantee for HD data can be provided, so all the streams are serviced equally or with a contention-based policy. That inevitably affects the video quality of HD video what results in failure to accomplish identification to a required extent.

We also show that some additional controlling information like GPS location of the object or camera map location can be easily transmitted together with video data, since WiMAX BS can delegate the rest of available bandwidth for such data communication as a BE service with no guarantee for latency and rate. These data can be transmitted during detection gaps, when most cameras are in state of LD video distribution.

3.3 Case study scenarios

In our simulation we first assume baseline scenario, when network topology involves 2 cameras with HD video streams and 1 additional flow served for delivery of SD data but with less important service requirements. Both HD flows are set with 8Mbsec data rate and treated as UGS connections with ensured bandwidth allocation. This scenario describes the situation when the standard approach is applied and the total video is received with superior quality from 2 cameras and standard quality with 4 Mbsec rate from 1camera for surveillance purpose. The total system load consists of video streams produced by installed cameras. Network bandwidth is constant and well sufficient for effective management of the incoming traffic for all scenarios.

For the second test we left the same full system load and bandwidth parameters, but apply the proposed technology and add more connections accounted to increased number of cameras. UGS connection with 8 Mbpsec corresponds to the camera with HD video, but 3 rtPS connections with 4 Mbs load each referred to 3 cameras enabling SD video transmission and imply no event details are tracked within their viewing sector.

Test №1	Rate, Mbsec	Total Data, MB	Total system bandwidth, MB	Simulation time, sec	Transmission	Channel Bandwidth, MHz	PHY mode	Number of MSs
UGS1	8	20	28	6	DL	5	512 S-OFDMA	1
UGS2	8							1
rtPS1	4							1
Test №2								
UGS1	8	20	28	6	DL	5	512 S-OFDMA	1
rtPS1	4							1
rtPS2	4							1
rtPS3	4							1
Test №3								
UGS1	8	20.2	28	6	DL	5	512 S-OFDMA	1
rtPS1	3							1
rtPS2	3							1
rtPS3	3							1
BE1	1.5							1
BE2	1.5							1
BE3	0.2							1

Table 2.1. Simulation parameters.

Finally the 3rd test embraces the same full load segmentation principles, as exemplified in the second test, but we reduce traffic on rtPS connections down to 3Mbsec, while 2 more BE connections with 1.5 Mb/sec load were set to simulate cameras with LD video flows. In this test we add new BE connection with 0.2 Mbsec load to imitate control data transmission, such as GPS location or image delivery. Thus, in the last scenario we admit that 3 rtPS cameras were switched to less consuming mode with rare frame/second rate video, or black/white color transmission, but the rest of the total system load was allocated for new 2 cameras with LD video traffic transported over BE connections respectively. Moreover, an additional 0.2 Mbs BE connection produces a slight increase in the total system load for adequate analysis. The main simulation parameters of the considered tests are provided in the Table 2.1.

It should be noted that we set the same values of the total system load and system bandwidth for most of the experiments, except the final scenario with a small load overcome. The total data amount is re-allocated between the varied number of transport connections of defined QoS classes to model the variations of quality-selected video streams to compare network performance for the considered test scenarios. Summary throughput comparison is illustrated in Figure 1.5. Every graph on this figure correlates to summarized throughput values of a particular test.

The whole simulation was carried out with support of WiMAX software module for NS-2 simulator designed by Chen , Wang, Tsai and Chang and proposed in (Chen et. al, 2006).

3.4 Simulation results analysis

With much attention to HD video streams we should note that the higher date rate of about 7 Mbsec for UGS connection corresponding to superior video transmission, levels out around the same value throughout the whole experiment. This fact intensely shows that for all cameras with higher level of QoS requirements, WiMAX provides with sufficient resources to deliver superior video in spite of a number of supplementary cameras generating traffic with lower QoS needs. This is explained by QoS scheduling policy in which UGS connections are given priority amid the rest and the required resources are first delegated to serve these traffic delivery. Thus, the experimental figures demonstrate that the most important video with HD selected quality is supplied at the requested level.

With gradual network expansion, the system is again capable of providing distribution with support of required QoS metrics for both UGS and rtPS connections, as exemplified in Figure 2.3. rtPS connections with date rates surrounding default parameters of 4 Mbs and 3 Mbs are illustrated in Figures 2.2, 2.3 and 2.4 respectively. Thus, the system is flexible to optimize available bandwidth in a way, when service needs for traffic with HD and SD level are properly satisfied. The similar tendency was revealed in (Markarian et. al, 2010).

In the final Test 3 the system extension to 3 new cameras have led to 40 % drop in rate values for BE connections, as described in Figure 2.4. To sustain data rates steady for connections of higher service categories, the system is slower to serve BE. Besides, no service guarantee is provided for BE connections and, therefore, exemplified as lower experimental indications in comparison with required ones.

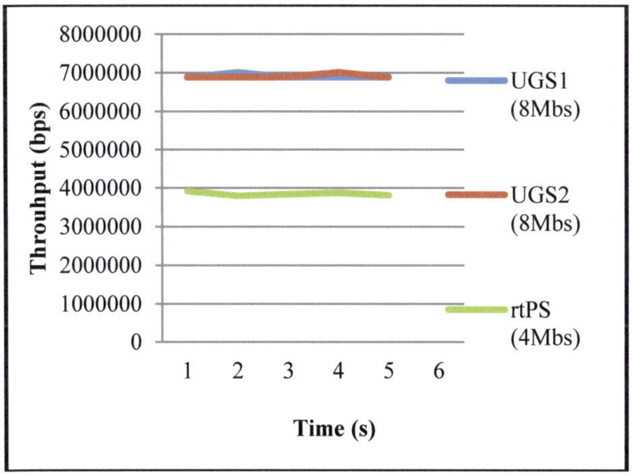

Fig. 2.2. Throughput results for Test 1.

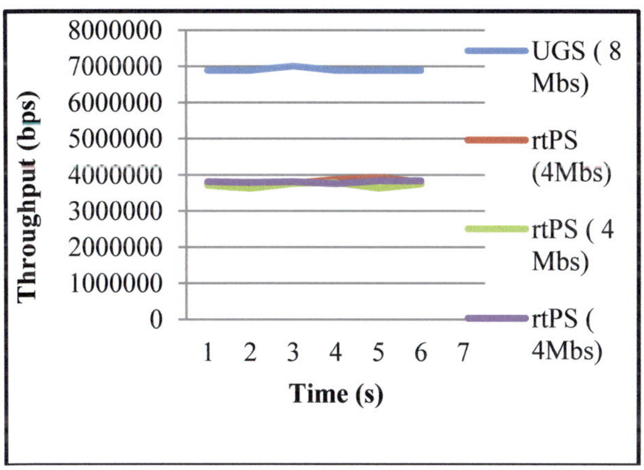

Fig. 2.3. Throughput results for Test 2.

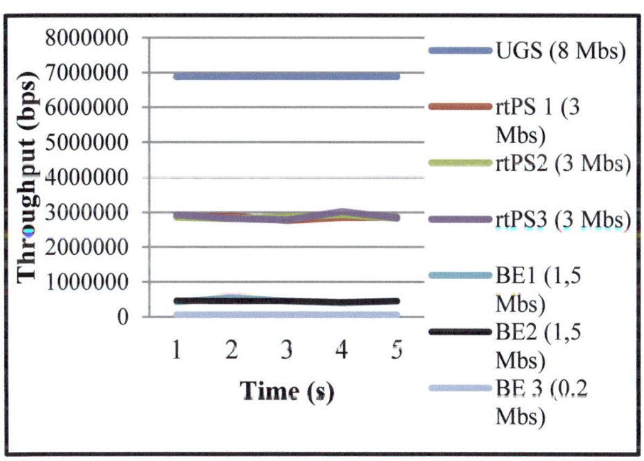

Fig. 2.4. Throughput indications for Test 3

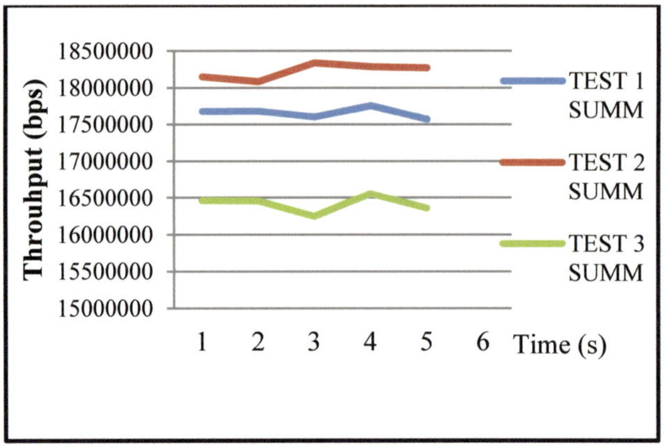

Fig. 2.5. Summary throughput comparison.

Nevertheless, with implementation to a real-life scenario, cameras with LD streaming transmit less timely-important information, therefore, the prioritized video uses UGS-based connection. Thus, lower data rate and higher delay are still justified by our introduced concept for selective video-quality in surveillance applications. Each time an alarm situation is detected, superior video quality is delivered along with rare frame/second rate video from LD network cameras enabling to properly react to emergency event and control the environment simultaneously. Based on summary throughput analysis, depicted in Figure 2.5, we observe that the lower value of around 16 Mbsec was obtained for the most complicated network topology comprising of 7 terminals. This throughput indication is 17 % less than maximum figure of 18.3 Mbsec achieved in Test 2 with only HD and SD traffic involved.

The minimal value of summary throughput, demonstrated in the Test 3, is a result of smaller resources allocated for BE connections with data rates well below default figures. In this case, the system provides low date rate to save additional bandwidth, as BE data can be delivered within longer period with higher latency, hence summary throughput dropped, illustrating 17 % bandwidth economy in comparison with an indication of Test 2.

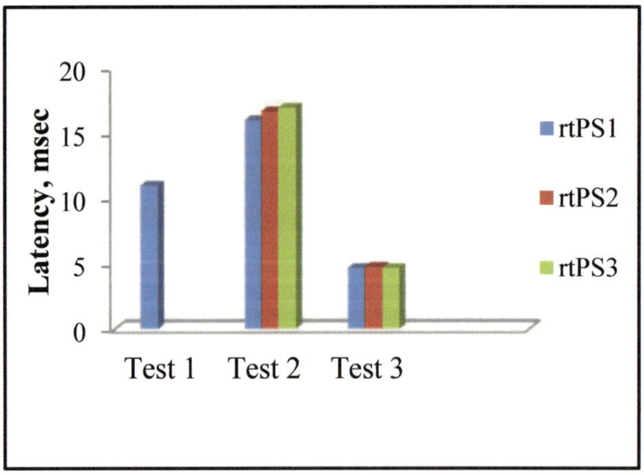

Fig. 2.6. Average latency for rtPS traffic.

Average latency values, depicted in Figure 2.6 for rtPS connections, demonstrate that the minimal figures were obtained for Test 3, in which the system resources were utilized in the best way, thanking to allocation of some of the total load for delay-tolerant BE connections of LD video and image/data traffic.

3.5 Simulation outcome

In this section we introduce an efficient distribution technique for multiple video streams over WiMAX-based monitoring and surveillance networks. We performed a computer simulation of the selected case-study scenarios which incorporate dynamic quality-based adaptation of video data entering the system and QoS categorized support for incoming traffic with HD, SD and LD quality.

The experimental results demonstrate that the introduced concept enables an optimized system resource utilization in case of network extension within the constant system bandwidth. The test results proves the feasibility of supplementary control data distribution with no service guarantee together with important HD video streams when the system is managed with help of video quality selection with integrated alarm-driven functionality.

The fulfilled experimement opens ways to theoretical foundation for successful implementation of QoS-supported 4G systems in surveillance application with traffic-consumed real-time video delivery.

4. Conclusions

In the provided chapter we have described an efficient methodology to support real-time video delivery in E-health and video surveillance applications over WiMAX systems. We have experimentally shown how WiMAX technology is able to satisfy stringent demands for bandwidth-consuming and delay-sensitive video traffic distribution in specified application areas. In overall, the developed technique demonstrates consideable achievments in system bandwidth optimization and ensures the reliable system performance under the selected cased-study scenarios. The proposed technique also reflects flexibility of the WiMAX QoS-supported concept in order to be successfully exploited for real-time video transmission across telemedicine and video surveillance multi-user networks.

5. Acknowledgement

This work was supported by the EU FP7 WiMAGIC Project and authors would like to express their gratitude to Rinicom Ltd for the opportunity to work on this project.

6. References

Article is available from encyclopedia Wikipedia, "eHealth". *Wikipedia the free encyclopaedia.* [Online] Available from: http://en.wikipedia.org/wiki/EHealth [Accessed: 5 March, 2010]

A.K. Zvikhachevskaya (2010). "Novel Wireless Communication Systems and Protocols for E-health Applications", PhD thesis, available at the Lancaster University PhD database, Lancaster, UK

H. A. Skinner, O. Maley, C. D. Norman (2006). "Developing Internet-Based eHealth Promotion Programs: The Spiral Technology Action Research (STAR) Model". Health Promot Pract 2006; 7; 406 originally published online Jul 13, 2006; DOI: 10.1177/1524839905278889. [Online]. Available from: http://hpp.sagepub.com/ cgi/content/abstract/7/4/406]

Bobadilla, P. Gomez, J. I. Godino, Mapaci (2007). A Real Time e-Health Application to Assist Throat Complaint Patients", iciw, pp.63, *Second International Conference on Internet and Web Applications and Services (ICIW'07)*

F. Hu, S. Kumar. (2003). "QoS Considerations in Wireless Sensor Networks for Telemedicine". *Proceedings of SPIE ITCOM Conference*, Orlando, FL, 2003.

Y. Lin, et al., (2004). "A Wireless PDA Based Physiological Monitoring System for Patient Transport". *IEEE Transactions on Information Technology in Biomedicine*, 2004, Vol. 8, issue 4, p. 439-447. [Online]. Available from: http://www.ncbi.nlm.nih.gov/ pubmed/15615034 [Accessed: September 2009].

F. Hu, S.Kumar, (2006). "The Integration of Ad hoc sensor networks and Cellular Networks for Multi-class Data Transmission". *Ad hoc Networks Journal (Elsevier)*, 2006. Volume 4(Issue 2): p. 254-282. ISSN:1570-8705. [Online]. Available: http://portal.acm.org/citation.cfm?id=1640928 [Accessed: September 2008].

N. Maglaveras, et al., (2002). "Home care delivery through the mobile telecommunications platform: the Citizen Health System (CHS) perspective". International Journal of Medical Informatics, 2002. Vol. 68: p. 99-111. [Online]. Available: http://linkinghub.elsevier.com/retrieve/pii/S1386505602000692[Accessed:September 2008].

P.Dudzik, et al., (2009). "Wireless ATM as a base for medical multimedia applications and telemedicine". *Computer Systems and Applications - CSA'98*, Irbid, Jordan. April 1998. [Online]. Available: http://en.scientificcommons.org/43302879 [Accessed: September 2009].

D. J. Vergados, D. D. Vergados., I. Maglogiannis, (2006). "Applying Wireless DiffServ for QoS Provisioning in Mobile Emergency Telemedicine". IEEE *Communications Society subject matter experts for publication in the IEEE GLOBECOM 2006 proceedings, 2006.*

Y. Liu, et al., (2006). "Dynamic Admission and Congestion Control for Real-time Traffic in IEEE 802.11e Wireless LANs". *2006 IEEE International Conference on Wireless and Mobile Computing, Networking and Communications*, ISBN: 1-4244-0494-0, June 19-June 21, 2006. [Online] Available: http://www.computer.org/portal/web/csdl/ doi/10.1109/WIMOB.2006.1696391 [Accessed: March 2009].

D. Gao, J. Cai, (2005). "Admission Control with Physical rate measurement for IEEE 802.11e Controlled Channel Access".*IEEECommunicationLetters,vol.9,no.8,August2005.* [Online]Availilable: http://ant.comm.ccu.edu.tw/course/94_WLAN/1_Papers/LTR,%20Admission%20control%20with%20physical%20rate%20measurement%20for%20IEEE%20802.11e%20controlled%20channel%20access.pdf

K. Noimanee, S. Noimanee, P. Khunja, and P. Keawfoonrungsie, (2010) "Medical Consult-based System for Diagnosis on WiMAX Technology", *International Journal on Applied Biomedical Engineering* Vol. 3, No. 1, p.51-55, 2010.

D.Tsitserov, G. Markarian, I. Manuilov, (2008) "Real-Time Video Distribution over WiMAX Networks", *Proceedings of the 9th Annual Postgraduate Symposium "The Convergence of Telecommunications, Networking and Broadcasting", PGnet Conference*, Liverpool, 23-24 July, 2008. Available from: http://www.cms.livjm.ac.uk/pgnet2008/Proceeedings/Papers/2008019.pdf. (URL).

G. Markarian, D. Tsitserov, A. Zvikhachevskaya, (2010). "Novel Technique for Efficient Video Distribution over WiMAX networks". *Proceedings of 21st Annual IEEE International Symposium on Personal, Indoor and Mobile Radio Communications (PIMRC 2010)*, 2010, Istanbul.

J. G. Andrews, (2007). *"Fundamentals of WiMAX: Understanding Broadband Wireless Networking", in Prentice Hall Communications Engineering and Emerging Technologies Series. Prentice Hall, 2007.*

D.Niyato, E.Hossain, J.Diamond, (2007). "IEEE 802.16/WiMAX-Based Broadband Wireless Access and Its Application for Telemedicine/e-Health Services". *IEEE Wireless Communications*, p. 1536-1284. February 2007.

D. Tsitserov, A. Zvikhachevskaya, (2010). "The Novel Cross-Layer Algorithm for Distribution of MPEG-4 ES-segmented Flows over IEEE 802.16". *Proceedings of 4th International Symposium on Broadband Communications, ISBC2010*. Malaysia. 2010.

R. S. H. Istepanian, S. Laxminarayn, C. S Pattichis, (2006). *"M-Health: Emerging Mobile Health Systems"*, Springer-Verlag, ISBN/ISSN 0387265589, London.

A. Zvikhachevskaya, G.Markarian, L. Mihaylova, (2009). "QoS consideration for wireless telemedicine/e-health services". *Proceedings of the IEEE Wireless Communications and Networking Conference WCNC*, Budapest, Hungary, 2009.

N. Y. Philip, (2008). "Medical Quality of Service for Optomised Ultrasound Streaming in Wireless Robotic Tele-Ultrasonography System ", PhD Thesis, Kingston University. 2008.

«Overview of the MPEG-4 Standard», (2002). ISO/IEC JTC1/SC29/WG11, N4668, March 2002.

G.Markarian, (2010 a). "Wireless Broadband Communications for Video Surveillance - Trends, Problems and Solutions" (invited tutorial). *Proceedings of the 13th International simposium on Wireless Personal Multimedia Communications, p.13.* October 11-14, Recife, Brazil. 2010.

J. Chen , C.-C. Wang, F. C.-Da Tsai, C.-Wei Chang, (2006). "The Design and Implementation of WiMAX Module for ns-2 Simulator", *Proceedings of the WNS2'06 conference*, October 10, 2006, Pisa, Italy.

M. Lalwani, S. Kulasekare, (2011)."Analysis of Video Surveillance over WiMAX Networks", Final Report, Simon Fraser University, April, 2011; available from: www.sfu.ca/-mla17/ENSC427.HTML

R. Henshaw, (2008). "The Wireless Video Surveillance Opportunity: Why WiMAX is not just for Broadband Wireless Access", WiMax.com Broadband Solutions, Inc.,2008; available from: http://www.wimax.com/features/the-wireless-video-surveillance-opportunity-why-wimax-is not just for-broadband-wireless-access

D. S. Ciochina, C.A. Condrachi, (2008). "Video Surbeillance Application Using WiMAX as a Wireless Technology", Orange Romania/Technical Department, 2008; available from: www.orange.ro.

I. Ahmad, D. Habibi, (2011). "A WiMAX Solution for Real-Time Video Surveillance in Public Transport", *International Journal of Computer Networks & Communications (IJCNC)* Vol.3, No.2, March 2011.

Dr. Emilio Maggio, Dr. Andrea Cavallaro. (2011). "*Video Tracking: Theory and Practice*", «Wiley», 292p., January 2011.

12

Cross-Layer Application of Video Streaming for WiMAX: Adaptive Protection with Rateless Channel Coding

L. Al-Jobouri and M. Fleury
University of Essex, United Kingdom

1. Introduction

Video streaming is an important application of broadband wireless access networks such as IEEE 802.16d,e (fixed and mobile WiMAX) (IEEE 802.16e-2005, 2005; Andrews et al., 2007; Nuaymi, 2007), as it essentially justifies the increased bandwidth compared to 3G systems, which bandwidth capacity will be further expanded in part 'm' of the standard (Ahmandi, 2011, written by Intel's chief technology officer). Broadband wireless access continues to be rolled out in many parts of the world that do not benefit from existing wired infrastructures or cellular networks. In particular, it allows rapid deployment of multimedia services in areas in the world unlikely to benefit from extensions to both 3G such as High Speed Downlink Packet Access (HSDPA) and UMTS such as Long-Term Evolution (Ekstrom et al., 2006). WiMAX is also cost effective in rural and suburban areas in some developed countries (Cicconetti et al., 2008). It is also designed to provide effective transmission at a cell's edge (Kumar, 2008), by allocation to a mobile user of sub-channels with separated frequencies to reduce co-channel interference. Time Division Duplex (TDD) through effective scheduling of time slots increases spectral efficiency, while the small frame size of 5 ms can reduce latency for applications such as video conferencing. The transition to the higher data rates of IEEE 802.16m indicates the competiveness of WiMAX.

Mobile WiMAX was introduced in 2007, as part *e* of the IEEE 802.16 standard, to strengthen the fixed WiMAX part d standard of 2004. Mobile WiMAX, IEEE 802.16e, specifies the lower two layers of the protocol stack. Like many recent wireless systems, part *d* utilized Orthogonal Frequency Division Multiplexing (OFDM) as a way of increasing symbol length to guard against multi-path interference. The sub-carriers inherent in OFDM were adapted for multi-user usage by means of Orthogonal Frequency Division Multiple Access (OFDMA), allowing subsets of the lower data-rate sub-carriers to be grouped for individual users. Sub-channel spectral allocation can range from 1.25 MHz to 20 MHz. Adaptive antenna systems and Multiple Input Multiple Output (MIMO) antennas can improve coverage and reduce the number of base stations. Basic Multicast and Broadcast Services (MBS) are supported by mobile WiMAX. IEEE 802.16m (Ahmandi, 2011) is expected to increase data rates to 100 Mbps mobile and 1 Gbps fixed delivery. However, 802.16m is not backwards compatible with 802.16e, though it does support joint operation with it.

One of the drivers of WiMAX's development is its suitability (because of centralized scheduling using TDD) for video streaming. Video streaming, as a part of Internet Protocol TV (IPTV) (DeGrande et al., 2008), can support time-shifted TV, start-again live TV, and video-on-demand. As an example, the UK's BBC iPlayer supports the former two of these unicast services, though using a form of block-based streaming in which differences in bandwidth capacity at the access network are accommodated by changes in spatial resolution. As the iPlayer's TV display is through a browser plug-in an alternative name for this service is Internet TV. Internet TV differs from what might be termed true IPTV as it uses 'best-effort' IP routing. The iPlayer is probably the best approximation to the type of video streaming considered in this Chapter. However, this Chapter does not utilize the chunk-based pseudo streaming of the BBC iPlayer but a packet-based streaming directly from the output of the codec or from pre-encoded stored video. It also does not use the Transmission Control Protocol (TCP) that underlies the Hyper Text Transport Protocol (HTTP) as this can lead to unacceptable delays across wireless networks, as TCP reacts to adverse channel conditions as if they were traffic congestion. IPTV as a service to set-top boxes or desk-top PCs generally includes TV channel multiplexing within a coded stream encapsulated in (say) MPEG-2 Transport System (TS) application-layer packets as well as an Electronic Program Guide (EPG) service. When transferred to a mobile system, this type of IPTV may well require the video service office (VSO) (DeGrande et al., 2008), as the last step in a content delivery network (CDN) overlay to respond to channel selection by the user rather than deliver all channels to the user (as occurs in fiber-to-the-home services). Such CDNs also have the important function of caching content nearer to users. It should be remarked that the BBC, provider of the iPlayer, acts as a public service and, hence, does not require a formal business model, whereas other IPTV services generally have a traditional business plan and may employ encryption and digital rights management .

It has become increasingly clear that Next Generation Networks (NGNs) will not be based on wireline devices as previously envisaged but on mobile devices. However, the volatile nature of the wireless channel (Goldsmith, 2005), due to the joint effect of fading, shadowing, interference and noise, means that an adaptive approach to video streaming is required. To achieve this exchange of information across the protocol layers is necessary, so that the application-layer can share knowledge of the channel state with lower protocol layers. Though a cross-layer application in general has its detractions, such as the difficulty of evolving the application in the future, because of the delay constraints of video streaming and multimedia applications in general, its use is justified.

This Chapter provides a case study, in which information from the PHYsical layer is used to protect video streaming over a mobile WiMAX link to a mobile subscriber station (MS). Protection is through an adaptive forward error correction (FEC) scheme in which channel conditions as reported by channel estimation at the PHY layer serve to adjust the level of application-layer FEC. This flexibility is achieved by use of rateless channel coding (MacKay, 2005), in the sense that the ratio of FEC to data is adjusted according to the information received from the PHY layer. The scheme also works in cooperation with PHY-layer FEC, which serves to filter out packet data in error, so that only correctly received data within a packet are passed up the layers to the video-streaming application. The 802.16e standard provides Turbo coding and hybrid Automatic Repeat request (ARQ) at the PHY layer with scalable transmission bursts depending on radio frequency conditions. However, application-layer forward error correction (Stockhammer et al., 2007) is still recommended for IPTV during severe error conditions.

Rateless channel coding allows the code rate to be adaptively changed according to channel conditions, avoiding the thresholding effect associated with fixed-rate codes such as Reed-Solomon. However, the linear decode complexity of one variant of rateless coding, Raptor coding (Shokorallahi, 2006), has made it attractive for its efficiency alone. For broadcast systems such as 3GPP's Multimedia Broadcast Multicast System (MBMS) (Afzal, 2006) , as channel conditions may vary for each receiver, the possibility of adapting the rate is not exploited, even with a rateless code. However, for unicast video-on-demand and time-shifted TV streaming it is possible to adaptively vary the rate according to measured channel conditions at the sender. These services are a commercially-attractive facility offered by IPTV as they add value to a basic broadcast service.

In addition to analysis of the cross-layer protection scheme, the Chapter demonstrates how source-coded error resilience can be applied by means of data-partitioning of the compressed video bitstream. This in turn encourages the use of duplicate data, as a measure against packet erasure. Packet erasure can still occur despite adaptive FEC provision for data within WiMAX packets, i.e. Medium Access Control (MAC) protocol data units (MPDUs). Assessment of the results of the adaptive protection scheme is presented in terms of packet drops, data corruption and repair, end-to-end delay introduced, and the dependency of objective video quality upon content type.

The remainder of this Chapter is organized as follows. Section 2 sets the context for the case study with discussion of WiMAX cross-layer design, IPTV for WiMAX, together with source and channel coding issues. Section 3 presents the simulation model for the case study with some sample evaluation results. Finally, Section 4 makes some concluding remarks.

2. Context of the case study

This Section now describes research into cross-level design for mobile WiMAX in respect to video streaming.

2.1 WiMAX cross-layer design

The number of cross-layer designs for wireless network video-streaming applications has considerably increased (Schaar & Shankar, 2005) with as much as 65% of applications in mobile ad hoc networks adopting such designs. This should not be a surprise, as source coding and streaming techniques in the application layer cannot be executed in isolation from the lower layers, which coordinate error protection, packet scheduling, packet dropping when buffers overflow, routing (in ad hoc and mesh networks), and resource management.

In WiMAX multicast mode, scheduling decisions for the real-time Polling Service (rtPS) queue, one of the WiMAX quality of service queues (Andrews et al., 2007), in particular are suspended. This can cause excessive delay to multimedia applications. To avoid this, in Chang & Chou (2007) knowledge of the application types and their delay constraints is conveyed to the datalink layer, where the scheduling mode is decided upon. The network layer can also benefit from communication with the datalink layer in order to synchronize

WiMAX and IP handoff management (Chen & Hsieh, 2007) and in that way reduce the number of control messages. For further general examples of cross-layer design in WiMAX, the reader should consult Kuhran et al. (2007).

Video applications using PHY layer information were targeted in Juan et al. (2009) and She et al. (2009). In Juan et al. (2009), layers of a scalable video stream were mapped onto different 802.16e connections. The base station (BS) periodically reports average available bandwidth to a collocated video server, which then dynamically allocates video packets to the connections. The base layer occupies one connection while the remaining enhancement layer(s) packets occupy the second connection. If base layer packets (and certain key pictures) are lost, then the BS only retransmits these if available bandwidth permits. In She et al. (2009), cross-layer design was applied to WiMAX IPTV multicast to guard against channel diversity between different receivers. The solution again utilized scalable video layers but, instead of a mapping onto different connections, superposition coding is employed. In such coding, more important data are typically modulated at Binary Phase Shift Keying (BPSK) whereas enhancement layers are transmitted at higher order modulation such as 16QAM (16-point Quadrature Amplitude Modulation). A cross-layer unit performs the superposition at the BS, whereas, at the subscriber stations, layers are selected according to channel conditions. Both these schemes fall into the class of wireless medium-aware video streaming. However, neither of these papers explained how signaling between lower and higher level protocols can take place.

In Neves et al. (2009) it was pointed out that IEEE 802.21 Media Independent Handover (MIH) services (IEEE 802.21, 2008) already provides a framework for cross-layer signaling that could be enhanced for more general purposes. In fact, another WiMAX specific set of standardized communication primitives is IEEE 802.16g. However, it could be that legacy WiMAX systems will need to be provided with a different interface. In 802.21, a layer 2.5 is inserted between the level 2 link layer and the level 3 network layer. Upper-layer services, known as MIH users or MIHU communicate through this middleware to the lower layer protocols. One of the middleware services, the Media Independent Event Service (MIES) is responsible for reporting events such as dynamic changes in link conditions, link status and quality, which appears suitable or at least near to the requirements of the adaptive scheme reported in this Chapter.

There are penalties in applying a cross-layer scheme (Kawadia & Kumar, 2003), namely it may result in a monolithic application that is hard to modify or evolve. However, for wireless communication (Srivastava & Motani, 2005) an adaptive scheme that leverages information across the layers can cope with the volatile state of the channel due to fading and shadowing and the constrained available bandwidth of the channel. It is not necessary to abandon layering altogether in a 'layerless' design but simply to communicate between the layers. Video applications break protocol boundaries with limited objectives in mind, though improvements in performance remain the goal. Performance may be defined variously in terms of reduction of delay, reduction of errors, throughput efficiency, and, in wireless networks, reduction of energy consumption. This list by no means exhausts the possible trade-offs that can be engineered through cross-layer exchange of information.

2.2 IPTV video streaming

The ability to provide TV over wireless (and digital subscriber line) access networks has undoubtedly been encouraged by the increased compression achievable with an

H.264/Advanced Video Coding (AVC) codec (Wiegand et al., 2003), for example reducing from at least 1.5 Mbps for MPEG-2 video to less than 500 kbps for equivalent quality TV using H.264/AVC compression. The density of subscribers is linked to the number of sub-channels allocated per user, which is a minimum of one per link direction. In a 5 MHz system, the maximum is 17 uplink and 15 downlink sub-channels. For a 10 MHz system (FFT size 1024) 35 downlink and 30 uplink sub-channels are available. For a mobile WiMAX (IEEE 802.16e) 10 MHz system, capacity studies (So-In et al., 2010) suggest between 14 and 20 mobile TV users per cell in a 'lossy' channel depending on factors such as whether simple or enhanced scheduling and whether a single antennas or 2×2 MIMO antennas are activated. However, given the predicted increase in data rates arising from IEEE 802.16m, the number of uni-cast video users (Oyeman et al., 2010) with 4×2 Multi User (MU)-MIMO antennas, will be 44 at 384 kbps and 22 at 768 kbps in an urban environment. For a similar configuration but using IEEE 802.16m 20 MHz (FFT size 2048) rather than IEEE 802.16m 80 MHz channels (4 FFT of size 2048 each) , the authors of Oyeman et al. (2010) reported the number of uni-cast video users to be 11 and 6 depending on data-rates. However, it should be born in mind that the capacity of a WiMAX cell can be scaled up by means of sectored antennas, whereas the above capacities for IEEE 802.16m are for a single sector. A typical arrangement (Jain et al., 2008) is to have three sectors per cell. It should be remarked that in Oyeman et al. (2010), the subscriber density of LTE-Advanced is assessed as very similar to that of IEEE 802.16m.

In Degrande et al. (2008), ways to improve IPTV quality were discussed with the assumption that intelligent content management would bring popular video content nearer to the end viewer. The typical IPTV architecture considered, Fig. 1a, assumes a super head-end (SHE) distributor of content across a core network to regional video hub offices (VHOs). VHOs are connected to video serving offices (VSOs) over a regional metro network. It is a VSO that interacts with users over an access network. While Degrande et al. (2008) have managed networks using IP framing but *not* 'best-effort' routing in mind, CDNs such as iBeam and Limelight originated for the unmanaged Internet. Microsoft TV IPTV Edition is probably the best known of the managed network proprietary solutions and this too can utilize WiMAX delivery (Kumar, 2008).

An overview of how an IPTV system with WiMAX fixed or mobile delivery is presented in Uilecan et al. (2007). The system takes advantage of WiMAX's point-to-multipoint (PMP) mode for the broadcast of TV channels. MPEG2-TS packets containing multiplexed TV channels are encapsulated in RTP/UDP/IP packets. Header suppression and compression techniques reduce the overhead. In Issa et al. (2010), IPTV streaming was evaluated on a WiMAX testbed for downlink delivery of TV channels and uplink delivery of either TV news reports or video surveillance; refer to Figure 1b. Broadly for streaming media WiMAX's application class 3 supports medium bandwidth between 0.5 and 2 Mbps and jitter less than 100 ms. In fact, the ITU-T's recommendations for IPTV (not mobile TV) are even more stringent with jitter less than 40 ms and packet loss rates less than 5%. Video conferencing (not covered in this Chapter) will require jitter less than 50 ms but probably much lower bandwidths and end-to-end latency less than 160 ms.

In a native Real-Time Protocol (RTP) solution for IPTV distribution, the Real-Time Protocol Streaming Protocol (RTSP) is available for TV channel selection and can support pseudo video cassette recorder functions such as PAUSE and REWIND. The Real-Time Control

Protocol (RTCP) is suitable for feedback that may be used to reduce the streaming rate for live video, or by stream switching or a bitrate transcoder if pre-encoded video is being streamed.

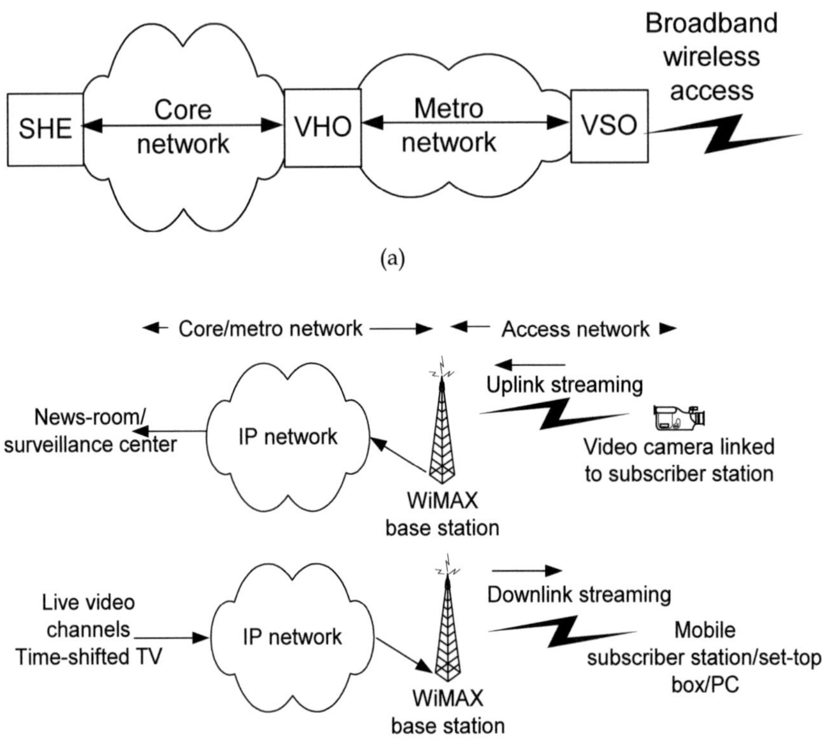

Fig. 1. (a) Schematic IPTV distribution network (b) Downlink and uplink streaming scenarios.

Originally, it was assumed (Kumar, 2008) that the IP networks involved would form "walled gardens", which would be managed by telecommunications companies ('telcos') and which might exclude competitors in the speech communication market such as Skype voice-over-IP and include traditional forms of mobile broadcast. Originally also it was thought that WiMAX's extended coverage would function as a backhaul service to IEEE 802.11 networks, which are limited in range by their access control mechanism, whereas WiMAX has been developed as a replacement for many smaller but isolated IEEE 802.11 hotspots. The IP Multimedia Subsystem (IMS) then allows roaming across networks with a common framing standard, outside the 'walled garden'. In the IMS view, WiMAX is an underlying network just as LTE would be. WiMAX's real-time Polling Service (rtPS) is the scheduling service class suited to IPTV video streaming.

2.3 Source coding for video streaming

Source coding issues are now briefly discussed. As mentioned in Section 1, data-partitioning was enabled for error resilience purposes. In an H.264/AVC codec (Wenger, 2003), when data-partitioning is enabled (Stockhammer & Bystrom, 2007), inter-coded slices are normally divided into three separate partitions according to decoding priority. These data are packed into different Network Abstraction Layer units (NALU's). Each NALU is encapsulated into an IP/RTP/UDP packet for possible IMS transport. Each partition is located in either of type-2 to type-4 NAL units. A NAL unit of type 2, also known as partition-A, comprises the most important information of the compressed video bit stream of P- and B-pictures, including the MB addresses, MVs, and essential headers. If any MBs in these pictures are intra-coded, their frequency transform coefficients are packed into the type-3 NAL unit, also known as partition B. Type 4 NAL, also known as partition-C, carries the transform coefficients of the motion-compensated inter-picture coded macroblocks. When motion-copy error concealment is enabled at a decoder, then receipt of a partition-A carrying packet is sufficient to enable a partial reconstruction of the frame. When the quantization parameter (QP) is appropriately set, the smaller size of partition-A results in smaller packet length and, hence, a reduced risk of error.

In adverse channel conditions, duplicate partition-A packets are transmitted. On the other hand, the duplicate partition-A stream should be turned off during favorable channel conditions. In an H.264/AVC codec, it is instead possible to send redundant pictures slices (Radulovic et al., 2007), which employ a coarser quantization than the main stream, but this can lead to encoder-decoder drift. Besides, for data-partitioning, replacing one partition with a redundant slice with a different QP to the other partitions would not permit reconstruction in an H.264/AVC codec.

In order to decode partition-B and -C, the decoder must know the location from which each MB was predicted, which implies that partitions B and C cannot be reconstructed if partition-A is lost. Though partition-A is independent of partitions B and C, Constrained Intra Prediction (CIP) should be set in the codec configuration (Dhondt et al., 2007) to make partition-B independent of partition-C. By setting this option, partition-B MBs are no longer predicted from neighboring inter-coded MBs. This is because the prediction residuals from neighboring inter-coded MBs reside in partition-C and cannot be accessed by the decoder if a partition-C packet is lost. There is a by-product of increasing overhead from extra packet headers in a reduction in compression efficiency but the overall decrease in packet size may be justified in error- prone environments.

2.4 Rateless channel coding for video streaming

Rateless or Fountain coding (MacKay, 2005), of which Raptor coding (Shokorallahi, 2006) is a subset, is ideally suited to a binary erasure channel in which either the error-correcting code works or the channel decoder fails and reports that it has failed. In erasure coding, all is not lost as flawed data symbols may be reconstructed from a set of successfully received symbols (if sufficient of these symbols are successfully received). A fixed-rate (n, k) Reed-Solomon (RS) erasure code over an alphabet of size $q = 2L$ has the property that if any k out of the n symbols transmitted are received successfully then the original k symbols can be decoded. However, in practice not only must n, k, and q be small but also the computational complexity of the decoder is of order $n(n - k) \log_2 n$. The erasure rate must also be estimated in advance.

The class of Fountain codes allows a continual stream of additional symbols to be generated in the event that the original symbols could not be decoded. It is the ability to easily generate new symbols that makes Fountain codes rateless. Decoding will succeed with small probability of failure if any of k (1 + ε) symbols are successfully received. In its simplest form, the symbols are combined in an exclusive OR (XOR) operation, according to the order specified by a random, low density generator matrix and, in this case, the probability of decoder failure is $\partial = 2^{-k\varepsilon}$, which, for large k, approaches the Shannon limit. The random sequence must be known to the receiver but this is easily achieved, through knowledge of the sequence seed.

Luby transform (LT) codes (Luby, 2002) reduce the complexity of decoding a simple Fountain code (which is of order k^3) by means of an iterative decoding procedure. The 'belief propagation' decoding relies on the column entries of the generator matrix being selected from a robust Soliton distribution. In the LT generator matrix case, the expected number of degree one combinations (no XORing of symbols) is $S = c \ln(k/\partial)\sqrt{k}$, for small constant c. Setting $\varepsilon = 2 \ln(S/\partial) S$ ensures that, by sending $k(1 + \varepsilon)$ symbols, these symbols are decoded with probability $(1 - \partial)$ and decoding complexity of order $k \ln k$.

The essential differences between Fountain erasure codes and RS erasure codes are that: Fountain codes in general (not Raptor codes) are not systematic; and that, even if there were no channel errors, there is a small probability that the decoding will fail. In compensation, they are completely flexible, have linear decode computational complexity, and generally their overhead is considerably reduced compared to fixed erasure codes. Apart from the startling reduction in computational complexity, a Raptor code (Shokorallahi, 2006) has the maximum distance separable property. That is, the source packets can be reconstructed with high probability from any set of k or just slightly more than k received symbols. A further advantage of Raptor coding is that it does not share the high error floors on a binary erasure channel (Palanki & Yedidai, 2004) of prior rateless codes. However, it is probably the combination of closeness to the ergodic capacity and the low rate of decoder error (Castura & Mao, 2006) that most determines the advantage of Raptor codes over other forms of rateless channel coding.

3. Case study

A video application can adopt at least three methods of protection for fragile video streams. The first method is application-layer channel coding. However, application coding is only effective to the extent that a packet actually reaches a wireless device and is not lost beforehand. Packets can be lost in a variety of ways: because of buffer overflow; or because the signal-level drops below the receiver's threshold; or because the physical-layer forward error correction is unable to reconstruct enough of the packet to be able to pass data up to the application layer. Therefore, the second method of protection is duplication of all or part of the original bitstream. The duplicated packets are sent alongside the original video stream. A third method is to anticipate errors at the source-coding stage through error resilience, with a good number of such techniques presented in Stockhammer & Zia (2007). Error resilience can act as an aid to reconstruction through error concealment. The scheme described in this Chapter's case study utilizes all three methods of protection. Simulations show that in particularly harsh channel conditions the scheme is able to protect the video stream against data loss and subsequently achieve reasonable video quality at the mobile device. Without the protection scheme the video quality would be poor.

In the protection scheme, application-layer channel coding takes advantage of rateless channel coding (MacKay, 2005) to dynamically adapt to channel conditions. Extra redundant data are 'piggybacked' onto a new packet so as to aid the reconstruction of a previous packet. To achieve adaptation (and also to turn off duplicate slices during favorable conditions) channel estimation is necessary. As an example, the IEEE 802.16e standard (IEEE 802.16e-2005, 2005) specifies that a mobile station or device should provide channel measurements, which can either be received signal strength indicators or may be carrier-to-noise-and-interference ratio measurements made over modulated carrier preambles. Therefore, to aid in this process the method assumes one of these methods is implemented.

Error resilience is provided by data partitioning (Stockhammer & Bystrom, 2007). Data-partitioning rearranges the video bitstream according to the reconstruction priority of the compressed data. There is less overhead than other forms of error resilience such as the popular Flexible Macroblock Ordering (Lambert et al., 2005). Consequently, data-partitioning can operate during favorable channel conditions, as well as unfavorable channel conditions. On the other hand, the duplicate stream protection mentioned previously should be turned off during favorable channel conditions, as its transmission involves a significant overhead. 'Redundant' data at coarser quantization levels can be sent instead of duplicated data but redundancy results in encoder-decoder drift, unless a memory-intensive, multiple-reference scheme (Zhu et al., 2006) is employed.

3.1 Implementing the protection scheme

In the adaptive channel coding scheme, the probability of channel byte loss through fast fading (*BL*) serves to predict the amount of redundant data to be added to the payload. In an implementation, *BL*, is found through measurement of channel conditions. If the original packet length is *L*, then the redundant data is given simply by

$$\begin{aligned} R &= L\times BL + (L\times BL^2) + (L\times BL^3)\ldots \\ &= L/(1-BL) - L, \end{aligned} \tag{1}$$

which adds successively smaller additions of redundant data, based on taking the previous amount of redundant data multiplied by *BL*.

Rateless code decoding in traditional form operates by a belief-propagation algorithm (MacKay, 2005) which is reliant upon the identification of clean symbols. This latter function is performed by PHY-layer forward error correction, which passes up correctly received blocks of data (checked through a cyclic redundancy check) but suppresses erroneous data. For example, in IEEE 802.16e (Andrews et al., 2007), a binary, non-recursive, convolutional encoder with a constraint length of 7 and a native rate of 1/2 operates at the PHY layer.

If a packet cannot be decoded, despite the provision of redundant data, extra redundant data are added or 'piggybacked' onto the next packet. In Figure 2, packet X is corrupted to such an extent that it cannot be immediately decoded. Therefore, in packet X+1 some extra redundant data are included up to the level that decode failure is no longer certain.

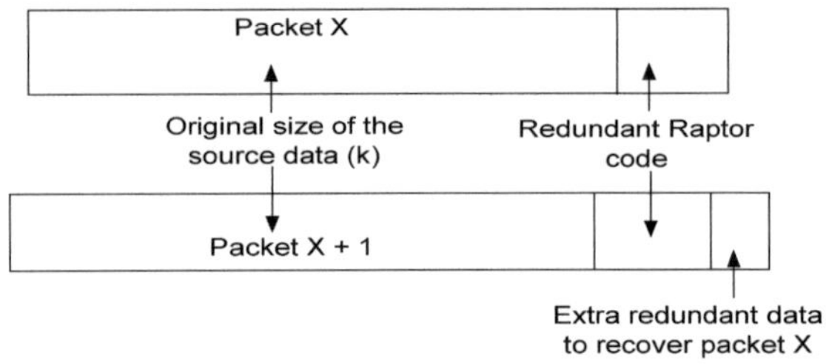

Fig. 2. Division of payload data in a packet (MPDU) between source data, original redundant data and piggybacked data for a previous erroneous packet.

3.2 Modeling the WiMAX environment

To evaluate the scheme, transmission over WiMAX was carefully modeled. The PHY-layer settings selected for WiMAX simulation are given in Table 1. The antenna heights are typical ones taken from the standard (IEEE 802.16e-2005, 2005). The antenna was modeled for comparison purposes as a half-wavelength dipole, whereas a sectored set of antenna on a mast might be used in practice to achieve directivity and, hence, better performance. The IEEE 802.16e Time Division Duplex (TDD) frame length was set to 5 ms, as only this value is supported in the WiMAX forum simplification of the standard. The data rate results from the use of one of the mandatory coding modes (IEEE 802.16e-2005, 2005) for a TDD downlink/uplink sub-frame ratio of 3:1. The base station (BS) was assigned more bandwidth capacity than the uplink to allow the WiMAX BS to respond to multiple mobile devices.

Parameter	Value
PHY	1024 OFDMA
Frequency band	5 GHz
Bandwidth capacity	10 MHz
Duplexing mode	TDD
Frame length	5 ms
Max. packet length	1024 B
Raw data rate (downlink)	10.67 Mbps
Modulation	16-QAM 1/2
Guard band ratio	1/16
MS transmit power	245 mW
BS transmit power	20 W
Approx. range to SS	1 km
Antenna type	Omni-directional
Antenna gains	0 dBD
MS antenna height	1.2 m
BS antenna height	30 m

OFDMA = Orthogonal Frequency Division Multiple Access,
QAM = Quadrature Amplitude Modulation, TDD = Time Division Duplex

Table 1. IEEE 802.16e parameter settings

Channel model

To establish the behavior of rateless coding under WiMAX, the ns-2 simulator augmented with a module or patch [12] that has proved an effective way of modeling IEEE 802.16e's behavior. Ten runs per data point were averaged (arithmetic mean) and the simulator was first allowed to reach steady state before commencing testing.

A two-state Gilbert-Elliott model served to simulate the channel model for WiMAX. In (Wang & Chang, 1996), it was shown that this model sufficiently approximates to Rayleigh fading, as occurs in urban settings during transmission from a base station to a mobile device. Moreover, in Jiao et al. (2002) it was shown that a first-order Markov chain can also model packet-level statistics. The main intention of our use of the twofold Gilbert-Elliott model was to show the response of the protection scheme to 'bursty' errors. These errors can be particularly damaging to compressed video streams, because of the predictive nature of source coding. Therefore, the impact of 'bursty' errors (Liang et al., 2008) should be assessed in video-streaming applications.

To model the effect of slow fading at the packet-level, the PGG (probability of being in a good state) was set to 0.95 and the PBB (probability of being in a bad state) = 0.96. The model has two hidden states which were modeled by Uniform distributions with PG (probability of packet loss in a good state) = 0.02 and PB (probability of packet loss in a bad state) = 0.01. The selection of a Uniform distribution is not meant to model the underlying physical process but to reflect the error patterns experienced at the application.

Additionally, it is still possible for a packet not to be dropped in the channel but, nonetheless, to be corrupted through the effect of fast fading. This byte-level corruption was modeled by a second Gilbert-Elliott model, with the same parameters (applied at the byte level) as that of the packet-level model except that PB (probability of byte loss) was increased to 0.165.

Assuming perfect channel knowledge of the channel conditions when the original packet was transmitted establishes an upper bound beyond which the performance of the adaptive scheme cannot improve. However, we have included measurement noise into the estimate of BL to test the robustness of the scheme. Measurement noise was modelled as a zero-mean Gaussian (normal) distribution and added up to a given percentage (5% in the evaluation) to the packet loss probability estimate.

In order to introduce sources of traffic congestion, an always available FTP source was introduced with TCP transport to a second mobile station (MS). Likewise, a CBR source with packet size of 1000 B and inter-packet gap of 0.03 s was also downloaded to a third MS. WiMAX has a set of quality-of-service queues at a BS. While the CBR and FTP traffic occupy the non-rtPS (non-real-time polling service) queue, rather than the rtPS queue, they still contribute to packet drops in the rtPS queue for the video, if the packet rtPS buffer is already full or nearly full, while the nrtPS queue is being serviced. Buffer sizes were set to fifty packets, as larger buffers lead to start-up delays and act as a drain upon MS energy.

The following types of erroneous packets were considered: packet drops at the BS sender buffer and packet drops through channel conditions; together with corrupted packets that were received but affected by Gilbert-Elliott channel noise to the extent that they could not be immediately reconstructed without a retransmission of piggybacked redundant data.

Notice that if the retransmission of additional redundant data still fails to allow the original packet to be reconstructed then the packet is simply dropped.

Raptor code model

In order to model Raptor coding, we employed the following statistical model (Luby et al., 2007):

$$\begin{aligned} P_f(m,k) &= 1 && \text{if} \quad m > k \\ &= 0.85 \times 0.567^{m-k} && \text{if} \quad m \geq k \end{aligned} \tag{2}$$

where $P_f(m,k)$ is the decode failure probability of the code with k source symbols if m symbols have been successfully received (and 1 - P_f is naturally the success probability). Notice that the authors of Luby et al. (2007) remark and show that for $k > 200$ the model almost perfectly models the performance of the code. In the experiments reported in this Chapter, the symbol size was set to bytes within a packet. Clearly, if instead 200 packets are accumulated before the rateless decoder can be applied (or at least equation (2) is relevant) there is a penalty in start-up delay for the video stream and a cost in providing sufficient buffering at the MSs. In the simulations, the decision on whether a packet can be decoded was taken by comparing a Uniformly-distributed random variable's value with that of the probability given by (2) for $k > 200$. The Uniform distribution was chosen because there is no reason to suppose that a more specific distribution is more appropriate.

It is implied from (2) that if less than k symbols (bytes) in the payload are successfully received then a further $k - m + e$ redundant bytes can be sent to reduce the risk of failure. In the evaluation tests, e was set to four, resulting in a risk of failure of 8.7 % in reconstructing the original packet if the additional redundant data successfully arrives. This reduced risk arises because of the exponential decay of the risk that is evident from equation (2) and that gives rise to Raptor code's low error probability floor.

Test video sequence

The test sequence was Paris, which is a studio scene with two upper body images of presenters and moderate motion. The background is of moderate to high spatial complexity. The sequences was variable bitrate encoded at Common Intermediate Format (CIF) (352 × 288 pixel/picture), with a Group of Pictures (GOP) structure of IPPP..... at 30 Hz, i.e. one initial Instantaneous Decoder Refresh (IDR)-picture followed by all predictive P-pictures. This structure removes the coding complexity of bi-predictive B-pictures at a cost in increased bit rate. Similarly, in H.264/AVC's Baseline profile, B-pictures are not supported to reduce complexity at the decoder of a mobile device. As a GOP structure of IPPP.... was employed, it is necessary to protect against temporal error propagation in the event of inter-coded P-picture slices being lost. To ensure higher quality video, 5% intra-coded MBs (randomly placed) (Stockhammer & Zia, 2007) were included in each frame (apart for the first IDR-picture) to act as anchor points in the event of slice loss. The JM 14.2 version of the H.264/AVC codec software was utilized, according to reported packet loss from the simulator, to assess the objective video quality (PSNR) relative to the input YUV raw video. Lost partition-C carrying packets were compensated for by error concealment at the decoder using the MVs in partition-A to predict the missing MB.

3.3 Evaluation results

Figure 3 shows the effect of the various schemes on packet drops when streaming Paris. 'Data-partition' in the Figure legend refers to sending no redundant packets. 'Duplicate X' refers to sending duplicate packets containing data-partitions of partition type(s) X, in addition to the data-partition packets. The proposed redundant schemes were also assessed for the presence of CIP or its absence. From Figure 3, the larger packet drop rates at quantization parameter (QP) = 20 will have a significant effect on the video quality. However, the packet size changes with and without CIP have little effect on the packet drop rate.

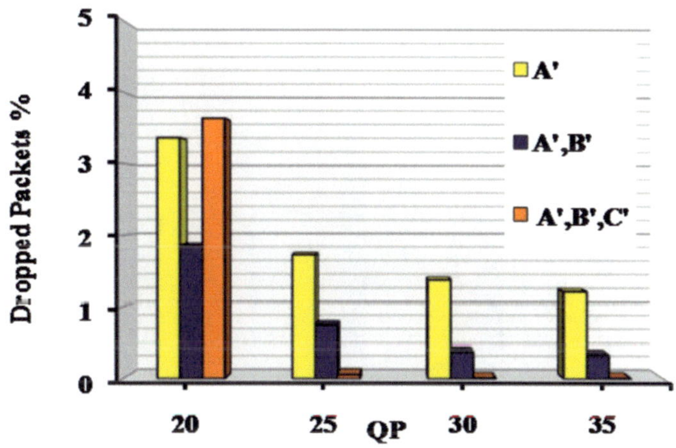

(a)

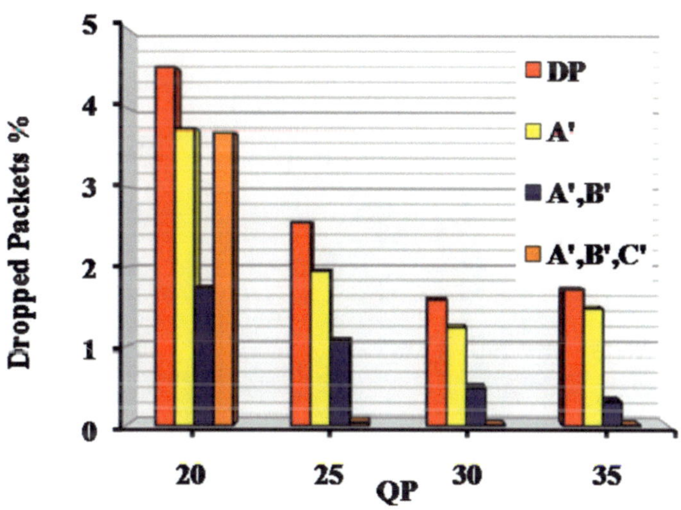

(b)

Fig. 3. *Paris* sequence protection schemes packet drops, (a) with and (b) without CIP. A` = duplicate partition-A; A`,B` = duplicate partitions A and B; A`, B`, C` = duplicate partitions A`, B`, and C`; DP = data-partitioning without duplication.

Figure 4 shows the pattern of corrupted packet losses arising from simulated fast fading. There is actually an increase in the percentage of packets corrupted if a completely duplicate stream is sent (partitions A, B, and C), though this percentage is taken from corrupted original and redundant packets. However, the effect of the corrupted packets on video quality only occurs if a packet cannot be reconstructed after application of the adaptive retransmission scheme.

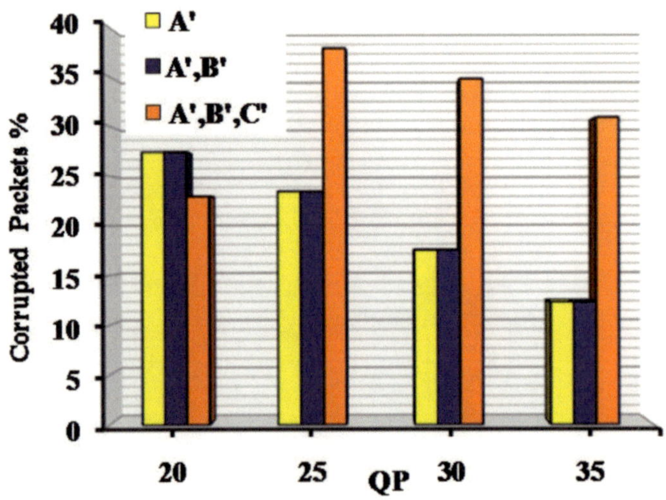

(a)

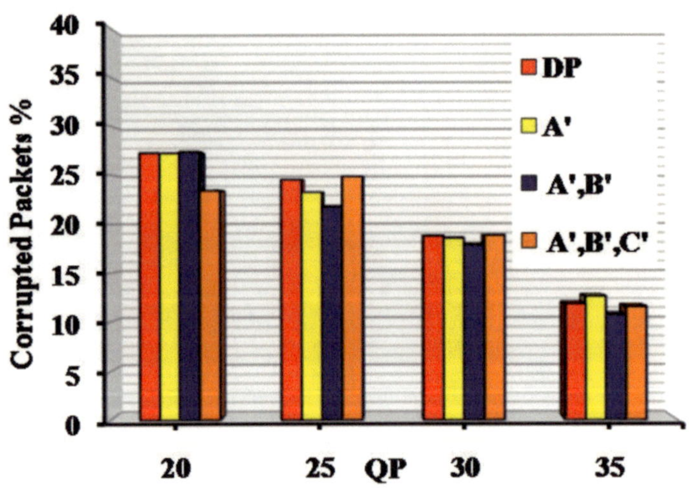

(b)

Fig. 4. *Paris* sequence protection schemes corrupted packets, (a) with and (b) without CIP. A` = duplicate partition-A; A`, B` = duplicate partitions A and B; A`, B`, C` = duplicate partitions A`, B`, and C`; DP = data-partitioning without duplication.

Examining Figure 5 for the resulting objective video quality, one sees that data partitioning with channel coding, when used without duplication, is insufficient to bring the video quality to above 31 dB that is to a good quality. PSNRs above 25 dB, we rate as of fair quality (depending on content and coding complexity). However, it is important to note that sending duplicate partition-A packets alone (without duplicate packets from other partitions) is also insufficient to raise the video quality to a good rating (above 31 dB). Therefore, to raise the video quality to a good level (above 31 dB) requires not only the application of the adaptive rateless channel-coding scheme but also the sending of duplicate data streams with duplication of more than just partition-A packets.

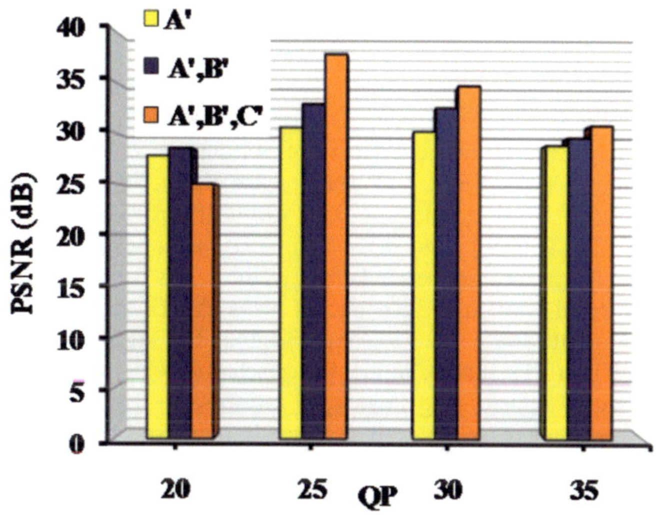

(a)

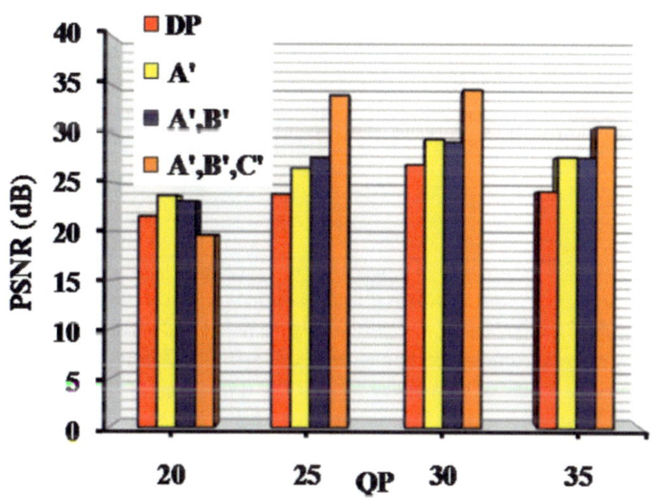

(b)

Fig. 5. *Paris* sequence protection schemes video quality (PSNR), (a) with and (b) without CIP. A` = duplicate partition-A; A`, B` = duplicate partitions A and B; A`, B`, C` = duplicate partitions A`, B`, and C`; DP = data-partitioning without duplication.

The impact of corrupted packets, given the inclusion of retransmitted extra redundant data, is largely seen in additional delay. There is an approximate doubling in per- packet delay between the total end-to-end delay for corrupted packets, about 20 ms with CIP and 17 ms without, and normal packet end-to-end delay. Normal packets do not, of course, experience the additional delay of a further retransmission prior to reconstruction at the decoder. Nevertheless, the delays remain in the tens of millisecond range, except for when QP = 20, when end-to-end delay for the scheme with a complete duplicate stream exceptionally is as high as 130 ms. It must be recalled that, for the duplicate stream schemes, there is up to twice the number of packets being sent. This type of delay range is acceptable even for interactive applications, but may contribute to additional delay if it forms part of a longer network path.

4. Concluding remarks

IEEE 802.16 and more narrowly the WiMAX Forum's simplification of the standards are well suited to video streaming but some form of application layer error protection will be necessary, of the type presented in this Chapter's case study. For severe channel conditions combined with traffic congestion, not only does forward error correction seem a necessary overhead, together with source-coded error resilience, but additional duplication of some part of the encoded bit-stream may be advisable. In the case study, data partitioning had the dual role of providing a way to reduce packet sizes (MPDUs) and a way to scale layer duplication. However, alternative schemes exist such as the MPEG-Pro COP #3 (Rosenberg & Schulzrinne) IP/UDP/RTP packet interleaving scheme which includes FEC as separate packets, and it is worth considering how application layer packet interleaving could be included in the presented scheme, though at a cost in increased latency. Such schemes have the advantage that they can be applied to multicast as well as unicast delivery, as there is no requirement for repair packets. However, the feedback implosion at a remote multicast server that results from repair packet requests from multiple video receivers can be avoided in the Chapter's scheme as the single request for extra 'piggybacked' redundant data can be turned off. This will require a determination of what level of adaptive FEC is necessary to support multicast delivery without repair packets. All the same in the Internet TV version of IPTV, multicast from a remote server prior to reaching the WiMAX access network is unlikely. This is because the Internet Group Management Protocol (IGMP) should be turned on at routers to support multicast, which is difficult to ensure.

5. References

Afzal, J.; Stockhammer, T.; Gasiba, T. & Xu, W. (2006). Video Streaming Over MBMS: A System Design Approach, *Journal of Multimedia*, Vol. 1, No. 5, pp. 25-35.

Ahmandi, S. (2011). *Mobile WiMAX: A Systems Approach to Understanding IEEE 802.16m Radio Access Technology*, Academic Press, ISBN978-0-12-374964-2, Elsevier, Amsterdam.

Andrews, J. G.; Ghosh, A. & Muhamed, R. (2007). *Fundamentals of WiMAX: Understanding Broadband Wireless Networking*, Prentice Hall, ISBN 0-13-222552-2, Upper Saddle River, NJ.

Castura, J. & Mao, Y. (2006). Rateless Coding over Fading Channels, *IEEE Communications Letters*, Vol. 10, No. 1, pp. 46-48.

Chang, B.-J.; & Chou, C.-M. (2007). Cross-layer Based Delay Constraint Adaptive Polling for High Density Subscribers in IEEE 802.16 WiMAX Networks, *Wireless Personal Communications*, Vol. 46, No. 3, pp. 285-304.

Chen, Y.-W. & Hsieh, F.-Y. (2007). A Cross-layer Design for Handoff in 802.16e Network with IPv6 Mobility, *IEEE Wireless Communications and Networking Conference*, pp. 3844-3849.

Cicconnetti, C.; Lenzini, L.; Mingozzi, E.; and Eklund, C. (2006). Quality of Service Support in IEEE 802.16 Networks, *IEEE Network*, Vol. 20, No. 2, pp. 50-55.

Degrande, N.; Laevens, K. & Vleeschauwer, D. De (2008). Increasing the User Perceived Quality for IPTV Services, *IEEE Communications Magazine*, Vol. 46, No. 2, pp. 94-100.

Dhondt, Y.; Mys, S.; Vermeirsch, K. & Walle, R. Van de (2007). Constrained Inter Prediction: Removing Dependencies between Different Data Partitions, *Proceedings of Advanced Concepts for Intelligent Visual Systems*, pp. 720-731.

Ekstrom, K. et al. (2006). Technical Solutions for the 3G Long-Term Evolution, IEEE Communications Magazine, Vol. 44, No. 3, pp. 38-45.

Lambert, P.; Neve, W. De; Dhondt, Y. & Wall, R. Van de (2005). Flexible Macroblock Ordering in H.264/AVC, *Journal of Visual Communication and Image Representation*, Vol. 17, No. 2, pp. 358-375.

Goldsmith, A. (2005) *Wireless communications*, Cambridge University Press, UK.

IEEE 802.16e-2005 (2005). IEEE Standard for Local and Metropolitan Area Networks. Part 16: Air Interface for Fixed and Mobile Broadband Wireless Access Systems.

IEEE 802.21-2008 (2008) 802.21 WG IEEE standard for local and metropolitan area networks, media independent handover services.

Issa, O.; Li, W.; & Liu, H. (2010). Performance Evaluation of TV over Broadband Wireless Access Networks, *IEEE Transactions on Broadcasting*, Vol. 56, No. 2, pp. 201–210.

Jain, R.; So-In, C. & Al-Tamimi, A.-K (2008) System-level Modeling of IEEE 802.16e Mobile WiMAX networks, *IEEE Wireless Communications*, Vol. 15, No. 5, pp. 73–79.

Jiao, Ch.; Schwiebert, L. & Xu, B. (2002). On Modeling the Packet Error Statistics in Bursty Channels, *IEEE Conference on Local Computer Networks*, pp. 534- 541.

Juan, H.-H.; Huang, H.-C.; Huang, C. & Chiang, T. (2009). Cross-layer Mobile WiMAX MAC Designs for the H.264/SVC Scalable Video Coding, *Wireless Networks*, Vol. 16, No. 1, pp. 113-123.

Kawadia, V. & Kumar, P.R. (2003). A Cautionary Perspective on Cross layer Design, *IEEE Wireless Communications Magazine*, Vol. 12, No. 1, pp. 3-11.

Kuhran, M.S.; Gür, G.; Togcu, T. & Alagöz, F. (2010). Applications of the Cross-layer Paradigm for Improving the Performance of WiMAX. *IEEE Wireless Communications*, Vol. 17, No. 3, pp. 86-95.

Kumar, A. (2008). *Mobile Broadcasting with WiMAX: Principles, Technology, and Applications*, Elsevier, ISBN 978-0-240-81040-9, Amsterdam.

Liang, Y.J.; Apostolopoulos, J.G. & Girod, B. (2008). Analysis of Packet Loss for Compressed Video: Effect of Burst Losses and Correlation Between Error Frames, *IEEE Transactions on Circuits and Systems for Video Technology*, Vol. 18, No. 7, pp. 861-874.

Lee, J. M.; Park, H.-J.; Choi, S. G.; & Choi, J.K. (2009). Adaptive Hybrid Transmission Mechanism for On-demand Mobile IPTV over WiMAX, *IEEE Transactions on Broadcasting*, Vol. 55, No. 2, pp. 468-477.

Luby, M. (2002) LT codes, *34rd Annual IEEE Symposium on the Foundations of Computer Science*, pp. 271–280.

Luby, M.; Gasiba, T.; Stockhammer, T. & Watson, M. (2007). Reliable Multimedia Download Delivery in Cellular Broadcast Networks, *IEEE Transactions on Broadcasting*, Vol. 53, No. 1, pp. 235-246.

MacKay, D.J.C. (2005). Fountain Codes, *IEE Proceedings: Communications*, Vol. 152, No. 6, pp. 1062–1068.

Neves, P.; Sargneto, S.; Pentikousis, K. & Fontes, F. (2009) WiMAX Cross-layer System for Next Generation Heterogeneous Environments, pp. 357-376, In *WiMAX, New Developments*, Dalal, U.D.D. & Kosta, Y.P. (eds.), In-Tech, ISBN 978-953-7619-53-4, Vienna, Austria.

Nuaymi, L. (2007) *WiMAX: Technology for Broadband Wireless Access*, Wiley & Sons, ISBN 978-0-470-02808-7, Chichester, UK.

Oyman, O.; Foerster, J.; Tcha, Y.-j.; & Lee, S.-C. (2010). Towards Enhanced Mobile Video Services over WiMAX and LTE, *IEEE Communications Magazine*, Vol. 48, No. 8, pp. 68–76.

Palanki, R. & Yedidai, J. (2004). Rateless Codes on Noisy Channels, *International Symposium on Information Theory*, p. 37.

Radulovic, J.; Wang, Y.-K.; Wenger, S.; Hallapuro, A.; Hannuksela, M.H. & Frossard, P. (2007). Multiple Description H.264 Video Coding with Redundant Pictures, *International Workshop on Mobile Video*, pp. 37-42.

Rosenberg, J. & Schulzrinne, H. (1999). An RTP Payload Format for Generic Forward Error Correction, Internet Engineering Task Force, RFC 2733.

Schaar, M. van der & Shankar, N.S. (2005). Cross-layer Wireless Multimedia Transmission: Challenges, Principles, and New Paradigms, *IEEE Transactions on Wireless Communications*, Vol. 12, No. 4, pp. 50–58.

She, J.; Yu, X.; Po, P.-H.; & Yang, E.H. (2009) A Cross-layer Design Framework for Robust IPTV Services over IEEE 802.16 Network, *IEEE Journal of Selected Areas in Communications*, Vol. 29, No. 2, pp. 235-245.

Shokorallahi, A. (2006) Raptor codes, *IEEE Transactions on Information Theory*, Vol. 52, No. 6, pp. 2551-2567.

So-In, C.; Jain, R. & Tamini, A.-K. (2010). Capacity Evaluation of IEEE 802.16e WiMAX, *Journal of Computer Systems, Networks, and Communications*, Vol. 2010 [online], 12 pages.

Srivastava, V. & Motani, M. (2005). Cross-layer Design: A Survey and the Road Ahead, *IEEE Communications Magazine*, Vol. 43, No. 12, pp. 712-721.

Stockhammer, T. & Bystrom, M. (2004). H.264/AVC Data Partitioning for Mobile Video Communication, *Proceedings of IEEE International Conference on Image Processing*, pp. 545-548.

Stockhammer, T. & Zia, W. (2007). Error-resilient Coding and Decoding Strategies for Video Communication, pp. 13-58, In Schaar, M. van der & Chou, P.A. (eds.) *Multimedia in IP and Wireless Networks*, Academic Press, ISBN 978-0-12-088480, Burlington, MA.

Stockhammer, T.; Luby, M. & Watson, M. (2008). Application Layer FEC in IPTV Services, *IEEE Communications Magazine*, Vol. 46, No. 5, pp. 94-101.

Uilecan, I.V.; Zhou, C. & Atkin, G.E. (2007). Framework for Delivering IPTV Services over WiMAX Wireless Networks, *IEEE International Conference on Electronics/Information Technology*, pp. 470–475.

Wang, H. & Chang, P. (1996). On Verifying the First-order Markovian Assumption for a Rayleigh Fading Channel Model, *IEEE Transactions on Vehicular Technology*, Vol. 45, No. 2, pp. 353–357.

Wiegand, T.; Sullivan, G.J.; Bjontegaard, G. & Luthra, A. (2003). Overview of the H.264/AVC Video Coding Standard, *IEEE Transactions on Circuits and Systems for Video Technology*, Vol. 13, No. 7, pp. 560-576.

Wenger, S. (2003). H.264/AVC over IP, *IEEE Transactions for Circuits and Systems for Video Techno*logy, Vol. 13, No. 7, pp. 645-655.

Zhu, C.; Wang, Y.K.; Hannuksela, M. & Li, H. (2006). Error Resilient Video Coding Using Redundant Pictures, *IEEE International Conference on Image Processing*, pp. 801–804.

13

Public Safety Applications over WiMAX Ad-Hoc Networks

Jun Huang[1], Botao Zhu[1] and Funmiayo Lawal[2]
[1]*Jiangsu University,*
[2]*University of Ottawa,*
[1]*China*
[2]*Canada*

1. Introductions

1.1 Special needs of public safety communications

Wireless communications in the public safety heavily depends on the robustness, reliability, availability and usability of the communication system. In the past decades this was achieved at the price of extremely high system cost, and was often based on specialized solution that lacked interoperability. Faced by severe cost constraints, the need to ensure interoperation of various agencies, and the desire to involve existing infrastructures available, the public safety community is increasingly attracted by the opportunity to utilize off-the-shelf technology in conjunction with both specialized and commercial communication systems.

The most basic communication need of the public safety is radio-based voice communications. This type of communication allows dispatchers to direct personnel to areas where incidents have occurred. The trend in this marketplace has been geared towards allowing for inter-agency communication in case of large-scale disasters. The most notable large-scale response effort occurred on September 11, 2001, when multiple agencies responded to the attacks in New York. The state of the most basic radio technology could not meet the increasing demand for radio communications that arose on that day. The crush of radio communications flooded the spectrum, and caused massive failures across the board with regard to the base station relaying of crucial information, led to more deaths of first responders. The most gripping issue regarding the state of the technology at that time was the fact that the same failures had occurred in 1993 and nothing had been done to address the issue. More focus had been put on developing faster and more lucrative consumer market, and the mainstream vendors had forgot this niche space.

Radio was the primary medium for the transmission of voice communications. Later developments allowed for the transmission of voice and data over the same radio spectrum. The problem was that the only people capable of receiving these transmissions were other first responders in the same department. There was an inability to communicate across different departments or agencies for coordination during a disaster. The conventional radio system typically had three segregated channels: car to station, station to car and car to car.

There was also a shortfall due to the fact that personnel must wait for a transmission to complete prior to being able to send their own transmissions, since the channel only allowed for one speaker at a time. A vehicular mesh network would have allowed for additional channel resources for voice communication. Further, a video channel could have been set up with real-time situational awareness, with a tie in to vehicle or body cameras. Short message service through the use of private messaging networks would also have been available in the event that a voice channel was unavailable, thus allowing for vital information to be relayed immediately rather than waiting for a chance to transmit. P25 group is addressing this issue for voice and data; here we focus more on video on-the-go.

When a fireman trying to rescue a people, the environment is harsh and noisy, some times voice is not that effective and live video or GPS (Global Positioning System) data is needed to assist the coordination's. The camera is normally mounted on firemen's helmet, and wirelessly transmitted to the fire-engines (service vehicles) on the spot, for the commander to see how are every team members doing; the goal is to keep firemen alive at the first place, and then to rescue as many people as possible. Comparing with voice or GPS and other sensor data such as temperature, CO density etc, video data is relative large and harder to get through wireless channel, however "a picture may worth a thousand words"; for this reason we focus on the evaluating video over Vehicular Ad-hoc Network (VANET) in this study.

Fig.1.1 shows video communication application of the techniques disclosed herein, for public safety authority usage. The system includes a national control centre at the gateway level, a police car and a fire engine incorporating mobile servers at the service truck level, and mobile terminals which are carried by public safety personnel. The terminals gather information which being transmitted to the servers and then on to the national control centre for subsequent access by client systems.

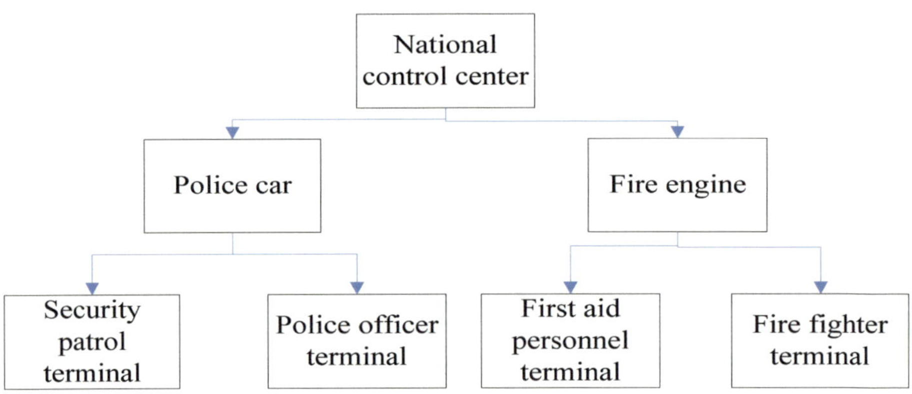

Fig. 1.1. A system architecture of public safety communications

Fig.1.2 is a typical Point-to-Multi-Point (PMP)/ Multi-Point-to-Point (MPP) and Peer-to-Peer (P2P) JXTA network, including fixed client systems operatively coupled to a gateway through a communication network. The gateway is operatively coupled to a mobile server through a satellite system, and also to a remote server. The mobile server is operatively coupled to mobile communication devices, including a mobile client system and mobile terminals. The remote server is operatively coupled to remote terminals.

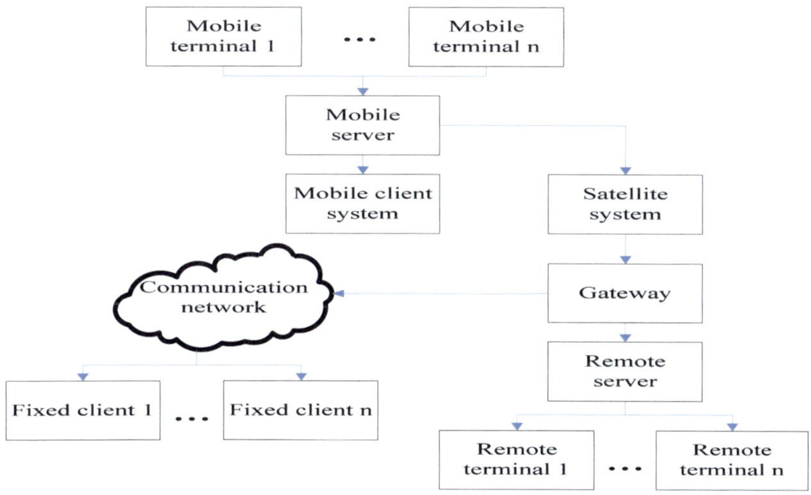

Fig. 1.2. PMP/MPP/P2P public safety networks

Note that any thing mobile must go through wireless here. Software defined radio is used to bridge the gaps, between each section of the network, while they are moved around.

Fig. 1.3. Public safety system road test scenes

Above are the streets views where communications between our mobile server and mobile client were interrupted for more than 20% of time, where end-to-end delay exceeded more than 10 seconds at the peaks, during the frequency and network switching. Those field tests have partially trigged our in depth studying.

1.2 Vehicular networks for road safety

Vehicle to vehicle communication needs a unique Ad-hoc communication scheme that is self-organizing, and it can function without a pre-existing cellular infrastructure network. This is an essential feature of VANET because when conventional communication towers are suffering outages or become non-existent, Ad-hoc communication can provide an effective way to transmit information. Due to the rapidly changing topology and the speed of the vehicles in Ad-hoc network, a number of issues become increasingly important to ensure the efficiency and stability of this network. Here we focus on the video traffic sizing challenge, which is the key to unlock the power of video applications. Like every other wireless environment, transmitting video signals in a VANET poses concerns. Handling congestion and packet loss becomes more difficult and delicate in a VANET environment where interference is inevitable. Interference such as electromagnetic waves from starting car engines with electronics, from Additive White Gaussian Noise (AWGN) wireless channel under critical weather conditions, can all affect the Quality of Service (QoS) as seen by the end user. The topology is constantly changing and vehicles could move out of sight from one another causing an outage in video transmission.

In addition, unlike every other network environment, VANET mobility has a peculiar and unique nature due to the randomness of human behaviour. In creating an effective mobility model, vehicle-to-vehicle interaction and vehicle to infrastructure interaction needs to be considered carefully and closely. One of the major research issues in VANET is the creation of an effective simulation platform that can integrate a network simulator with a realistic vehicular traffic simulation model. According to (Sommer & Dressler, 2008), the effect of having a realistic mobility model is evident. In integrating a network model with a VANET mobility model, two approaches are identified: an open-loop integration approach and a closed-loop integration approach. The latter entails integrating traces generated from a mobility simulator to a network simulator while the former runs the two simulators concurrently. In other words, in the closed-loop approach, the traffic simulator and the external VANET mobility simulator are connected using High Level Architecture (HLA) design for distributed computer simulation systems, so that the two components feed the most recent information back to each other. The closed-loop approach is more effective as it allows the effect of the wireless signals to govern the mobility patterns of drivers. It also models driver reactions to certain wireless signals as detailed in (Sommer & Dressler, 2008).

1.3 WiMAX made for VANET

WiMAX (WiMa, 2009) is a 4G equivalent technology standardized by IEEE802.16 that enables the delivery of last mile wireless broadband access. The name WiMAX was created by the WiMAX forum, which was formed in June of 2001 to promote conformity and interoperability of the standard (Brit, 2010). The WiMAX technology (Ghosh, 2007) provides ease deployment as it eliminates the use of cables and can save investment when used in remote and rural areas. The technology is scalable and has a flexible frequency re-use scheme because it can use Orthogonal Frequency Division Multiplexing (OFDM) technology. WiMAX implements full Multiple-Input and Multiple-Output (MIMO) setting, which is a good fit for mobile and car applications, by enhancing timely information delivery to save lives and improve quality of life.

A comparison of these physical layer technologies that could be used for VANET is shown in Table1.1 (Morgan, 2010). The '$$' in the table was used to denote the cost per bit for each technology where '$' represents the least expensive and '$$$$' represents the most expensive. Through comparison, one can see that WiMAX is the most cost effective approach by providing a data rate that can satisfy the needs of our mobile multimedia users (low latency and high coverage) at high speed and at an affordable cost.

One of the major challenges in VANET design is the development of an effective platform that can bring all issues described earlier under one umbrella – a complete simulation model. Since it is safer and more cost efficient to simulate possible solutions rather than field experimenting of driving at 140km/hr, creating an effective VANET simulation platform

Items	*WiMAX*	*Satellite*	*DSRC*	*FM Radio*	*GSM*	*CDMA*
Max Range km	<50	1000s	< 1	100s	<10	<10
Data Rate mbps	70	100	10	0.01	0.1	2
Cost per bit	$$	$$$$	$	$	$$$	$$$
Average Latency	Lo	Lo	Very Lo	Hi	Lo	Lo
Connectivity	Hi	Very Hi	Lo	Lo	Hi	Very Hi
Sustain km/hr	180	100	80	120	140	110

Table 1.1. Comparison of related wireless technologies for video on the go application

has become of pertinent importance in research and industry. One of the major challenges faced is integrating an effective mobility model that puts vehicle to vehicle interaction and vehicle to infrastructure interaction into consideration, along with platform possessing the full functionalities of a communication device with effective receiving, processing and transmitting capabilities, thus emulating a real world situation. Human behavioural modelling are also some of the other issues to be modelled as close to reality as possible, to produce conclusions that can be used in the real world. Although (Wegener et al., 2008) have worked on creating a similar platform, no specific work have been done using OPNET as a popular network simulation tool. In addition, customizing the platform for real-time video traffic is a specific area we explored using different traffic level scenarios.

1.4 WiMAX Ad-hoc network

WiMAX is a broadband wireless technology that can sustain voice, video and data services at high moving speed while maintaining high data rates. Mobile WiMAX is based of OFDMA physical layer of the 802.16e-2005 standard, which is a revision of the fixed WiMAX standard. IEEE 802.16e provides functionalities such as BS handoffs, MIMO transmit/receive diversity, and scalable Fast Fourier Transform sizes (Li, 2006). WiMAX is considered one of the most promising technologies in the rural area today. Ad-hoc network (Song & Oliver, 2004) has emerged, for instance, wireless mesh network, and it rapidly gained acceptance and interest from both academic and industrial communities for the advantages of low up-front cost, easy network maintenance, good robustness, usability, reliable service and larger coverage. Thus, the mesh mode was defined in the IEEE 802.16 standard as an additional architecture to the previous Point to Multi-Point (PMP) mode. In the PMP mode, nodes are organized into a cellular like structure consisting of a Base Station (BS) and some Subscriber Stations (SS). All the SSs must be within the transmission range of the BS, and traffic only occurs directly between BS and SS. Mesh SS communication without going through the Mesh BS, network traffic can through other Mesh SS, two Mesh SS communicate in direct. Comparing with PMP mode, the mesh mode can provide better coverage, survivability, flexibility and scalability, thus a great deal of research works have been done focusing on WiMAX (Zhou & Ji, 2010) mesh networks for performance improvement. Many of the works concentrated on the construction of routing trees (Chen et al., 2008) and link or packet scheduling with spatial reuse, aiming to maximize the throughput, maximize the number of concurrent transmission links, minimize the end-to-end delay, and provide better fairness. The Ad-hoc mode of VANET for public safety is a special mesh mode; the focus is more on survivability and usability rather than increased bandwidth.

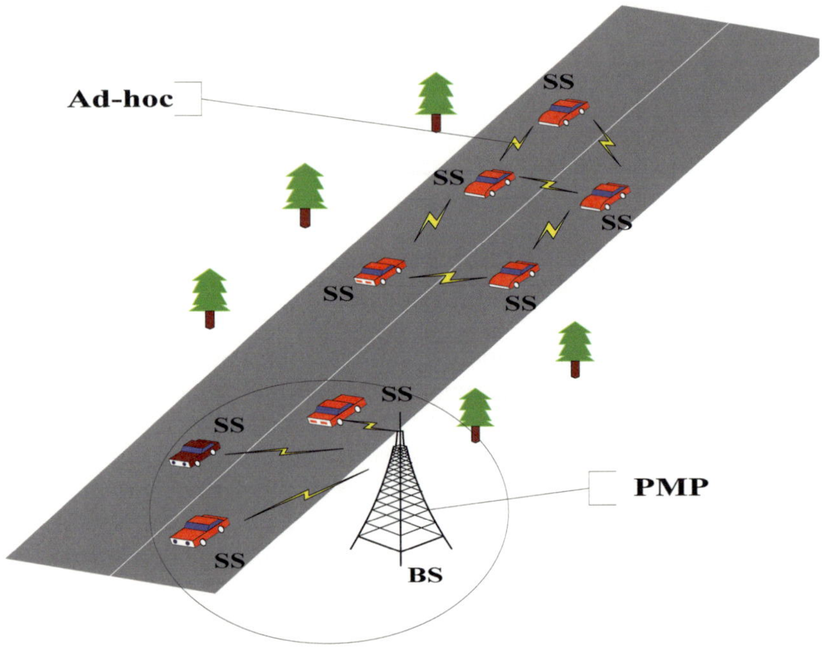

Fig. 1.4. WiMAX Ad-hoc vehicle networks

2. Public safety networks operation, models and assumptions

2.1 Safety network operation

This section describes the network layout of VANET with WiMAX technology along with their operation that are of interest to this research.

2.1.1 General network layout of VANET

In the VANET we envisioned, each vehicle has the ability to communicate with any neighbouring vehicles. Depending on the nature of the message, the information either remains within the VANET or venture out to the backhaul network via the Road Side Unit (RSU). For instance, brake warning sent from preceding cars, tailgate and collision warnings are messages that can remain in the VANET network. In the sensor application (Li et al, 2009), video messages are forwarded from the point of interest (which could be a traffic congestion area, camera view from unmanned car, road block, accident scene etc), to the backhaul network via the RSU to aid traffic personals, emergency agents or any other party to respond to such situations more effectively.

To study the traffics generated within the network, we consider a VANET consisting of N cars communicating with each other and with the Internet via RSUs. The network topology is shown in Fig.2.1. The RSU (BS1 or BS2) has the capability to handle up to 100 cars simultaneously. Each car is associated with the RSU depending on their distance to one another. The video packets are routed and given priority due to the service class name associated with them and the scheduling type, which handles the bandwidth request/grant mechanism. The silver service class and the Real-time Polling Service (RTPS) scheduling are used. Maximum sustainable traffic and reserved traffic rates are set to 384kbps for this service class. The minimum rate between cars is set to 96kbps.

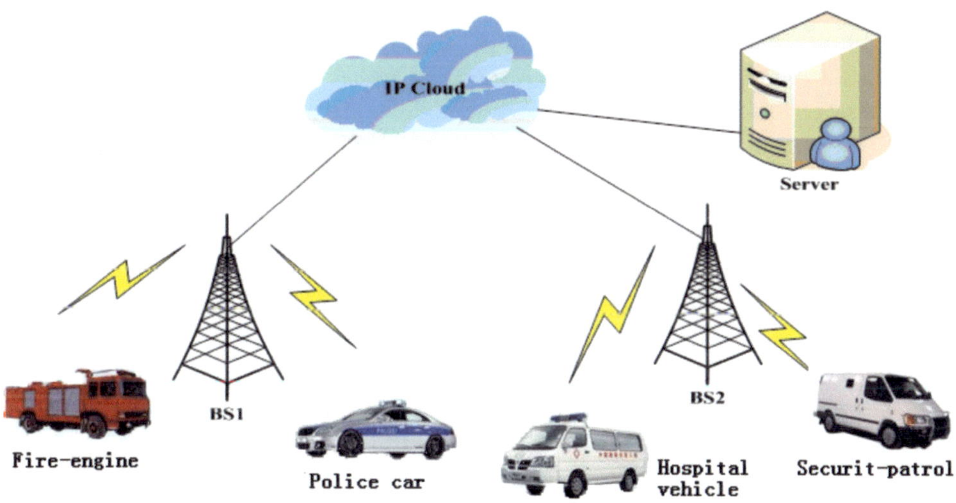

Fig. 2.1. Public safety network topology model

At the SS station, over the low sub-layer Air Interface, the average Service Data Unit (SDU) size is less than 768 bytes, such that the entire packet can survive the wireless transmission. The larger packet is very vulnerable to interference of all kinds. Each video arriving from the higher layer is expected to be broken down to this size range. Any packet size greater than this shall be segmented before encapsulated into a Protocol Data Unit (PDU) and transmitted with appropriate header information, any packet less than this shall be merged with previous leftover or next small packet if possible before encapsulated for Air Interface. When a SS wants to transmit video, the video is generated from the application layer using our traffic generation model. The packet is sent to the RSU and the RSU forwards the packet accordingly. The IP cloud is set to its default values and acts as a router. The server is configured to accept packets generated by our model.

WiMAX is known for its data rates up to 128Mbps downlink and 56Mbps uplink using its MIMO antenna techniques. In our case, we used Simple Input Simple Output (SISO) antenna technique, which supports up to 1Mbps uplink and downlink. It defines service flows that can be mapped into gradual IP sessions to enable end-to-end IP based QoS. Scalability, Security and mobility management are the other major features of WiMAX technology.

In our OPNET model, WiMAX does not support network-assisted handover, base station-initiate periodic ranging and power management. A sub-channel is allocated to each user thereby reducing the channel interference in the frequency domain. OFDMA is the scheme used allowing multiple accesses to every user on our network. At the Network layer, IPv4 is used for addressing and Routing Information Protocol (RIP) is used as the routing protocol. RTSP is a real-time streaming protocol designed for streaming video.

2.2 Public safety network models and assumptions

2.2.1 VANET Video model

Fig.2.2 shows a diagram summarizing the various components of our model. The video VANET OPNET model, consist mainly of the Video model and the VANET model. By first analyzing a live video trace, characterizing the trace and modeling the characterized trace then feed it into our simulator, to obtain the final Video model. On the other hand, the VANET model consists of the VANET mobility model and a communication model.

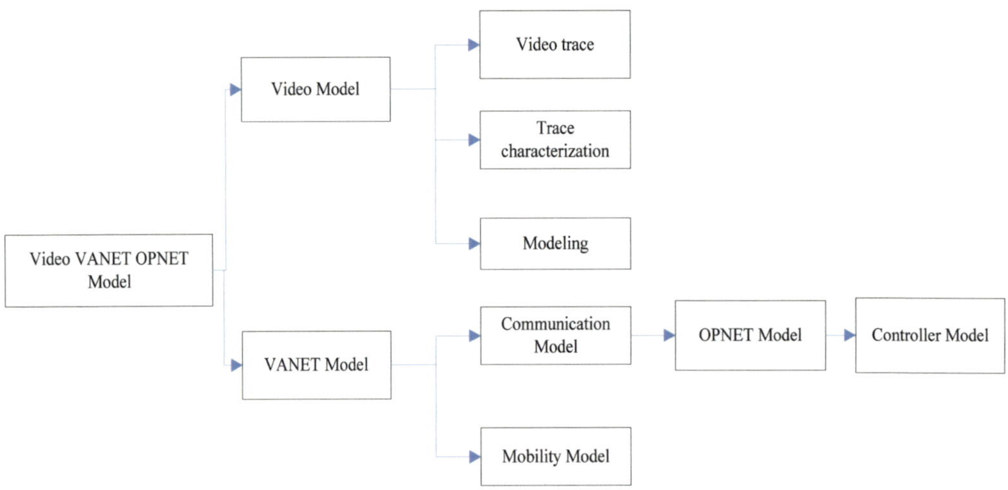

Fig. 2.2. Video VANET OPNET model tree structure

OPNET modeller provided the platform for the communication model and allowed for the integration of the various components of the Video VANET OPNET model.

a. VANET model

From our survey, Table 2.1 shows a summary of the findings.

The result of this analysis presents VanetMobiSim as the only mobility model found as of the time of development that could be integrated into OPNET consequently influencing our choice. VanetMobiSim's ability to integrate into OPNET comes with its flexible to manipulate its output file by coding its output generator file to produce a desired format.

Besides its adaptable output abilities, VanetMobiSim incorporates both microscopic and macroscopic models to allow the modelling of vehicle-to-vehicle and vehicle-to-infrastructure interaction. Traffic light integration, stop signs, human mobility dynamics

Items	*OPNET*	*ns2*	*QualNet*
MoVES	No	No	No
STRAW	No	No	No
VanetMobiSim	Yes	Yes	Yes
SUMO	No	Yes	Yes
SHIFT	No	No	No
GMSF	No	Yes	Yes

Table 2.1. Mobility model summaries

and safe inter-distance management are all modeled in this tool. The different forms of topology are shown in Fig.2.3 (Fiore et al., 2007). VanetMobiSim provides a flexible platform in which the user can configure the path used during a trip between Dijkstra shortest-path, road-speed shortest path and a density-based shortest path. The trip could either be generated by random source-destination or activity-based (Fiore et al., 2007).

a) User- defined topology b) Randomly defined topology c) GDF map topology

Fig. 2.3. Typical mobility topologies

The RSU and car communication are the major communication nodes in VANET. Our RSU is a simplified WiMAX BS. Each car is equipped with proper communication tools to enable car to car and car to infrastructure (RSU in our case) interaction. The design of each RSU is robust and non-application sensitive so that every car can send and receive a wide range of information. Table 2.2 shows the basic essential characteristics of our model along with some typical settings.

Parameter	*Value*
Physical layer	IEEE 802.16e
BS TX power (W)	5
Number of TX	SISO
BS Antenna Gain (dBi)	15
Minimum Power Density (dBm/Hz)	-80
Maximum Power Density (dBm/Hz)	-30
Link bandwidth (MHz)	20
Base Frequency (GHz)	5.8
Physical layer Profile	OFDM

Table 2.2. Typical RSU parameters

b. IEEE802.16 video model

The video model is one of the main components of our VANET OPNET model as our research focuses on real-time video communication in a VANET environment. In creating our video model, we put certain factors into consideration to measure the usefulness of the model. According to (Huang, 2001) factors like parsimony, analytic correctness, flexibility, implement ability and absolute accuracy was considered with MOS (Mean Opinion Score) method, on a scale of 1 to 3, using the factors mentioned above, 1 being the least and 3 the greatest. As common sense, each model has its pros and cons. With respect to our application, we choose parsimony and implement ability as our highest priorities.

Items	*Mini Pareto*	*FBM*	*TCP*
Parsimony	2	3	1
Analytical	2	1	1
Flexibility	1	1	1
Implemental	3	2	1
Accuracy	2	2	3

Table 2.3. Traffic model methodology comparisons

Table 2.3 shows other models and their MOS rating with respect to the factors described above. We have taken a systematic approach in developing our mini-Pareto model. Video traffic trace was collected using the same camera used for a car-to-car road test. The traces were analyzed and stochastically represented and plugged into our simulation platform.

2.2.2 Modeling assumptions

Unless otherwise stated, the following are assumptions taken throughout the chapter:

1. Every vehicle in the network is equipped with necessary radio. Every vehicle on the road has the capability to receive from and send video data to other vehicles via the RSUs.
2. BS is a "stationary" node. This is required due to the limitation of our OPNET model and we need it to act as an intermediate node for packet forwarding to the destination.
3. No disruption in a communication channel because one can use dedicated channel allocation once the node is in the communication range of a RSU.
4. Finite buffer size for each transmitter: this is a more realistic assumption, which would also allow us to find the trade-off between buffer size and end-to-end delay.
5. The RSU use OFDMA for multiplexing and their is always a slot available for each SS sending video traffic, Media Access Control (MAC) layer stress test will be studied later.

3. Laboratory set-up and trace collection

This section presents the experimental set-up of our model. It discusses the trace collection process and the initial analysis done on the trace. The later sections then describe the simulation environment, scenarios and performance measures used in this work.

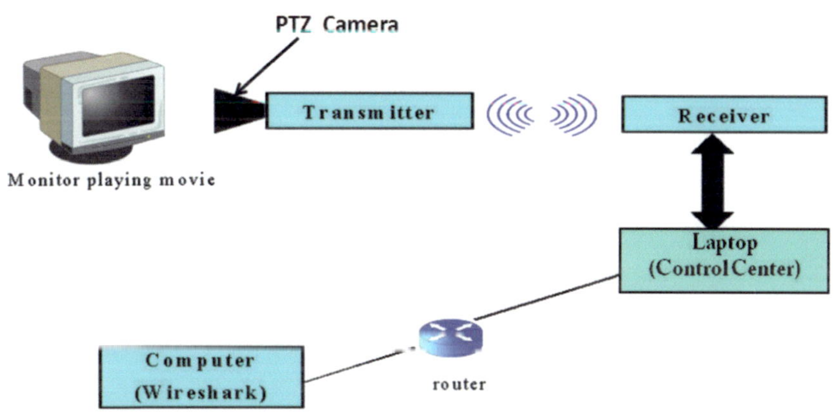

Fig. 3.1. Set-up for taking wireless video traces

3.1 Experimental conditions

We need to first collect video traces in order to model a video characteristic that is as close to reality as possible. Traces from a live camera were obtained using WireShark software on a monitoring PC. The set-up is shown in Fig.3.1. The monitor was used to play a series of video clips for 10 minutes each. The transmitter sends the compressed (in a ratio of 250:1) and encrypted video images to the control centre via a car-to-car radio system. The receiver decodes and decompresses the received video frames and plays the image at the control centre at about 20fps. The control centre laptop and the receiver are connected using a USB port. The control centre is then connected through a router to a computer hosting the packet trace-capturing tool - WireShark. Once the system is turned on, the computer with the WireShark software is set to access the "capture" folder in the control centre before streaming the video. The WireShark software is turned on and the trace capture begins. The video clips were chosen based on the activity rate in the clip. Three types of video clips are chosen and described in the following.

1. Action movies. This type of movies has a lot of movements and hence more variations in frame sizes. "The Prisoner" by Jackie Chan was chosen for study here.
2. Drama: This type of movies has an average movement and hence, it's a mixture of frame sizes. "The Game Plan" with Dwayne Johnson was chosen for this study.
3. A romantic: This type of movies has very slow scenes hence, little or no variation in frame sizes. "28 Day" with Sandra Bullock was chosen for this study.

3.2 Initial analysis and detail parameter matching

The next challenge was to analyse the trace and create a video traffic model. Fig.3.2 shows the schematics of our traffic model. The number of sources, N, was to be chosen bearing in mind the trade-off between parsimony and accuracy (Parsimony refers to the provision of the simplest and most frugal available solution to a certain problem). Each mini-source represents each set of video object sub-stream with the switch being regulated. The switch is configured to form traffic with a long-range dependency. We modelled the on-time switch by a Pareto distribution and the off-time by an exponential process, since we believe the memory between action sequences is negligible, but within the same Action Unit (AU), the sequences are strongly correlated (Gu & Ji, 2004).

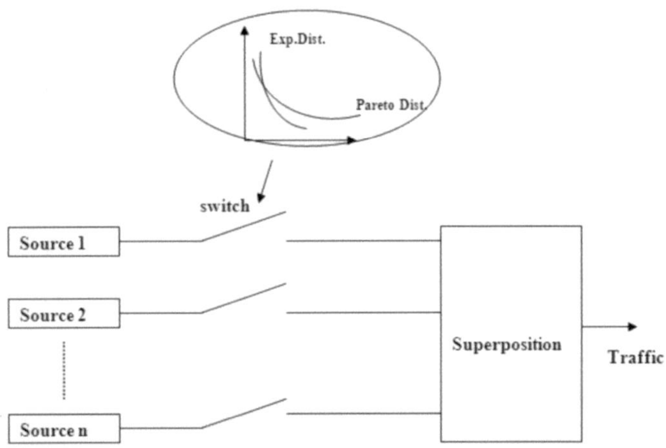

Fig. 3.2. Mini source model for video traffic generation

The problem with previous standardized 4IPP model is the matching process with Index of Dispersion for Counts (IDC) curve from measured data is complicated, especially when we need to scale up or down for different data rate and different applications. Also the distribution is tied down on the traditional exponential, lacks the flexibility to include the more general distribution such as Pareto, or Weibull on-off latterly proven has property of large deviation (Duffy & Sapozhnikov, 2007); most engineers also believed that Weibull can model wide range of WWW traffic, and Pareto is good for video. Here is a quick review of distributions used in OPNET tool:

Pareto definition:

$$f(x) = c_p \alpha_p^{c_p} / x^{c_p+1}, \quad \alpha_p \le x < \infty \tag{1}$$

Mean: $E(x) = c_p\alpha_p / (c_p - 1)$, $c_p > 1$. Arg1: location $= \alpha_p > 0$, Arg 2: shape $= c_p > 0$.

Exponential definition:

$$f_x(x_o) = \begin{cases} \alpha_e e^{-\alpha_e x_o} & x_o > 0 \\ 0 & \text{otherwise} \end{cases}, \ \alpha > 0 \tag{2}$$

Mean: $E(x) = \alpha_e^{-1}$.

Weibull definition:

$$f(x) = \begin{cases} 1 - e^{-\left(\frac{x}{\beta_w}\right)^{c_w}} & x > 0 \\ 0 & \text{otherwise} \end{cases} \tag{3}$$

Arg: shape $= c_w$.

Lognormal definition:

$$f(x) = \int \begin{cases} \frac{1}{x\sqrt{2\pi\sigma^2}} \exp \frac{-(\ln x - \mu)^2}{2\sigma^2} & \text{if} x > 0 \\ 0 & \text{otherwise} \end{cases} \tag{4}$$

Mean period: $E(x) = e^{\mu + \sigma^2/2}$

In this new proposed baseline extension model, we have included both Pareto and Weibull distribution, which are available now in OPNET library due to our previous suggestions to the tool vendor. The detail matching steps from WireShark measurement to the OPNET parameters are offered as followings:

Here is the procedure deciding the number of mini sources (N_s).

$$N_S = 1 + \overline{R_I}/\overline{R_T} \tag{5}$$

$$R_I = P_I/T_I \tag{6}$$

$$\overline{R_T} = \overline{P}/\overline{T_I} \tag{7}$$

R_I is instant rate, R_I is average instant rate, P_I is instant packet size, $\overline{P}$ is average packet size, T_I is instant inter-arrival time, $\overline{T_I}$ is average inter-arrival time, $\overline{R_T}$ is average rate total.

We recommend 9 on-off mini sources, if you wish to skip above step, however above matching process is not limited to 9, can be more or less, depends on the trace characteristics and how accurate or how fast you want the model be, more mini sources, simulation will run slower, but relatively more accurate. Once the number of mini source is chosen, we are ready to find out the corresponding histogram from the WireShark trace of packet size.

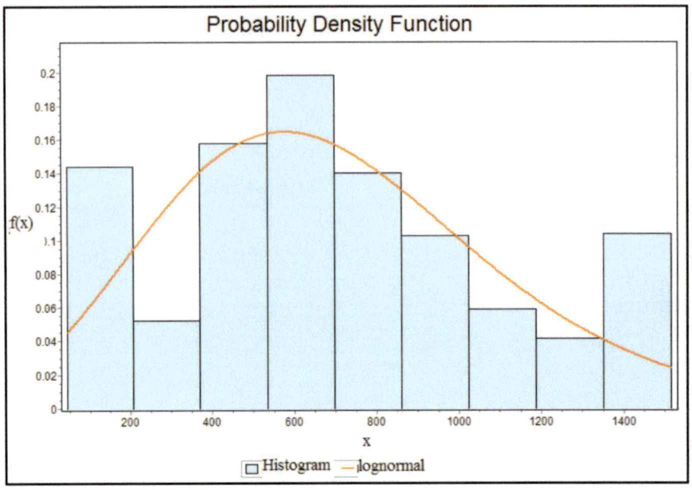

Fig. 3.3. Typical PDF of wireless video trace

Having obtained the i=9 bin pdf(i) (Fig.3.3) of the video trace calculated using a tool called EasyFit, the program decided that the lognormal distribution as the best distribution to fit the given data. The orange curve is the result automatically generated using the lognormal distribution. Matlab or Excel Macro can also be used to fulfil this task of finding the best-fit analytic curve for the histogram. Once the relative strength of each mini source is identified, we need to find out the fundamental Hurst parameter as follows:

IDC formula is defined below:

$$F(T)=\frac{E[S_T-E[S_T]]^2}{E[S_T]}=\frac{E[(S_T)^2]-E^2[S_T]}{E[S_T]}, \quad S_T=X_1+X_2...+X_T \tag{8}$$

$$F(T)-1=(T/T_0)^{\lambda}, \lambda=2H-1, \ \lambda\in(0,1) \tag{9}$$

$$\log(F(T)-1)=(2H-1)\log(T/T_0) \tag{10}$$

We can obtain the slope λ from different points $(\log(F(T)-1),\log(T/T_0))$ on the IDC curve.

The shape parameter, c, of the Pareto distribution is related to the Hurst parameter as shown in equation (11) below. The slope of the IDC curve gives the Hurst parameter from equation (12). As shown from the curve, the fractal effect calms down, but does not disappear, for this reason we call it persistent Hurst phenomenon (Mehrvar et al., 1996).

$$H=\frac{1}{2}(3-c) \tag{11}$$

$$H = \frac{1}{2}(1+\lambda) \quad (12)$$

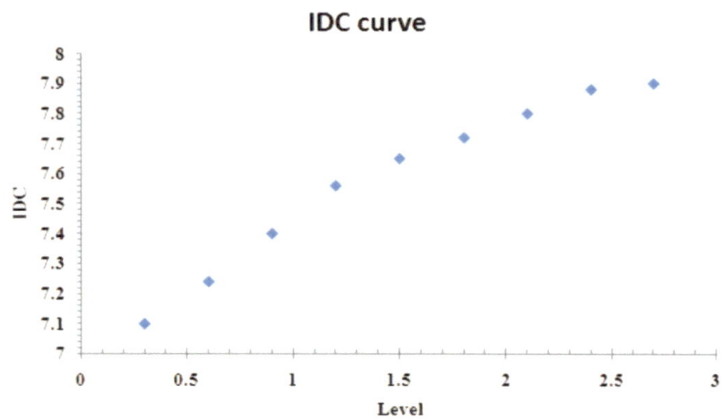

Fig. 3.4. Entire trace IDC curve

Again IDC curve can be either calculated by Matlab program or Excel Spreadsheet by calling standard deviation function recursively. When video is compressed with little loss of information, the peak to average ratio decreases. The Hurst parameter can be used to reflect this invariance phenomenon of entropy conservation property (Hong et al., 2001). Since our video is highly compressed but still preserves the original entropy of the information, we use the Hurst parameter to accurately capture this scaling invariance, modeling it with on-time distribution. On the other hand, an exponential distribution was used to model the off-time, which represents the time between each object action scene. This distribution was chosen with the observation that the action scene sequence is relatively memory less. And thus we have:

$$T_{on} = P(i)/\left(\overline{R}_T/N_s\right),\ 1 \le i \le 9 \quad (13)$$

$$\alpha_p = T_{on} * \left(C_p - 1\right)/C_p \quad (14)$$

P(i) is the packet size for ith mini-source, $\overline{R_T}$ was calculated from WireShark trace, 46.13kbps for our case. The Pareto shape is $C_p = 1.6$, and for Weibull $C_w = C_p - 1 = 0.6$, in our situation.

Now we have on-time calculated, to obtain off-time, we need to find out the frame Correlation, the formula is below:

$$R_X(t_1,t_2) = E\{X(t_1)X(t_2)\} = \int_{-\infty}^{+\infty}\int_{-\infty}^{+\infty} x_1x_2 f(x_1,x_2,t_1,t_2)dx_1dx_2 \quad (15)$$

$$T_{off} + T_{on} = L/pdf(i),\ 1 \le i \le 9\,. \quad (16)$$

L is the Correlation Length, depends on Frame FP, which is calculated by WireShark, 54.4ms for us, $L = L_c \times FP$.

$$\int_{\tau_0}^{\infty} r_X(\tau)d\tau < 0.25, \tau = t_1 - t_2 \tag{17}$$

Correlation lags, $L_c = \tau_0 = 20$ here. Table 3.1 shows a summary of each mini Pareto source and its characterization values mapped according to above steps.

Bins	*Location (ms)*	*Packet Size P(i) (Bytes)*	*Mean Off-Time (ms)*	*Mean On-Time (ms)*	*pdf(i)*
mini1	*56*	*96*	*13450*	150	0.08
mini2	*168*	*288*	*9442*	449	0.11
mini3	*253*	*432*	*6126*	674	0.16
mini4	*365*	*624*	*5071*	973	0.18
mini5	*449*	*768*	*6056*	1198	0.15
mini6	*561*	*960*	*7570*	1497	0.12
mini7	*646*	*1104*	*10367*	1722	0.09
mini8	*758*	*1296*	*16112*	2021	0.06
mini9	*842*	*1440*	*19514*	2246	0.05

Table 3.1. Mini sources with characterization values for OPNET

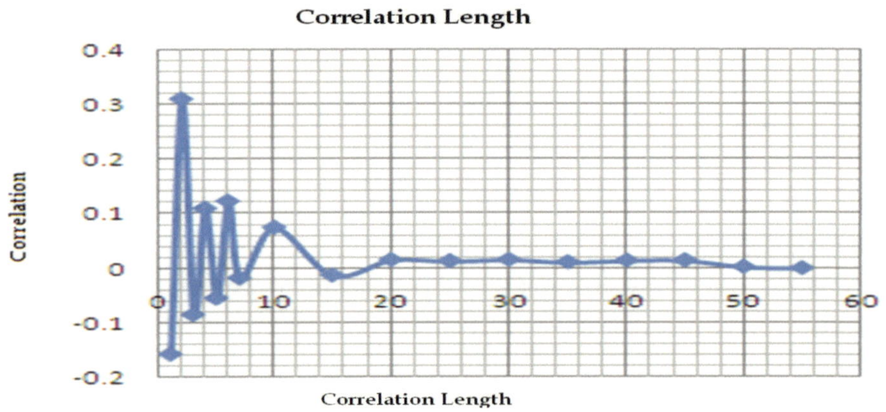

Fig. 3.5. Frame correlation length

In summary, for our Pareto mini source, the shape is obtained from IDC slope, the location is obtained from the mean value of the inter-arrival time and the shape, and finally the mean off time is derived from the correlation length of the trace and the lognormal distribution of packet size. The correlation curve is shown in Fig.3.5. Correlations show a predictive relationship in a sequence of data. Fig.3.5 shows that our frames are correlated since actions are correlated, however, as one can see, there is not much correlation beyond a length of 20 lags; consequently, our correlation length can be either visually chosen for simplicity, or go through integration of the formula (17) for a more tightly bounding. Note that curve fitting to lognormal can be skipped; we do it for the purpose of the easier scaling of the model later on, without refitting for every trace. When scaling for different video rate, all we need to do is simply set the constant bit rate of each mini source in OPNET to the desired video rate directly. More important, with the above-mentioned distributions, the mini on-off model can be readily mapped into BMAP/D/1/K queue or $M^X/G/1/PS$ queue (Feng & Misra, 2003), where we can obtain the numerical solution or analytical bound below.

$$\text{Mean(JobSojournTime)} \le \frac{\text{Variance(Batch)} \times \text{Variance(Service)} \times \text{Load}}{\text{Mean(Batch)} \times \text{Mean(Service)} \times (1 - \text{Load})} \tag{18}$$

With our carefully matched Pareto and Lognormal distributions, the quick calculation shows that the actual delay bound could be as 69 times higher than a simple M/M/1 queue estimation.

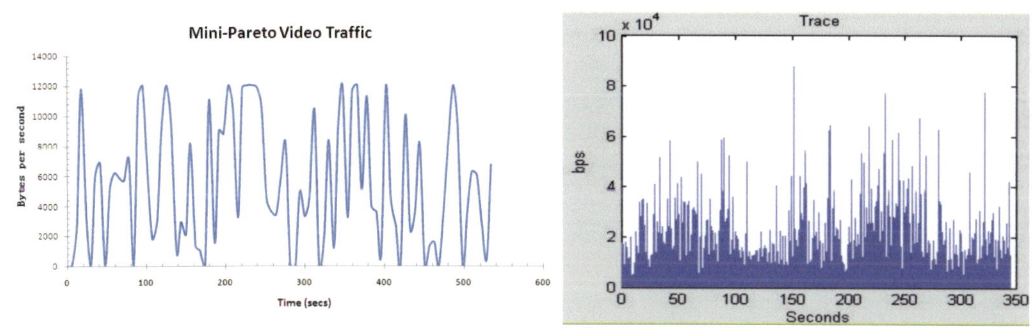

Fig. 3.6. Mini-Pareto traffic sample reproduced from video trace

The original bps sample and generated mini-Pareto video traffic bytes per second are shown in Fig.3.6; these kinds of large deviation from the video trace can never be reproduced by matching with the traditional Poisson process, neither Interrupted Poisson process.

4. Public safety simulation and overall performance evaluation

4.1 Simulation environment setting

As discussed earlier, OPNET can provide a platform to create and test an analytic and practical video model; it can also provide the ability to integrate the model into a VANET environment.

The common simulation parameters for each scenario are shown in Table 4.1 below, in places where parameters were changed for specific purposes, it will be indicated and discussed. Each simulation was simulated for a simulation time of 3600secs. The terrain dimensions vary slightly from scenario to scenario. They are an average of 1300 X 1250 m in area. The relative (x, y) position on the terrain is used to integrate the VANET mobility model trajectories and to obtain the initial positions of the vehicles. Vehicular environment for the path loss parameter is modeled according to the description in the "Radio Tx Technologies for IMT2000" white paper of the ITU. The shadow fading standard deviation was set to 10dB. The trajectories vary from scenario to scenario and will be discussed below.

4.2 Safety simulation scenarios

Two scenarios were chosen to simulate: school zone scenario, highway scenario. The goal was to study different geographic areas with varying traffic congestion, varying wireless interference and varying traffic speed limit.

We discuss below the specifications of model components in each scenario. In general, each scenario consists of N mobile nodes (Mobile Station on vehicle) and two BS to cover the geographical area represented. Performance measures such as end-to-end delay; usability (outage) will be evaluated and discussed for each scenario.

Parameter	*Value*
Physical layer	IEEE802.16e (WiMax)
Data rate	10Mbps
BS TX power	5 W
MS Tx power	1 W
Antenna type	Omni-directional
BS antenna gain	15dBi
MS antenna gain	9dBi
Link bandwidth	20MHz
Modulation scheme	16-QAM
Path loss parameter	Vehicular environment
Number of vehicles	10
Mobility model	VanetMobiSim
Number of RSU's	2
Simulated time	3600secs
Seeds	127
Terrain dimensions	1300x1250m

Table 4.1. OPNET simulation parameters

We study a network of 10 cars and 2 RSU's with each car has a maximum sustainable traffic rate of 384 Kbps except where otherwise stated. The link between the RSU and the backhaul network was a DS3 link with a capacity of 44.736 Mbps as shown in Fig.2.1. The buffer size at each SS was set to 256 KB except where otherwise stated. The simulation was run for 60 mins simulated time and the number of packets generated per node is about 10,000.

4.2.1 School safety scenario

This scenario was used to simulate a school zone with lots of stop signs and obstructions that can let children safely cross the road. The maximum speed in this scenario is 30 km/hr. The trajectory in this scenario was generated using specific formatting in VanetMobiSim. Fig. 4.1 show a screen shot of this scenario.

In this scenario, a mobility model with clusters was used to generate the trajectory for each node. The clusters are programmed to populate the scenario at random times during the simulation process to mimic the behaviour of a School zone environment. The path loss model to be applied to signals being received at the WiMAX MAC in this scenario was the "Vehicular Environment" model with shadow fading of 10 dB.

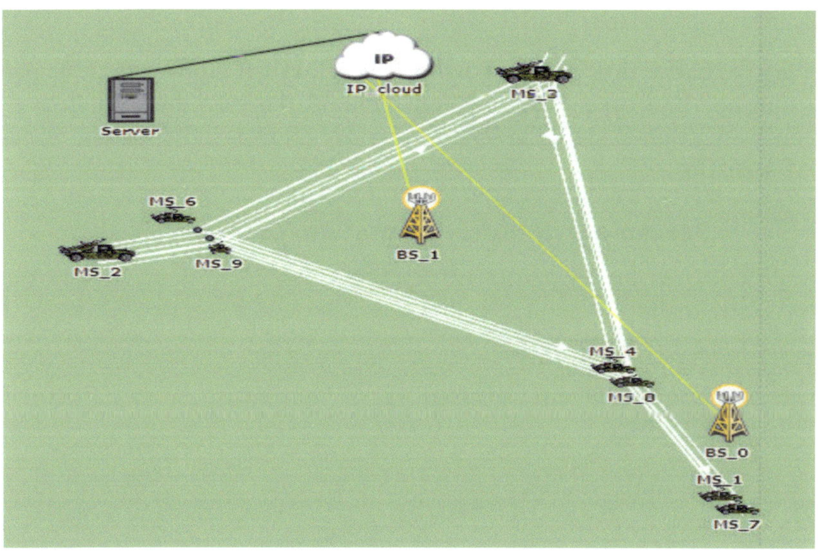

Fig. 4.1. School zone scenario with mobile trajectory

a. Mean end-to-end delay

Fig.4.2 shows the mean end-to-end delay performance as it varies with different buffer sizes and service rates. The mean end-to-end delay is seen to increase as the buffer size increases and decrease as the service rate increases. The curve is an increasing curve with a positive slope. This corresponds with the behaviour of a traffic that follows the Pareto distribution. The difference in the mean end-to-end delay for the service rates of 0.5 Mbps and 1 Mbps is not substantial. This is due to the fact that the school zone scenario has light traffic which implying that the delay is more and more dominated by other factors such as CPU speed different from the buffer capacity.

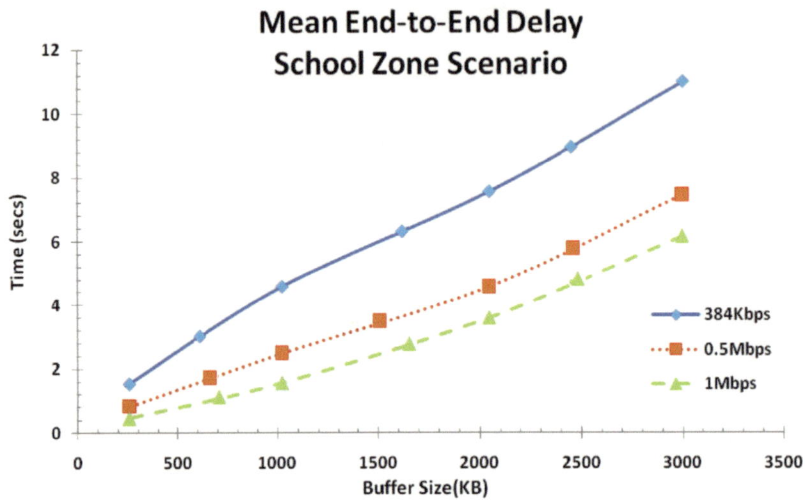

Fig. 4.2. Mean end-to-end delay performance for school zone scenario

b. Buffer overflow percentage

The percentage of buffer overflow is shown in Fig.4.3. When buffer overflow, packet is lost or delay is long and thus considered to be outage time. It is seen that at higher service rates the buffer does not saturate since the school zone scenario has lighter traffic as compared to the highway scenario.

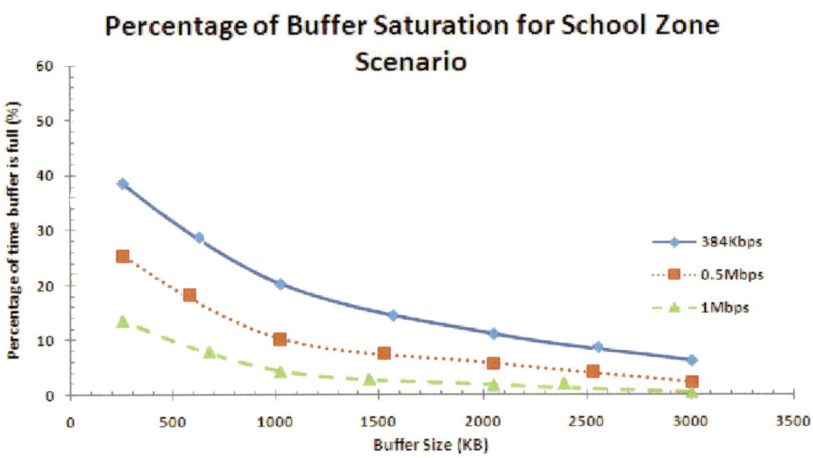

Fig. 4.3. Percentage of buffer saturation performance for school zone scenario

4.2.2 Highway safety scenario

This scenario was used to simulate the highway with a minimum speed of 60 km/hr and a maximum speed of 100 km/hr. In the trajectory of this scenario, the maximum number of traffic lights is one, just before the cars enter the highway. Fig.4.4 shows this scenario with the trajectory represented by the white lines shown.

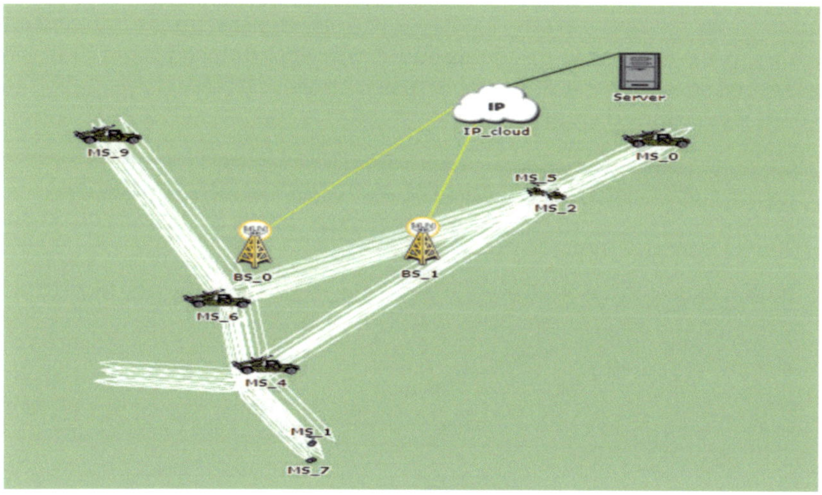

Fig. 4.4. Highway scenario with vehicle moving trajectory

The Highway scenario simulates an area where cars move at high speed of 100 km/hr.

a. Mean end-to-end delay

The mean end-to-end delay performance for the highway scenario is shown in Fig.4.5. As expected, as the service rate increases, the end-to-end delay reduces. However, one can see that the speed of the vehicles is a large factor here, compare with school zone, the situation is much worse for the same buffer size and the same service rate. Increasing the service rate reduces the end-to-end delay and increasing the buffer size increases the end-to-end delay.

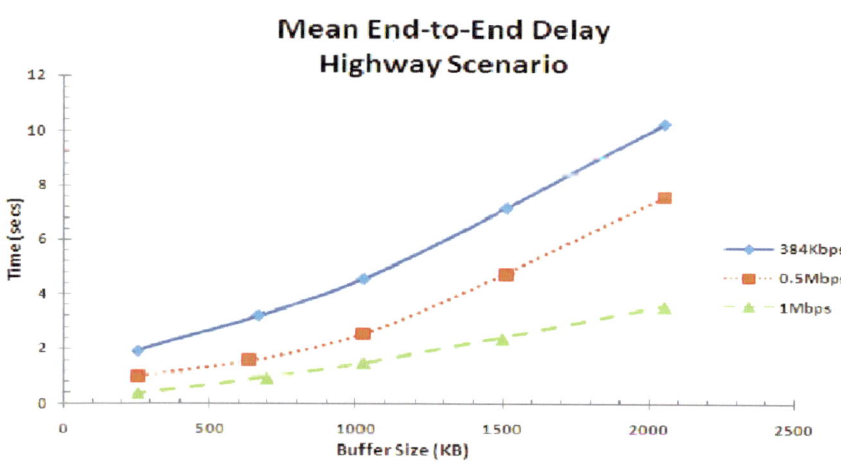

Fig. 4.5. Mean end-to-end delay performance for highway scenario

b. Buffer overflow (service outage) percentage

The percentage of buffer saturation of the Highway scenario is shown in Fig.4.6. The normal trend is followed in this case, i.e., as the buffer size increases, the percentage of time for which the buffer is full decreases. It is important to note that the reduction in percentage of buffer saturation as the service rate increases in this scenario might be caused by other factors such as, packet loss due to connection drops and reduced bandwidth. Again it is worse than School zone. By the way, the actual road tests we have conducted with police cars agree with this usability observation.

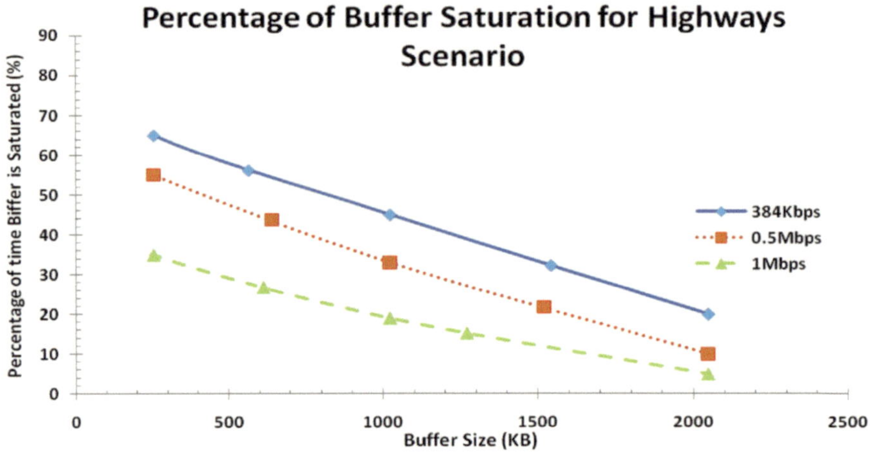

Fig. 4.6. Percentage of buffer saturation performance for highway scenario

5. Current implementation of the public safety wireless video system

With the past surge of the commercialization of the Internet, the continuing expansion of wireless services, and the increasing usage of multimedia applications, communication traffic demand has seen a steady increase. Researchers are diligently working towards disruptive technology that has not previously been given substantial attention narrowband wireless video applications to public safety.

Today's Internet does not provide the necessary QoS guarantees that are needed to support high-quality, real-time video transmission. Multimedia data transmitted over the Internet often suffers from delay, jitter, and data loss. Data loss, in particular, can be extremely damaging to compressed video since the intra-frame dependencies needed to achieve high-compression rates in video exacerbate the data loss when primary frames are lost. Unlike data applications, video applications can tolerate some short loss. A small gap in a video stream may not significantly impair media quality, and may not even be noticeable to users. However, long loss can result in unacceptable media quality, or service outage.

A number of techniques exist to repair packet loss in a media stream. These techniques have proven to be effective for audio stream data loss but may have yet to be applied to video, but in a significant different way. In particular, we propose a video interleaving approach to reduce the damage to a video stream from packet loss. Interleaving assumes that better perceptual quality can be achieved by spreading out bursty packet losses in a media flow. In other words, several unnoticeable short gaps degrade quality way less than a long gap in a multimedia flow. The basic idea of interleaving is to uniformly spread out long gaps in the video stream into several short gaps. In this way the effect of the loss of multiple consecutive frames is ameliorated, and the perceptual quality will be increased dramatically.

At the sender, frames in a video stream are first interleaved, with the original consecutive frames being separated by a specific distance that is given by the interleaving algorithm. After arriving at the receiver, frames are then reconstructed back to their original order. If consecutive loss occurs in the interleaved stream during transmission or as a result of single loss propagation, after reconstruction at the receiver, a long gap in the stream caused by the consecutive loss or propagated loss will be spread out into several unnoticeable short gaps.

This is different from audio interleaver in that no complicated Forward Error Correction is needed. Fig.5.1 shows the diagram of the wireless video streaming system. Video information is collected by the input video source, processed by the video encoder, interleaving system, channel encoder, modulator and transmitted through the transceiver to a destination. In the transmit chain explicitly shown in Fig.5.1, an interleaving system is used to interleave collected information. Video information received by the receiver is processed by the demodulator and the channel decoder. The de-interleaving system is employed to reverse the interleaving, which may be bit/byte/packet interleaving for example, applied to the received video information by an interleaving system at a transmitting device. De-interleaved video information is decoded by the video decoder, and output the video output device, which may be a display screen, for example. The particular structure and operation of the encoder may be different for different formats of video information, and the channel encoder, the modulator, and the transmitter will similarly be dependent upon communication protocols and media using which information is to be transmitted.

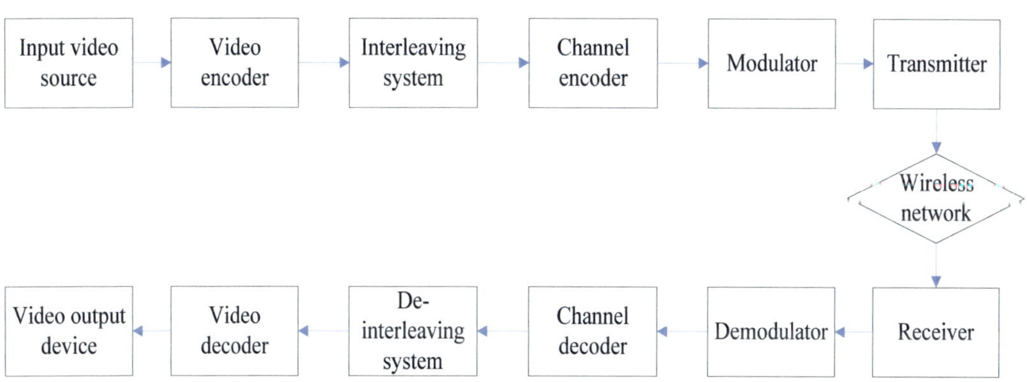

Fig. 5.1. System structure for breaking up long outage into short one

The interleaving system of Fig.5.2 implements an interleaving path which includes multiple interleavers, a packet interleaver, a frame interleaver, a byte interleaver, and a bit interleaver, each having a respective interleaving length. Also it includes a controller to control which interleavers are active in the interleaving path and thus the aggregate interleaving length at any time, a memory for storing information during interleaving and mappings between information types, operating conditions, and interleaving lengths.

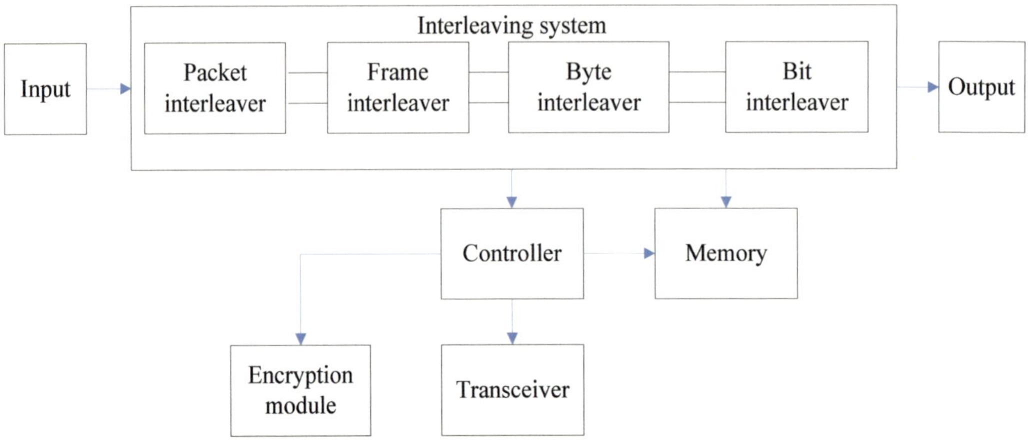

Fig. 5.2. Architecture of the degrading concealment interleaving system

Combining interleaving with encryption and watermark, instead of adding a stand-alone device, represents brand new thinking for lightweight all-in-one design philosophy. The hash key may be used to simply encrypt and mark the information itself, or to determine the position of original information after interleaving, rather than the complicated encrypting the actual information. Security information, a key for instance, can be combination of numerical number and alphabetical mark. We can pick a number from a password, if the password is "1326" and the frame interleaver is used for combined interleaving and encrypt marking, the first frame is swapped with the third frame in position, the second and the sixth frames are swapped, so on so forth, when the group of picture (frame) is set to 10. If the group of picture is set to 60, the key of "1646" will swap 16th frame with 46th frame. These rules could be exchanged using standard secured key exchange protocol as well.

Fig.5.3 shows a burst error reduction algorithm with adaptive control. It changes the interleaver size according to information provided by a run-time algorithm. Each sender and receiver receives video packets from each other, then analyzes the received video packet, and in particular video packet headers according to particular implementation, determines whether the sequence number of RTSP is damaged, and if the sequence number is changed, then the number of hops that the video packet passed is calculated. If the sequence number is not changed, then a current interleaving size is not changed, as indicated.

After calculating the number of hops, and also the number of errors reported on different layers at checkpoint for error, a determination is made, as to whether the overall error is above a threshold. If so, then interleaver size and thus interleaving length for an interleaving path is adjusted. If the number of hops for a packet is greater than 1, a runtime check for congestion on a communication link is performed. Illustrative examples of runtime checks are described in further detail below. If congestion is above a predetermined, selected, or remotely specified threshold, as determined, then interleaver dimension is changed, by enabling or disabling one or more additional interleavers.

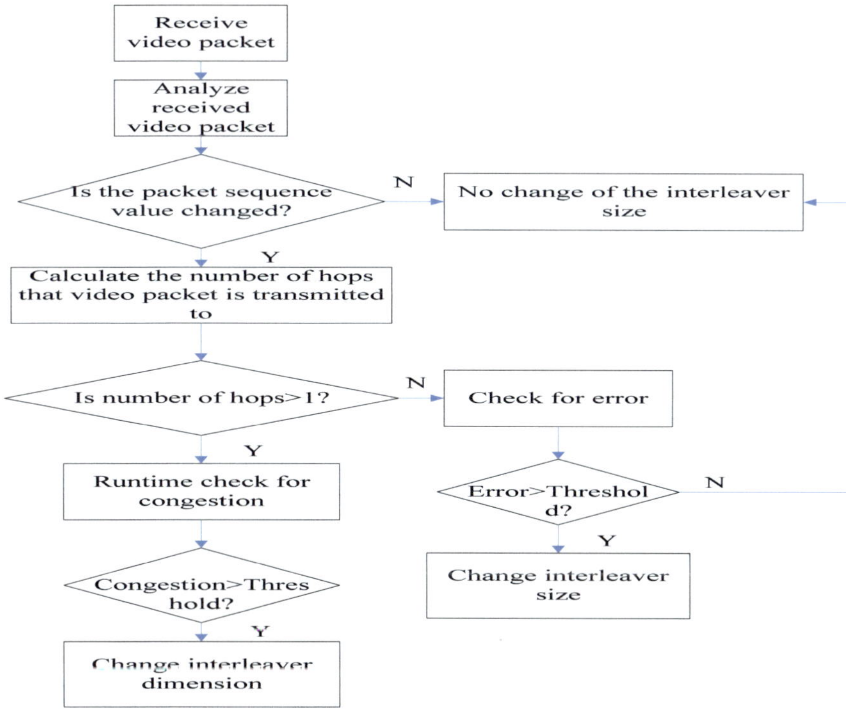

Fig. 5.3. Long burst impairment reduction algorithm

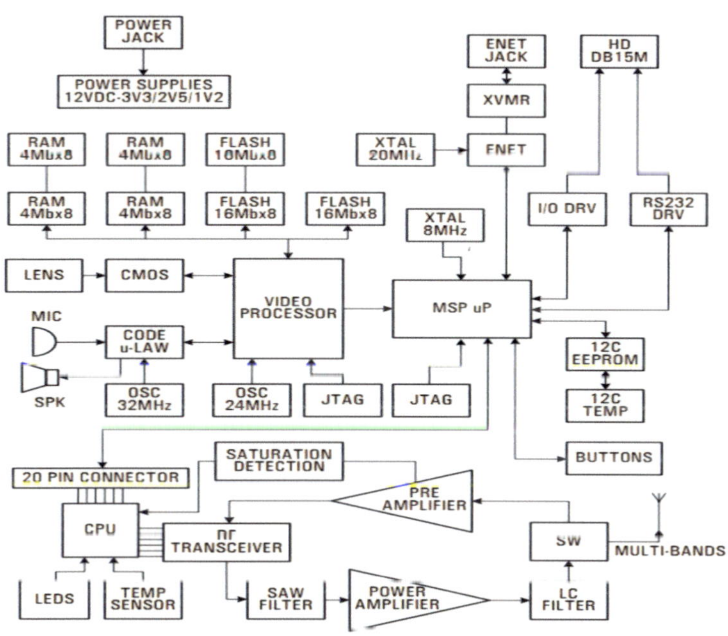

Fig. 5.4. Mobile terminal detail structure used in the experiment

Fig.5.4 is a block diagram of an example mobile terminal, including both wireless and video parts. Interleaving/integrity/and down sampling are mostly done in the Video processor, with some assistant from MSP Micro Processor for network layer related process, e.g. packet header filtering to distinguish other sensor signal from the video signal, and the third CPU for physical layer process, such as power amplifier saturation warning.

Deinterleaving/security and up sampling are done in Personal Computer for current product. And will be in the video processor or Field Programmable Gate Array (FPGA) chip in future product.

6. Conclusion

In this work, we were able to create an effective public safety WiMAX Ad-hoc video simulation platform with which other researchers can develop and test various public safety applications. This is done by designing our video model to work as individual sub-streams mini-Pareto model, matched with real trace from the actual car-to-car public safety camera. The platform was built using the OPNET simulation tool which allowed for the integration of all models - developed from statistically analysing a live video trace, real world map, trajectory simulations; and provided a complete tier of communication layers for proper performance analysis. This methodology does not limit itself to symmetrical wireless ad-hoc video traffic other situation applies as well.

Integrating a VANET mobility model into our platform created challenges for us. So far, due to the limitation of the OPNET library available, we have to run mobility model first before integrating the results into our communication model. It would have been best to run the simulation model and the mobility model concurrently to allow the wireless communication affect the mobility of vehicles so that more quick results can be obtained.

It is observed on the field that the scenario in which the application is deployed agrees with the performance obtained here in principal.

In summary, we were able to extend and implement a standard IEEE802.16 theoretical video model, integrate this new video model matched from WireShark real trace together with a VANET mobility model into the OPNET simulation tool, assisted by Matlab calculations.

7. Acknowledgement

The authors are grateful to Mr. Yanjun Chen and Prof. Yi Zhu of Jiangsu University, Prof. Oliver Yang of University of Ottawa, Prof. Lianfeng Shen of Southeast University, Prof. Zhiyong Zhang of Tamkang University, Prof. Tony Bailetti and Dr. Helen Tang of Carleton University, Prof. Zhen Liu of Jiangsu S&T University, Dr. Qiubo Ye and Hong Qian of CRC, Mike Zhou of RIM, Kathy Miao of IBM and Matt Woods of IP Unwired for their valuable contributions, comments, related discussions and constant encouragements during the research work lasted a decade long; also a number of public safety agencies in Canada and China for their diligent supporting of many unforgettable road tests!

8. References

Brit, E.(2010). WiMax. *Encyclopædia Britannica.* Accessed on 27 Jul. 2010. Available from http://www.britannica.com/EBchecked/ topic/1017801/WiMax

Chen, X. J., Chen, Z. J. & Xu, G. H. (2008). The research on routing protocol of sense wireless network in environmental monitoring. *Proceeding of the 4th International Conference on Wireless Communications, Networking and Mobile Computing*, pp.1-4, ISBN 978-1-4244-2107-7, Dalian, Oct 12-14, 2008

Duffy, K. R. & Sapozhnikov, A. (2007). The large deviation principle for the on/off weibull sojourn process. Journal of Applied Probability. Volume 45, pp.107-117

Fiore, M., Harri, J., Filali, F.& Bonnet, C.(2007). Vehicular Mobility Simulation for VANETs. *Simulation Symposium, 2007. ANSS '07. 40th Annual*, pp.301-309, ISSN 1080-241X Norfolk, vol., no., March 26-28, 2007

Feng, H. & Misra, V.(2003). Asymptotic bounds for $M^X/G/1$ processor sharing queues. *Technical report CUCS-006-04*, July 2003. http://www.cs.columbia.edu/~hanhua/

Ghosh, A. & Ghosh, M. (2007). *Fundamentals of WiMAX: Understanding Broadband Wireless Networking*. Prentice Hall, 2007

Gu, H. & Ji, Q. (2004). An automated face reader for fatigue detection. *Proceeding of Automatic Face and Gesture Recognition, Sixth IEEE International Conference*, pp. 111-116, ISBN 0-7695-2122-3, May 17-19, 2004

Huang, J. (2001). Presentation for Generalizing 4IPP Traffic Model for IEEE 802.16.3. *IEEE 802.16 Broadband Wireless Access Working Group*, Available from http://www.ieee802.org/16/tg3/contrib/802163p-00_58.pdf

Hong, S. H., Park, R. & Lee, C. B. (2001). Hurst parameter estimation of long-range dependent VBR MPEG Video traffic in ATM Networks. *Journal of Visual Communication and Image Representation*, Vol. 12, issue 1, March 2001, pp 44-65

Li, X., Huang, H. & Shu, W. (2009). VStore: Towards Cooperative Storage in Vehicular Sensor Networks for Mobile Surveillance. *Proceedings of IEEE Wireless Communication and Networking Conference*, pp.1-6, ISSN 0163-6804, Budapest, Hungary, April, 2009

Li, K.(2006). IEEE 802.16e- 2005 Air Interface Overview. *WiMAX Solutions Division Intel Mobility Group*, June 05, 2006

Morgan, Y.L. (2010). Managing DSRC & Wave Standards Suite Operations: In a V2V Scenario. *International Journal of Vehicular Technology*, Volume 2010, Available from http://www.hindawi.com/journals/ijvt/

Mehrvar, H., Le-Ngove, T. & Huang, J. (1996). Performance Evaluation of Bursty Traffic Using Neural Networks. *Proceeding of Canadian Conference Electrical and Computer Engineers*, pp.995-958, ISBN 0-7803-3143-5, Calgary, May 26-29, 1996

Song, G. & Oliver, Y. (2004). Minimum energy multicast routing for wireless Ad-hoc networks with adaptive antennas. *Proceeding of the 12th IEEE International Conference on Network Protocols*. ISSN 1092-1648, Oct 5-8, 2004

Sommer, C. & Dressler, F. (2008). Progressing toward Realistic Mobility Models in VANET Simulations, *IEEE Communication magazine*, vol. 46, no. 11, Nov. 2008, pp. 132-137, ISSN: 0163-6804

WiMa. (2009). IEEE WirelessMan 802.16. *The 802.16 WirelessMAN MAC: It's Done, but What Is It?* Available from
http:/ieee802.org/16/docs/01/80216-01_58rl/pdf, Accessed in 2009

Wegener, A., Hellbruck, H. & Wewetzer, C. (2008). VANET Simulation Environment with Feedback Loop and its Application to Traffic Light Assistance. *Proceedings of IEEE*

Globecom workshop, pp.1-7, *2008 IEEE,* ISBN 978-1-4244-3061-1, New Orleans,vol. 3, Nov. 30, Dec. 4, 2008

Zhou, L. Y. & Ji, W. W. (2010). A study on the application of WiMAX access technology in IPTV system Source. *Proceedings of the 2th International Conference on Networks Security, Wireless Communications and Trusted Computing,* pp.183-186, ISBN 978-0-7695-4011-5, April 2010

14

Multihop Relay-Enhanced WiMAX Networks

Yongchul Kim and Mihail L. Sichitiu
Department of Electrical and Computer Engineering
North Carolina State University
Raleigh
USA

1. Introduction

The demand for high speed data service has been increasing dramatically since the Internet has become a part of people's lives. Most broadband wireless service providers have boosted data service rates by adopting recently developed technologies such as OFDM, MIMO, and smart antennas. However, in practice there are still problems such as coverage holes due to shadowing, and poor signal to interference and noise ratio (SINR) for the subscriber stations (SSs) that are far away from the base station (BS). A simple solution for this problem is to add more BSs, but it is a very inefficient solution especially when there are few SSs to be served (e.g., in rural areas.) As an alternative to adding BSs, deploying low-cost relay stations (RSs) provides a cost-effective way to overcome the above problem (RSs are a simplified version of a full BS resulting in with lower upfront cost than BS; additionally, RSs do not require backhaul connections, thus reducing operating costs). The WiMAX specification was amended (802.16j, 2009) to include multihop relays, an extension which has gained much attention and proved to be an attractive technology for the next-generation of wireless communications. Furthermore, the currently evolving Long Term-Evolution Advanced (LTE-A) standard considers multihop relaying as an essential feature (3GPP, 2009). In this chapter, we study the effect of using RSs on both capacity and coverage enhancements.

The IEEE 802.16j amendment focuses on the deployment of RSs in such a way that the network capacity can be enhanced or coverage of the network can be extended. Accordingly, two different types of RSs are specified in the amendment: *transparent* RS (T-RS) mode and *non-transparent* RS (NT-RS) mode. In T-RS mode, the framing information is always transmitted by the BS to the SSs directly, while data traffic can be relayed via RSs. Therefore, in cells with T-RSs, the SSs associated with an RS have to be located within the coverage of the BS. Conversely, in NT-RS mode, the framing information is transmitted along the same path as data traffic to the SSs, and thus the RS operates effectively as a BS for connected SSs. Therefore, T-RSs allow throughput enhancement only, whereas NT-RSs can extend the coverage as well as increase the throughput. Both types of relays can serve unmodified SSs (i.e., the SSs do not distinguish between genuine BSs and RSs). In this chapter, we analyze the benefits of using RSs for the capacity enhancement scenario with T-RSs and the coverage extension scenario with NT-RSs respectively.

In the first part of this chapter, we focus on improving cell capacity by deploying T-RSs inside a cell, and consider the placement of RSs that maximizes cell capacity. According to the location of RSs, the network capacity will vary significantly, i.e., when the RS is either very close to the BS or far away from the BS, only a few SSs will benefit from the RS. In order to maximize the number of SSs that can achieve greater throughput through the T-RSs, we need to find the optimal placement of T-RSs such that the cell throughput is maximized. We also show how various network parameters such as reuse factor, terrain types, RS antenna gain, and the number of RSs affect the optimal placement of RSs and the capacity gain compared to the conventional scenario (i.e., without using RSs).

In the second part, we focus on deploying NT-RSs for the purpose of coverage extension. We explore three different issues in this part. First, we study several scheduling schemes such as orthogonal, overlapped, and optimal schemes in order to maximize cell throughput while serving the SSs in a fair manner. Second, we analyze cell coverage extension by varying both the location and number of RSs from a cost efficiency perspective. Finally, we explore an extension of the optimal scheme to a general multihop relaying scenario, and analyze the network throughput degradation due to the increase of relay hops under the optimal scheduling scheme.

The rest of this chapter is organized as follows. In the next section, we discuss the system model including system description, SINR analysis, fading channel, and relay strategy. In Section III, we present the capacity enhancement scenario with T-RSs. In Section IV, we analyze cost effective coverage extension scenarios with NT-RSs. The optimal scheduling scheme is also presented and compared with the orthogonal and overlapped schemes. Section V concludes the chapter.

2. System model

2.1 System description

In this chapter, we consider a single WiMAX cell consisting of a central BS and several RSs for the purpose of capacity enhancement, coverage extension, or both. The SSs are uniformly distributed throughout the cell and only N SSs are randomly chosen to be active at a time for each scenario. The BS is responsible for allocating resources for the SSs to be served and is connected to the backhaul network, while the RSs have no backhaul links but are wirelessly connected to the BS. The main responsibility of the RS is to relay data between the BS and SSs. All RSs and the BS are referred to as *service nodes* in the rest of this chapter. The RS to RS connection is also allowed for the coverage extension scenario. However, in the capacity enhancement scenario, we assume only two-hop relaying since more than two-hop relaying without extending coverage reduces the efficiency of using RSs. The one-hop links BS to RS and RS to RS are referred to as *relay links*, and the links BS to SS and RS to SS as *access links*.

We assume that every node in a cell has a single omni-directional antenna and operates in half-duplex mode, hence, no terminal can receive and transmit data simultaneously. The frequency reuse scheme is not considered in the scenario for the optimal placement of T-RSs, i.e., only one node can be active at a time. In contrast, for the coverage extension scenario, the proposed optimal scheduling scheme allows for frequency reuse in order to maximize the bandwidth efficiency, hence, the throughput degradation due to coverage extension can be minimized. The standard allows for two types of duplex methods to separate the uplink (UL) and downlink (DL) channels: time division duplex (TDD) and frequency division duplex (FDD); we assume TDD in this chapter since TDD makes more efficient use of the spectrum by dynamically allocating the amount of time slots to each direction. The system parameters used for the analysis are listed in Table 1.

System Parameters		OFDMA Parameters	
Operating Frequency	3.5 GHz	FFT Size	1024
Duplex	TDD	Sub-carrier Frequency Spacing	10.94 kHz
Channel Bandwidth	10 MHz	Useful Symbol Time	91.4 μs
BS/RS Height	50 m	Guard Time	11.4 μs
SS Height	1.5 m	OFDM Symbol Duration	102.9 μs
BS/RS Antenna Gain	17 dBi	Data Sub-carriers(DL / UL)	720 / 560
SS Antenna Gain	0 dBi	Pilot Sub-carriers(DL / UL)	120 / 280
BS/RS Power	20 W	Null Sub-carriers(DL / UL)	184 / 184
SS Power	200 mW	Sub-channels(DL / UL)	30 / 35
BS/RS Noise Figure	3 dB		
SS Noise Figure	7 dB		

Table 1. Simulation Parameters

2.2 SINR analysis

The adaptive modulation and coding scheme (AMC) is the primary method of maintaining the quality of wireless transmission. WiMAX supports AMC by defining seven combinations of modulation and coding rate that can be used to achieve various data rates specified by the standard (Andrews et al., 2007). Table 2 shows the modulation and coding rates and the corresponding achievable data rate; the last column represents the minimum required threshold values of SINR computed by a bit error rate expression for M-QAM (Foschini & Salz, 1983) when bit error rate is 10^{-6}. For example, when the received SINR is between 9.1 and 11.73dB, the achievable data rate will be 5.25Mbps. In general, a higher order modulation tends to be used close to the base station, whereas lower order modulations tend to be used at longer ranges.

The IEEE 802.16 working group has recommended the Erceg-Greenstein path loss model for fixed wireless application systems. Since we did not consider the mobility of the SSs in this analysis, we used this model to calculate path loss. The Erceg-Greenstein models are divided into three types of terrains, namely A, B and C. Terrain type A has the highest path loss, and is applicable to hilly terrains with moderate to heavy foliage densities. Type C has the lowest path loss and applies to flat terrains with light tree densities. Type B is suitable for intermediate terrains. The basic path loss equation with correction factors is presented in (Erceg & Hari, 2001):

$$L = 20\log_{10}\left(\frac{4\pi d_0}{\lambda}\right) + 10\alpha\log_{10}\left(\frac{d}{d_0}\right) - 10.8\log_{10}\left(\frac{h}{2}\right) + s + 6\log_{10}\left(\frac{f}{2000}\right), \quad (1)$$

where d_0=100m, α is the path loss exponent dependent on terrain type, d is the distance between the transmitter and receiver, h is the receiver antenna height, f and λ are the frequency and wavelength of the carrier signal, and s is a zero mean shadow fading component. When the path loss value between the transmitter and receiver is computed by using (1), the received signal power P_r can be calculated by:

$$P_r = \frac{G_t G_r P_t}{L}, \quad (2)$$

Modulation & Coding rate	Downlink Data Rate [Mbps]	Spectral Efficiency [bps/Hz]	Threshold [dB]
QPSK 1/2	5.25	1.0	9.1
QPSK 3/4	7.87	1.5	11.73
16 QAM 1/2	10.49	2.0	13.87
16 QAM 3/4	15.74	3.0	17.55
64 QAM 2/3	20.99	4.0	20.86
64QAM 3/4	23.61	4.5	22.45
64 QAM 5/6	26.23	5.0	24.02

Table 2. SINR Threshold Set for a BER of 10^{-6}

where G_t, G_r, and P_t represent the transmitting antenna gain, receiving antenna gain, and the transmission power. Once the received signal power is computed, the SINR value at the receiver can be easily determined by:

$$SINR = \frac{P_r}{P_N + \beta P_I}, \tag{3}$$

where β is the number of co-channel cells of the first tier, P_N is the noise power, and P_I is the interference signal power from a neighboring cell on the same frequency as the current cell. In a similar way to the received signal power, the interference signal power can be computed by using (1) and (2) with the information of co-channel distance. We assume that the co-channel distance is $R\sqrt{3\tau}$ (Rappaport, 2001), where τ is the reuse factor and R is the cell size. For the capacity enhancement scenario with T-RSs, only inter-cell interference is considered since no concurrent transmissions are allowed inside a cell, while for the coverage extension scenario with NT-RSs, intra-cell interferences is also considered as multiple nodes can transmit simultaneously.

2.3 Fading channel

In a wireless communication system, a signal can travel from transmitter to receiver over multiple paths, and hence the received signal can fade. This phenomenon is referred to as multipath fading. In a fading environment the received signal power varies randomly over time due to multipath fading. We assume well known fading channels such as the Rayleigh fading and Rician fading channel for the access link and the relay link respectively. The Rayleigh fading channel is most applicable when there is no propagation along the line of sight between the transmitter and receiver, while the Rician fading channel is more appropriate when there is a dominant line of sight component at the receiver. For an access link, the amplitude distribution of the received signal is accurately modeled by a Rayleigh distribution:

$$p(\rho) = \frac{\rho}{\sigma^2} \exp\left(-\frac{\rho^2}{2\sigma^2}\right), \rho \geq 0, \tag{4}$$

where ρ is the amplitude of the received signal and σ^2 is the local mean power. From the probability density function (pdf) of the received signal amplitude (4), the pdf of received signal power, γ, can be derived and has the exponential pdf (Zhang & Kassam, 1999):

$$p(\gamma) = \frac{1}{\gamma^*} \exp\left(-\frac{\gamma}{\gamma^*}\right), \gamma \geq 0, \tag{5}$$

where γ^* is the mean power of the received signal. In a similar way, for a relay link, the amplitude distribution of the received signal is more accurately modeled by a Rician distribution:

$$p(\rho) = \frac{\rho}{\sigma^2} \exp\left(-\frac{\rho^2 + \nu^2}{2\sigma^2}\right) I_0\left(\frac{\rho\nu}{\sigma^2}\right), \rho \geq 0, \nu \geq 0, \tag{6}$$

where $I_n(\cdot)$ is the nth-order modified Bessel function of the first kind, and $\frac{1}{2}\nu^2$ and σ^2 are the power of the LOS component and the power of all other scattered components respectively. Thus, the total mean power of the received signal, γ^*, can be expressed as $\gamma^* = \frac{1}{2}\nu^2 + \sigma^2$. The ratio between the signal power in the dominant component and the local mean scattered power is defined as Rician K-factor (Erceg & Hari, 2001), where $K = \frac{\nu^2}{2\sigma^2}$. When the K-factor is equal to zero, the Rician distribution becomes a Rayleigh distribution; from the pdf of received signal envelope (6), the pdf of received signal power, γ, can be derived by transforming random variable ρ into γ by considering the relationship between amplitude and power of the signal, $\gamma = \frac{1}{2}\rho^2$. The pdf of received signal power can be expressed as:

$$p(\gamma) = \frac{(1+K)e^{-K}}{\gamma^*} \exp\left(\frac{(1+K)\gamma}{\gamma^*}\right) I_0\left(\sqrt{\frac{4K(1+K)\gamma}{\gamma^*}}\right), K \geq 0, \gamma \geq 0, \tag{7}$$

where γ^* is the mean power of the received signal.

2.4 Relay strategy

The SSs that are outside of the transmission range of the BS have to be served via RSs, while the SSs located inside the transmission range of the BS can be served either directly from the BS or via RSs. In general, the SSs transmit/receive data to/from the BS via RSs only if the achievable relay data rate (BS-RS-SS) is greater than direct data rate (BS-SS). Since we assume that every node has a single antenna and operates in half duplex mode, the relay data rate is influenced by the link capacities and time durations of both hops involved. It is unlikely that the SSs having a good direct link capacity from the BS will use an RS to achieve a higher data rate. To compute the relay data rate in this chapter, we assume that the incoming data and outgoing data at the RSs should be equal. Let C_{BS-RS} and C_{RS-SS} be the capacities of BS to RS and RS to SS links respectively when each link is given the whole bandwidth, and also let t_{BS-RS} and t_{RS-SS} be the time durations of BS to RS and RS to SS links respectively. Focusing on the DL transmission, $t_{BS-RS} + t_{RS-SS}$ should be equal to the total duration of DL subframe. The amount of data transferred from the BS to an RS is equal to the amount of data transferred from the RS to an SS:

$$C_{BS-RS} \cdot t_{BS-RS} = C_{RS-SS} \cdot t_{RS-SS}. \tag{8}$$

Thus, the average data rate of an SS using an RS is equal to the amount of data received divided by the time required to receive it:

$$C_{BS-SS} = \frac{C_{BS-RS} \cdot t_{BS-RS}}{t_{BS-RS} + t_{RS-SS}}, \tag{9}$$

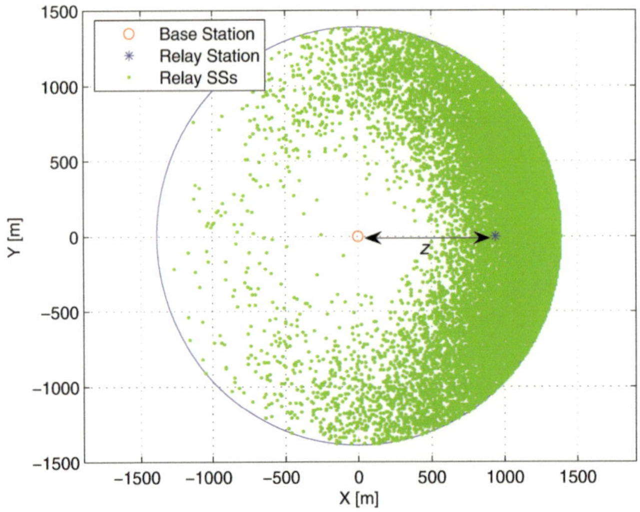

Fig. 1. A scenario with a single RS optimally placed at distance z from the BS showing the nodes that benefit from the RS.

as the RS cannot receive from the BS while transmitting to the SS. Consequently, using (8), the relay data rate of an SS can be rewritten as:

$$\frac{1}{C_{BS-SS}} = \frac{1}{C_{BS-RS}} + \frac{1}{C_{RS-SS}}. \tag{10}$$

3. Capacity enhancement

In this section, we focus on improving cell capacity by deploying T-RSs (transparent relays) inside a cell, and consider the placement of RSs that maximizes cell capacity. There has been a great deal of research directed toward improving the capacity of wireless networks at the physical layer. However, the achievable bit rate is still limited by the received signal strength due to the fact that wireless signal attenuates severely as it propagates between transmitter and receiver. Especially, the SSs located at the edge area of a cell achieve very limited data rates. To mitigate this problem, deploying RSs inside a cell can increase the achievable bit rate between a transmitter and a receiver leading to capacity enhancement.

3.1 Optimal placement of transparent relay

A network operator always desires the most cost effective solution with a minimal deployment expenditure for the provision of satisfactory service. Therefore, in order to provide efficient multi-hop relay networks, we need to know the optimal location of RSs for maximizing network capacity. In this subsection, we show how to find the optimal location of a T-RS that maximizes the cell capacity; the results can be easily extended to scenarios with multiple RSs. We assume that multiple RSs are deployed in a circular pattern inside a cell. Thus, the optimal distances from the BS to the RSs are consistent, however the optimal distance could vary as the number of RSs changes.

As shown in Fig. 1, the BS is located at the center of the cell and the SSs are distributed uniformly in the coverage area of the BS with constant grid size 10m. The location of an SS can be uniquely identified by its coordinates. We define the set of coordinates of the SSs inside a cell, $\mathcal{Q} = \{(x_1, y_1), (x_2, y_2), ..., (x_N, y_N)\}$ where N is the total number of SSs inside a cell. When a T-RS is located at (x, y) and an SS i is located at (x_i, y_i), $i \in \{1, 2, ..., N\}$, the distances of the links BS to RS, RS to SS, and BS to SS can be expressed as:

$$\begin{aligned} d_{BS-SS} &= \sqrt{x_i^2 + y_i^2}, \\ d_{BS-RS} &= \sqrt{x^2 + y^2}, \\ d_{RS-SS} &= \sqrt{(x_i - x)^2 + (y_i - y)^2}. \end{aligned} \tag{11}$$

Let us denote with $C_{SS}(x_i, y_i)$ the achievable data rate of an SS that is located at (x_i, y_i). We also denote with $C_{SS}^{Direct}(x_i, y_i)$ and $C_{SS}^{Relay}(x_i, y_i)$ the achievable direct data rate (BS-SS) and relay data rate (BS-RS-SS) of an SS located at (x_i, y_i) respectively when the whole channel bandwidth is used for that SS; with the distances computed by equation (11), the path losses and average received signal powers are calculated by using equations (1), (2). After that, the random values of the received signal powers can be generated by distributions (5), (7), and then the instantaneous SINR values can be computed by equation (3). Consequently, using the threshold SINR values in Table 2 and the equation (10) for the relay case, the achievable direct and relay data rates of an SS can be determined. To maximize the cell capacity, every SS will choose the best achievable data rate between direct and relay data rates:

$$C_{SS}(x_i, y_i) = \max\left(C_{SS}^{Direct}(x_i, y_i), C_{SS}^{Relay}(x_i, y_i)\right). \tag{12}$$

Due to the fading channel effect (i.e., received signal power is a random variable), an SS that is close to the BS does not always achieve a higher data rate than an SS that are further away from the BS. Likewise, as shown in Fig. 1 a few SSs located on the left side of the cell can benefit from the RS placed on the right side of the cell. We denote with *relay SSs* and *direct SSs* the SSs whose relay data rate is higher than direct data rate and the SSs whose direct data rate is higher than relay data rate respectively. We also define the mean cell capacity, C_{cell}, as the summation of every SS's achievable data rate divided by the total number of SSs in a cell. Therefore, the optimal placement of an RS can be determined in such a way that the mean cell capacity is maximized:

$$\arg\max_{(x,y)} \frac{1}{N} \sum_{(x_i,y_i)\in\mathcal{Q}} C_{SS}(x_i, y_i; x, y). \tag{13}$$

3.2 Impact of network parameters

In this subsection, we show how various network parameters such as reuse factor, terrain types, RS antenna gain, and the number of RSs affect the optimal placement of RSs and the cell capacity. To evaluate the performance of using RSs in comparison to the conventional scenario (without using RSs), we define the capacity gain parameter. Let C_{cell}^{Relay} and C_{cell}^{Direct} be the cell capacity with RS and without RS respectively, the relative capacity gain can be defined

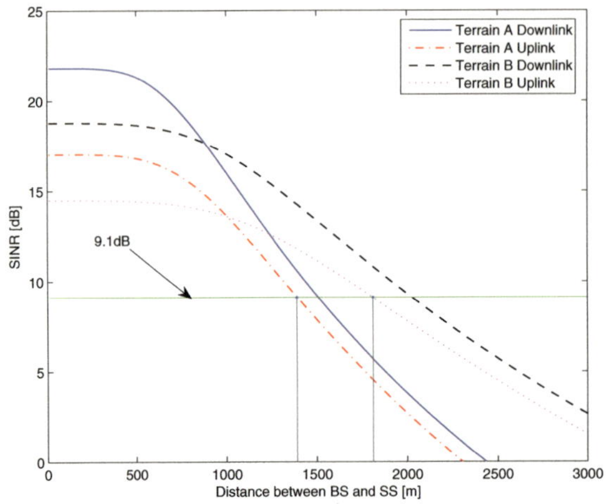

Fig. 2. Downlink and uplink edge SINR as a function of cell size for two different terrain types.

as:

$$Capacity\ Gain = \frac{C_{cell}^{Relay} - C_{cell}^{Direct}}{C_{cell}^{Direct}}. \tag{14}$$

We assume that our basic system has a reuse factor of seven, one sector per cell, terrain type A, 17dBi RS antenna gain, and four RSs deployed to enhance cell capacity. The system is analyzed by varying one specific parameter without changing the rest of basic system parameters. For each scenario, the cell size (cell radius) will vary since the received SINR value at the SS is affected by the different network parameters. Fig. 2 shows how cell size can be determined for a specific scenario. The SINR value of an SS located at the edge of the cell decreases as the distance between BS and SS increases on both DL and UL regardless of the terrain types. The decrease in SINR is more significant for terrain type A. However, when the edge SS is close to the BS, the SINR values under terrain type A are even higher than that of terrain type B because the co-channel interference values are significant when the received signal powers are similar in both terrain types. Once either the DL or UL SINR value reaches the minimum threshold value 9.1dB, that distance from the BS is considered as a cell size.

The first scenario is to evaluate the capacity gain and the optimal placements of RSs by varying the reuse factor. The reuse factor represents the number of cells grouped in a cluster. When the number of cells in a cluster increases, the co-channel distance also increases. The co-channel interference is a function of co-channel distance, hence, the larger the reuse factor the smaller the interference power from neighboring cells. Moreover, the cell size is affected by the reuse factor. When the reuse factor decreases, the cell size also decreases as the co-channel interference increases. When the cell size is small, most of the SSs do not benefit from the RSs since they are close enough to the BS leading to the smaller capacity gain. In other words, it is not very useful to deploy RSs in scenarios with a smaller reuse factor. Fig. 3(a) shows the cell capacity gain as a function of the location of RS in each reuse factor scenarios four, seven, nine, and twelve. The maximum cell capacity gain of 31.54% is achieved for a reuse factor of twelve, while only a capacity gain of 13.22% is obtained in the scenario with a reuse factor of four. Although RSs are optimally placed in a cell, only small number of SSs can benefit from

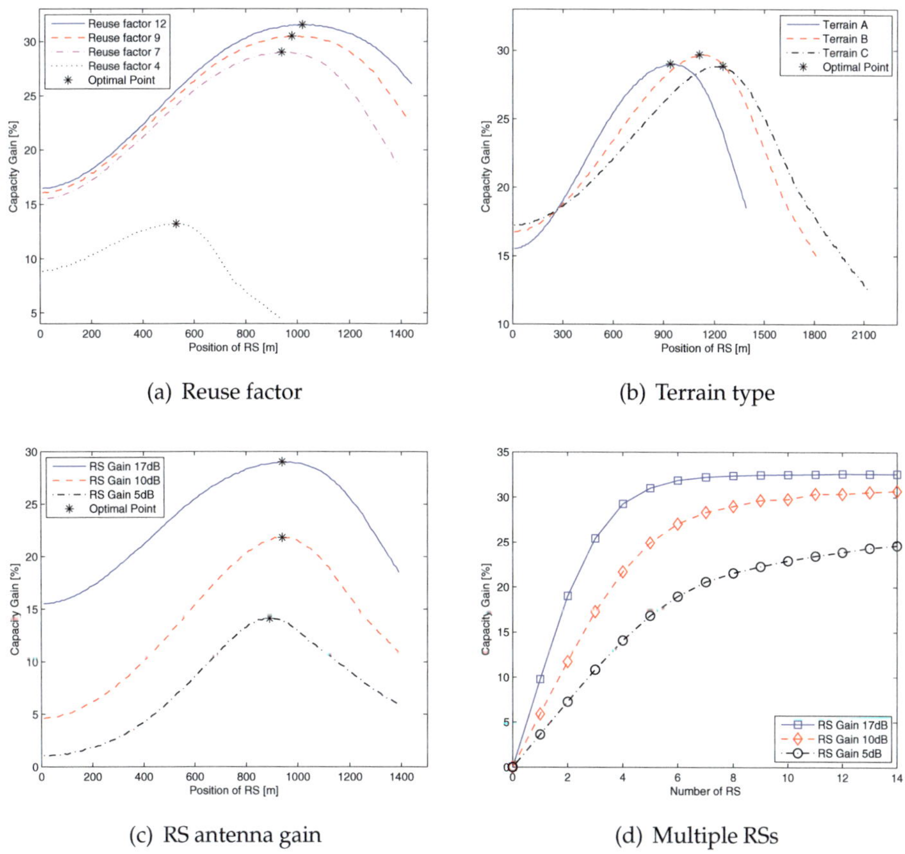

(a) Reuse factor

(b) Terrain type

(c) RS antenna gain

(d) Multiple RSs

Fig. 3. Cell capacity gains for different network parameters when using four T-RSs.

RSs in the scenario with the reuse factor four. It is interesting to note that RSs are increasing the capacity of the cell even if placed near the base station due to Rayleigh fading that may result in SSs preferring the link to the RSs to the link to the BS. The exact cell capacities and relay locations are listed in Table 3.

In the second scenario we study the impact of terrain types mentioned in section 2.2 on the capacity gain and the optimal placements of RSs. When terrain type changes from A to C, the path loss decreases between transmitter and receiver. Thus, the received signal mean power increases leading to a bigger cell size as terrain types changes from A to C. The cell sizes for terrain type A, B, and C are 1390m, 1810m, and 2120m respectively. Fig. 3(b) shows achievable capacity gains with respect to the location of RS for different terrain type scenarios. The maximum capacity gains for each scenario are very similar to each other, 29.7% is achieved for terrain type B and 28.87% is achieved for terrain type C. However, the ratio of optimal location of RS to each cell radius decreases as terrain type changes from A to C. When terrain type is A, the optimal location of RS is 67.63% of the cell radius, while optimal location of RS for terrain type C is 58.96%.

In the third scenario we analyze the capacity gain and the optimal placements of RSs by varying the RS antenna gain. The cost of an RS is assumed to be much less than a BS since

Scenario	Radius [m]	C_{cell}^{Direct} [Mbps]	Optimal Location	C_{cell}^{Relay} [Mbps]	Capacity Gain[%]
Reuse factor 4	930	16.6132	530	18.8094	13.22
Reuse factor 7	1390	12.9280	940	16.6804	29.03
Reuse factor 9	1420	12.6865	980	16.5553	30.50
Reuse factor 12	1440	12.5138	1020	16.4602	31.54
Terrain B	1810	12.4314	1110	16.1237	29.70
Terrain C	2120	12.2409	1250	15.7743	28.87
RS Gain 10dB	1390	12.9272	940	15.7502	21.84
RS Gain 5dB	1390	12.9266	890	14.7527	14.13

Table 3. Capacity and optimal relay position for the basic scenario and variations of the reuse factor, terrain type, and RS antenna gain.

an RS does not have a backhaul link, but the antenna gain of an RS could be as good as the BS antenna gain. We use 17dBi RS antenna gain for our basic system, and change that to 10dBi and 5dBi to analyze the impact of RS antenna gain on the system. Fig. 3(c) shows the cell capacity gain results for different RS antenna gains. The maximum capacity gain 29.03% is achieved when RS antenna gain is 17dBi and the lowest capacity gain 14.13% is achieved when RS antenna gain is 5dBi. It is clear that the cell capacity gain is significantly impacted by the antenna gain of RS. That is, the lower the RS antenna gain the lower the capacity gain achieved. It is also shown that the different antenna gains of RS had no significant impact on the optimal placement of RS.

In the last scenario of this subsection we explore the impact of the number of RSs on cell capacity. In general, the more RSs, the higher the cell capacity. However, the network cost will also increase as the number of RS increases. Fig. 3(d) shows how much capacity gain can be achieved with respect to the number of RS for different RS antenna gain scenarios. The increase rate of capacity gain for each scenario is clearly decreasing as the number of RS increases. When the RS antenna gains are 17dBi, 10dBi, and 5dBi, the capacity gains are limited to approximately 33%, 31%, and 25% respectively. Therefore, it is clear that the addition of RSs after a certain number of RSs does not further improve the cell capacity, e.g., deploying more than six RSs in 17dBi antenna gain scenario is not useful. The capacity gain with 14 RSs of antenna gain 5dBi is lower than that of a system with five RSs of antenna gain 10dBi or three RSs of antenna gain 17dBi. That is, the RS antenna gain has more significant impact on the capacity gain than the number of RSs.

4. Coverage extension

In this section, we focus on deploying NT-RSs for coverage extension. In order to analyze the benefits of using RSs from a coverage extension perspective, we need to examine how deploying RSs affects cost and throughput as well as coverage increase. To analyze the variation of throughput under the influence of RSs, the scheduling scheme has to be taken into account since the achievable throughput could differ significantly according to the scheduling schemes used. We first present three scheduling schemes: orthogonal, overlapped, and optimal that can be used in two-hop relaying networks. We then analyze the cost effective coverage extension problem by varying both the location and number of RSs (Kim & Sichitiu, 2010a). Finally, we explore an extension of the optimal scheduling scheme to a general multihop relaying scenario, and examine the impact of an increased number of relay hops on the network performance (Kim & Sichitiu, 2011a). Fig. 4 shows an example coverage

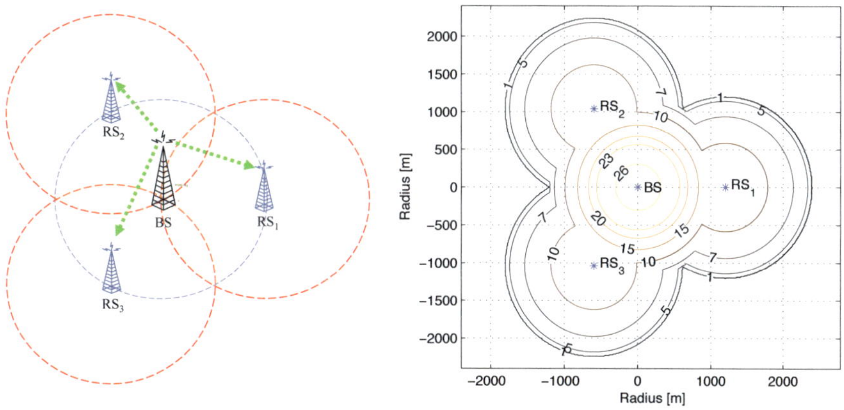

(a) Coverage extension scenario (b) Contour graph

Fig. 4. (a) Coverage extension scenario and (b) Contour graph of achievable average data rate when three RSs are deployed at the edge of the cell.

extension scenario where three NT-RSs are deployed at the edge of the BS transmission range. Using this scenario, we evaluate the performance of the scheduling schemes presented in the following subsection. The cell radius for the coverage extension scenario is assumed to be 1200m (determined by the condition that the cell coverage probability under the Rayleigh fading channel is greater than 90% (Erceg & Hari, 2001)), and the rest of network parameters are the same as the basic system in Section 3.

4.1 Scheduling schemes

According to the standard (802.16j, 2009), it is possible for the NT-RSs and BS to transmit and receive simultaneously to/from the associated SSs during the access zone period. Therefore it is challenging to schedule transmission opportunities to each link in the network, especially for serving SSs in a fair manner. We first present two existing scheduling schemes, namely the orthogonal and overlapped schemes (Park et al., 2009), and then introduce an optimal scheme (Kim & Sichitiu, 2011b). The orthogonal scheme minimizes interference by disallowing frequency reuse; however it can lead to lower throughput performance, whereas the overlapped scheme can achieve higher throughput by maximizing frequency reuse, but the outage rate is also increased due to significant intra-cell interference. In (Park et al., 2009), the boundary between access and relay zones was not dynamically selected according to the traffic load but statically determined for each scenario. To overcome this problem, we formulate the optimization problem for both orthogonal and overlapped schemes such that the cell throughput is maximized by optimally determining the boundary between the access and relay zones. Furthermore, we introduce an optimal scheduling scheme to combine the advantages of orthogonal and overlapped schemes. The optimal scheme maximizes the frequency reuse efficiency while avoiding outage events due to interference. The formulated optimization problems can be solved by using linear programming.

OFDMA based WiMAX networks allow scheduling to be done in the tiling frame structure (two-dimensional time×frequency) to attain frequency selectivity and multiuser diversity. However, due to the fact that the original tile scheduling problem is NP-hard (Deb et al., 2008), we will not consider multiuser scheduling over the frequency domain. Thus the entire

Notation	Description
$\mathcal{R}$	Set of RSs
$\mathcal{R}^+$	Set of service nodes (BS and RSs)
$\mathcal{U}$	Set of possible transmission subsets of service nodes
$\mathcal{V}$	Set of possible transmission subsets of relay links
$\mathcal{L}$	Set of possible relay links ($l_{ij} \in \mathcal{L}$)
l_{ij}	Relay link between service node i and j ($i \in \mathcal{R}^+, j \in \mathcal{R}$)
$\mathcal{S}$	Set of SSs
$\mathcal{S}_b$	Set of SSs associated with BS
$\mathcal{S}_R$	Set of SSs associated with RSs
$\mathcal{S}_r^u$	Set of SSs associated with RS $r \in \mathcal{R}$ in the subset u
$\mathcal{S}_{r^+}^u$	Set of SSs associated with $r^+ \in \mathcal{R}^+$ in the subset u
C_s^u	Achievable data rate of SS $s \in \mathcal{S}$ in the subset u
C_r	Achievable data rate of RS $r \in \mathcal{R}$ from the BS
C_{ij}^v	Achievable data rate for a relay link l_{ij} in the subset v
λ_s^u	Time fraction allocated to SS $s \in \mathcal{S}$ in the subset u
λ_r	Time fraction allocated to RS $r \in \mathcal{R}$ in relay zone
λ^u	Time fraction of subset $u \in \mathcal{U}$ in access zone
λ^v	Time fraction of subset $v \in \mathcal{V}$ in relay zone
λ_{ij}^v	Time fraction allocated for a relay link l_{ij} in the subset v
T_s	Throughput of SS $s \in \mathcal{S}$

Table 4. Notations used

spectrum is allocated to each node whenever they are allowed to transmit, i.e., scheduling is done by assigning time slots to nodes. For scheduling, fairness is an important issue. The well known fairness schemes such as max-min and proportional fairness for the relay enhanced WiMAX system are evaluated in our previous paper (Kim & Sichitiu, 2010b). The goal of max-min fairness is to allocate network resources in such a way that none of the active SSs can achieve more throughput than other active SSs without decreasing the throughput of other SSs. We formulate the optimization problem for each scheduling scheme under the max-min fairness constraints (Tassiulas & Sarkar, 2002). The notations used for the formulations are listed in Table 4.

4.1.1 Orthogonal scheduling scheme

The orthogonal scheme is an extreme solution for schedule resources since it does not allow frequency reuse in order to minimize intra-cell interference during the access zone period. In this scheme, only one service node can be active at a time to transmit to the associated SSs. Hence, there is no outage problem due to interference, but the achievable network throughput will be degraded due to inefficient radio resource utilization. To determine the time duration of each transmission for the orthogonal scheme, we formulate the optimization problem such that the cell throughput is maximized under max-min fairness constraints. Since it is impossible to increase the throughput of an SS without decreasing that of other SSs in the orthogonal scheme, the max-min fairness is equivalent to absolute fairness, i.e., every active SS achieves an equal throughput.

We denote with $\mathcal{R}$ and $\mathcal{S}$ the sets of RSs and SSs respectively, and the set of service nodes (the BS and RSs) is denoted as $\mathcal{R}^+$ (i.e., $|\mathcal{R}^+| = |\mathcal{R}| + 1$). Let $\mathcal{U}$ be the set of possible simultaneously active service nodes during the access zone period. Each element $u \in \mathcal{U}$ represents one subset of $\mathcal{R}^+$ with all the service nodes in that subset able to transmit concurrently to their associated SSs. In the orthogonal scheme, only single element subsets are possible in $\mathcal{U}$ since simultaneous transmission among service nodes is not allowed, hence, the two sets $\mathcal{U}$ and $\mathcal{R}^+$ have the same cardinality (i.e., $|\mathcal{U}| = |\mathcal{R}^+|$) in the orthogonal scheme. Let C_s be the achievable data rate of an SS from the service node that the SS is associated with. In other words, C_s represents the instantaneous link capacity between an SS and its service node. Similarly, the achievable data rate of an RS from the BS is denoted by C_r. Also, let λ_s and λ_r be the time fraction allocated to an SS and the time fraction allocated to an RS respectively. When the transmission subset $u \in \mathcal{U}$ changes, an SS can be associated with a different service node. Thus, the achievable data rate and time fraction of an SS in each transmission subset u are denoted by C_s^u and λ_s^u respectively. The ultimate throughput of an SS, T_s, during the current DL subframe can be expressed as the summation of throughputs received in each transmission subset u when the SS was allocated the time fraction λ_s^u:

$$T_s = \sum_{u \in \mathcal{U}} C_s^u \lambda_s^u, \qquad \forall s \in \mathcal{S}. \tag{15}$$

Our goal of maximizing cell throughput corresponds to maximizing the throughput, T_s, of any subscriber $s \in \mathcal{S}$ in the orthogonal scheme. Finding the maximum achievable throughput of an SS can be formulated as a linear programming problem as follows:

$$\max_{s \in \mathcal{S}} \; T_s. \tag{16}$$

Subject to:

$$T_{s_1} = T_{s_2}, \qquad \forall s_1, s_2 \in \mathcal{S} \; (s_1 \neq s_2). \tag{17}$$

$$C_r \lambda_r = \sum_{s \in \mathcal{S}_r} C_s \lambda_s, \qquad \forall r \in \mathcal{R}. \tag{18}$$

$$\lambda^u = \sum_{s \in \mathcal{S}^u_{r^+}} \lambda_s^u, \qquad \forall r^+ \in u, \forall u \in \mathcal{U}, |\mathcal{S}^u_{r^+}| > 0. \tag{19}$$

$$\sum_{u \in \mathcal{U}} \lambda^u + \sum_{r \in \mathcal{R}} \lambda_r \leq 1. \tag{20}$$

$$0 \leq \lambda_s^u, \lambda_r \leq 1, \qquad \forall u \in \mathcal{U}, \forall s \in \mathcal{S}, \forall r \in \mathcal{R}. \tag{21}$$

Here, $\mathcal{S}_r$ and $\mathcal{S}_{r^+}$ denote the set of SSs associated with RS $r \in \mathcal{R}$ and service node $r^+ \in \mathcal{R}^+$ respectively. In each transmission subset u, the sets $\mathcal{S}_r$ and $\mathcal{S}_{r^+}$ are denoted by $\mathcal{S}_r^u$ and $\mathcal{S}_{r^+}^u$. The first constraint ensures that every active SS in a cell achieves an equal throughput. The second constraint states that there is no data loss at the RSs, the data transferred from BS to RS $r \in \mathcal{R}$ is equal to the data transferred from RS r to the associated SSs. The third constraint ensures that resources within the duration of each transmission subset u are fully utilized by the associated SSs. Thus, the time fraction of each transmission subset u, λ^u, is equal to the summation of time fractions allocated to SSs associated with r^+ when r^+ is the element of subset u. The forth constraint captures the fact that the DL subframe consists of an access zone and a relay zone. The summation of time fractions of every transmission subset will be equivalent to the access zone time fraction, and the summation of time fractions allocated to RSs is the same as the relay zone time fraction. The sum of access and relay zone time fractions should be less than or equal to one. The final constraint restricts the amount of each time fraction allocated to SSs

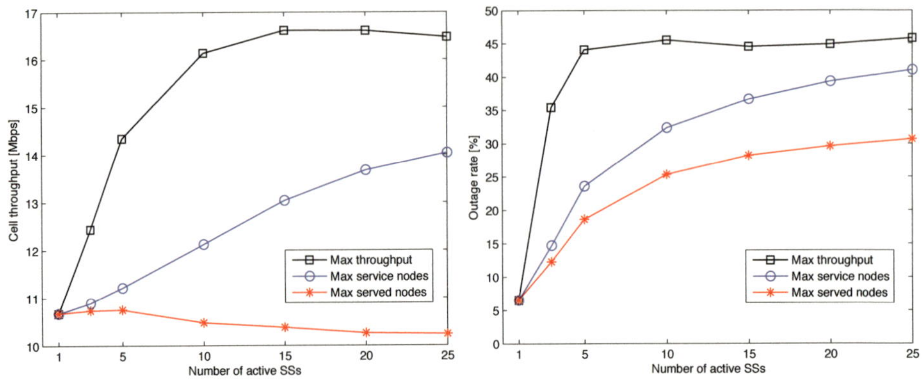

(a) Cell throughput

(b) Outage probability

Fig. 5. (a) Cell throughput and (b) outage probability as a function of the number of active SSs within a cell for different subset selection objectives for the overlapped scheme.

and RSs to be positive and smaller than one. By using these scheduling constraints (17)-(21), the objective function (16) can be maximized. Once the throughput of an SS (16) is maximized, the cell throughput can be computed by:

$$\text{Cell Throughput} = \sum_{s \in \mathcal{S}} T_s. \tag{22}$$

4.1.2 Overlapped scheduling scheme

The main goal of the overlapped scheme is to fully reuse radio resources during the access zone period. Multiple service nodes can transmit data to the associated SSs simultaneously, thereby enhancing network throughput, but at the same time outage events increase due to significant intra-cell interference. All the service nodes in a cell can be active simultaneously to maximize the chance of frequency reuse, however according to the number of active SSs and distributions of the SSs, some of the service nodes may not have to be active since it is possible that none of the active SSs are associated to that service node. Therefore, the decision regarding which service node should be active over a frame needs to be made at the beginning of a frame before scheduling. In other words, the overlapped scheme will consider only one transmission subset of the active service nodes over one frame duration. That is, once the subset of active service nodes is determined at the beginning of a frame, it will last for the duration of the entire current frame. For determining the set of active service nodes, we should also consider an additional *subset selection objective* (Kim & Sichitiu, 2011b):

- maximizing cell throughput,
- maximizing the number of active service nodes, or
- maximizing the number of served SSs.

According to which of these objectives are selected, a different subset of active service nodes is determined for the overlapped scheme. Now, we can formulate the optimization problem as in Section 4.1.1 to maximize cell throughput for the overlapped scheme by considering only one subset $u \in \mathcal{U}$ instead of the entire set $\mathcal{U}$. When the service nodes of the selected subset are active simultaneously, each active SS can be associated only to the service node that has the strongest link capacity to that SS over the entire duration of that frame. However, not every active SS can fully use resources during the access zone period because we assume max-min fairness and also the relay links (BS to RSs) have to share the resources orthogonally to transfer data from the BS to the RSs. Therefore, for the SSs that are served via RSs, the SSs with low link capacity are allocated large fractions of time while the SSs with high link capacity may have smaller time fractions, hence, the absolute fairness still holds for the SSs associated with RSs, but the SSs associated with the BS directly may achieve a higher throughput under max-min fairness since these larger throughputs do not affect the throughput of the rest of SSs associated with the RSs. Consequently, the objective of maximizing the throughput of an SS does not correspond to maximizing the cell throughput for the overlapped scheme since the equivalence of the absolute and max-min fairness constraint does not hold. Therefore, the objective of the optimization problem for the overlapped scheduling scheme is expressed as:

$$\max \sum_{s \in \mathcal{S}} T_s. \tag{23}$$

Let $\mathcal{S}_b$ and $\mathcal{S}_R$ be the set of SSs associated with the BS and RSs respectively over a frame period. The max-min fairness ensures that every SS achieves an equal throughput in each subset $\mathcal{S}_b$ and $\mathcal{S}_R$. However, the throughput of an SS in $\mathcal{S}_b$ could be higher than the throughput of an SS in $\mathcal{S}_R$. Therefore, the first constraint (17) in the orthogonal scheme is modified for the overlapped scheme as follows:

$$\begin{aligned} T_{s_1} &= T_{s_2}, \quad \forall s_1, s_2 \in \mathcal{S}_b \ (s_1 \neq s_2) \\ T_{s_3} &= T_{s_4}, \quad \forall s_3, s_4 \in \mathcal{S}_R \ (s_3 \neq s_4) \\ T_{s_5} &\geq T_{s_6}, \quad \forall s_5 \in \mathcal{S}_b, \ \forall s_6 \in \mathcal{S}_R. \end{aligned} \tag{24}$$

The second and fifth constraints (18), (21) do not change for the overlapped scheme as there is no data loss at the RSs, but the third and forth constraints (19), (20) are modified because there may be wasted resources in the access zone due to fairness and only one subset of $\mathcal{R}^+$ is considered in the overlapped scheme:

$$\lambda^u \geq \sum_{s \in \mathcal{S}^u_{r^+}} \lambda^u_s, \ \forall r^+ \in u, \forall u \in \mathcal{U}, |\mathcal{S}^u_{r^+}| > 0 \tag{25}$$

$$\lambda^u + \sum_{r \in \mathcal{R}} \lambda_r \leq 1. \tag{26}$$

Consequently, the cell throughput for the overlapped scheme can be maximized by solving linear programming with the objective (23) under constraints (18), (21), (24), (25), (26). Any subset could be chosen at the beginning of a frame based on the subset selection objectives, and the selected service nodes in that subset will be optimally scheduled to maximize cell throughput by using the optimization problem formulated above. However, the cell throughput and outage rate can vary according to the subset of active nodes chosen. Fig. 5 shows the cell throughput and outage rate as a function of the number of active SSs for different subset selection objectives for the overlapped scheme. The max throughput objective achieves the highest cell throughput, while the max served nodes objective attains the lowest cell throughput. In contrast, the outage rate performance of the max served nodes objective case is the best among three objective cases, while the outage rate of the max throughput objective case is the worst.

4.1.3 Optimal scheduling scheme

The key drawback of the orthogonal scheme is the bandwidth inefficiency because it prevents frequency reuse during the access zone period. For the overlapped scheme, the high outage probability is a serious problem. To eliminate these problems, an optimal scheduling scheme is proposed. The main task of an optimal scheme is to maximize bandwidth efficiency by allowing frequency reuse while avoiding outage events due to intra-cell interference. In order to accomplish the main task of the optimal scheme, we need to consider all possible transmission subsets of service nodes during the access zone period, i.e., the set $\mathcal{U}$ is the power set of $\mathcal{R}^+$ excluding the empty set in the optimal scheme scenario ($|\mathcal{U}| = 2^{|\mathcal{R}^+|} - 1$). In each subset of service nodes, an active SS can be either outage due to interference or associated with the active service node that has the highest link capacity to that SS. The achievable data rate of an SS, C_s^u, varies for each subset of active service nodes because intra cell interference changes according to the number of active service nodes in each subset.

Similar to the orthogonal scheme, the objective of the optimization problem for the optimal scheme is to maximize the throughput of any active SS in a cell since the equivalence of the absolute and max-min fairness holds for the optimal scheme. In this scheme, the whole bandwidth should be fully utilized without wasting resources, thus none of the active SSs can achieve more throughput without decreasing the throughput of the other SSs. The first, forth and fifth constraints (17), (20), (21) in the orthogonal scheme do not change for the optimal scheme, but the second and third constraints (18), (19) are modified to take into account every possible subset of service nodes for the optimal scheme. In the orthogonal and overlapped schemes, the set of SSs associated with RS r, $\mathcal{S}_r$, does not change over one DL subframe interval because an RS r can be active only one time to transfer data to the SSs associated to that RS, i.e., an RS r can not be included in more than two subsets. However, in the optimal scheme, an RS r can be active more than once as part of different subsets, i.e., an RS r can be an element of multiple subsets. Thus, the second constraint can be rewritten as:

$$C_r \lambda_r = \sum_{u \in \mathcal{U}} \sum_{s \in \mathcal{S}_r^u} C_s^u \lambda_s^u, \quad \forall r \in \mathcal{R}. \tag{27}$$

To ensure that resources within the duration of each transmission subset u are fully utilized by the associated SSs in the optimal scheme, there should not be any wasted resources. For example, when r_1^+ and r_2^+ are active service nodes in the subset u, the summation of time fractions allocated to SSs associated with r_1^+ should be equal to the summation of time fractions allocated to SSs associated with r_2^+:

$$\begin{aligned} \lambda^u = \sum_{s \in \mathcal{S}_{r_1^+}^u} \lambda_s^u = \sum_{s \in \mathcal{S}_{r_2^+}^u} \lambda_s^u, \\ \forall r_1^+, r_2^+ \in u, \forall u \in \mathcal{U}, |\mathcal{S}_{r_1^+}^u| > 0, |\mathcal{S}_{r_1^+}^u| > 0. \end{aligned} \tag{28}$$

Therefore, the cell throughput for the optimal scheme can be maximized by solving the linear programming problem with the objective (16) under constraints (17), (20), (21), (27), (28).

To evaluate the performance of the optimal scheduling scheme, we compare its performance with the orthogonal and overlapped schemes. Fig. 6 shows the cell throughput and outage rates as a function of the number of active SSs in a cell. To obtain the average cell throughput value, the simulation is repeated 10,000 times for each scenario with N active SSs randomly placed in the cell with a uniform distribution. For the overlapped scheme, the max served

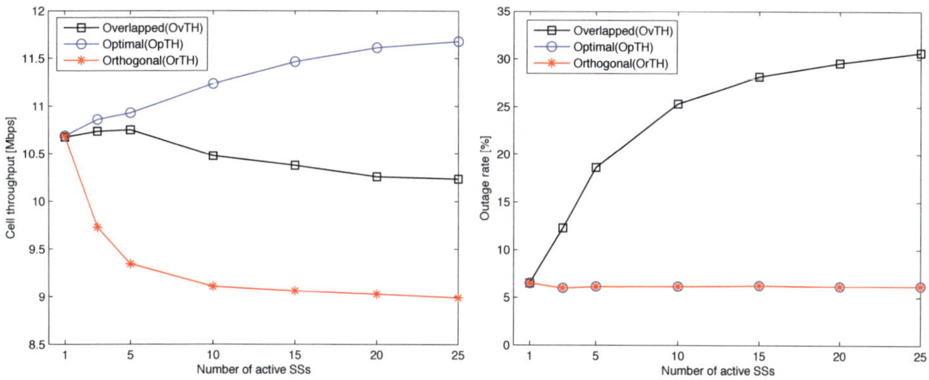

(a) Cell throughput (b) Outage probability

Fig. 6. (a) Cell throughput and (b) outage probability as a function of the number of active SSs within a cell for different two-hop scheduling schemes.

nodes subset selection objective is assumed. When there is only one active SS in a cell, there is no difference between the three scheduling schemes on both the cell throughput and outage rate since there is no frequency reuse and intra-cell interference. However, the differences becomes significant as the number of active SSs increases. The cell throughput achieved by the orthogonal scheme decreases because it is more likely to have SSs with low link capacities consuming large fractions of the time in order to preserve fairness, while the cell throughput for the optimal scheme grows as the number of active SSs increases since the optimal scheme maximizes frequency reuse without increasing outage rates. As shown in Fig. 6(b), the outage rate from the optimal scheme is identical to the result from the orthogonal scheme, while the outage rate for the overlapped scheme continues to rise significantly as more SSs join the cell. Although there is no interference between service nodes in the orthogonal scheme, about 6% of active SSs still encounter outage due to the Rayleigh fading channels. Overall, the cell throughput and outage rate performance can be dramatically enhanced by using the optimal scheduling scheme.

4.2 Cost effective coverage extension

In this subsection we analyze a cost effective coverage extension scenario by varying both the location and number of RSs. We use the optimal scheduling scheme presented in the previous subsection to compute the cell throughput for each coverage extension scenario and analyze the impact of varying the location and number of RSs on network throughput as well as network cost. We assume that each cell has between one and six RSs arranged in a circular pattern at the same distance from the BS to extend the cell coverage, and the cost of an RS is assumed to be 40% of the cost of a BS (Upase & Hunukumbure, 2008). To compare the network cost enhanced with RSs with the network cost without RSs, we define the relative cost parameter:

$$Relative\ Cost[\%] = \frac{\text{Cost of network with RS}}{\text{Cost of network without RS}} \times 100. \quad (29)$$

For example, assume that the total network service area is covered by 100 BSs without RSs (i.e., 100 cells); when two RSs are deployed in each cell to extend the cell coverage, the total

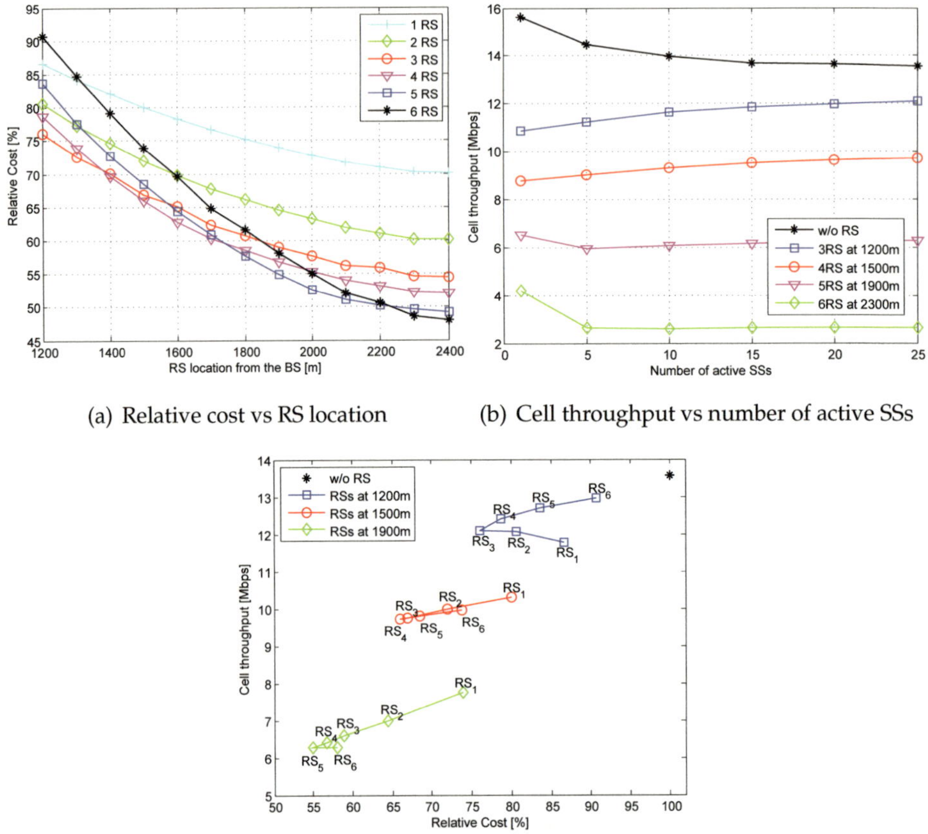

(a) Relative cost vs RS location (b) Cell throughput vs number of active SSs

(c) Cell throughput vs relative cost

Fig. 7. Cost effective coverage extension scenario results

number of cells needed to cover the service area will be decreased. Assume that the new number of cells is 50 (i.e., 50 BSs and 100 RSs); then the relative cost of this relay enhanced network is 90% ($50 + 100 \times 0.4$). We consider this relative cost as the network cost.

Figure 7(a) shows relative costs as a function of the RS location for a different number of RSs per each cell. The distance between the BS and RSs varies from 1200m to 2400m, which is twice the cell size. Since the RSs have higher antenna gains and transmission power than those of the SSs, the location of RS could be outside of the BS coverage. As the location of RSs from the BS increases, the relative costs decrease because the cell coverage extends when the RSs are further away from the BS leading to fewer required cells to cover the network area, hence, lower cost. When the number of RSs increases from one to three, the relative cost decreases regardless of the location of RSs, however, when the number of RSs is greater than four, the relative cost of the higher number of RSs is not necessarily lower than that of a smaller number of RSs. For example, the minimum relative cost is achieved by the three RSs case, i.e., deploying more than four RSs at 1200m is not desirable as they are more costly. From 1400m to 1700m deploying four RSs is the optimal case, and from 1800m to 2200m five RSs is better, and six RSs is better for distances from 2300m to 2400m.

To evaluate the effect of varying locations and numbers of RSs on the cell throughput, the optimal scheduling scheme is used such that every active SS can achieve the same throughput and a reduced outage rate. Fig. 7(b) shows cell throughput as a function of the number of active SSs for different locations and numbers of RSs. In the case without-RSs, the cell coverage area is the minimum, while the cases with-RSs increase the cell coverage as the placement of the RS is further away from the BS. Thereby, the cell throughput decreases as the cell coverage increases. However, the cell throughput of the case without-RSs decreases as the number of active SSs increases, while some of the cases with RSs show that cell throughput tends to increase with the number of active SSs. This change in cell throughput is due to the fact that it is more likely to have SSs with small link capacities consuming a large fraction of the time in order to preserve fairness as the number of active SSs increase, but in the cases with RSs the frequency reuse efficiency can overcome this tendency, hence, it is clear that the frequency reuse scheme has a positive impact on the cell throughput. Especially, when the location of RSs is less than 1500m, the achievable cell throughput continues to grow with the number of active SSs.

To be able to determine the cost effective coverage extension scenario, we need to simultaneously examine the effects of using RSs on both cost and throughput. Fig. 7(c) shows the cell throughput as a function of relative cost when the number of active SS is 25. Cell throughputs for three different RS locations and one to six RSs are plotted in this graph. Overall, the relative cost decreases as the location of RSs is further away from the BS and the number of RSs increases, but at the same time cell throughput also decreases. In other words, lower network cost can be achieved at the expense of the cell throughput. However, when the location of RSs is 1200m, the cell throughput continues to increase as the number of RSs increases due to the increase in frequency reuse. Especially, when the number of RSs changes from one to three, the relative cost decreases significantly. That is, deploying up to three RSs at 1200m is beneficial from both throughput and cost points of view, but deploying more than four RSs at 1200m is enhancing throughput at more cost. Therefore, cost effective coverage extension without significant throughput degradation is always feasible by carefully choosing both the location and number of RSs.

4.3 Generalization for multihop scenario

We explore an extension of the optimal scheduling scheme to a general multihop relaying scenario by allowing more than three-hop relaying in a cell. Although the two-hop scenario is technically a multihop scenario, we will use the term "multihop" to refer to scenarios with more than one relay tier. Many researchers have focused on two-hop relaying networks since this scenario has the largest throughput gain and more hops per connection cause a greater delay. On the other hand, it is clear that the cell coverage will be significantly extended by increasing the number of relay hops. Therefore, it is of interest to explore how increasing the number of relay hops can affect the network performance under an optimal scheduling scheme. The optimal scheme presented in the previous subsection is aimed at two-hop relaying networks where the relay zone is orthogonally shared by relay links (BS to RSs). In a multihop relaying scenario, the RSs should be able to relay data to/from other RSs, hence, frequency reuse between relay links is also possible during the relay zone period. Therefore, the extension of the optimal scheduling scheme should take into account frequency reuse in both the access and relay zone periods. We present an extended optimal scheduling scheme in this subsection, and evaluate its performance using two example three-hop relaying scenarios.

4.3.1 Optimal scheme for multihop

In the two-hop relaying scenario, the SSs are receiving data from the BS either directly or via only one RS, while in the multihop scenario, the SSs can receive data through multiple RSs. Also, multiple paths from the BS to each SS exist in a cell. If we assume that there is no frequency reuse in the relay zone, as it is in two-hop scenario, the optimal path to each SS will be the path that has the highest achievable relay data rate of the SS. However, to minimize throughput degradation due to the increase of number of hops, frequency reuse must be considered. Thereby, it is possible that a lower capacity relay link can also be scheduled for RSs to relay data, since multiple relay links can be active simultaneously during the relay zone period. To formulate the optimization problem for the multihop optimal scheduling scheme, we need to consider every possible link between service nodes as relay links, and then consider every possible set of relay links that can be active simultaneously.

Let l_{ij} be a relay link from the service node $i \in \mathcal{R}^+$ to $j \in \mathcal{R}$, and let $\mathcal{L}$ be the set of all possible relay links (i.e., $l_{ij} \in \mathcal{L}$). To consider every possible simultaneous transmissions between relay links, we denote with $\mathcal{V}$ the power set of $\mathcal{L}$ excluding the empty set and any sets of links that cannot be active at the same time. Thus, each element $v \in \mathcal{V}$ is a subset of relay links that can be active at the same time. Also, we denote with C_{ij} and λ_{ij} the achievable data rate and time fraction allocated for a relay link l_{ij} respectively. For the simultaneous transmission subset v, the achievable data rate and time fraction allocated for a relay link could vary due to intra-cell interference, hence, the C_{ij} and λ_{ij} when the relay links in the subset v are active are denoted by C_{ij}^v and λ_{ij}^v respectively. To simplify the notation, let T_r be the total amount of data transferred from an RS r to the associated SSs during the DL subframe interval as shown in (27):

$$T_r = \sum_{u \in \mathcal{U}} \sum_{s \in \mathcal{S}_r^u} C_s^u \lambda_s^u, \quad \forall r \in \mathcal{R}. \tag{30}$$

The objective of the multihop optimal scheme is to maximize the throughput of any active SS in a cell since the equivalence of the absolute and max-min fairness holds for the multihop optimal scheme. Finding the maximum achievable throughput of an SS can be formulated as a linear program as follows:

$$\max_{s \in \mathcal{S}} T_s. \tag{31}$$

Subject to:

$$T_{s_1} = T_{s_2}, \qquad \forall s_1, s_2 \in \mathcal{S}\ (s_1 \neq s_2). \tag{32}$$

$$\sum_{v \in \mathcal{V}} \sum_{l_{ij} \in v} C_{ij}^v \lambda_{ij}^v = \sum_{v \in \mathcal{V}} \sum_{l_{jk} \in v} C_{jk}^v \lambda_{jk}^v + T_j, \qquad \forall i \in \mathcal{R}^+, \forall j,k \in \mathcal{R}. \tag{33}$$

$$\lambda^u = \sum_{s \in \mathcal{S}^u_{r_1^+}} \lambda_s^u = \sum_{s \in \mathcal{S}^u_{r_2^+}} \lambda_s^u, \qquad \forall r_1^+, r_2^+ \in u, \forall u \in \mathcal{U}, |\mathcal{S}^u_{r_1^+}| > 0, |\mathcal{S}^u_{r_1^+}| > 0. \tag{34}$$

$$\sum_{u \in \mathcal{U}} \lambda^u + \sum_{v \in \mathcal{V}} \lambda^v \leq 1. \tag{35}$$

$$\lambda^v = \lambda_{l_1}^v = \lambda_{l_2}^v, \qquad \forall l_1, l_2 \in v, \forall v \in \mathcal{V}. \tag{36}$$

$$0 \leq \lambda_s^u, \lambda_r \leq 1, \qquad \forall u \in \mathcal{U}, \forall s \in \mathcal{S}, \forall r \in \mathcal{R}. \tag{37}$$

The first constraint ensures that every active SS in the cell achieves an equal throughput. The second constraint states that there is no data loss at the RSs: the data transferred from any of service node $i \in \mathcal{R}^+$ to an RS j is equal to the sum of data transferred from the RS j to any

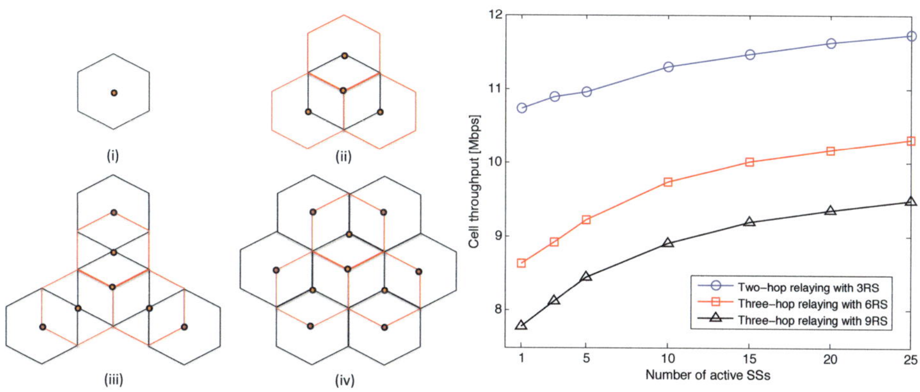

(a) Coverage extension scenarios

(b) Cell throughput results

Fig. 8. (a) A coverage extension scenario with (i) no RS, (ii) three RSs, (iii) six RSs, (iv) nine RSs and (b) cell throughput results for different multihop scenarios.

of service nodes $k \in \mathcal{R}$ and data transferred from the RS j to the associated SSs. The third constraint ensures that resources within the duration of each transmission subset u are fully utilized by the associated SSs. The forth constraint captures the fact that the sum of access and relay zone time fractions should be less than or equal to one. The summation of every time fraction of subset λ^v is equivalent to the relay zone time fraction. The fifth constraint ensures that resources within the duration of each transmission subset v are fully utilized by the associated relay links. The final constraint restricts the amount of each time fraction allocated to SSs and RSs to be positive and smaller than one. By using these scheduling constraints (32)-(37), the objective function (31) can be maximized and the cell throughput with multihop optimal scheduling scheme can be easily computed.

To evaluate the performance of the multihop optimal scheduling scheme, we show two three-hop relaying scenarios with six and nine RSs respectively. Fig. 8(a) shows how the RSs are deployed to extend the cell coverage for the two-hop and three-hop scenarios. We do not assume a hexagonal cell shape in this work, but use it to demonstrate how much cell coverage can be extended with the increase in the number of hops. For example, when three RSs are deployed in the two-hop scenario, the extended cell coverage is three times lager than the cell without RSs. Similarly, the coverage of three-hop scenarios with six and nine RSs can be five and seven times larger than that of a cell with no RSs. Fig. 8(b) shows the cell throughput results for two-hop and three-hop relaying scenarios. It is clear that the cell throughput decreases as the number of hops increases. The throughput increase rates of three-hop cases are slightly higher than that of the two-hop case as the number of active SSs increases. When the number of active SSs is 25, the throughput degradations from two-hop to three-hop with six and nine RSs are approximately 12% and 19% respectively, which is surprisingly low considering the significant increase in cell coverage.

5. Conclusion

In this chapter we studied the impact of deploying RSs on both capacity and coverage aspects in relay enhanced WiMAX networks. In particular, this chapter is composed of two main parts: Section 3 is targeted at optimizing the placement of transparent RSs that maximize the cell capacity; Section 4 is focused on cost effective coverage extension in the non-transparent

RS mode. In Section 3, we present the optimal placement of transparent RSs in WiMAX networks. The results show how various network parameters such as reuse factor, terrain types, RS antenna gain, and the number of RSs affect the optimal placement of RSs. In Section 4, we explore three different issues with regard to coverage extension scenario. First, we present three scheduling schemes called orthogonal, overlapped, and optimal to maximize cell throughput while preserving fairness. The results show that the cell throughput and outage rate performance can be dramatically enhanced by using the optimal scheduling scheme. Second, we suggest some design guidelines allowing network operators to achieve cost-effective coverage extensions without significant throughput degradation. In general, the lower the relative cost the lower the cell throughput; however, a higher cell throughput can be achieved with lower relative cost by carefully choosing both location and number of RSs. Finally, we extend our optimal scheduling scheme to a general multihop relaying scenario in order to show the impact of an increased number of relay hops on the network performance.

6. References

3GPP (2009). Relay advancements for E-UTRA (LTE-Advanced), *TR 36.806*, 3rd Generation Partnership Project (3GPP).

802.16j (2009). IEEE standard air interface for broadband wireless access systems: Multihop relay specification.

Andrews, J. G., Ghosh, A. & Muhamed, R. (2007). *Fundamentals of WiMAX: Understanding Broadband Wireless Networking*, Prentice Hall PTR, Upper Saddle River, NJ, USA.

Deb, S., Mhatre, V. & Ramaiyan, V. (2008). WiMAX relay networks: opportunistic scheduling to exploit multiuser diversity and frequency selectivity, *MOBICOM*, ACM, pp. 163–174.

Erceg, V. & Hari, K. V. S. (2001). Channel models for fixed wireless applications, *IEEE 802.16 Broadband Wireless Access Working Group*, Technical Report.

Foschini, G. J. & Salz, J. (1983). Digital communications over fading radio channels, *Bell System Tech. J.* pp. 429–456.

Kim, Y. & Sichitiu, M. L. (2010a). Cost effective coverage extension in 802.16j mobile multihop relay networks, *WCNC*, IEEE, pp. 1–6.

Kim, Y. & Sichitiu, M. L. (2010b). Fairness schemes in 802.16j mobile multihop relay networks, *GLOBECOM 2010, 2010 IEEE Global Telecommunications Conference*, pp. 1 –5.

Kim, Y. & Sichitiu, M. L. (2011a). Optimal max-min fair resource allocation in multihop relay-enhanced WiMAX networks, *IEEE Trans. Veh. Technol.* accepted. URL: *http://www4.ncsu.edu/~mlsichit/Research/Publications/wimaxOptimalSchedulingTVT.pdf*

Kim, Y. & Sichitiu, M. L. (2011b). Optimal resource allocation in multihop relay-enhanced WiMAX networks, *WCNC*, IEEE, pp. 1–6.

Park, K., Ryu, H. S., Kang, C. G., Chang, D., Song, S., Ahn, J. & Ihm, J. (2009). The performance of relay-enhanced cellular OFDMA-TDD network for mobile broadband wireless services, *EURASIP J. Wirel. Commun. Netw.* 2009: 1–10.

Rappaport, T. (2001). *Wireless Communications: Principles and Practice*, 2nd edn, Prentice Hall PTR, Upper Saddle River, NJ, USA.

Tassiulas, L. & Sarkar, S. (2002). Max-min fair scheduling in wireless networks, *INFOCOM*, Vol. 2, pp. 763–772.

Upase, B. & Hunukumbure, M. (2008). Dimensioning and cost analysis of multihop relay-enabled WiMAX networks, *Fujitsu Sci. Tech. J.* 44(3): 303–317.

Zhang, Q. & Kassam, S. (1999). Finite-state Markov model for Rayleigh fading channels, *IEEE Trans. Commun.* 47(11): 1688–1692.

15

Cost Effective Coverage Extension in IEEE802.16j Based Mobile WiMAX Systems

Se-Jin Kim, Byung-Bog Lee, Seung-Wan Ryu,
Hyong-Woo Lee and Choong-Ho Cho
*Dept. of Computer and Information Science, Korea University,
Republic of Korea*

1. Introduction

Recently, there have been numerous standardization activities for making the IEEE802.16e system highly effective in supporting mobile users based on orthogonal frequency division multiple access (OFDMA) with Time Division Duplex (TDD) mode in (IEEE Standard 802.16e-2005, 2005) and (WiMAX Forum). Standardization activities of the IEEE802.16e system were completed in 2005, and currently enhancements of the IEEE802.16e standard are under discussion. One such enhancement effort is cell coverage extension and link throughput enhancement which was studied in IEEE802.16j TG in (IEEE802.16j-2008, 2008). In order to achieve these goals, the IEEE802.16j TG introduces mobile multi-hop relay technology to the IEEE802.16e system. The other enhancement effort is being discussed in IEEE802.16m TG. The IEEE802.16m system, called the gigabit WiMAX, makinly aims to enhance system throughput up to 1Gbps. Besides, this system aims to support legacy IEEE802.16 standards including the IEEE802.16j, and to interworking with other wireless systems such as 3GPP LTE and IMT-advanced system in (WiMAX Forum) and (Lee et al., 2006).

In particular, in Korea, the IEEE802.16e standard-based wireless broadband (WiBro) system was developed in 2005 and WiBro service was launched in 2006 covering isolated areas in Seoul. Since the initial launch, WiBro service providers have been trying to extend service areas from Seoul to regions nationwide. In order to cover the entire region of Korea, however, deployment cost for the traditional infrastructure consisting of only Base Stations (BSs) is estimated to be astronomical. For this reason, it is necessary to adopt a cost-effective service coverage extension method, and the mobile multi-hop relay (MMR) WiBro system is considered as a strong candidate for possible implementation.

In this chapter, we investigate various issues on multi-hop relay based WiBro/WiMAX systems, i.e., the IEEE802.16j system, with a focus on cost-effective cell coverage extension under various deployment situations. Since the coverage extension problem may occur in both of metropolitan areas and rural areas when the user-traffic density is relatively moderate or low, we first introduce an overview of IEEE802.16j standard. Secondly, we discuss several topologies and the resulting cost-effective coverage extension methods for each case. Also, we propose and analyze two sectored cellular based cost-effective coverage extension methods, the Narrow-Beam Trisector Cell (NBTC) and Wide-Beam Trisector Cell (WBTC) system based approaches. Finally, we present a practical deployment scenario consisting of three phases depending on the user-traffic density and the number of traffic relaying hops.

The remainder of this chapter is organized as follows. In section 2, we introduce standardization and research issues in MMR WiBro/WiMAX system. In section 3, we analyze cost-effective coverage extension methods under various MMR topologies that may occur in metropolitan and rural areas. Then we present sectored cellular-based coverage extension approaches in section 4. Based on the above analysis results, we propose a practical deployment scenario in section 5. Finally, we conclude this chapter by suggesting future research issues in section 6.

2. An overview of the MMR WiBro/WiMAX system

2.1 Standardization and research issues in the MMR WiBro/WiMAX system

In general, the concept of multi-hop relaying is adopted to extend cell coverage, to enhance link throughput per user, and to increase network reliability, while supporting quality of service requirements. One approach to enhance system capacity, throughput per user, and system reliability is adoption of Relay Stations (RSs) in low SINR cell boundary areas. In addition, by using demodulation/decode and forward functionalities, RSs could substitute higher data rate SINR links for existing lower SINR links. This spectrum efficiency results in system capacity enhancement. In fact, a part of the enhanced capacity could be used for multi-hop relay function while the other part of the capacity could be used to provide subscribers with increased throughput.

The purpose of cell-coverage extension is three-fold. First, it can be used to remove coverage holes that can be found around high-rise buildings or other wave propagation obstacles. Second, it can be used to give safe communication coverage to subscribers who are far away from or isolated from a BS. Third, it can be used to provide reliable communication links to mobile vehicles on which an RS is installed.

IEEE802.16j TG aims to enhance the IEEE802.16e (mobile WiMAX or WiBro) system in terms of throughput per user and cell coverage by introducing multi-hop based RSs in (IEEE 802.16j-2008, 2008) and (Siemens and ETRI, 2007). Coverage extension in the legacy WiBro/WiMAX system can be achieved based on the existing PMP (Point to Multi-Point) mode, while throughput enhancements can be achieved by higher-order modulation schemes based on the higher quality signal strength between BS and RS, and RS and Mobile Station (MS). However, without introducing MMR technology, even all the enhancements, which are related to the PHY (Physical) and the MAC (Medium Access Control) layers, are restricted to two hops, from BS to RS and RS to MS.

Fig.1 shows the overall MMR system's configuration studied in IEEE802.16j TG. The RS being considered in IEEE802.16j TG can be categorized into 3 types: the fixed RS, the nomadic RS, and the mobile RS. In addition, direct communication between MSs is not allowed. Therefore, the full mesh-networking approach support connections and multiple hops are beyond the scope of IEEE802.16j TG in (IEEE802.16j-2008, 2008).

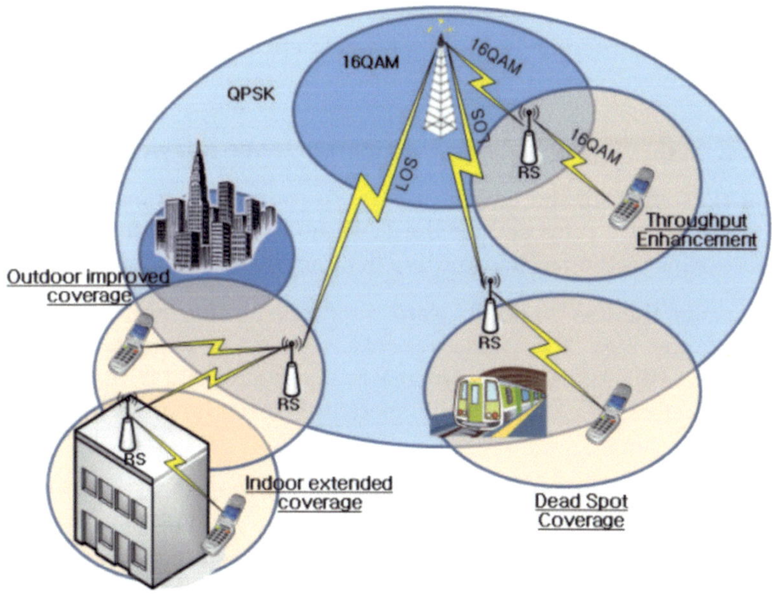

Fig. 1. IEEE802.16j TG MMR system.

2.2 Technical requirements for relay stations of the MMR WiBro/WiMAX

There are several technical requirements for the RSs of the MMR WiBro/WiMAX system (Siemens and ETRI, 2007). In this subsection, we summarize major technical functionalities of RSs to improve network throughput and to extend cell coverage.

- Support of legacy WiBro/WiMAX services: An RS should be able to support WiBro/WiMAX services to MSs transparently. An RS should register its existence to the network, and relay uplink and downlink traffic between the BS and an MS based on the WiBro/WiMAX protocols in (Lee et al., 2006) and (Siemens and ETRI, 2007).
- Network Entry functionalities: An RS could be operated in both MS-mode and BS-mode. When an RS is turned on, the RS can be operated in MS-mode to obtain synchronization information from the BS. After the synchronization process, the RS negotiates with the BS for the capability of an RS. Then, the RS operates in MS-mode for communication with the BS, and in BS-mode for communication with MSs. Therefore, the RS should include both MS-mode and BS-mode functionalities.
- Handover Support: Since an RS must provide legacy WiBro/WiMAX services, it should be able to support MS mobility. RS handover procedures that are similar to MR-BS handoff procedures should be performed for the usage models which can be found in guideline document in (IEEE802.16j-06/015, 2006).
- Cell-throughput enhancement: In order to enhance cell throughput per user, an RS should support Adaptive Modulation Coding (AMC) technology in hop-by-hop links, such as relay links and access links. Also cooperative relay technology is considered to achieve cell-throughput enhancement.
- Cell-coverage extension: In order to extend cell coverage by multi-hop links, RSs should relay the traffic from MR-BS/RS to RS/MS and vice versa. There could be

several kinds of frame structures that can support such multi-hop links. A transparent frame structure and a non-transparent frame structure were adopted in IEEE802.16j standard.

2.3 Transparent mode relaying vs. non-transparent mode relaying

In IEEE802.16j TG standardization, two types of RS modes, the transparent mode and the non-transparent mode, are proposed for the TDD mode. The key difference between these two relay modes is the method of transmitting the framing information.

In transparent mode, the RSs do not forward the frame header information, and hence do not extend the cell coverage of the wireless access system. Consequently, the main use case for transparent mode RSs is to facilitate increases in capacity within the cell coverage. This type of RS is of lower complexity, and only operates in a centralized scheduling mode for a topology of up to two hops.

On the other hand, the non-transparent mode RS is generally located in the cell edge to extend BS coverage using the remaining capacity of the BS. The RSs generate their own framing information or forward those provided by the BS depending on the scheduling approach (i.e., distributed or centralized). In the centralized scheduling, RSs relay MAP information received from the BS to its neighboring offspring RSs and MSs. Therefore, the centralized scheduling scheme is simple and efficient, and thus used in smaller grids such as metropolitan-area topologies. Meanwhile, in the distributed scheduling, an RS receives MAP information from the BS similar to the centralized mode. However, it creates its own MAP information and transmits it to its neighboring offspring RSs and MSs. Therefore, the distributed scheduling scheme can provide fault tolerance and scalability for large-scale topologies. However, the transmission of the framing information can result in high interference between neighboring RSs. Also, non-transparent mode RSs can operate in topologies larger than two hops in either centralized or distributed scheduling mode.

As with the former 802.16e frame structure in (IEEE802.16e-2005, 2005), the 802.16j frame is divided into the DL and UL subframes. However, unlike the previous frame structure, these subframes are further divided into zones to support BS-RS/MS communication (the Access Zone (AZ)) and RS-MS communication (the Relay Zone (RZ)). Thus, the Relay Receive/transmit Transition Interval (R-RTI) and Relay Transmit/receive Transition Interval (R-TTI), have to be inserted between AZ and RZ for both the DL and UL subframes in order to give the wireless devices sufficient time to switch from transmission mode to reception mode, or vice versa.

In transparent mode, the BS transmits data to both of the MSs and RSs in AZ, and then the RSs retransmit the received data to MSs located in its coverage. In RZ, the BS operates either in a silent mode or in a cooperative transmission mode. The frame structure of the non-transparent mode is similar to that of the transparent mode, but a BS and RSs can transmit data at the same time in AZ due to lower interference. An RS retransmits the received data to MSs located in its coverage. In RZ, a BS transmits data to RSs and all RSs operate in receiving mode.

The coverage extension, therefore, can be achieved using the non-transparent mode RSs. For the remainder of this chapter, we will concentrate on the non-transparent mode RSs.

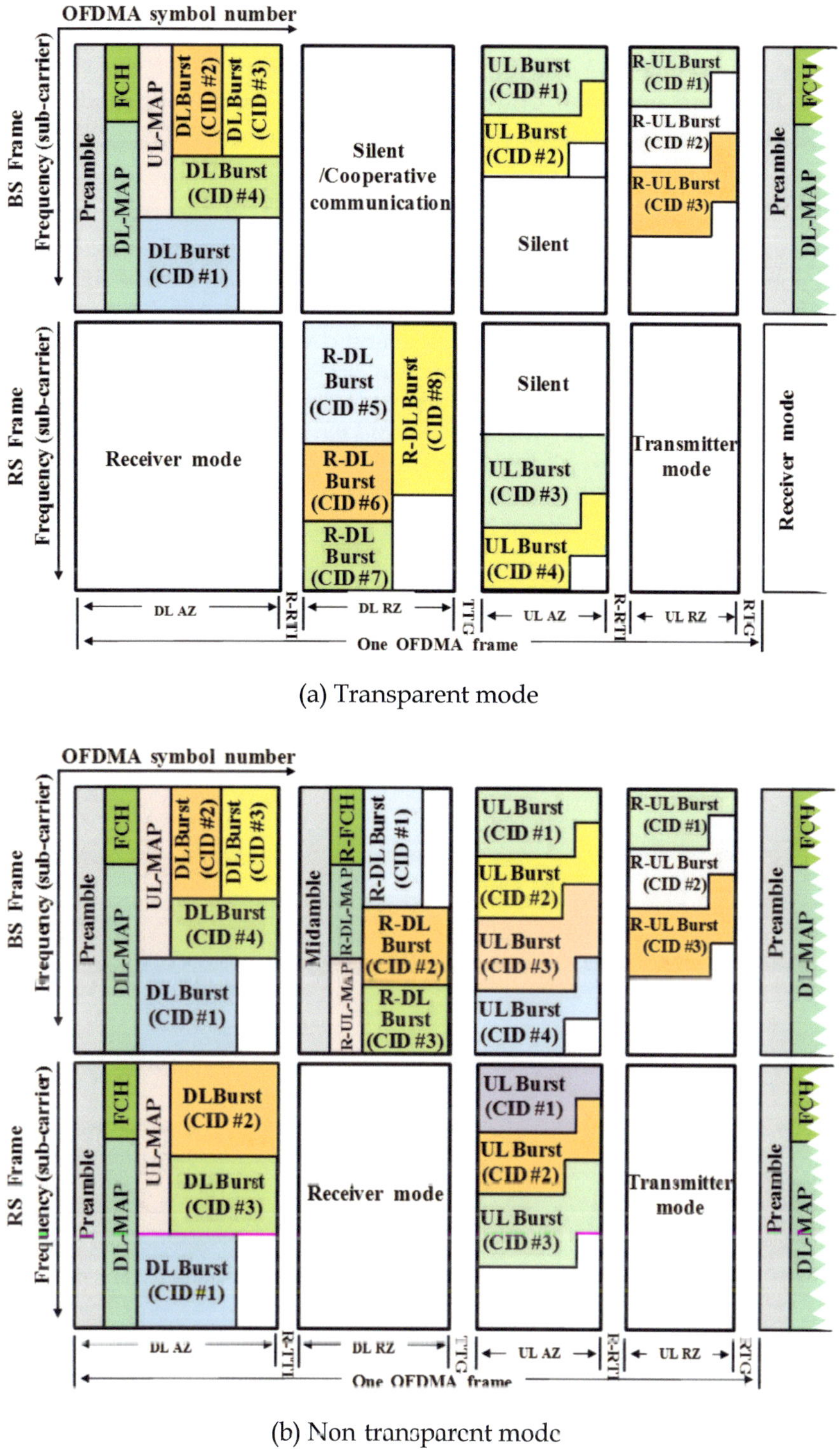

Fig. 2. Frame structures in IEEE802.16j.

3. Coverage extension with minimal deployment cost

In this section, we propose and analyze various MMR topologies for cell-coverage extension using omni-directional antennas. Under such topologies, we investigate deployment cost and optimal numbers of BSs and RSs with respect to various user traffic densities.

3.1 Omni-directional antenna based multi-hop relay topologies

We propose various MMR topologies for both metropolitan and rural areas. We assume that omni-directional antennas are used in both a BS and RSs. We also assume that the maximum number of hops is 4 which means that there could be at most 3-tiers of RSs from the BS. For the convenience of analysis, we also assume that the unit coverage areas in the metropolitan area topologies and the rural area topologies are a square and a hexagon respectively. A BS is assumed to have different sizes of coverage area according to its transmission power. In other words, in terms of the unit coverage area, a BS could have a different number of unit areas for its coverage. In general, RSs have less transmission power compare to the BS, and thus, in this chapter, we assume that each RS covers one unit of coverage area regardless of topologies. Figure 3 shows three types of different topologies for each case of the metropolitan and rural areas, i.e., Type-A, B, and C, for the metropolitan areas and Type-D, E, and F are for the rural areas. For the metropolitan areas, we use grid topology models which are generally used for metropolitan topology. An RS coverage is a square but the BS coverages are one, four, five squares for Type-A, B, C. On the other hand, for the rural areas, we use hexagonal topology models. An RS coverage is a hexagon but the BS coverages are one, seven, three hexagons for Type-D, E, F. The fill patterns and colors denote the number of tiers for RSs.

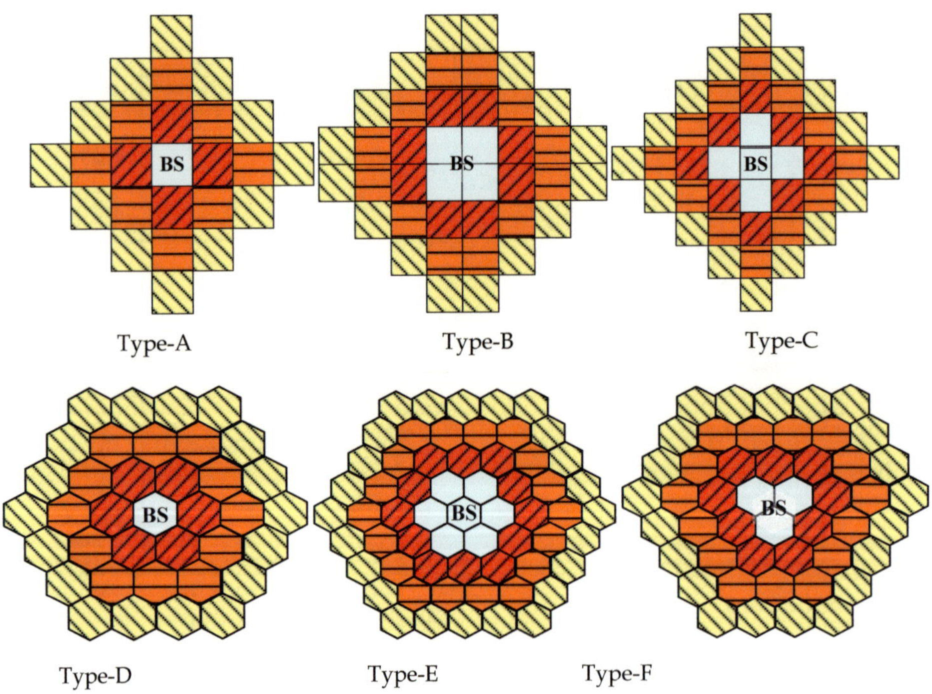

Fig. 3. Multihop-based topologies in metropolitan (Type-A, B, and C) and rural (Type-D, E, and F) area.

3.2 Modeling of a cost-effective coverage extension problem

When a target service area is given, the cost-effective coverage enhancement problem of the MMR WiBro/ WiMAX system is generally solved with respect to the traffic density per unit area. Since each BS and RS has its own coverage area in terms of unit service area and the user traffic is generated proportional to the size of service area, each BS and RS has a maximum number of accommodating MSs for its coverage area. In addition, since a BS has limited capacity, a limited number of RSs can be connected to a BS for coverage extension. From the characteristics of each proposed MMR topology, the number of RSs in each tier is also limited. For example, in case of type B, there could be at most 8, 12, or 18 RSs in each tier, respectively.

As a result, the cost-effective coverage extension problem can be formulated as a cost minimization problem with several constraints for a given user-traffic density using an integer Linear programming. Then, it can be expressed in terms of the number of BSs and RSs that are required to cover the target service area for a given user-traffic density. The object function and constraints for the minimum deployment cost are evaluated by the following optimization problem:

Minimize

$$C_T = C_{BS}X_{BS} + C_{RS}X_{RS\_1} + C_{RS}X_{RS\_2} + C_{RS}X_{RS\_3} \tag{1}$$

Subject to

$$A_{BS}X_{BS} + A_{RS}X_{RS\_1} + A_{RS}X_{RS\_2} + A_{RS}X_{RS\_3} \geq A_T \tag{2}$$

$$\begin{aligned}(C-\rho * A_{BS})X_{BS} - \rho * A_{RS}X_{RS\_1} \\ -\rho * A_{RS}X_{RS\_2} - \rho * A_{RS}X_{RS\_3} \geq 0\end{aligned} \tag{3}$$

$$\left.\begin{aligned}&X_{RS\_1} - N_1 \cdot X_{BS} \leq 0 \\ &N_1 \cdot X_{RS\_2} - N_2 \cdot X_{RS\_1} \leq 0 \\ &N_2 \cdot X_{RS\_3} - N_3 \cdot X_{RS\_2} \leq 0\end{aligned}\right\} \tag{4}$$

$$X_{BS} \in \{1,2,\cdots\} \tag{5}$$

$$X_{RS\_j} \in \{0,1,2,\cdots\}, j \in \{1,2,3\} \tag{6}$$

where, C_T is total deployment cost for a given user traffic density, X_{BS_i} is the number of BSs and X_{RS_i} is the number of RSs at i-th tier. C_{BS} and C_{RS} are the costs of a BS and an RS, respectively. There are four constraints in the optimization problem; constraints for seamless covering of the target service area in (2) where A_{BS} and A_{RS} are the coverages of a BS and an RS, constraints for the capacity limitation of a BS in (3) where C is the capacity of a BS and ρ is a given user traffic density, constraints for the maximum number of RSs (N_1~N_3) at each tier in (4), and constraints for the integrality of the number of BSs and RSs in (5) and (6).

3.3 Performance analysis of coverage extension and deployment cost

In order to analyze the cost-effective coverage extension problem for each topology under a given user-traffic density, several modeling parameters shown in Table 1 are used. We used CPLEX for optimization tool. In the analysis, for the sake of modeling convenience, it is assumed that user traffic is uniformly distributed, no Adaptive Modulation Coding (AMC) option, no resource allocation for BS and RSs, and no interference models are used. In addition, the cost of an RS is assumed to be a portion of the cost of a BS. In this chapter, three different RS cost levels, 10%, 20% and 30% of a BS cost, are used

Parameter	Value			
Total area (A_T)	300 Km^2			
BS coverage (A_{BS})	Metropolitan Area		Rural Area	
	Type-A	1 Km^2	Type-D	1 Km^2
	Type-B	4 Km^2	Type-E	7 Km^2
	Type-C	5 Km^2	Type-F	3 Km^2
RS coverage (A_{RS})	1 Km^2 (a square)		1 Km^2 (a hexagon)	
Traffic density (ρ)	0.1~10bps/m^2 (0.1~10Mbps/km^2)			
RS cost (C_{RS})	10%, 20%, 30% of BS cost per RS			
BS total capacity (C)	50 Mbps			

Table 1. Model Parameters for Metropolitan and Rural area systems.

Figure 4 shows the efficiency of the MMR-based coverage extension approach in terms of the coverage extension and the total deployment cost for the case of metropolitan areas and rural areas, respectively. Figure 4-(a) shows that the coverage area with the type-C approach is bigger than with type-A when the same number of BSs is used. However, as the traffic density increases, the difference between each MMR approach and the difference between MMR approaches and BS-only approach tend to vanish due to the limited capacity of a BS. Figure 4-(b) shows the total deployment cost of three MMR approaches and BS-only approaches for seamless covering of the target metropolitan service area with respect to the traffic density. As shown in Figure 4-(b), type-C topology covers the target area with the lowest cost while type-A covers with the highest cost. The reason is that because of small coverage of a BS, type-A requires more number of BSs to cover the target area, and thus results higher deployment cost. Meanwhile, when RS costs of type-B and type-C are 30%, and 20% and 30% of BS cost, deployment of such MMR topologies are not beneficial than the BS-only approach because of the higher RS cost compared to the coverage extension effect.

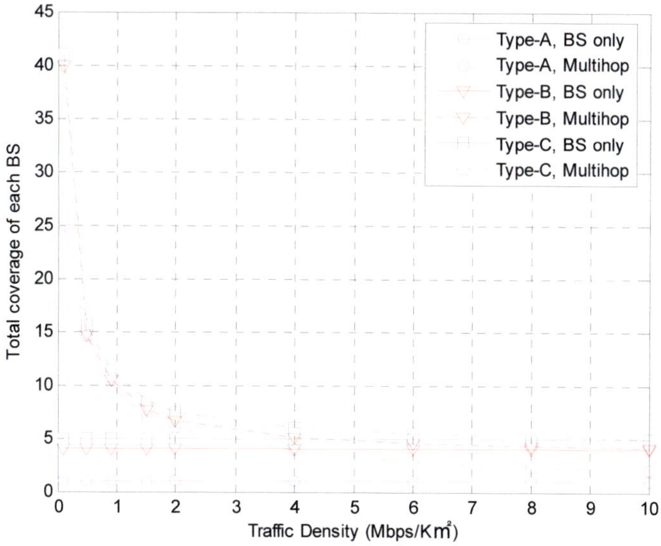

(a) coverage extension

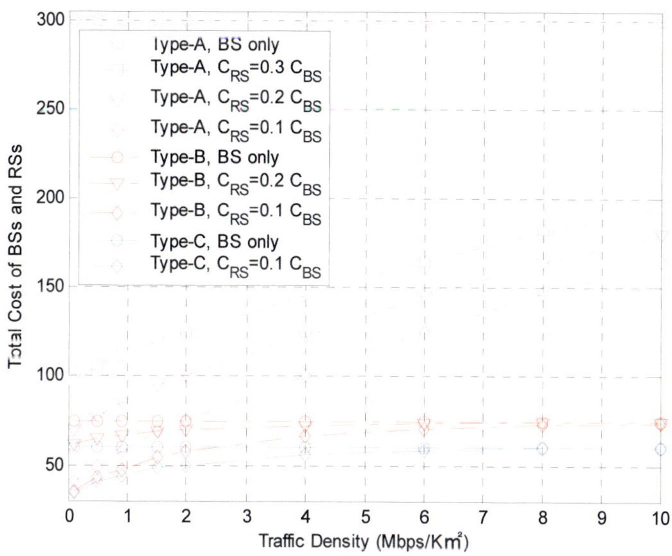

(b) Deployment cost

Fig. 4. Analysis results on coverage extension (a) and deployment cost (b) under various topologies for metropolitan area.

Figure 5-(a) also shows that the coverage area with type-E approach is bigger than with type-D when the same number of BSs is used. However, similar to the result of figure 4-(a), as the traffic density increases, the difference between each MMR approach types and the difference between MMR approaches and BS-only approach are tend to vanish due to the limited capacity of a BS. Figure 5-(b) shows the total deployment cost of three MMR approaches and

BS-only approaches for seamless covering of the target rural service area with respect to the traffic density. As shown in Figure 5-(b), type-E topology covers the target area with the lowest cost while type-D covers with the highest cost due to the same reason explained in figure 4-(b). In addition, similar to the figure 4-(b), when RS costs of type-E are 20% and 30% of BS cost, deployment of such MMR topologies are not beneficial than the BS only approach because of the same reason explained in figure 4-(b).

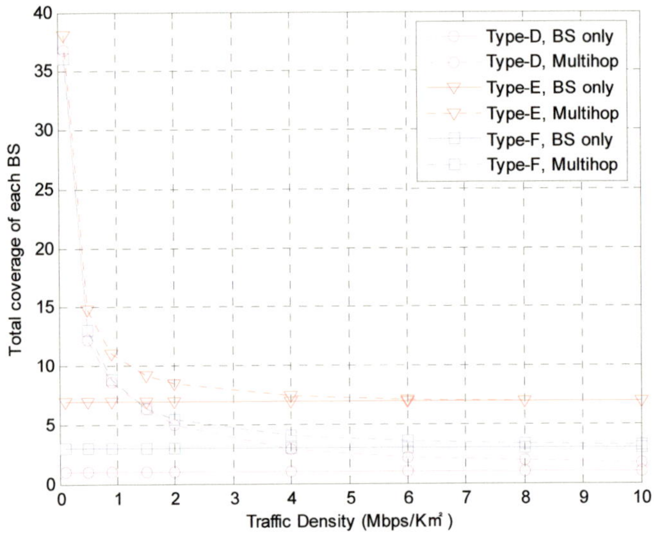

(a) coverage extension

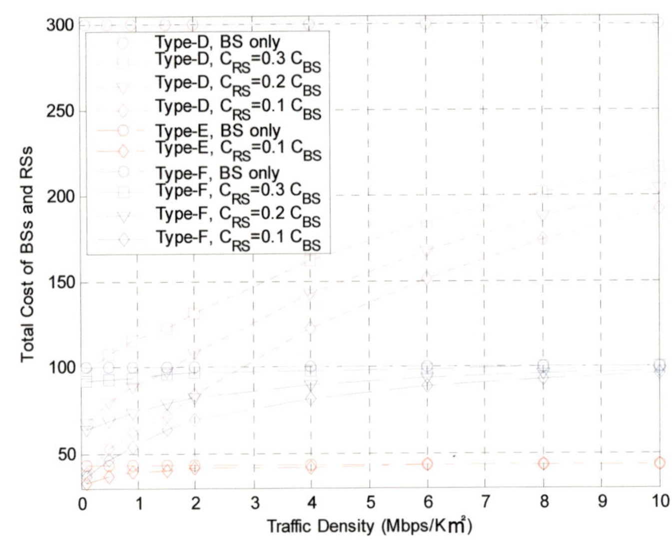

(b) Deployment cost

Fig. 5. Analysis results on coverage extension (a) and deployment cost (b) under various topologies for rural area.

4. Coverage extension with sectorized approaches

In this section, we propose and analyze various MMR topologies for cell-coverage extension using directional-antennas. Under such topologies, we investigate deployment cost and optimal numbers of BSs and RSs with respect to various user traffic densities.

4.1 Sectored BS based Multi-hop relay approaches

In the conventional cellular system with omni-directional antenna, the contour of a cell for equal received power from the antenna is approximately a circular, and generally assumed

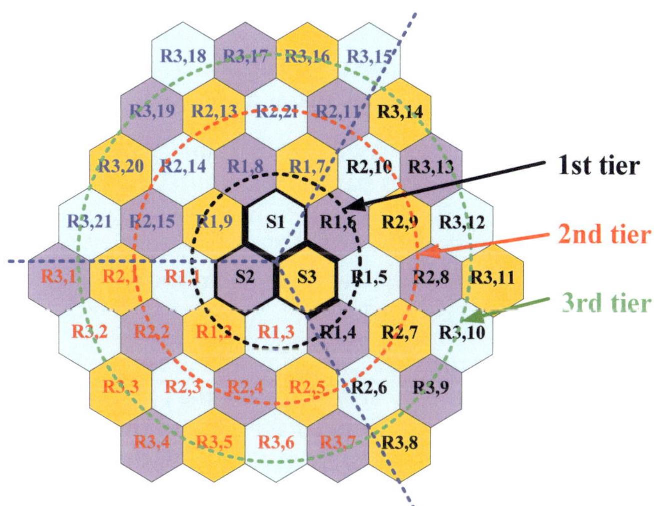

(a) Multihop-based NBTC

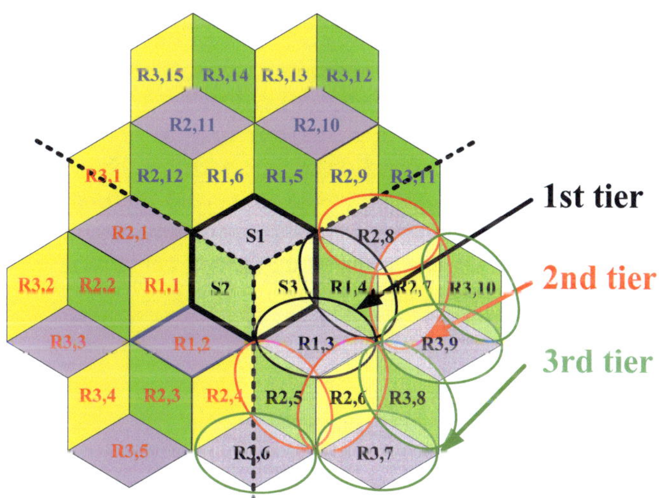

(b) Multihop-based WBTC

Fig. 6. Multihop-based NBTC and WBTC systems.

to be a hexagonal shape in (IEEE 802.16j-06/015, 2006). Consequently, in order to increase a BS capacity through decreasing radio frequency interference, sectorized cell structures such as the WBTC and NBTC systems were introduced in (Noh, 2006) and (Choi et al., 2006). The number of sectors in a cell is depend on the characters of antenna but three sectored (120°) or six sectored(60°) antenna is normally used. In general, the representative three sector cell structures are the WBTC and NBTC. A BS is covered with three 60° directional antennas in the NBTC. The coverage area of a sector in the NBTC assumed to be a hexagonal shape because the narrow beam radiation pattern matches well a hypothetical hexagonal shape. With three such antennas, the coverage contour of the NBTC cell composed of three sectors is therefore like a clover leaf. On the other hand, the coverage area of a sector in the WBTC assumed to be a rhombus shape because the wide beam radiation pattern with 120° antennas matches well a rhombus shape, and thus the coverage area of the WBTC is a hexagonal shape consisting of three rhombus shape sectors.

We propose two multi-hop approaches based on NBTC and WBTC BS sectorization. Figure 6 shows two multi-hop relay approaches based on the NBTC and WBTC BS structures. As shown in figure 6-(a), an NBTC BS is assumed to have three hexagonal shape sectors (S_i, i=1,2,3), and the shape of each RS is assumed to be a hexagon. On the other hand, as shown in figure 6-(b), a WBTC BS is assumed to have three rhombus shape sectors (S_i, i=1,2,3), and the shape of each RS is assumed to be a rhombus.

In this chapter, we only consider the NBTC and the WBTC based MMR approaches having up to 3 tiers. S_i means i-th sector of a BS and $R_{i,j}$ means the j-th RS at i-th tier. For example, every sector in NBTC based MMR system may have at most three RSs at the first tier, five RSs at the second tier, and seven RSs at the third tier as illustrated in figure 6-(a). On the other hand, every sector in WBTC based MMR system may have at most two RSs at the first tier, four RSs at the second tier, and five RSs at the third tier as illustrated in figure 6-(b).

4.2 Problem formulation and performance analysis

In order to analyze the cost-effective coverage extension problem for each sectorized BS based MMR approaches under a given user traffic density, we formulate similar integer Linear optimization problem introduced in the previous section where omni-directional antenna is assumed to use. The object function of the sectorized BS based MMR approach is slightly different from the previous model. In the sectorized BS based MMR approach, the BS cost is divided into two parts, one for BS tower construction cost and the other for sector antenna related cost. Therefore, the objective function and constraints can be expressed as below.

Minimize

$$\begin{aligned} C_T &= C_{BS}X_{BS} + C_{RS}\sum_{i=1}^{3}\sum_{j=1}^{l} X_{RS_{i,j}} \\ &= C_{BS}^{*}X_{BS} + \sum_{i=1}^{3} C_{BS}^{i} X_{RS_{i,0}} + C_{RS}\sum_{i=1}^{3}\sum_{j=1}^{l} X_{RS_{i,j}} \end{aligned} \tag{7}$$

Subject to

$$A_{BS}X_{BS} + A_{RS}\sum_{i=1}^{3}\sum_{j=1}^{l} X_{RS_{i,j}} \geq A_T \tag{8}$$

$$C \cdot X_{BS} - \rho\left(\sum_{i=1}^{3}\sum_{j=0}^{l} X_{RS_{i,j}}\right) \geq 0 \tag{9}$$

$$\left.\begin{aligned} & X_{RS_{i,0}} - N_1 \cdot X^{i}{}_{BS} \leq 0, \\ & N_1 \cdot X_{RS_{i,2}} - N_2 \cdot X_{RS_{i,1}} \leq 0, \\ & N_2 \cdot X_{RS_{i,3}} - N_3 \cdot X_{RS_{i,2}} \leq 0, \\ & N_3 \cdot X_{RS_{i,4}} - N_4 \cdot X_{RS_{i,3}} \leq 0, \end{aligned}\right\} for\ \forall i \in \{1,2,3\} \tag{10-1}$$

$$\begin{aligned} & X_{BS} - (X_{RS_{1,0}} + X_{RS_{2,0}} + X_{RS_{3,0}}) \leq 0, \\ & 3X_{BS} - (X_{RS_{1,0}} + X_{RS_{2,0}} + X_{RS_{3,0}}) \geq 0, \end{aligned} \tag{10-2}$$

$$X_{BS} = X_{RS_{1,0}} \Rightarrow \begin{cases} X_{BS} - X_{RS_{1,0}} \leq 0, \\ X_{BS} - X_{RS_{1,0}} \geq 0, \end{cases} \tag{10-3}$$

$$X_{RS_{1,0}} \geq X_{RS_{2,0}} \geq X_{RS_{3,0}} \Rightarrow \begin{cases} X_{RS_{1,0}} - X_{RS_{2,0}} \geq 0, \\ X_{RS_{2,0}} - X_{RS_{3,0}} \geq 0 \end{cases} \tag{10-4}$$

$$X_{BS} \in \{1,2,\cdots\} \tag{11}$$

$$X_{RS_{i,j}} \in \{0,1,2,\cdots\}\ for\ \ \forall i \in \{1,2,3\}, \forall j \in \{0,1,2,3,\cdots\} \tag{12}$$

where, C_T, C_{BS} and C_{RS} are the same cost values used in (1), and X_{BS} is the number of BS. In order to consider sectorized BS based MMR approach, several variables are introduced. $X_{RS_{i,j}}$ is the number of RS at i-th tier of j-th BS sector, and $X_{RS_{i,0}}$ is the number of sector antenna in BSs. C*BS is the BS tower construction costs, and CiBS is the i-th sector antenna related cost in a BS. Constraints of the optimization problem and assumptions are similar to the previous optimization model in sub-section 3.2. The constraint for seamless covering of the target service area is in (8) where AiBS is the coverage of a directional-antenna in a BS and (9) is the constraints for the capacity limitation of a BS. (10-1) is the constraints for the maximum number of RSs at each tier where N1~ N4 are variables and (10-2) is the constraint about that a BS can have three directional-antennas. (10-3) is the constraint about a BS must have a sectored antenna and (10-4) is the order of sectored antennas in a BS. The constraints of (11) and (12) are the integrality of the number of BSs and RSs.

Figure 7 shows analytical results on coverage extension and deployment cost for NBTC and WBTC BS-based MMR systems. Modeling parameters are the similar to the previous model shown in Table 2.

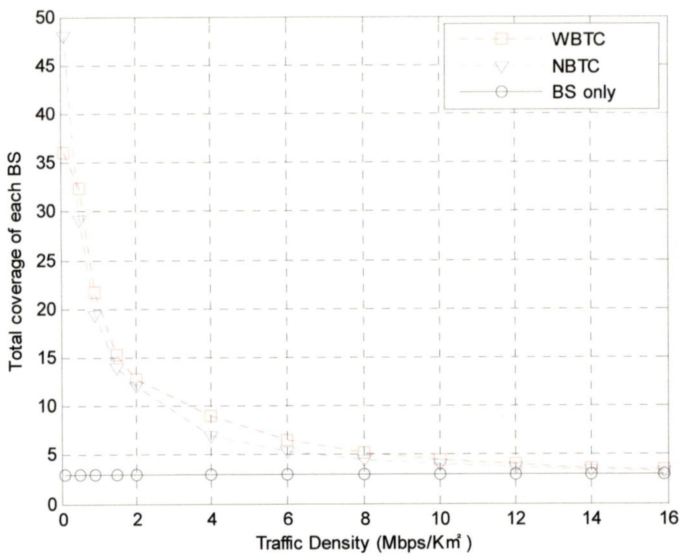

(a) coverage extension

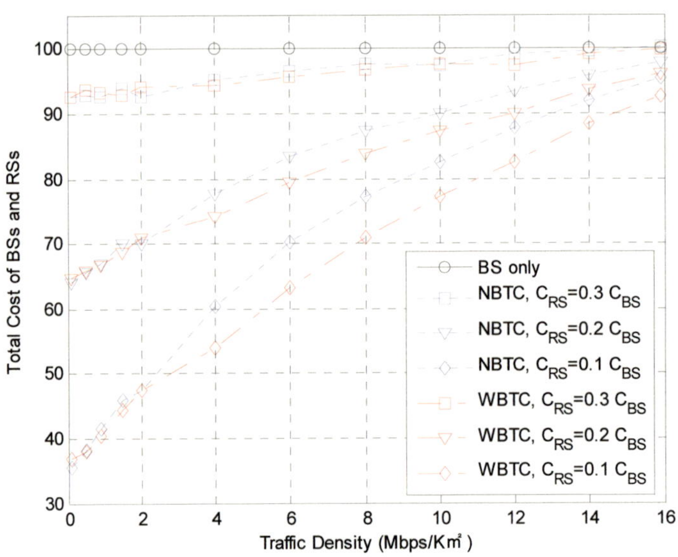

(b) Deployment cost

Fig. 7. Analysis results on coverage extension (a) and deployment cost (b) under WBTC and NBTC based MMR systems.

Parameter	Value
Total area (A_T)	300 Km^2
BS coverage (A_{BS})	3 Km^2 (Three sector antennas)
RS coverage (A_{RS})	1 Km^2
Traffic density (ρ)	0.1~15.9bps/m^2 (0.1~15.9Mbps/km^2)
RS cost (C_{RS})	10%, 20%, 30% of BS cost per RS
BS total capacity (C)	60 Mbps (Each antenna has 20 Mbps)

Table 2. Model Parameters for NBTC and WBTC Systems.

Figure 7 shows the analytical results of coverage extension and deployment cost for the NBTC and the WBTC BS based MMR systems with respect to the user traffic density under 300Km^2 of the target service area. As shown in figure 7-(a), the coverage of the WBTC BS-based MMR system is shown to be slightly larger than that of the NBTC BS-based MMR system while, the coverage of the NBTC BS-based MMR system is shown to be larger than that of the WBTC BS-based MMR system when the traffic density is less than 0.4 Mbps/Km^2. However, there is no significant difference between those two approaches in terms of coverage extension.

As shown in figure 7-(b), the deployment cost of the WBTC BS based MMR approach is shown to be lower than that of the NBTC BS based MMR system when the RS cost is 10% or 20% of the BS cost. However, there is no difference between two approaches when the RS cost is 30% of BS. Therefore, we could conclude that the WBTC BS-based MMR system is better than the NBTC BS-based MMR system in terms of the coverage extension and deployment cost except for the case when the traffic density is less than 0.4 Mbps/Km^2.

5. Multi-hop system deployment scenario

In previous two sections, we analyze various MMR WiBro/WiMAX systems for cost-effective coverage extension under various topologies and BS sectorizations. From the above analytical results, we can conclude that the multi-hop relay based WiBro/WiMAX system can be deployed in several phases according to the user-traffic density and resulting system performance.

Figure 8 shows a practical deployment scenario of the MMR WiBro/WiMAX system consisting of three deployment scenarios in terms of the user-traffic density and the number of tiers. In this scenario we assume that the minimum required data rate per user 128kbps. Phase 1 corresponds to the situation when the traffic density per unit area is less than 1.5 Mbps. In this phase, due to the low traffic density a BS has enough capacity to accommodate many users, and thus the MMR system could have more than 2 tiers to cover large service area. Phase 2 corresponds to the situation when the traffic density is 1.5~5 Mbps/Km^2. In this case, a BS could have 1 or 2 tiers to cover a service area. Phase 3 corresponds to the moderate traffic density situation having more than 5 Mbps/Km^2. In this phase, there could be less than 1 tier RSs connected to a BS, and thus RSs are used to extend BS coverage toward a specific direction from a BS or to cover radio shadowing areas.

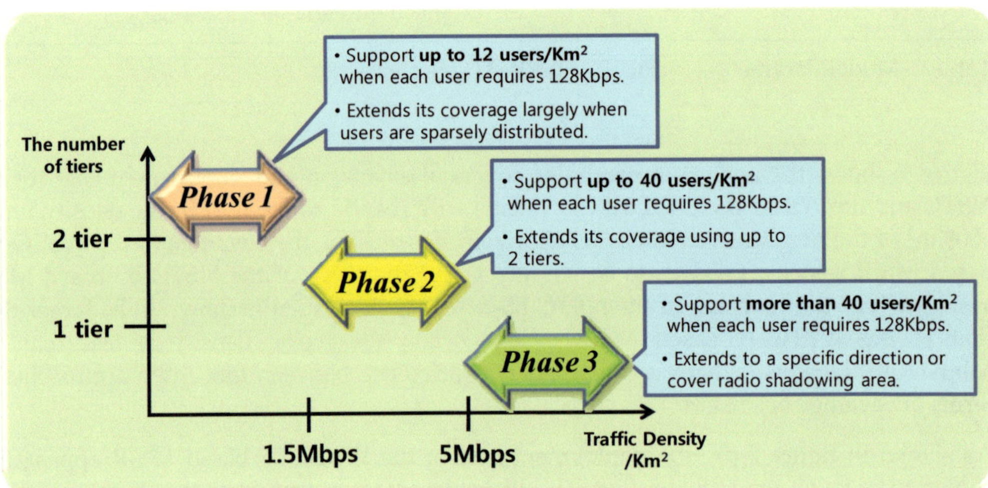

Fig. 8. Deployment Scenario for Multihop-based WiBro/WiMAX systems.

6. Conclusions

In this chapter, we first obtained the optimum numbers of BSs and RSs for minimizing total system cost using optimization problem given maximum system capacity of WiBro/ WiMAX systems. We proposed various multi-hop based network topologies using omni-directional antenna for metropolitan area network and rural area network. We, also, proposed multi-hop based network topologies using directional antenna, the NBTC BS-based and the WBTC BS-based MMR systems. Through analytical methods, we obtained the optimal network configuration given traffic density for various topologies and compared the cost with the traditional topology. Finally, we presented practical deployment scenario consisting of three phases depending on user-traffic density and the number of traffic relaying hops. We presented the basic studies of the multi-hop based NBTC BS-based and WBTC BS-based MMR systems without considering interference in this work. However, the WBTC system has generally the bigger interference than the NBTC system. We are planning to investigate and compare the performance of various topologies taking interference into consideration in MMR WiBro / WiMAX systems.

7. Acknowledgment

This research was supported by Basic Science Research Program through the National Research Foundation of Korea(NRF) funded by the Ministry of Education, Science and Technology(2010-0025125).

8. References

IEEE Standard 802.16e-2005, (2005). IEEE Standard for Local and Metropolitan Area Networks Part 16: Air Interface for Fixed and Mobile Broadband Wireless Access Systems, Feb. 2006.

IEEE 802.16j-2008, (2007). Baseline Document for Draft standard for Local and Metropolitan Area Networks Part 16: Air Interface for Fixed and Mobile Broadband Wireless Access Systems, June. 2007

WiMAX Forum, http://www.wimaxforum.org

IEEE802.16 TGj PAR, http://grouper.ieee.org/groups/802/16/relay

S. Lee, et al., (2006). "The Wireless Broadband (WiBro) System for Broadband Wireless Internet Services," IEEE Communications Magazine, Vol.44, No.7, (July 2006), pp.106-112, ISSN 0163-6804

Siemens and ETRI, (2007). "IR 1.3 Multi-hop Capabilities of WiMAX," Technical paper, (September 2007).

IEEE 802.16j-06/015, (2007). Guideline documents: Usage Models, IEEE802.16j TG

V. H. MacDonald, (1979). AMPS: The Cellular Concept, Bell System Technical Journal, Vol.58, No.1, (Jan 1979) pp.15-41

Sun-Kuk Noh, (2006). A Study on the WBTC and NBTC for CDMA Mobile Communications Networks, ICCSA 2006, (May 2006) pp.582-590

Dong You Choi, Kwan Houng Lee, (2006). Analysis on the Cell Sectorization Using the Wide-beam and Narrow-beam for CDMA Mobile Communication Channels, International Conference on Computational Intelligence & Security, Vol.1, (Nov. 2006) pp.882-885

Chang-Hoi Koo, Yong-Woo Chung, Dynamic Cell Coverage Control for Power Saving in IEEE802.16 Mobile Multihop Relay Systems, Systems and Networks Communication 2006, (Oct. 2006) pp.60

IST WINNER II Project, Deliverables-D3.5.1 v1.0, https://www.ist-winner.org

IST FIREWORKS Project, http://fireworks.intranet.gr
IST ROMANTIK Project, http://www.ist-romantik.org

16

A WiMAX Network Architecture Based on Multi-Hop Relays

Konstantinos Voudouris[1], Panagiotis Tsiakas[1], Nikos Athanasopoulos[1], Iraklis Georgas[1], Nikolaos Zotos[2,3] and Charalampos Stergiopoulos[1,2]

[1]*Department of Electronics, Technological Educational Institution of Athens, Attiki,*
[2]*Future Intelligence Ltd, Attiki,*
[3]*NCSR Demokritos, Attiki,*
Greece

1. Introduction

It has become apparent in the recent years that in order for the next generation of wireless technology (whether this is WiMAX, LTE or any other 4G implementation) to be able to deliver ubiquitous broadband content, the network is required to provide excellent coverage, both outdoor and indoor, and significantly higher bandwidth per subscriber [Voudouris *et al.*, 2009]. In order to achieve that at frequencies above 2 and 3 GHz, which are targeted for future wireless technologies, network architecture must reduce significantly the cell size or the distance between the network and subscribers' antennas. While micro, pico and femto Base Transceiver Station (BTS) technologies reduce the cost of base-station equipment, they still rely on a dedicated backhaul. One solution introduced with the WiMAX 802.16j standard is the wireless Multi-hop Relay Station (MRS), intended to overcome these challenges. On one hand, it should be small, cost-effective and easy to install for enabling mass deployment in indoor and outdoor environments and creating relatively small areas with excellent coverage and high capacity availability. On the other hand, it does not require any dedicated backhaul equipment as it receives its capacity from centralized base-stations via the same resources used for the access service. In a setting where a MRS exists, enabling MIMO transmission, the link referred to needs to be specified. This means that, when a 2x2 setting is mentioned, there can be either two transmit antennas on the base station and two receive antennas on the relay station, or two transmit antennas on the relay station and two receive antennas on the subscriber's device [Chochliouros *et al.*, 2009].Wireless Multi-hop Relay Stations (MRSs), when deployed in various sights, result in increased throughput or coverage. A general case, where a relay station can be used, is in situations with coverage constraints such as areas where there is presence of physical obstacles (e.g. buildings, forests), or in indoor coverage cases. Some examples are large office buildings, University campuses, and villages in unreachable areas on rockier uplands etc. Another scenario, where MRSs can be used, is for high mobility users with increased bandwidth requirements, such as trains with a great number of wireless users. Such a mobile subscriber will more likely have data rate degradations due to non-fixed position. In this case, a relay station can be considered as the most feasible solution in terms of cost and easiness of installation in every public transport vehicle, providing increased coverage and throughput to mobile WiMAX users. In order to achieve certain bit error rate levels on the data transmitted to the subscribers, WiMAX uses adaptive modulation. In case the subscriber is far from the base station or the environment introduces a lot of interference, the modulation used will be adapted accordingly, reducing the available data rate of the user. The use of a relay station can improve the provided service to the end-user, since it can transcode the received signal from the base station increasing the data rate provided to that user. This scenario is applicable in suburban environments, where users are usually away from the base stations, as well as in environments with increased interference.

2. IEEE 802.16j protocol stack overview

In WiMAX the IEEE only defined the Physical (PHY) and Media Access Control (MAC) layers in 802.16 [IEEE 802.16, 2004]. This approach has worked well for technologies such as Ethernet and WiFi, which rely on other bodies such as the IETF (Internet Engineering Task Force) to set the standards for higher layer protocols such as TCP/IP (Transport Control Protocol / Internet Protocol), SIP (Session Initiation Protocol), VoIP (Voice over Internet Protocol) and IPSec (Internet Protocol Security). The main objective of the IEEE 802.16 standards is to develop a proper set of specifications of the air interface, while the WiMAX Forum defines the system profile, which is a list of selected functionalities for a particular usage scenario and overall wireless network architectures [Nakamura, 2008]. The sector continues to innovate in many ways; one of these is the development/promotion of new standards to solve "open" problems. Such an initiative (which is currently of extreme interest) is the development of the Multi-hop Relay Standard, IEEE 802.16j. This is being developed to provide low cost coverage in the initial stages of network deployment and increased capacity when there is high utilisation of the network. The standard has been identified with a better feasibility and efficiency due to the similarities in the MAC and PHY layers and the support of fast route change [Harmonized Contribution 802.16j]. The standard is expected to have significant impact in new 802.16 rollouts.

When deploying an IEEE 802.16j based WiMAX network, this can be considered as a cellular network since they both have the same design principles [Lin *et al., 2007*]. The standard specifies a set of technical issues in order to enhance previous standards with the main objective of supporting relay concepts [Zeng *et al.*, 2008].

The 802.16j standard defines an air interface between a Multi-hop Relay-Base Station (MR-BS) and a Relay Station (RS) [Okuda *et al.*, 2008]. Its most important technical issues are listed as follows: Centralized vs. distributed control; Scheduling; Radio resource management; Power control; Call admission and traffic shaping policies; QoS based on network wide load balancing and congestion control; Security, and; Management. Besides these issues, when it comes to network deployment the main objective remains the optimal placement of the RSs. Operators are mainly concerned about operating costs, revenues and the pay-off periods for their investments; but the quality of services offered, is also an important issue for them. The IEEE 802.16j protocol layering for simple RS is shown in Fig.1. Principles of layers are briefly discussed as follows: Principles of layers are briefly discussed as follows: The MAC Converge Sub-layer (CS) is primarily used to "map" any data received from the upper layers (e.g. IP, Ethernet) to an appropriate MAC connection, to manage data

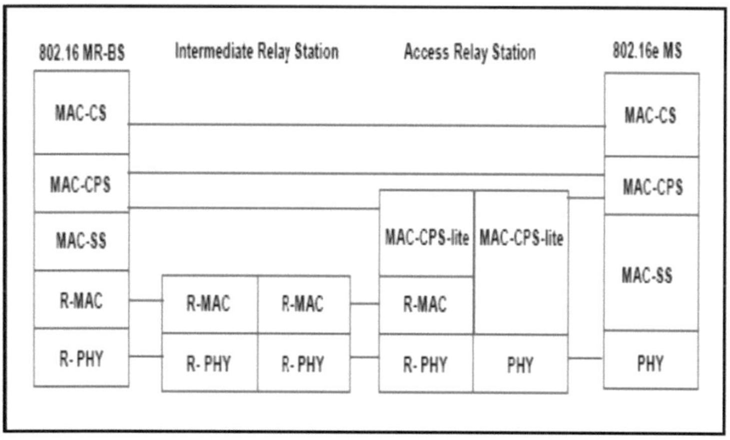

Fig. 1. IEEE 802.16j Protocol layering for simple RS

flow from the upper layer and to ensure that QoS requirements are fulfilled. More specifically, the CS accepts data from the network layer through the CS Service Access Point (SAP) and performs data classification into the appropriate MAC Service Data Units (SDUs). A "classifier" is an entity which selects packets based on the content of packet headers and categorises them according to a set of "matching criteria" (such as destination IP address). The external higher-layer Protocol Data Units (PDUs) entering the CS are checked against those criteria and accordingly delivered to a specific MAC connection. The MAC Common Part Sub-layer (CPS) is a connection oriented protocol. Contrarily to previous wireless network technologies (such as IEEE 802.11), it does provide QoS guarantees. It receives the MAC SDUs from MAC SAP; next, it delivers the MAC PDUs to a peer MAC SAP according to the requested QoS, to perform various transport functions (i.e. packing, fragmentation and concatenation). Each MAC PDU is identified by a unique connection identifier (CID). In the scope of past IEEE 802.16 standard versions, in 2004, an operation has been defined where multiple MAC PDUs could be concatenated and comprised into a single burst for transmission purposes. These PDUs had to be encoded and modulated, by using the same PHY (i.e. using the same burst profile); however they could be associated with subscriber stations. The position of each burst into the DL (downlink) frame has been specified by DL-MAP2, which contained additional Information Elements (IEs). The IEs specify the CID of the receiver, the burst profile used, the start time of the burst and a bit indicating whether an optional preamble I present. The IEEE 802.16e standard has further extended the DL MAP IE of legacy IEEE 802.16, in order to carry the identifiers of multiple connections (CIDs) in a single IE. However, the last missing link for enabling efficient MAC PDU concatenation on relay link is the capability of supporting multiple connections using one uplink (UL) information element. Several approaches have been made, until now, to extend the UL MAP3 IE for relay link. The MAC Security Sub-layer (MAC-SS) handles security issues such as authentication, key exchange and privacy by encrypting the connections between BS and subscriber station. It is based on the Private Key Management (PKM) protocol, which has been enhanced to "fit" the IEEE 802.16 standard. At the time a subscriber connects to the BS, they perform mutual authentication with public-key cryptography using X.509 certificates [Chokhani *et al.*, 1999]. The payloads themselves are encrypted by using a symmetric-key system, which may be either DES (data encryption standard) with cipher block chaining or triple DES with two keys. The Relay MAC (R-MAC) Sub-layer has been introduced in the IEEE 802.16j standard. It provides efficient MAC PDU relaying/ forwarding and control functions, (such as scheduling, routing, and flow control). It is applicable to the links between MR-BS and RSs and between RSs. The Relay Physical (R-PHY) layer provides definition of physical layer design, (i.e. sub-channelization, modulation, coding, etc.), for links between MR-BS and RS and between RSs. The IEEE 802.16j standard has extended the past IEEE 802.16e frame structure to support in-band BS-to-RS communication. A high level diagram of the 802.16j frame structure in TDD (Time Division Duplex) OFDMA PHY mode is shown in Fig.5. The frame structure supports a typical two-hop relay-enhanced communication, where some MSs are attached to a RS and communicating with a BS via the RS, and some MSs connected directly to the BS.

In Fig.2, the horizontal dimension denotes time and the vertical dimension denotes frequency. Frame sections in grey denote receive (Rx) operation, whereas sections in white denote transmit (Tx) operation. The BS and RS frames are subdivided into DL and UL subframes in order to support TDD operation. Both DL and UL subframes are further subdivided into MS and RS zones4. The MS zones, supported at both the BS and RS, are backwards compatible with the 802.16e standard. The RS transmits to MSs in its coverage in the DL MS zone and receives control and data from the BS in the adjacent DL RS zone5. Each MR-BS frame begins with a preamble followed by a FCH (Frame Control Header) and the DL-MAP and possibly UL-MAP. The DL subframe shall include at least one DL access zone and may include one or more DL relay zones. The UL subframe may include one or more UL access zones and it may include one or more UL relay zones. A relay zone may be utilized for either transmission or reception, but the MRBS shall not be required to support both modes of operation within the same zone.

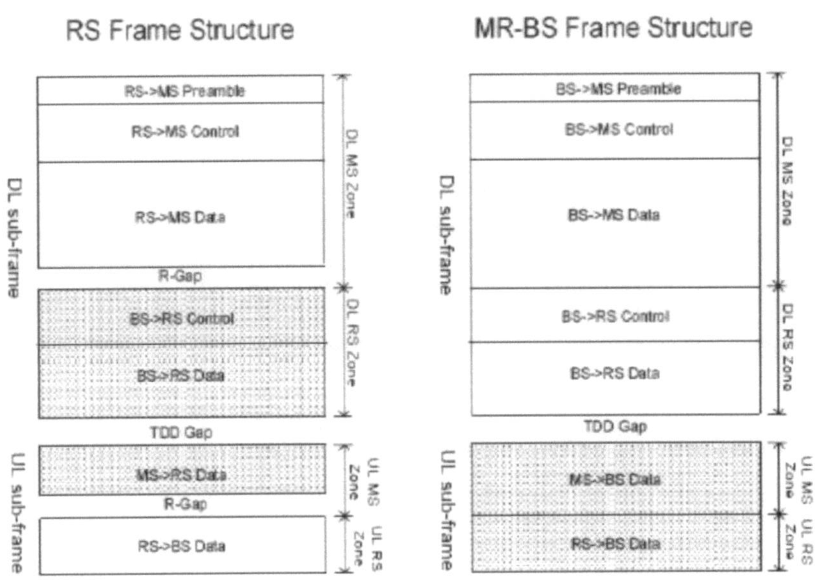

Fig. 2. OFDMA 802.16j frame structure.

3. Relay network overview

The following section provides an overview of a Relay Network in WiMAX Network build-out. It provides a general overview of the Relay network applications and benefits, an overview of the network architecture and the resulting requirements for the Relay node to fit into this architecture, and an overview of the required Protocol Stack to be run on the Relay node in the network.

3.1 Relay network general overview

There are various choices available for operators deploying WiMAX networks to improve indoor or outdoor coverage or to increase network capacity. These choices include various types of base stations: marcocells, microcells, or picocells in an outdoor environment, picocells in public indoor locations or within enterprise buildings, and femtocells for residential. The primary difference between these cells (performance-wise) is the size of coverage. Macrocells are the base stations with longest range, but are also the most expensive to purchase, deploy and maintain. Micro, pico and femto base stations are used to fill in coverage gaps and establish coverage in buildings where the macrocell signals can hardly penetrate. A significant side-effect of placing a large number of base stations in a region is that each needs a dedicated broadband backhaul connection. Micro, pico, and femto cells can use either wireline or wireless links for their backhaul, depending on the cost, availability and scalability of different solutions. In particular, they can support in-band backhaul to enable operators to use their spectrum holdings to carry backhaul traffic to the nearest macro base station or to the nearest microcell or picocell with wireline backhaul.

The IEEE 802.16j Mobile Multi-hop Relay (MMR) specifications are aimed to extend base station reach and coverage for WiMAX networks, while minimizing wireline backhaul requirement. The relay architecture will allow operators to use in-band wireless backhaul while retaining all the standard WiMAX functionality and performance.

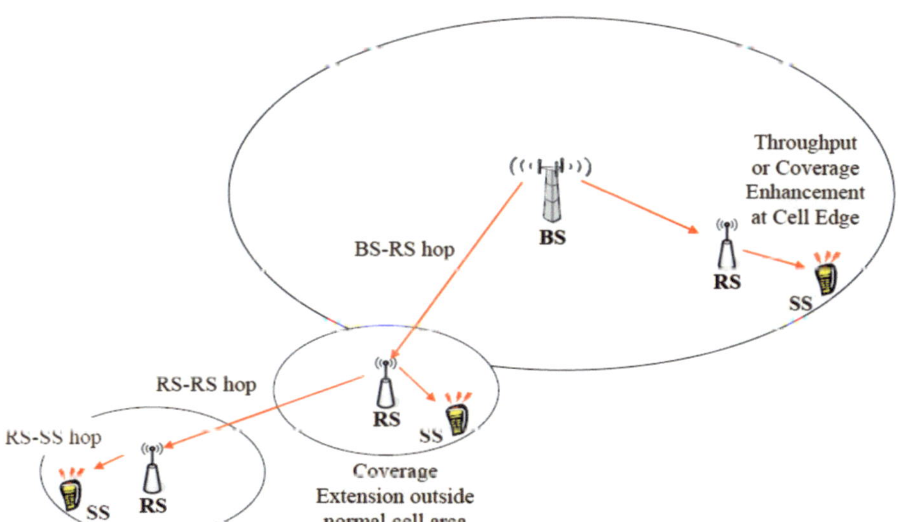

Fig. 3. Throughput or coverage enhancement

For example, in the previous figure, the MMR base station provides the primary area of coverage. It also has a backhaul connection, such as leased copper, fiber optics, or microwave radio link. The relay station extends the base station coverage. A mobile subscriber station (SS) can connect to a base station, an MMR base station (MRBS) or a Relay Station (RS).

The figure below shows additional usage scenarios for BS+RS deployment. These include coverage hole elimination within cell area (right); and temporary deployment for disaster/emergency situations and for special events (left).

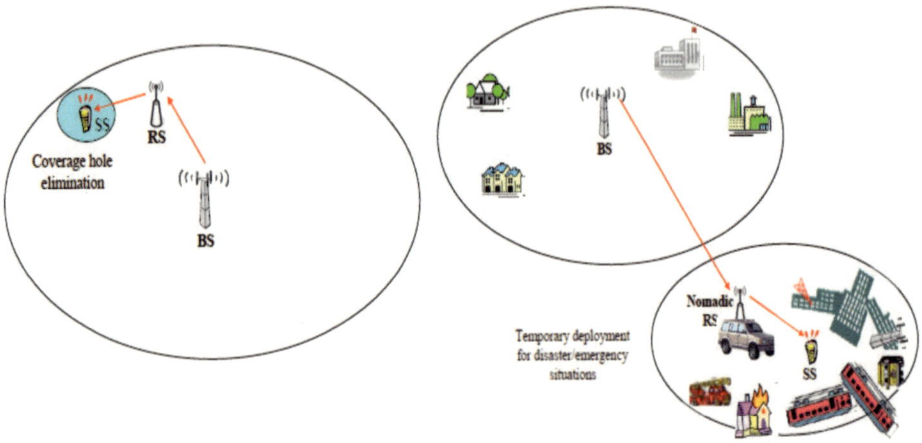

Fig. 4. Coverage hole elimination; Temporary deployment

Relay can be used in WiMAX networks for improving the coverage at the edge of the cell or for providing coverage to an area outside the coverage area of the cell. Relay can also be used for solving specific coverage problem within the coverage area of the cell ("hole filler") or provide coverage to tunnels and roads. Relay can also be used for adding throughput to certain "hot spot" areas within the coverage area of a cell. One important usage that has drawn growing attention in recent years is "In-building" or "indoor coverage". The growing traffic generated by indoor users can be met by the use of several types of relays and system configurations ranging from dedicated outdoor relays to distributed indoor relays with a multi hop configuration. A relay mounted on a vehicle can provide continuous service to users located in the vehicle while communicating and keeping continuity by performing continuous handover with base stations along the way. Relay can also be used on temporary basis, providing additional capacity to certain locations where a heavy traffic is expected for a limited time such as sporting events, concerts, and other events where a large crowd is expected to gather. Relay can also be deployed by first responders in order to provide coverage in an area where rescue operations take place.

3.2 Relay network architecture overview

Multi-hop Relay (MR) is a deployment that may be used to provide additional coverage or performance advantage in an access network. In MR networks, the Multi-hop Relay BS (MR-BS) is connected to several Relay Stations (RS), in a multi-hop topology, in order to enhance the network coverage and capacity density. Traffic and signalling between the SS and MR-

BS are relayed by the RS thereby extending the coverage and performance of the system in areas where RSs are deployed. Each RS is under the supervision of an MR-BS. In a system with more than two hops, traffic and signalling between an access RS and MR-BS may also be relayed through intermediate RSs. The RS is fixed in location. The SS may also communicate directly with the MR-BS. The following figure illustrates the MR-BS and two-hop RS deployment. For each of the RS there is an ACCESS link that covers the current cell and a BACKHAUL link to the next cell.

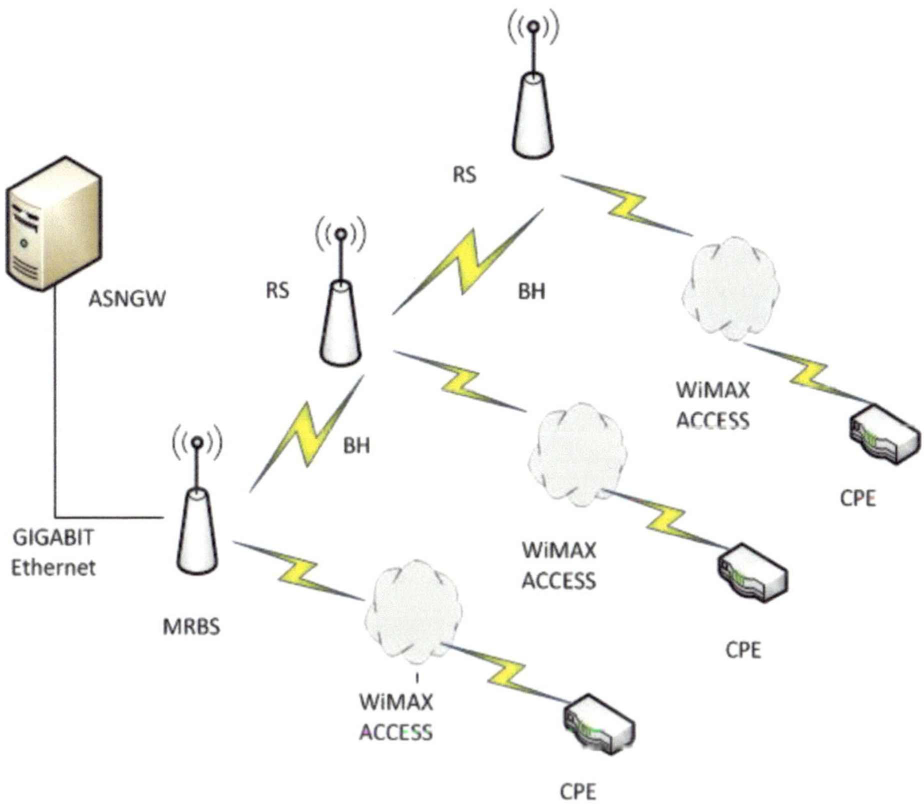

Fig. 5. MR-BS and RSs deployment

When considering relay stations, a distinction can be made between transparent and non-transparent relay stations. The main difference between these RS types is the way that the MS regards its serving or super-ordinate RS. When the super-ordinate RS of the MS is a transparent RS, then the MS regards the MRBS as its serving BS. From the MS point of view the transparent RS in the middle does not exist. The transparent RS assists the MRBS-MS communication link without any visibility of the RS to the MS. In case the super-ordinate RS of the MS is a non-transparent RS, then the MS regards its super-ordinate RS as if it is its serving 802.16e BS. Non-transparent relay stations can employ either distributed scheduling or centralized scheduling. When using distributed scheduling, the RS makes decisions about resource allocation, CID allocation, and MAP building regarding subscribers (and RSs) in its control. When using centralized scheduling, the MRBS makes those decisions for all the RSs under its supervision, and communicates these decisions to all RSs, which execute them.

4. Relay networking architecture options

This section lists several options for modifying the WiMAX architecture to enable support for relay. Each option is presented and its main cons and pros are listed. It begins with a discussion regarding MAC-based relay, as suggested by 802.16j, and then proceeds to network-based relay.

4.1 MAC-based relay

The first architecture option to consider is called "MAC-based relay". MAC-based relay is the approach used by 802.16j non-transparent mode. In this approach, the backhaul link between the Relay Station (RS) and the Multi-hop Relay Base Station (MRBS) is contained within the MAC (L2) layer. In most aspects, the ASN-GW is unaware of the fact that a MS is connected to the BS via a RS. Rather, the MRBS behaves as if the MS is registered with it directly. In this approach, no networking modifications are required to support relay traffic (though some modifications may be needed for RRM and other aspects, see below). From a networking perspective, the RS does not behave like a BS. The GRE tunnel used to carry the Service Flow is terminated at the MRBS and not at the RS. 802.16j MAC-based relay can support multi-hop (3-hop and more) relay without network modifications.

Pros:

- No network modifications.
- Lower latency for relay transport
- Easier to implement QoS constraints, because RS-MRBS link uses dedicated MAC, which can contain more information than available over R6.

Cons:

- RS-MRBS link not compliant to 802.16e MAC
- RS functionality does not build over existing WiMAX capabilities. Mobility, RRM needs to be modified to support relay.

4.2 Network-based relay

The MAC-based relay has one important drawback: The air interface between the RS and the MRBS in not 802.16e compliant. While modifying an existing BS networking capabilities is relatively painless, modifying as existing 802.16e BS to support new MAC features (such as 802.16j) is much more complex. Likewise, upgrading a deployed 802.16e network to support relay is much easier if it does not involve radio modifications. In addition, there is also decreased availability of existing test equipment. This makes development costs much higher for MAC variants. It is also important to note that 802.16j standard was not yet presented to NWG. Expected NWG activities may include RRM, network-assisted distributed scheduling and handover support. Therefore, we need to consider Network-based relay options, where the RS is composed of interconnected WiMAX BS and a WiMAX MS. The link between the RS and the MRBS is 802.16e-compliant, and networking elements are used to enable relay. Using 802.16e compliant air interface simplifies development and deployment. A RS with two 802.16e-compliant air interfaces (one for access and one for backhaul) is called Simultaneous Transmit and Receive (STR) Relay. Network-based relay options are presented in the following sections.

4.3 QoS over relay

WiMAX defines QoS parameters - most importantly, maximum tolerable latency and jitter - over the air interface connections. When combined with QoS mechanisms over the wireline backbone, or when the wireline backbone is good enough, the WiMAX QoS can provide predictable end-to-end QoS. When Multi-hop Relay is introduced, QoS must be applied to two independent wireless links - the link between the MS and the RS, and the link between the RS and the MRBS. For MAC-based relay, one can envision a cooperative QoS mechanism in which downlink packets which accumulated high latency in the MRBS/RS link are prioritized over other packets on the RS/MS link (and vice versa). However, such a mechanism is bound to be complicated and bandwidth-consuming, and is not included in 802.16j MAC-based relay. Having abandoned cooperative QoS, the multi-hop QoS problem becomes a problem of assigning Service Flows individual QoS parameters for each of the wireless links. For MAC-based relay, this can be done by observing the CID in each PDU and associating each CID/SF with QoS parameters over the RS/MRBS link. For network-based relay, the CID is not available but the SF can be deduced from the GRE tunnel used to transport the data path. In this sense, MAC-based relay and network-based provide equivalent end to end QoS.

802.16j MAC-based relay also defines Tunnel CIDs (T-CIDs). A T-CID is a connection which traverses all the hops between a RS and the MRBS (See 6.5.3.3). Several user-facing Service Flows are aggregated into a single T-CID, and this single T-CID has a set QoS parameters. Resources for the T-CID are scheduled as if it was a single connection with given QoS parameters, ignoring the individual SFs. Aggregating SFs into a T-CID simplified scheduling and can reduce bandwidth requirements. For example, several MAC PDUs which have a 32-bit CRC each, can be combined to a single Relay MAC PDU which only has a single 32-bit CRC. This mechanism of tunneling different SFs in a single connection can be used for network-based relay easily, though not as efficiently.

4.4 Radio Resource Management (RRM) and relay

Radio Resource Management (RRM) is the control of MSs and BSs aimed to increase the bandwidth capacity of the complete network. The two primary tools used for RRM are: *Power Control* - lowering the power used by transmitter to reduce interference to receivers *Handoffs* changing the BS with which a station communicates in a way which would allow lower transmit power (and less interference). For example, A MS can be handed off from one BS to another if the MS significantly approaches the target BS, so the signal loss between the MS and the target BS is lower, in comparison to the original BS. Additionally, a MS might be handed off from a "crowded" BS to a less utilized BS. This way, the under-utilized BS becomes more utilized, thus wasting less spectrum. The over-utilized BS now has room for new subscribers, and until they arrive, it can either increase bandwidth allocation for the remaining subscribers, or use more robust modulations to reduce transmit power. When relay stations are introduced to the network, RRM is affected in several aspects. First, the new RSs fill the airwaves as if they were BSs: They cause interferences to MSs trying to reach other RSs or BSs. In this aspect, deploying a RS is similar to deploying a BS (or a Pico- BS). Once can assume that the mechanism used to control inter-BS interferences can be used to control interferences between RSs and BSs.

The second aspect in which relay complicate RRM is the fact that each relayed MS now takes part in more than a single wireless link, and RRM aims to optimize all of them simultaneously. Refer to Figure 35 and consider that the MS can reach both RS A2 and B1 equally well. Deciding which RS will serve the MS is a function of load over the various air interfaces. Which RS is better suited to handle the MS depends on how many subscribers and how much traffic is relayed not only thorough A2 and B1 (as would have been for the non-relay case), but also through the MRBSs A and B and RSs A1 and B2. For example, if there is a high traffic demand from MSs attached to A1, and little demand from other MSs. If the demand is high enough, then the air interface used by MRBS A to provide the backhaul becomes highly utilized. In this case, moving the MS to RS B1 is preferred as it can use the under-utilized air interface of MRBS B. It is therefore clear that in order to have effective RRM, handover decision-making must take into account the load on relevant backhaul links.

The third aspect in which relay affects RRM is the introduction of Mobile RS (MRS). An MRS is a RS which can change its super-ordinate MRBS, e.g., a RS mounted inside a train, providing access to passengers. In terms of mobility, MRS largely functions as a MS. An efficient network should take into account the special requirements of MRS when considering MRS handoffs. The WiMAX networking architecture defines two Profiles. Profile A defines distributed RRM, in which every BS has a Radio Resource Agent (RRA) which collects radio utilization and interference metrics, and a Radio Resource Controller (RRC) which communicates with the RRA and with other RRCs to initiate handovers. Profile C defines the RRC to reside in the ASN-GW. In Profile C, the RRC polls (RRAs) inside every BS to decide when a handover is needed. When using network-based relay, each RS is in essence a BS, together with an RRA and possibly an RRC. If, for the initial stages of deployment, traffic demands are moderate, then using the existing WiMAX RRM, which does not take into account the load over the backhaul links, with little or no modifications, is probably sufficient for the initial stages of deployment. Building RRM for an optimal network can be delayed to when the network is more mature. When using MAC-based relay, a RS does not inherently have RRM capabilities. It is therefore mandatory to implement RRM for the RS in the MRBS.

5. Multi-hop Relay Scenarios

A RS can be adapted at several mobility levels, i.e. fixed, nomadic, mobile, and can be used in the Next Generation Networks for improving the communication quality by several aspects, depending on the user needs and the environment conditions and constraints. In this section, the most common relay scenarios are described.

5.1 Hole filler

A RS can be used inside the service area of the cell in order to improve link quality to those specific areas that do not have sufficient link quality due to excessive link attenuation from the BS. This attenuation can be caused among other factors due to shadowing of buildings or due to a given hilly topography. Such a scenario is shown in Fig. 6.

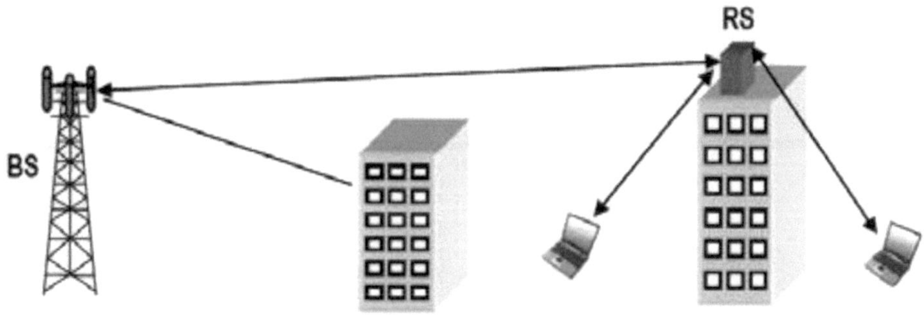

Fig. 6. Hole Filler Relay Station Scenario

5.2 Cell extension

In this case, a RS is used to increase the coverage area of a cell. A RS can extend the coverage area in a certain location at the edge of the cell as shown in Fig. 7a or cover an area separated from the coverage area of the cell as shown in Fig. 7b. This latter configuration is sometimes called "remote sector". Cell extension can be used in a more strategic manner where multiple RSs are deployed around the perimeter of a cell to achieve higher coverage area with a single BS, as shown in Fig. 7c. This concept may be used as a part of the network design strategy in order to provide coverage with as few as possible base stations.

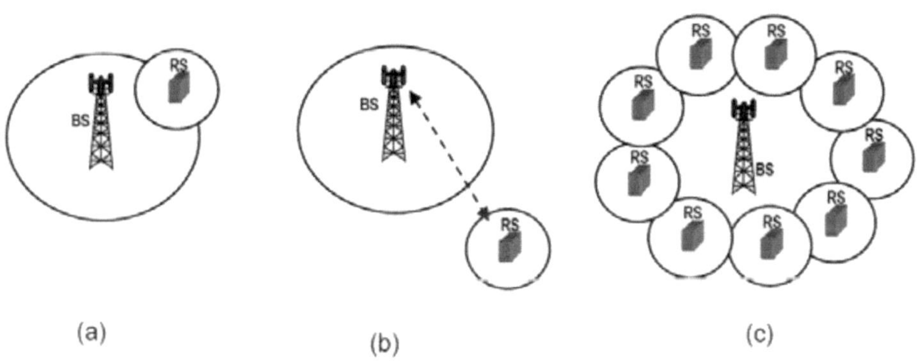

Fig. 7. Cell Extension Relay Station Scenario

5.3 Capacity and throughput

In most scenarios, the use of an RS can increase the per-MS throughput, system capacity and QoS. A single link between the BS and RS with high Signal to Interference plus Noise Ratio (SINR) (and as a result with high order modulation and coding scheme) can be replaced with multiple RS to BS links with low SINR. The result is an increase in spectral efficiency which produces a capacity increase. This additional capacity can be used for providing higher throughput to individual MSs or to support more MSs within the coverage area of the RS. In addition, the link reliability is enhanced due to improved SINR. Fig. 8 depicts a scenario where four RSs are deployed in the service area of a sector. The capacity of the original coverage area may increase by a factor of four due to the fact that each RS provides its own capacity to the area and MSs which had low quality link to the BS may have better link quality to the serving RS. This increase in link quality is translated by the inherent link adaptation process to higher throughput and therefore to a higher capacity.

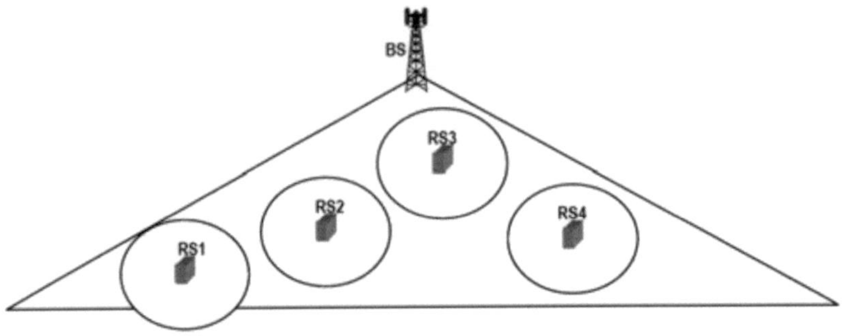

Fig. 8. Capacity and Throughput Improvement Relay Station Scenario

5.4 Indoor usage scenarios

The majority of cellular traffic is generated from buildings. Providing service to indoor MSs by the same BS that provides service to outdoor MSs has several major disadvantages. First, due to the building walls introduced attenuation, BS-MS link might be marginal or of a low quality thus limiting the data rate and consuming excessive time-frequency resource from the BS. MSs which reside in high floors of a building, pose another issue. They are exposed to multiple BSs arriving signals and as a result, two problems may occur: First, signals may interfere with each other, hence degrading the SINR of the MS. Second, MS may enter into an undesired hand off process and as a result, excessive handover processes might occur. This may result in power consumption of the radio, backhaul and computational resources. Relaying technologies in indoor environments can be proved challenging in order to improve the communication. The major methods used for providing dedicated coverage for indoor MSs are described below.

5.4.1 Fixed and nomadic RS with direct connection to the BS

A fixed RS is mounted in a way that its antenna maintains good link quality concurrently with the BS and with the MSs which reside in the building. Alternatively, the RS may be lightweight, nomadic, and similar to a WiFi router.

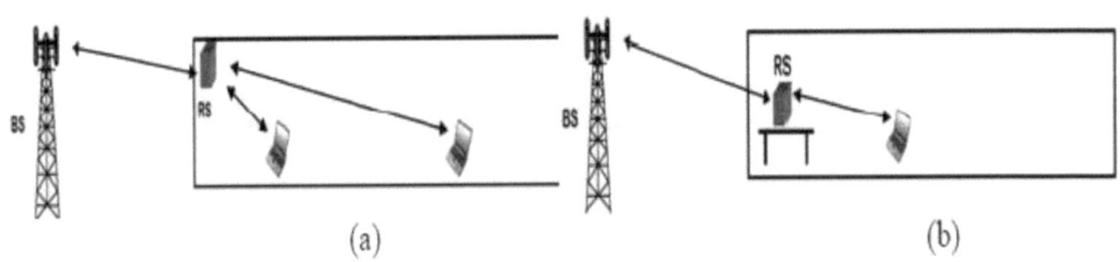

Fig. 9. A Relay Station with direct connection to the Base Station in an indoor scenario. In (a) the Relay Station is Fixed, while in (b) the Relay Station is Nomadic.

5.4.2 Multi-hop RS

When large area floors need to be covered, a single RS may not be sufficient. In such cases, multiple RSs can be distributed over the floor connected to each other using the multi-hop capability. The internal RSs can be chained to a RS mounted at the edge of the floor with a good link to the BS as depicted in Fig. 10.

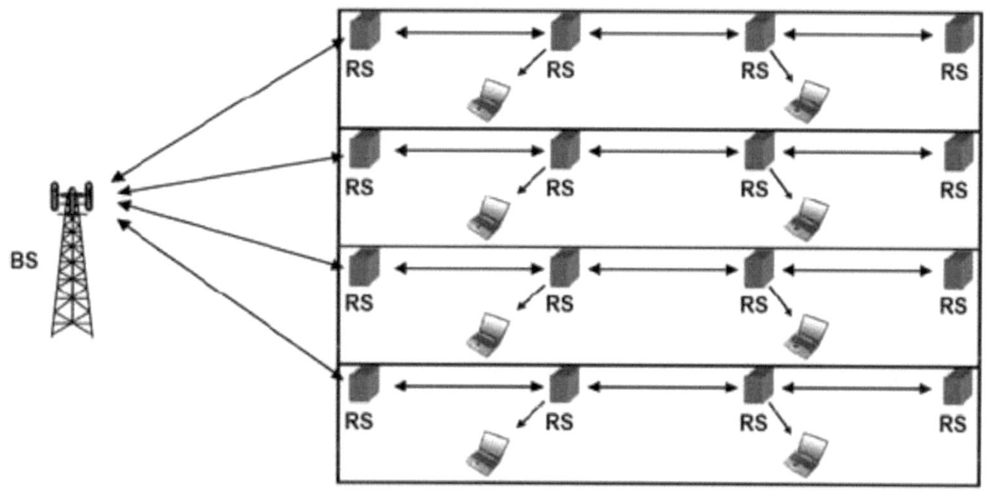

Fig. 10. A Multi-hop Relay Station indoor scenario.

5.4.3 Coverage with dedicated external RSs

In some cases it may possible to provide in-building coverage with external RSs "illuminating" the building from outside as depicted in Fig. 11. The advantage of this model is the elimination of the excruciating need for installing equipment and cabling inside the building

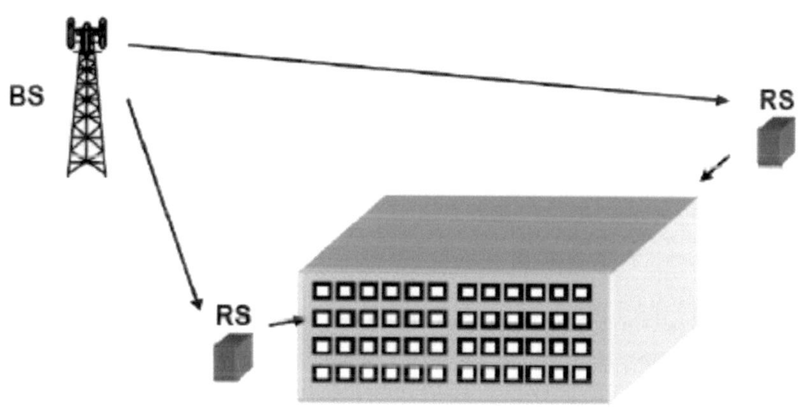

Fig. 11. Two external Relay Stations illuminating a building.

5.5 Road and tunnel

The common requirement for roads and tunnels is the need to provide coverage along a linear path. Since high mobility MSs are served, it is recommended to allow the move without handover between the coverage area of each RS. Fig. 12 depicts a scenario where a BS feeds in parallel three transparent RSs. This configuration allows continues high SINR connection along the road without handover.

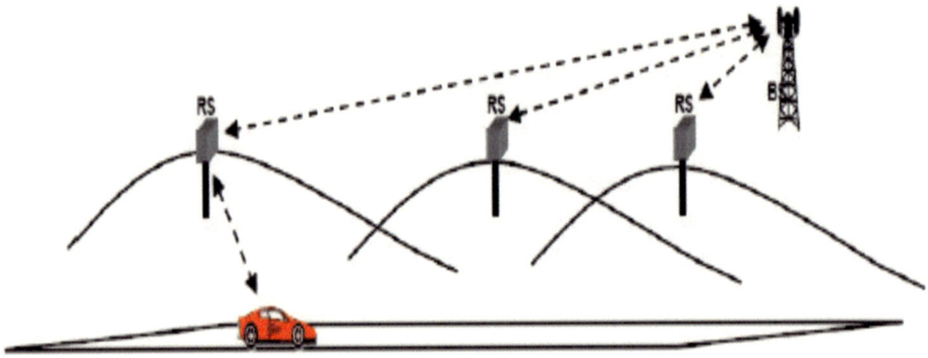

Fig. 12. A Base Station feeds three Relay Stations in parallel.

Fig. 13 shows a tunnel covered by RSs. The RSs inside the tunnel have no connection with the external BSs and therefore multi-hop RSs must be used. In order to minimize the number of hops, the internal RSs can be split to two groups each fed from a BS at each side of the tunnel.

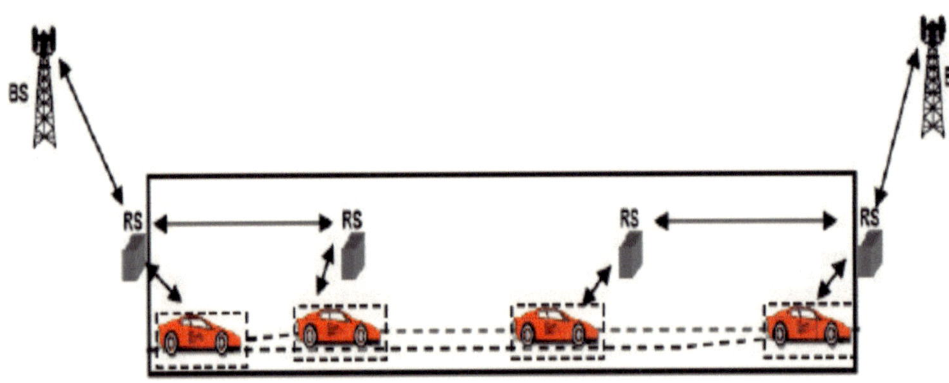

Fig. 13. A tunnel coverage through a multi-hop Relay Station configuration.

5.6 Temporary relay stations

RSs may be deployed temporarily to provide coverage or additional capacity. An example would be in events where a heavy load on the network is expected only in certain predefined time interval such as sporting events, concerts or other events where a large crowd is expected to gather. In the case of adding capacity to an area where an event takes place, usually reasonable coverage from the macro BS already exist. In another usage model a relay can be used for providing coverage in areas where network coverage is needed temporarily for an incidental reason. An example would be an emergency incident or a disaster recovery effort where fixed infrastructure network has been destroyed overloaded or never existed before. Service to those cases can be provided by a deployment of temporary RS usually suitably installed on a vehicle.

6. Field tests of the REWIND prototype system

REWIND project examines backhaul-less relay station implementations for WiMAX and supports the relevant standardisation process with interoperability, lab and field information on possible implementations of the WiMAX relay. The main scientific and technological objectives of the REWIND project are in the area of design, development and integration of advanced wireless relay stations based on OFDM technology, in order to foster and exemplify the WiMAX technology by delivering to the subscriber broadband multimedia content. It aims to increase the bandwidth and quality of service to the end users. REWIND prototype system brings a new and promising technology of using WiMAX Relay stations at certain areas that are facing significant performance problems. Performance and coverage tests for both MRBS and RS systems are described below. Two topologies were tested in terms of coverage (with or without MIMO functionality enabled) and total throughput the system can achieve.

6.1 System and measurements equipment

Fig. 14 illustrates the multi-hop relay base station (MRBS), the relay station (RS) system and a van on which the mobile station (MS) was installed and has been used for the field measurements.

Fig. 14. MRBS Antenna, transceiver, GPS-RS system-MS.

6.2 Topologies

The topologies that were used are shown in the following figures. Fig. 15 illustrates the network topology when a mobile station (MS) is directly connected to the MRBS (one hop), while Fig. 16 shows the network topology where the MS is connected to MRBS via RS (two hops topology).

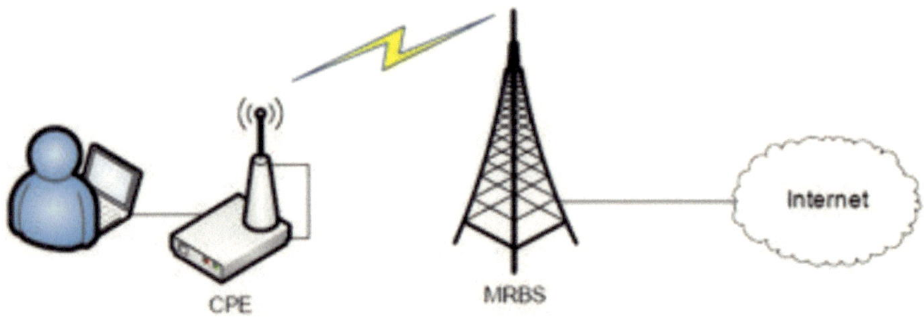

Fig. 15. Topology with MRBS active only.

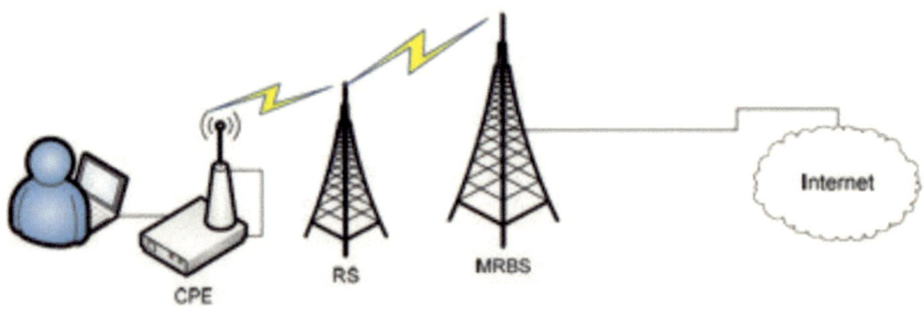

Fig. 16. Two hops topology.

6.3 Performance and coverage tests

In order to test the MRBS coverage area the following procedure was followed: An MS that was registered at MRBS, was placed at the roof of a van. The van started moving towards the cell boundary keeping the connection between the MS and the MRBS. The cell coverage of MRBS was the area that covered till the connection was cut off. The maximum distance where the connectivity of the CPE with the MRBS was alive, was at 440m. The same procedure was followed in order to define the coverage area with the RS as well. The results have shown that the usage of the RS extends the cell range to 570m by fully supporting all the QoS levels of 802.16j standard.

The use of MIMO was also tested, in order to clarify and evaluate the communication performance improvement for the MRBS. MIMO antenna technique was expected to improve throughput measurements in both nLOS/LOS conditions. Initially, MIMO was enabled at the MRBS and a van with a mobile station which was registered at the MRBS was moving along a predefined path, where measurements were taken at several places. In the next step, the MIMO functionality was disabled at the MRBS and the van followed the same predefined path, in which several measurements were taken. The obtained results showed that the use of MIMO functionality at the MRBS provided higher throughput values. The same procedure was performed for the MIMO functionality of the RS. The results in this case also showed higher throughput values and better that the ones of the MRBS. So, even in this scenario, the use of the RS provided improved results.

7. Conclusions

Relay technology has focused significant attention due to several important and quite identifiable reasons, i.e. simplicity, flexibility, deployment efficiency and cost effectiveness, as relays can permit a faster network rollout. Conformant to the scope of the actual European REWIND Research Project aiming to develop an effective relay station implementation for WiMAX technology, we have discussed several architecture principles and benefits for relay-based networks and we have presented an essential description of the related RS software architecture design (PHY and MAC layers architecture) for the realization of a proper relay node required for the relay station functionality. Our approach was based on the "core concept" of a multi-hop relay network architecture, adopting the IEEE 802.16j standard.

8. Acknowledgements

The present work has been performed in the scope of the REWIND ("RElay based WIireless Network and StandarD") European Research Project and has been supported by the Commission of the European Communities - Information Society and Media Directorate General (FP7, ICT-The Network of the Future, Grant Agreement No.216751).

9. References

Choclliouros, I. .; Mor, A. ; Voudouris, K.; Agapiou, G.; Aloush, A. ; Belesioti, M.; Sfakianakis, E. & Tsiakas, P. (2009). A Multi-Hop Relay Station Software Architecture Design, on the Basis of the WiMAX IEEE 802.16j Standard, *Proceedings of VTC 2009 69th IEEE International Conference on Vehicular Technology*, pp. 132-135, Barcelona, Spain, April 26-29, 2009

Chokhani, S. & Ford, W. (1999). Internet X.509 Public Key Infrastructure: Certificate Policy and Certification Practices Framework. *RFC 2527. Internet Engineering Task Force (IETF)*, Sterling,VA, USA (1999)

Harmonized Contribution 802.16j (Mobile Multihop Relay) Usage Models, Available from http://grouper.ieee.org/groups/802/16/relay/ocs/80216j-06-015.pdf

IEEE 802.16-2004, IEEE Standard for Local and metropolitan area networks. Part 16: Air Interface for Fixed and Mobile Broadband Wireless Access Systems, Available from http:// standards.ieee.org/getieee802/download/802.16e-2005.pdf

Lin, B.; Ho, H. ; Xie, L-L & Shen, X. (2007). Optimal relay station placement in IEEE 802.16j networks, *Proceedings of IWCMC'07 International Conference on Wireless Communications and Mobile Computing*, pp. 25-30, Honolulu, Hawai, USA, March, 2007

Nakamura, M. (2008). Standardization Activities for Mobile WiMAX. *Fujitsu Journal of Science & Technology Robotic Systems*, Vol.44, No.3, (2008), pp. 285-291

Okuda, C.; Zhu,C. & Viorel, D. (2008). Multihop Relay Extension for WiMAX Networks - Overview and Benefits of IEEE 802.16j standard. *Fujitsu Journal of Science & Technology Robotic Systems,* Vol.44, No.3, (2008), pp. 292-302

Voudouris, K.; Chocliouros, I. ; Tsiakas, P. ; Mor, A. ; Agapiou, G. ; Aloush, A. ; Belesioti, M. & Sfakianakis, E. (2009). Developing an Innovative Multihop Relay Station Software Architecture in the scope of the REWIND European Research Program, *Proceedings of MOBILIGHT 2009 1st International Conference on Mobile Lightweight Wireless Systems*, pp. 144-154, Athens, Greece, May 18-20, 2010

Zeng, H. & Zhu, C. (2008). System-Level Modeling and Performance Evaluation of 802.16j Multi-hop Relay Systems, *Proceedings of IWCMC, Next Generation Mobile Networks Symposium*, pp. 25-30, Crete Island, Greece, August, 2008